Fundamental Constants

Quantity	Symbol	Approximate Value	Current Best Value[†]
Speed of light in vacuum	c	3.00×10^8 m/s	2.99792458×10^8 m/s
Gravitational constant	G	6.67×10^{-11} N·m^2/kg^2	$6.67259(85) \times 10^{-11}$ N·m^2/kg^2
Avogadro's number	N_A	6.02×10^{23} mol^{-1}	$6.0221367(36) \times 10^{23}$ mol^{-1}
Gas constant	R	8.315 J/mol·K $= 1.99$ cal/mol·K $= 0.082$ atm·liter/mol·K	$8.314510(70)$ J/mol·K
Boltzmann's constant	k	1.38×10^{-23} J/K	$1.380658(12) \times 10^{-23}$ J/K
Charge on electron	e	1.60×10^{-19} C	$1.60217733(49) \times 10^{-19}$ C
Stefan-Boltzmann constant	σ	5.67×10^{-8} W/m^2·K^4	$5.67051(19) \times 10^{-8}$ W/m^2·K^4
Permittivity of free space	$\epsilon_0 = (1/c^2\mu_0)$	8.85×10^{-12} C^2/N·m^2	$8.854187817\ldots \times 10^{-12}$ C^2/N·m^2
Permeability of free space	μ_0	$4\pi \times 10^{-7}$ T·m/A	$1.2566370614\ldots \times 10^{-6}$ T·m/A
Planck's constant	h	6.63×10^{-34} J·s	$6.6260755(40) \times 10^{-34}$ J·s
Electron rest mass	m_e	9.11×10^{-31} kg $= 0.000549$ u $= 0.511$ MeV/c^2	$9.1093897(54) \times 10^{-31}$ kg $= 5.48579903(13) \times 10^{-4}$ u
Proton rest mass	m_p	1.6726×10^{-27} kg $= 1.00728$ u $= 938.3$ MeV/c^2	$1.6726231(10) \times 10^{-27}$ kg $= 1.007276470(12)$ u
Neutron rest mass	m_n	1.6749×10^{-27} kg $= 1.008665$ u $= 939.6$ MeV/c^2	$1.6749286(10) \times 10^{-27}$ kg $= 1.008664904(14)$ u
Atomic mass unit (1 u)		1.6605×10^{-27} kg $= 931.5$ MeV/c^2	$1.6605402(10) \times 10^{-27}$ kg $= 931.49432(28)$ MeV/c^2

[†]Reviewed 1993 by B. N. Taylor, National Institute of Standards and Technology. Numbers in parentheses indicate one standard deviation experimental uncertainties in final digits. Values without parentheses are exact (i.e., defined quantities).

Other Useful Data

Joule equivalent (1 cal)		4.186 J
Absolute zero (0 K)		-273.15°C
Earth:	Mass	5.97×10^{24} kg
	Radius (mean)	6.38×10^3 km
Moon:	Mass	7.35×10^{22} kg
	Radius (mean)	1.74×10^3 km
Sun:	Mass	1.99×10^{30} kg
	Radius (mean)	6.96×10^5 km
Earth-sun distance (mean)		149.6×10^6 km
Earth-moon distance (mean)		384×10^3 km

The Greek Alphabet

Alpha	A	α	Nu	N	ν
Beta	B	β	Xi	Ξ	ξ
Gamma	Γ	γ	Omicron	O	o
Delta	Δ	δ	Pi	Π	π
Epsilon	E	ε	Rho	P	ρ
Zeta	Z	ζ	Sigma	Σ	σ
Eta	H	η	Tau	T	τ
Theta	Θ	θ	Upsilon	Υ	υ
Iota	I	ι	Phi	Φ	ϕ, φ
Kappa	K	κ	Chi	X	χ
Lambda	Λ	λ	Psi	Ψ	ψ
Mu	M	μ	Omega	Ω	ω

Values of Some Numbers

$\pi = 3.1415927$	$\sqrt{2} = 1.4142136$	$\ln 2 = 0.6931472$	$\log_{10}e = 0.4342945$
$e = 2.7182818$	$\sqrt{3} = 1.7320508$	$\ln 10 = 2.3025851$	$1 \text{ rad} = 57.2957795$°

Mathematical Signs and Symbols

$\propto$	is proportional to	$\leq$	is less than or equal to
$=$	is equal to	$\geq$	is greater than or equal to
$\approx$	is approximately equal to	Σ	sum of
$\neq$	is not equal to	$\bar{x}$	average value of x
$>$	is greater than	Δx	change in x
$\gg$	is much greater than	$\Delta x \to 0$	Δx approaches zero
$<$	is less than	$n!$	$n(n-1)(n-2)\ldots(1)$
$\ll$	is much less than		

Unit Conversions (Equivalents)

Length

1 in. = 2.54 cm
1 cm = 0.394 in.
1 ft = 30.5 cm
1 m = 39.37 in. = 3.28 ft
1 mi = 5280 ft = 1.61 km
1 km = 0.621 mi
1 nautical mile (U.S.) = 1.15 mi = 6076 ft = 1.852 km
1 fermi = 1 femtometer (fm) = 10^{-15} m
1 angstrom (Å) = 10^{-10} m
1 light-year (ly) = 9.46×10^{15} m
1 parsec = 3.26 ly = 3.09×10^{16} m

Volume

1 liter (L) = 1000 mL = 1000 cm^3 = 1.0×10^{-3} m^3 = 1.057 quart (U.S.) = 54.6 in.^{3}
1 gallon (U.S.) = 4 qt (U.S.) = 231 in.^{3} = 3.78 L = 0.83 gal (Imperial)
1 m^3 = 35.31 ft^3

Speed

1 mi/h = 1.47 ft/s = 1.609 km/h = 0.447 m/s
1 km/h = 0.278 m/s = 0.621 mi/h
1 ft/s = 0.305 m/s = 0.682 mi/h
1 m/s = 3.28 ft/s = 3.60 km/h
1 knot = 1.151 mi/h = 0.5144 m/s

Angle

1 radian (rad) = 57.30° = 57°18′
1° = 0.01745 rad
1 rev/min (rpm) = 0.1047 rad/s

Time

1 day = 8.64×10^4 s
1 year = 3.156×10^7 s

Mass

1 atomic mass unit (u) = 1.6605×10^{-27} kg
1 kg = 0.0685 slug
[1 kg has a weight of 2.20 lb where g = 9.81 m/s^2.]

Force

1 lb = 4.45 N
1 N = 10^5 dyne = 0.225 lb

Energy and Work

1 J = 10^7 ergs = 0.738 ft·lb
1 ft·lb = 1.36 J = 1.29×10^{-3} Btu = 3.24×10^{-4} kcal
1 kcal = 4.18×10^3 J = 3.97 Btu
1 eV = 1.602×10^{-19} J
1 kWh = 3.60×10^6 J = 860 kcal

Power

1 W = 1 J/s = 0.738 ft·lb/s = 3.42 Btu/h
1 hp = 550 ft·lb/s = 746 W

Pressure

1 atm = 1.013 bar = 1.013×10^5 N/m^2 = 14.7 lb/in.^{2} = 760 torr
1 lb/in.^{2} = 6.90×10^3 N/m^2
1 Pa = 1 N/m^2 = 1.45×10^{-4} lb/in.^{2}

SI Derived Units and Their Abbreviations

Quantity	Unit	Abbreviation	In Terms of Base Units[†]
Force	newton	N	kg·m/s^2
Energy and work	joule	J	kg·m^2/s^2
Power	watt	W	kg·m^2/s^3
Pressure	pascal	Pa	kg/(m·s^2)
Frequency	hertz	Hz	s^{-1}
Electric charge	coulomb	C	A·s
Electric potential	volt	V	kg·m^2/(A·s^3)
Electric resistance	ohm	Ω	kg·m^2/(A^2·s^3)
Capacitance	farad	F	A^2·s^4/(kg·m^2)
Magnetic field	tesla	T	kg/(A·s^2)
Magnetic flux	weber	Wb	kg·m^2/(A·s^2)
Inductance	henry	H	kg·m^2/(s^2·A^2)

[†]kg = kilogram (mass), m = meter (length), s = second (time), A = ampere (electric current).

Metric (SI) Multipliers

Prefix	Abbreviation	Value
exa	E	10^{18}
peta	P	10^{15}
tera	T	10^{12}
giga	G	10^9
mega	M	10^6
kilo	k	10^3
hecto	h	10^2
deka	da	10^1
deci	d	10^{-1}
centi	c	10^{-2}
milli	m	10^{-3}
micro	μ	10^{-6}
nano	n	10^{-9}
pico	p	10^{-12}
femto	f	10^{-15}
atto	a	10^{-18}

PHYSICS

for

SCIENTISTS & ENGINEERS

Part 2
PHYSICS
for
SCIENTISTS & ENGINEERS
Third Edition

DOUGLAS C. GIANCOLI

PRENTICE HALL
Upper Saddle River, New Jersey 07458

Editor-in-Chief: Paul F. Corey
Production Editor: Susan Fisher
Executive Editor: Alison Reeves
Development Editor: David Chelton
Director of Marketing: John Tweedale
Senior Marketing Manager: Erik Fahlgren
Assistant Vice President of Production and Manufacturing: David W. Riccardi
Executive Managing Editor: Kathleen Schiaparelli
Manufacturing Manager: Trudy Pisciotti
Art Manager: Gus Vibal
Director of Creative Services: Paul Belfanti
Advertising and Promotions Manager: Elise Schneider
Editor in Chief of Development: Ray Mullaney
Project Manager: Elizabeth Kell
Photo Research: Mary Teresa Giancoli
Photo Research Administrator: Melinda Reo
Copy Editor: Jocelyn Phillips
Editorial Assistant: Marilyn Coco
Cover photo: Onne van der Wal/Young America
Composition: Emilcomp srl / Preparé Inc.

 © 2000, 1989, 1984 by Douglas C. Giancoli
Published by Prentice Hall
Upper Saddle River, NJ 07458

Photo credits appear at the end of the book, and
constitute a continuation of the copyright page.

Printed in the United States of America

10 9 8 7 6 5 4 3 2 1

ISBN 0-13-029095-5

Prentice-Hall International (UK) Limited, *London*
Prentice-Hall of Australia Pty. Limited, *Sydney*
Prentice-Hall Canada Inc., *Toronto*
Prentice-Hall Hispanoamericana, S.A., *Mexico City*
Prentice-Hall of India Private Limited, *New Delhi*
Prentice-Hall of Japan, Inc., *Tokyo*
Prentice-Hall (*Singapore*) Pte. Ltd.
Editora Prentice-Hall do Brasil, Ltda., *Rio de Janeiro*

CONTENTS

PREFACE xvii

SUPPLEMENTS xxviii

NOTES TO STUDENTS AND INSTRUCTORS ON
THE FORMAT xxx

USE OF COLOR xxxi

PART 1

1 INTRODUCTION, MEASUREMENT, ESTIMATING 1

1-1 The Nature of Science 2
1-2 Models, Theories, and Laws 3
1-3 Measurement and Uncertainty; Significant Figures 4
1-4 Units, Standards, and the SI System 6
1-5 Converting Units 8
1-6 Order of Magnitude: Rapid Estimating 9
*1-7 Dimensions and Dimensional Analysis 12
SUMMARY 13 QUESTIONS 13
PROBLEMS 14 GENERAL PROBLEMS 15

2 DESCRIBING MOTION: KINEMATICS IN ONE DIMENSION 16

2-1 Reference Frames and Displacement 17
2-2 Average Velocity 18
2-3 Instantaneous Velocity 20
2-4 Acceleration 23
2-5 Motion at Constant Acceleration 26
2-6 Solving Problems 28
2-7 Falling Objects 31
*2-8 Use of Calculus; Variable Acceleration 36
SUMMARY 38 QUESTIONS 38
PROBLEMS 39 GENERAL PROBLEMS 42

3 KINEMATICS IN TWO DIMENSIONS; VECTORS 45

3-1 Vectors and Scalars 45
3-2 Addition of Vectors—Graphical Methods 46
3-3 Subtraction of Vectors, and Multiplication of a Vector by a Scalar 48
3-4 Adding Vectors by Components 48
3-5 Unit Vectors 52
3-6 Vector Kinematics 53
3-7 Projectile Motion 55
3-8 Solving Problems in Projectile Motion 58
3-9 Uniform Circular Motion 63
3-10 Relative Velocity 66
SUMMARY 68 QUESTIONS 69
PROBLEMS 70 GENERAL PROBLEMS 74

4 DYNAMICS: NEWTON'S LAWS OF MOTION 77

4-1 Force 77
4-2 Newton's First Law of Motion 78
4-3 Mass 79
4-4 Newton's Second Law of Motion 80
4-5 Newton's Third Law of Motion 82
4-6 Weight—the Force of Gravity; and the Normal Force 85
4-7 Solving Problems with Newton's Laws: Free-Body Diagrams 88
4-8 Problem Solving—A General Approach 96
SUMMARY 97 QUESTIONS 97
PROBLEMS 98 GENERAL PROBLEMS 103

5 **FURTHER APPLICATIONS OF NEWTON'S LAWS** **106**

5-1 Applications of Newton's Laws
Involving Friction 106
5-2 Dynamics of Uniform Circular Motion 114
5-3 Highway Curves, Banked
and Unbanked 118
*5-4 Nonuniform Circular Motion 121
*5-5 Velocity-Dependent Forces; Terminal
Velocity 122
SUMMARY 124   QUESTIONS 124
PROBLEMS 125   GENERAL PROBLEMS 129

6 **GRAVITATION AND NEWTON'S SYNTHESIS** **133**

6-1 Newton's Law of Universal Gravitation 133
6-2 Vector Form of Newton's Law
of Universal Gravitation 136
6-3 Gravity Near the Earth's Surface;
Geophysical Applications 137
6-4 Satellites and "Weightlessness" 139
6-5 Kepler's Laws and Newton's Synthesis 143
6-6 Gravitational Field 146
6-7 Types of Forces in Nature 147
*6-8 Gravitational Versus Inertial Mass;
the Principle of Equivalence 148
*6-9 Gravitation as Curvature of Space;
Black Holes 149
SUMMARY 150   QUESTIONS 150
PROBLEMS 151   GENERAL PROBLEMS 153

7 **WORK AND ENERGY** **155**

7-1 Work Done by a Constant Force 156
7-2 Scalar Product of Two Vectors 159
7-3 Work Done by a Varying Force 161
7-4 Kinetic Energy and the
Work–Energy Principle 164
*7-5 Kinetic Energy at Very High Speed 169
SUMMARY 170   QUESTIONS 170
PROBLEMS 171   GENERAL PROBLEMS 174

8 **CONSERVATION OF ENERGY** **176**

8-1 Conservative and Nonconservative
Forces 177
8-2 Potential Energy 178
8-3 Mechanical Energy and Its
Conservation 182
8-4 Problem Solving Using Conservation
of Mechanical Energy 184
8-5 The Law of Conservation of Energy 189
8-6 Energy Conservation with Dissipative
Forces: Solving Problems 190
8-7 Gravitational Potential Energy
and Escape Velocity 192
8-8 Power 195
*8-9 Potential Energy Diagrams; Stable
and Unstable Equilibrium 197
SUMMARY 198   QUESTIONS 199
PROBLEMS 200   GENERAL PROBLEMS 204

9 LINEAR MOMENTUM AND COLLISIONS 206

9-1 Momentum and Its Relation to Force 206
9-2 Conservation of Momentum 208
9-3 Collisions and Impulse 211
9-4 Conservation of Energy and Momentum in Collisions 214
9-5 Elastic Collisions in One Dimension 214
9-6 Inelastic Collisions 217
9-7 Collisions in Two or Three Dimensions 219
9-8 Center of Mass (CM) 221
9-9 Center of Mass and Translational Motion 225
*9-10 Systems of Variable Mass; Rocket Propulsion 227
SUMMARY 230 QUESTIONS 230
PROBLEMS 231 GENERAL PROBLEMS 236

10 ROTATIONAL MOTION ABOUT A FIXED AXIS 239

10-1 Angular Quantities 240
10-2 Kinematic Equations for Uniformly Accelerated Rotational Motion 243
10-3 Rolling Motion (without Slipping) 244
10-4 Vector Nature of Angular Quantities 246
10-5 Torque 247
10-6 Rotational Dynamics; Torque and Rotational Inertia 249
10-7 Solving Problems in Rotational Dynamics 250
10-8 Determining Moments of Inertia 254
10-9 Angular Momentum and Its Conservation 256
10-10 Rotational Kinetic Energy 260
10-11 Rotational Plus Translational Motion; Rolling 262
*10-12 Why Does a Rolling Sphere Slow Down? 268
SUMMARY 268 QUESTIONS 269
PROBLEMS 270 GENERAL PROBLEMS 276

11 GENERAL ROTATION 279

11-1 Vector Cross Product 279
11-2 The Torque Vector 280
11-3 Angular Momentum of a Particle 281
11-4 Angular Momentum and Torque for a System of Particles; General Motion 283
11-5 Angular Momentum and Torque for a Rigid Body 285
*11-6 Rotational Imbalance 287
11-7 Conservation of Angular Momentum 288
*11-8 The Spinning Top 290
*11-9 Rotating Frames of Reference; Inertial Forces 291
*11-10 The Coriolis Effect 292
SUMMARY 294 QUESTIONS 294
PROBLEMS 295 GENERAL PROBLEMS 298

12 STATIC EQUILIBRIUM; ELASTICITY AND FRACTURE 300

12-1 Statics—The Study of Forces in Equilibrium 300
12-2 The Conditions for Equilibrium 301
12-3 Solving Statics Problems 303
12-4 Stability and Balance 308
12-5 Elasticity; Stress and Strain 309
12-6 Fracture 312
*12-7 Trusses and Bridges 315
*12-8 Arches and Domes 319
SUMMARY 321 QUESTIONS 321
PROBLEMS 322 GENERAL PROBLEMS 328

PART 2

13 FLUIDS 332

13-1	Density and Specific Gravity	332
13-2	Pressure in Fluids	333
13-3	Atmospheric Pressure and Gauge Pressure	337
13-4	Pascal's Principle	337
13-5	Measurement of Pressure; Gauges and the Barometer	338
13-6	Buoyancy and Archimedes' Principle	340
13-7	Fluids in Motion; Flow Rate and the Equation of Continuity	343
13-8	Bernoulli's Equation	345
13-9	Applications of Bernoulli's Principle: From Torricelli to Sailboats, Airfoils, and TIA	347
*13-10	Viscosity	350
*13-11	Flow in Tubes: Poiseuille's Equation	351
*13-12	Surface Tension and Capillarity	351
*13-13	Pumps	353

SUMMARY 354 QUESTIONS 354
PROBLEMS 356 GENERAL PROBLEMS 360

14 OSCILLATIONS 362

14-1	Oscillations of a Spring	363
14-2	Simple Harmonic Motion	364
14-3	Energy in the Simple Harmonic Oscillator	369
14-4	Simple Harmonic Motion Related to Uniform Circular Motion	371
14-5	The Simple Pendulum	371
*14-6	Physical Pendulum and Torsion Pendulum	373
14-7	Damped Harmonic Motion	374
14-8	Forced Vibrations; Resonance	378

SUMMARY 380 QUESTIONS 381
PROBLEMS 381 GENERAL PROBLEMS 386

15 WAVE MOTION 388

15-1	Characteristics of Wave Motion	389
15-2	Wave Types	391
15-3	Energy Transported by Waves	395
15-4	Mathematical Representation of a Traveling Wave	396
*15-5	The Wave Equation	399
15-6	The Principle of Superposition	401
15-7	Reflection and Transmission	402
15-8	Interference	404
15-9	Standing Waves; Resonance	405
*15-10	Refraction	408
*15-11	Diffraction	410

SUMMARY 410 QUESTIONS 411
PROBLEMS 412 GENERAL PROBLEMS 415

16 SOUND 417

16-1	Characteristics of Sound	417
16-2	Mathematical Representation of Longitudinal Waves	419
16-3	Intensity of Sound; Decibels	420
16-4	Sources of Sound: Vibrating Strings and Air Columns	424
*16-5	Quality of Sound, and Noise	429
16-6	Interference of Sound Waves; Beats	429
16-7	Doppler Effect	432
*16-8	Shock Waves and the Sonic Boom	435
*16-9	Applications; Ultrasound and Ultrasound Imaging	437

SUMMARY 438 QUESTIONS 438
PROBLEMS 439 GENERAL PROBLEMS 443

17 TEMPERATURE, THERMAL EXPANSION, AND THE IDEAL GAS LAW 445

17-1	Atomic Theory of Matter	446
17-2	Temperature and Thermometers	447
17-3	Thermal Equilibrium and the Zeroth Law of Thermodynamics	449
17-4	Thermal Expansion	450
*17-5	Thermal Stresses	454
17-6	The Gas Laws and Absolute Temperature	454
17-7	The Ideal Gas Law	456
17-8	Problem Solving with the Ideal Gas Law	457
17-9	Ideal Gas Law in Terms of Molecules: Avogadro's Number	459
*17-10	Ideal Gas Temperature Scale—A Standard	460

SUMMARY 461   QUESTIONS 461
PROBLEMS 462   GENERAL PROBLEMS 464

18 KINETIC THEORY OF GASES 466

18-1	The Ideal Gas Law and the Molecular Interpretation of Temperature	466
18-2	Distribution of Molecular Speeds	470
18-3	Real Gases and Changes of Phase	473
*18-4	Vapor Pressure and Humidity	474
*18-5	Van der Waals Equation of State	477
*18-6	Mean Free Path	478
*18-7	Diffusion	479

SUMMARY 481   QUESTIONS 481
PROBLEMS 482   GENERAL PROBLEMS 484

19 HEAT AND THE FIRST LAW OF THERMODYNAMICS 485

19-1	Heat as Energy Transfer	485
19-2	Internal Energy	487
19-3	Specific Heat	488
19-4	Calorimetry—Solving Problems	489
19-5	Latent Heat	490
19-6	The First Law of Thermodynamics	493
19-7	Applying the First Law of Thermodynamics; Calculating the Work	495
19-8	Molar Specific Heats for Gases, and the Equipartition of Energy	498
19-9	Adiabatic Expansion of a Gas	502
19-10	Heat Transfer: Conduction, Convection, Radiation	503

SUMMARY 508   QUESTIONS 509
PROBLEMS 510   GENERAL PROBLEMS 514

20 SECOND LAW OF THERMODYNAMICS; HEAT ENGINES 516

20-1	The Second Law of Thermodynamics—Introduction	516
20-2	Heat Engines	517
20-3	Reversible and Irreversible Processes; The Carnot Engine	520
20-4	Refrigerators, Air Conditioners, and Heat Pumps	525
20-5	Entropy	528
20-6	Entropy and the Second Law of Thermodynamics	529
20-7	Order to Disorder	533
20-8	Energy Availability; Heat Death	534
*20-9	Statistical Interpretation of Entropy and the Second Law	535
*20-10	Thermodynamic Temperature Scale; Absolute Zero and the Third Law of Thermodynamics	537

SUMMARY 539   QUESTIONS 539
PROBLEMS 540   GENERAL PROBLEMS 543

PART 3

21 ELECTRIC CHARGE AND ELECTRIC FIELD 545

21-1	Static Electricity; Electric Charge and Its Conservation	546
21-2	Electric Charge in the Atom	547
21-3	Insulators and Conductors	547
21-4	Induced Charge; the Electroscope	548
21-5	Coulomb's Law	549
21-6	The Electric Field	554
21-7	Electric Field Calculations for Continuous Charge Distributions	558
21-8	Field Lines	561
21-9	Electric Fields and Conductors	562
21-10	Motion of a Charged Particle in an Electric Field	564
21-11	Electric Dipoles	565

SUMMARY 567   QUESTIONS 568
PROBLEMS 569   GENERAL PROBLEMS 572

22 GAUSS'S LAW 575

22-1	Electric Flux	576
22-2	Gauss's Law	578
22-3	Applications of Gauss's Law	580
*22-4	Experimental Basis of Gauss's and Coulomb's Law	586

SUMMARY 586   QUESTIONS 587
PROBLEMS 587   GENERAL PROBLEMS 590

23 ELECTRIC POTENTIAL 591

23-1	Electric Potential and Potential Difference	591
23-2	Relation Between Electric Potential and Electric Field	595
23-3	Electric Potential Due to Point Charges	597
23-4	Potential Due to Any Charge Distribution	599
23-5	Equipotential Surfaces	600
23-6	Electric Dipoles	601
23-7	E Determined from V	602
23-8	Electrostatic Potential Energy; the Electron Volt	603
*23-9	Cathode Ray Tube: TV and Computer Monitors, Oscilloscope	605

SUMMARY 607   QUESTIONS 607
PROBLEMS 608   GENERAL PROBLEMS 611

24 CAPACITANCE, DIELECTRICS, ELECTRIC ENERGY STORAGE 613

24-1	Capacitors	613
24-2	Determination of Capacitance	614
24-3	Capacitors in Series and Parallel	617
24-4	Electric Energy Storage	620
24-5	Dielectrics	621
*24-6	Molecular Description of Dielectrics	624

SUMMARY 627   QUESTIONS 627
PROBLEMS 628   GENERAL PROBLEMS 632

25 ELECTRIC CURRENTS AND RESISTANCE   634

25-1	The Electric Battery	635
25-2	Electric Current	636
25-3	Ohm's Law: Resistance and Resistors	638
25-4	Resistivity	640
25-5	Electric Power	642
25-6	Power in Household Circuits	644
25-7	Alternating Current	645
25-8	Microscopic View of Electric Current: Current Density and Drift Velocity	647
*25-9	Superconductivity	650
25-10	Electric Hazards; Leakage Currents	651

SUMMARY 653   QUESTIONS 653
PROBLEMS 654   GENERAL PROBLEMS 656

26 DC CIRCUITS   658

26-1	EMF and Terminal Voltage	659
26-2	Resistors in Series and in Parallel	660
26-3	Kirchhoff's Rules	664
26-4	Circuits Containing Resistor and Capacitor (RC Circuits)	669
*26-5	DC Ammeters and Voltmeters	674
*26-6	Transducers and the Thermocouple	676

SUMMARY 678   QUESTIONS 678
PROBLEMS 679   GENERAL PROBLEMS 683

27 MAGNETISM   686

27-1	Magnets and Magnetic Fields	686
27-2	Electric Currents Produce Magnetism	689
27-3	Force on an Electric Current in a Magnetic Field; Definition of **B**	689
27-4	Force on an Electric Charge Moving in a Magnetic Field	692
27-5	Torque on a Current Loop; Magnetic Dipole Moment	695
*27-6	Applications: Galvanometers, Motors, Loudspeakers	697
27-7	Discovery and Properties of the Electron	699
*27-8	The Hall Effect	701
*27-9	Mass Spectrometer	702

SUMMARY 702   QUESTIONS 703
PROBLEMS 704   GENERAL PROBLEMS 707

28 SOURCES OF MAGNETIC FIELD   709

28-1	Magnetic Field Due to a Straight Wire	710
28-2	Force Between Two Parallel Wires	710
28-3	Operational Definitions of the Ampere and the Coulomb	712
28-4	Ampère's Law	712
28-5	Magnetic Field of a Solenoid and a Toroid	716
28-6	Biot-Savart Law	719
*28-7	Magnetic Materials—Ferromagnetism	722
*28-8	Electromagnets and Solenoids	723
*28-9	Magnetic Fields in Magnetic Materials; Hysteresis	724
*28-10	Paramagnetism and Diamagnetism	725

SUMMARY 727   QUESTIONS 727
PROBLEMS 728   GENERAL PROBLEMS 732

29 ELECTROMAGNETIC INDUCTION AND FARADAY'S LAW 734

29-1	Induced EMF	734
29-2	Faraday's Law of Induction; Lenz's Law	736
29-3	EMF Induced in a Moving Conductor	739
29-4	Electric Generators	740
*29-5	Counter EMF and Torque; Eddy Currents	742
29-6	Transformers and Transmission of Power	744
29-7	A Changing Magnetic Flux Produces an Electric Field	747
*29-8	Applications of Induction: Sound Systems, Computer Memory, the Seismograph	749

SUMMARY 750 QUESTIONS 750
PROBLEMS 751 GENERAL PROBLEMS 754

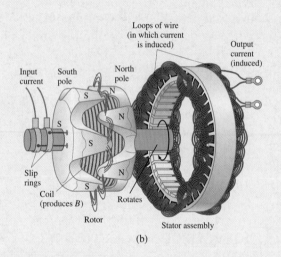

Loops of wire
(in which current
is induced)

Output
current
(induced)

Input
current

South
pole

North
pole

N

S

S

N

N

N

S

Slip
rings

Coil
(produces *B*)

Rotor

Rotates

Stator assembly

(b)

30 INDUCTANCE; AND ELECTROMAGNETIC OSCILLATIONS 756

30-1	Mutual Inductance	756
30-2	Self-Inductance	758
30-3	Energy Stored in a Magnetic Field	760
30-4	*LR* Circuits	762
30-5	*LC* Circuits and Electromagnetic Oscillations	764
30-6	*LC* Oscillations with Resistance (*LRC* Circuit)	766

SUMMARY 767 QUESTIONS 768
PROBLEMS 768 GENERAL PROBLEMS 770

31 AC CIRCUITS 772

31-1	Introduction: AC Circuits	772
31-2	AC Circuit Containing only Resistance *R*	773
31-3	AC Circuit Containing only Inductance *L*	773
31-4	AC Circuit Containing only Capacitance *C*	774
31-5	*LRC* Series AC Circuit	776
31-6	Resonance in AC Circuits	780
*31-7	Impedance Matching	781
*31-8	Three-Phase AC	782

SUMMARY 783 QUESTIONS 783
PROBLEMS 784 GENERAL PROBLEMS 785

32 MAXWELL'S EQUATIONS AND ELECTROMAGNETIC WAVES 787

32-1	Changing Electric Fields Produce Magnetic Fields; Ampère's Law and Displacement Current	788
32-2	Gauss's Law for Magnetism	791
32-3	Maxwell's Equations	792
32-4	Production of Electromagnetic Waves	792
32-5	Electromagnetic Waves, and Their Speed, from Maxwell's Equations	794
32-6	Light as an Electromagnetic Wave and the Electromagnetic Spectrum	798
32-7	Energy in EM Waves; the Poynting Vector	800
*32-8	Radiation Pressure	802
*32-9	Radio and Television	803

SUMMARY 806 QUESTIONS 807
PROBLEMS 807 GENERAL PROBLEMS 809

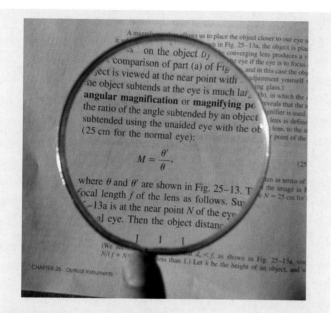

PART 4

33 LIGHT: REFLECTION AND REFRACTION 810

33-1	The Ray Model of Light	811
33-2	The Speed of Light and Index of Refraction	811
33-3	Reflection; Image Formation by a Plane Mirror	812
33-4	Formation of Images by Spherical Mirrors	816
33-5	Refraction: Snell's Law	822
33-6	Visible Spectrum and Dispersion	824
33-7	Total Internal Reflection; Fiber Optics	826
*33-8	Refraction at a Spherical Surface	828

SUMMARY 830   QUESTIONS 831
PROBLEMS 832   GENERAL PROBLEMS 835

34 LENSES AND OPTICAL INSTRUMENTS 836

34-1	Thin Lenses; Ray Tracing	837
34-2	The Lens Equation	840
34-3	Combinations of Lenses	843
34-4	Lensmaker's Equation	845
*34-5	Cameras	848
34-6	The Human Eye; Corrective Lenses	850
34-7	Magnifying Glass	853
*34-8	Telescopes	854
*34-9	Compound Microscope	856
*34-10	Aberrations of Lenses and Mirrors	858

SUMMARY 860   QUESTIONS 860
PROBLEMS 861   GENERAL PROBLEMS 864

35 WAVE NATURE OF LIGHT; INTERFERENCE 866

35-1	Huygens' Principle and Diffraction	867
35-2	Huygens' Principle and the Law of Refraction	867
35-3	Interference—Young's Double-Slit Experiment	870
35-4	Coherence	873
35-5	Intensity in the Double-Slit Interference Pattern	874
35-6	Interference in Thin Films	877
*35-7	Michelson Interferometer	881
*35-8	Luminous Intensity	882

SUMMARY 883   QUESTIONS 883
PROBLEMS 884   GENERAL PROBLEMS 885

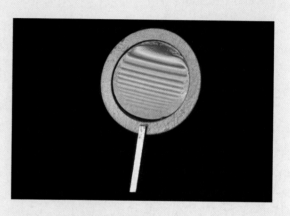

36 DIFFRACTION AND POLARIZATION 887

36-1	Diffraction by a Single Slit	888
36-2	Intensity in Single-Slit Diffraction Pattern	890
*36-3	Diffraction in the Double-Slit Experiment	893
36-4	Limits of Resolution; Circular Apertures	896
36-5	Resolution of Telescopes and Microscopes; the λ Limit	898
*36-6	Resolution of the Human Eye and Useful Magnification	899
36-7	Diffraction Grating	900
*36-8	The Spectrometer and Spectroscopy	901
*36-9	Peak Widths and Resolving Power for a Diffraction Grating	903
*36-10	X-Rays and X-Ray Diffraction	905
36-11	Polarization	907
*36-12	Scattering of Light by the Atmosphere	911

SUMMARY 911   QUESTIONS 912
PROBLEMS 913   GENERAL PROBLEMS 915

37 SPECIAL THEORY OF RELATIVITY   916

37-1   Galilean–Newtonian Relativity   917
*37-2 The Michelson–Morley Experiment   919
37-3   Postulates of the Special Theory
       of Relativity   922
37-4   Simultaneity   924
37-5   Time Dilation and the Twin Paradox   926
37-6   Length Contraction   930
37-7   Four-Dimensional Space–Time   932
37-8   Galilean and Lorentz Transformations   932
37-9   Relativistic Momentum and Mass   936
37-10 The Ultimate Speed   938
37-11 Energy and Mass; $E = mc^2$ 938
37-12 Doppler Shift for Light   942
37-13 The Impact of Special Relativity   943
       SUMMARY 944   QUESTIONS 945
       PROBLEMS 945   GENERAL PROBLEMS 947

PART 5

38 EARLY QUANTUM THEORY AND MODELS OF THE ATOM   949

38-1   Planck's Quantum Hypothesis   949
38-2   Photon Theory of Light and
       the Photoelectric Effect   952
38-3   Photons and the Compton Effect   955
38-4   Photon Interactions; Pair Production   957
38-5   Wave-Particle Duality; The Principle of
       Complementarity   958
38-6   Wave Nature of Matter   959
*38-7 Electron Microscopes   961
38-8   Early Models of the Atom   962
38-9   Atomic Spectra: Key to the Structure
       of the Atom   963
38-10 The Bohr Model   965
38-11 De Broglie's Hypothesis Applied to Atoms 971
       SUMMARY 972   QUESTIONS 973
       PROBLEMS 974   GENERAL PROBLEMS 976

39 QUANTUM MECHANICS   977

39-1   Quantum Mechanics—A New Theory   978
39-2   The Wave Function and
       Its Interpretation; the Double-Slit
       Experiment   979
39-3   The Heisenberg Uncertainty Principle   981
39-4   Philosophic Implications; Probability
       Versus Determinism   984
39-5   The Schrödinger Equation in One
       Dimension—Time-Independent Form   985
*39-6 Time-Dependent Schrödinger Equation   988
39-7   Free Particles; Plane Waves and Wave
       Packets   989
39-8   Particle in an Infinitely Deep Square
       Well Potential (a Rigid Box)   990
*39-9 Finite Potential Well   994
39-10 Tunneling through a Barrier   996
       SUMMARY 999   QUESTIONS 1000
       PROBLEMS 1000   GENERAL PROBLEMS 1002

40 QUANTUM MECHANICS OF ATOMS   1003

40-1   Quantum-Mechanical View of Atoms   1003
40-2   Hydrogen Atom: Schrödinger Equation
       and Quantum Numbers   1004
40-3   Hydrogen Atom Wave Functions   1007
40-4   Complex Atoms; the Exclusion Principle   1010
40-5   The Periodic Table of Elements   1012
40-6   X-Ray Spectra and Atomic Number   1013
*40-7 Magnetic Dipole Moments; Total
       Angular Momentum   1015
*40-8 Fluorescence and Phosphorescence   1019
*40-9 Lasers   1019
*40-10 Holography   1023
       SUMMARY 1025   QUESTIONS 1025
       PROBLEMS 1026   GENERAL PROBLEMS 1028

41 MOLECULES AND SOLIDS 1030

41-1	Bonding in Molecules	1030
41-2	Potential-Energy Diagrams for Molecules	1033
41-3	Weak (van der Waals) Bonds	1036
41-4	Molecular Spectra	1037
41-5	Bonding in Solids	1044
41-6	Free-Electron Theory of Metals	1045
41-7	Band Theory of Solids	1048
41-8	Semiconductors and Doping	1051
*41-9	Semiconductor Diodes	1052
*41-10	Transistors and Integrated Circuits	1054
	SUMMARY 1055 QUESTIONS 1056	
	PROBLEMS 1057 GENERAL PROBLEMS 1059	

42 NUCLEAR PHYSICS AND RADIOACTIVITY 1061

42-1	Structure and Properties of the Nucleus	1061
42-2	Binding Energy and Nuclear Forces	1064
42-3	Radioactivity	1067
42-4	Alpha Decay	1068
42-5	Beta Decay	1070
42-6	Gamma Decay	1072
42-7	Conservation of Nucleon Number and Other Conservation Laws	1073
42-8	Half-Life and Rate of Decay	1073
42-9	Decay Series	1077
42-10	Radioactive Dating	1078
42-11	Detection of Radiation	1080
	SUMMARY 1081 QUESTIONS 1081	
	PROBLEMS 1082 GENERAL PROBLEMS 1083	

43 NUCLEAR ENERGY: EFFECTS AND USES OF RADIATION 1085

43-1	Nuclear Reactions and the Transmutation of Elements	1085
43-2	Cross Section	1088
43-3	Nuclear Fission; Nuclear Reactors	1090
43-4	Fusion	1095
43-5	Passage of Radiation through Matter; Radiation Damage	1100
43-6	Measurement of Radiation—Dosimetry	1101
*43-7	Radiation Therapy	1104
*43-8	Tracers	1104
*43-9	Imaging by Tomography: CAT Scans, and Emission Tomography	1105
*43-10	Nuclear Magnetic Resonance (NMR) and Magnetic Resonance Imaging (MRI)	1107
	SUMMARY 1109 QUESTIONS 1110	
	PROBLEMS 1110 GENERAL PROBLEMS 1113	

44 ELEMENTARY PARTICLES 1114

44-1	High-Energy Particles	1115
44-2	Particle Accelerators and Detectors	1115
44-3	Beginnings of Elementary Particle Physics—Particle Exchange	1121
44-4	Particles and Antiparticles	1124
44-5	Particle Interactions and Conservation Laws	1124
44-6	Particle Classification	1126
44-7	Particle Stability and Resonances	1127
44-8	Strange Particles	1129
44-9	Quarks	1130
44-10	The "Standard Model": Quantum Chromodynamics (QCD) and the Electroweak Theory	1132
44-11	Grand Unified Theories	1134
	SUMMARY 1136 QUESTIONS 1137	
	PROBLEMS 1137 GENERAL PROBLEMS 1138	

45 ASTROPHYSICS AND AND COSMOLOGY **1140**

45-1 Stars and Galaxies   1141
45-2 Stellar Evolution; the Birth and Death of Stars   1145
45-3 General Relativity: Gravity and the Curvature of Space   1151
45-4 The Expanding Universe   1156
45-5 The Big Bang and the Cosmic Microwave Background   1159
45-6 The Standard Cosmological Model: The Early History of the Universe   1161
45-7 The Future of the Universe?   1165
SUMMARY 1168   QUESTIONS 1169
PROBLEMS 1170   GENERAL PROBLEMS 1171

APPENDICES

A   MATHEMATICAL FORMULAS   A–1
A-1   Quadratic Formula   A–1
A-2   Binomial Expansion   A–1
A-3   Other Expansions   A–1
A-4   Areas and Volumes   A–1
A-5   Plane Geometry   A–2
A-6   Trigonometric Functions and Identities   A–2
A-7   Logarithms   A–3

B   DERIVATIVES AND INTEGRALS   A–4

C   GRAVITATIONAL FORCE DUE TO A SPHERICAL MASS DISTRIBUTION   A–6

D   SELECTED ISOTOPES   A–9

ANSWERS TO ODD-NUMBERED PROBLEMS   A–14
INDEX   A–33
PHOTO CREDITS   A–57

PREFACE

A Brand New Third Edition

It has been more than ten years since the second edition of this calculus-based introductory physics textbook was published. A lot has changed since then, not only in physics itself, but also in how physics is presented. Research in how students learn has provided textbook authors new opportunities to help students learn physics and learn it well.

This third edition comes in two versions. The standard version covers all of classical physics plus a chapter on special relativity and one on the early quantum theory. The extended version, with modern physics, contains a total of nine detailed chapters on modern physics, ending with astrophysics and cosmology. This book retains the original approach: in-depth physics, concrete and nondogmatic, readable.

This new third edition has many improvements in the physics and its applications. Before discussing those changes in detail, here is a list of some of the overall changes that will catch the eye immediately.

Full color throughout is not just cosmetic, although fine color photographs do help to attract the student readers. More important, full color diagrams allow the physics to be displayed with much greater clarity. We have not stopped at a 4-color process; this book has actually been printed in 5 pure colors (5 passes through the presses) to provide better variety and definition for illustrating vectors and other physics concepts such as rays and fields. I want to emphasize that color is used pedagogically to bring out the physics. For example, different types of vectors are given different colors—see the chart on page xxxi.

Many more diagrams, almost double the number in the previous edition, have all been done or redone carefully using full color; there are many more graphs and many more photographs throughout. See for example in optics where new photographs show lenses and the images they make.

Marginal notes have been added as an aid to students to (i) point out what is truly important, (ii) serve as a sort of outline, and (iii) help students find details about something referred to later that they may not remember so well. Besides such "normal" marginal notes, there are also marginal notes that point out brief *problem solving* hints, and others that point out interesting *applications*.

The great laws of physics are emphasized by giving them a marginal note all in capital letters and enclosed in a rectangle. The most important equations, especially those expressing the great laws, are further emphasized by a tan-colored screen behind them.

Chapter opening photographs have been chosen to illustrate aspects of each chapter. Each was chosen with an eye to writing a caption which could serve as a kind of summary of what is in that chapter, and sometimes offer a challenge. Some chapter-opening photos have vectors or other analysis superimposed on them.

Page layout: complete derivations. Serious attention has been paid to how each page was formatted, especially for page turns. Great effort has been made to keep important derivations and arguments on facing pages. Students then don't have to turn back to check. More important, readers repeatedly see before them, on two facing pages, an important slice of physics.

Two new kinds of Examples: Conceptual Examples and **Estimates.**

New Physics

The whole idea of a new edition is to improve, to bring in new material, and to delete material that is verbose and only makes the book longer or is perhaps too advanced and not so useful. Here is a brief summary of a few of the changes involving the physics iself. These lists are selections, not complete lists.

New discoveries:
- planets revolving around distant stars
- Hubble Space Telescope
- updates in particle physics and cosmology, such as inflation and the age of the universe

New physics topics added:
- new treatment of how to make estimates (Chapter 1), including new Estimating Examples throughout (in Chapter 1, estimating the volume of a lake, and the radius of the Earth)
- symmetry used much more, including for solving problems
- new Tables illustrating the great range of lengths, time intervals, masses, voltages
- gravitation as curvature of space, and black holes (Chapter 6)
- engine efficiency (Chapter 8 as well as Chapter 20)
- rolling with and without slipping, and other useful details of rotational motion (Chapter 10)
- forces in structures including trusses, bridges, arches, and domes (Chapter 12)
- square wave (Chapter 15)
- using the Maxwell distribution (Chapter 18)
- Otto cycle (Chapter 20)
- statistical calculation of entropy change in free expansion (Chapter 20)
- effects of dielectrics on capacitor connected and not (Chapter 24)
- grounding to avoid electric hazards (Chapter 25)
- three phase ac (Chapter 31)
- equal energy in **E** and **B** of EM wave (Chapter 32)
- radiation pressure, EM wave (Chapter 32)
- photos of lenses and mirrors with their images (Chapter 33)
- detailed outlines for ray tracing with mirrors and lenses (Chapters 33, 34)
- lens combinations (Chapter 34)
- new radiation standards (Chapter 43)
- Higgs boson, supersymmetry (Chapter 44)

Modern physics. A number of modern physics topics are discussed in the framework of classical physics. Here are some highlights:
- gravitation as curvature of space, and black holes (Chapter 6)
- planets revolving around distant stars (Chapter 6)
- kinetic energy at relativistic speeds (Chapter 7)
- nuclear collisions (Chapter 9)
- star collapse (Chapter 10)
- galaxy red shift, Doppler (Chapter 16)
- atoms, theory of (Chapters 17, 18, 21)
- atomic theory of thermal expansion (Chapter 17)
- mass of hydrogen atom (Chapter 17)
- atoms and molecules in gases (Chapters 17, 18)
- molecular speeds (Chapter 18)
- equipartition of energy; molar specific heats (Chapter 19)
- star size (Chapter 19)
- molecular dipoles (Chapters 21, 23)
- cathode ray tube (Chapters 23, 27)
- electrons in a wire (Chapter 25)
- superconductivity (Chapter 25)
- discovery and properties of the electron, e/m, oil drop experiment (Chapter 27)
- Hall effect (Chapter 27)

- magnetic moment of electrons (Chapter 27)
- mass spectrometer (Chapter 27)
- velocity selector (Chapter 27)
- electron spin in magnetic materials (Chapter 28)
- light and EM wave emission (Chapter 32)
- spectroscopy (Chapter 36)

Many other examples of modern physics are found as Problems, even in early chapters. Chapters 37 and 38 contain the modern physics topics of Special Relativity, and an introduction to Quantum Theory and Models of the Atom. The longer version of this text, "with Modern Physics," contains an additional seven chapters (for a total of nine) which present a detailed and extremely up-to-date treatment of modern physics: Quantum Mechanics of Atoms (Chapters 38 to 40); Molecules and Condensed Matter (Chapter 41); Nuclear Physics (Chapter 42 and 43); Elementary Particles (Chapter 44); and finally Astrophysics, General Relativity, and Cosmology (Chapter 45).

Revised physics and reorganizations. First of all, a major effort has been made to not throw everything at the students in the first few chapters. The basics have to be learned first; many aspects can come later, when the students are more prepared. Secondly, a great part of this book has been rewritten to make it clearer and more understandable to students. Clearer does not always mean simpler or easier. Sometimes making it "easier" actually makes it harder to understand. Often a little more detail, without being verbose, can make an explanation clearer. Here are a few of the changes, big and small:

- new graphs and diagrams to clarify velocity and acceleration; deceleration carefully treated.
- unit conversion now a new Section in Chapter 1, instead of interrupting kinematics.
- circular motion: Chapter 3 now gives only the basics, with more complicated treatment coming later: non-uniform circular motion in Chapter 5, angular variables in Chapter 10.
- Newton's second law now written throughout as $m\mathbf{a} = \Sigma\mathbf{F}$, to emphasize inclusion of all forces acting on a body.
- Newton's third law follows the second directly, with inertial reference frames placed earlier. New careful discussions to head off confusion when using Newton's third law.
- careful rewriting of chapters on Work and Energy, especially potential energy, conservative and nonconservative forces, and the conservation of energy.
- renewed emphasis that $\Sigma\tau = I\alpha$ is not always valid: only for an axis fixed in an inertial frame or if axis is through the CM (Chapters 10 and 11).
- rolling motion introduced early in Chapter 10, with more details later, including rolling with and without slipping.
- rotating frames of reference, and Coriolis, moved later, to Chapter 11, shortened, optional, but still including why an object does not fall straight down on Earth.
- fluids reduced to a single chapter (13); some topics and details dropped or greatly shortened.
- clearer details on how an object floats (Chapter 13).
- distinction between wave interference in space, and in time (beats) (Chapter 16).
- thermodynamics reduced to four chapters; the old chapters on Heat and on the First Law of Thermodynamics have been combined into one (19), with some topics shortened and a more rational sequence of topics achieved.
- heat transfer now follows the first law of thermodynamics (Chapter 19).
- electric potential carefully rewritten for accuracy (Chapter 23).
- CRT, computer monitors, TV, treated earlier (Chapter 23).
- use of Q_{encl} and I_{encl} for Gauss's and Ampère's laws, with subscripts meaning "enclosed".
- Ohm's law and definition of resistance carefully redone (Chapter 25).
- sources of magnetic field, Chapter 28, reorganized for ease of understanding, with some new material, and deletion of the advanced topic on magnetization vector.
- circuits with L, C, and/or R now introduced via Kirchhoff's loop rule, and clarified in other ways too (Chapters 30, 31).
- streamlined Maxwell's equations, with displacement current downplayed (Chapter 32).
- optics reduced to four chapters; polarization is now placed in the same chapter as diffraction.

New Pedagogy

All of the above mentioned revisions, rewritings, and reorganizations are intended to help students learn physics better. They were done in response to contemporary research in how students learn, as well as to kind and generous input from professors who have read, reviewed, or used the previous editions. This new edition also contains some new elements, especially an increased emphasis on conceptual development:

Conceptual Examples, typically 1 or 2 per chapter, sometimes more, are each a sort of brief Socratic question and answer. It is intended that students will be stimulated by the question to think, or reflect, and come up with a response—before reading the Response given. Here are a few:

- using symmetry (Chapters 1, 44, and elsewhere)
- ball moving upward: misconceptions (Chapter 2)
- reference frames and projectile motion: where does the apple land? (Chapter 3)
- what exerts the force that makes a car move? (Chapter 4)
- Newton's third law clarification: pulling a sled (Chapter 4)
- free-body diagram for a hockey puck (Chapter 4)
- advantage of a pulley (Chapter 4), and of a lever (Chapter 12)
- to push or to pull a sled (Chapter 5)
- which object rolls down a hill faster? (Chapter 10)
- moving the axis of a spinning wheel (Chapter 11)
- tragic collapse (Chapter 12)
- finger at top of a full straw (Chapter 13)
- suction cups on a spacecraft (Chapter 13)
- doubling amplitude of SHM (Chapter 14)
- do holes expand thermally? (Chapter 17)
- simple adiabatic process: stretching a rubber band (Chapter 19)
- charge inside a conductor's cavity (Chapter 22)
- how stretching a wire changes its resistance (Chapter 25)
- series or parallel (Chapter 26)
- bulb brightness (Chapter 26)
- spiral path in magnetic field (Ch. 27)
- practice with Lenz's law (Chapter 29)
- motor overload (Chapter 29)
- emf direction in inductor (Chapter 30)
- photo with reflection—is it upside down? (Chapter 33)
- reversible light rays (Chapter 33)
- how tall must a full-length mirror be? (Chapter 33)
- diffraction spreading (Chapter 36)

Estimating Examples, roughly 10% of all Examples, also a new feature of this edition, are intended to develop the skills for making order-of-magnitude estimates, even when the data are scarce, and even when you might never have guessed that any result was possible at all. See, for example, Section 1–6, Examples 1–5 to 1–8.

Problem Solving, with New and Improved Approaches

Learning how to approach and solve problems is a basic part of any physics course. It is a highly useful skill in itself, but is also important because the process helps bring understanding of the physics. Problem solving in this new edition has a significantly increased emphasis, including some new features.

Problem-solving boxes, about 20 of them, are new to this edition. They are more concentrated in the early chapters, but are found throughout the book. They each outline a step-by-step approach to solving problems in general, and/or specifically for the material being covered. The best students may find these separate "boxes" unnecessary (they can skip them), but many students will find it helpful to be reminded of the general approach and of steps they can take to get started; and, I think, they help to build confidence. The general problem solving box in Section 4–8 is placed there, after students have had some experience wrestling with problems, and so may be strongly motivated to read it with close attention. Section 4–8 can, of course, be covered earlier if desired.

Problem-solving Sections occur in many chapters, and are intended to provide extra drill in areas where solving problems is especially important or detailed.

Examples. This new edition has many more worked-out Examples, and they all now have titles for interest and for easy reference. There are even two new categories of Example: Conceptual, and Estimates, as described above. Regular Examples serve as "practice problems". Many new ones have been added, some of the old ones have been dropped, and many have been reworked to provide greater clarity and detail: more steps are spelled out, more of "why we do it this way", and more discussion of the reasoning and approach. In sum, the idea is "to think aloud with the students", leading them to develop insight. The total number of worked-out Examples is about 30% greater than in the previous edition, for an average of 12 to 15 per chapter. There is a significantly higher concentration of Examples in the early chapters, where drill is especially important for developing skills and a variety of approaches. The level of the worked-out Examples for most topics increases gradually, with the more complicated ones being on a par with the most difficult Problems at the end of each chapter, so that students can see how to approach complex problems. Many of the new Examples, and improvements to old ones, provide relevant applications to engineering, other related fields, and to everyday life.

Problems at the end of each chapter have been greatly increased in quality and quantity. There are over 30% more Problems than in the second edition. Many of the old ones have been replaced, or rewritten to make them clearer, and/or have had their numerical values changed. Each chapter contains a large group of Problems arranged by Section and graded according to difficulty: level I Problems are simple, designed to give students confidence; level II are "normal" Problems, providing more of a challenge and often the combination of two different concepts; level III are the most complex, typically combining different issues, and will challenge even superior students. The arrangement by Section number means only that those Problems depend on material up to and including that Section: earlier material may also be relied upon. The ranking of Problems by difficulty (I, II, III) is intended only as a guide.

General Problems. About 70% of Problems are ranked by level of difficulty (I, II, III) and arranged by Section. New to this edition are General Problems that are unranked and grouped together at the end of each chapter, and account for about 30% of all problems. The average total number of Problems per chapter is about 90. Answers to odd-numbered Problems are given at the back of the book.

Complete Physics Coverage, with Options

This book is intended to give students the opportunity to obtain a thorough background in all areas of basic physics. There is great flexibility in choice of topics so that instructors can choose which topics they cover and which they omit. Sections marked with an asterisk can be considered optional, as discussed more fully on p. xxv. Here I want to emphasize that topics not covered in class can still be read by serious students for their own enrichment, either immediately or later. Here is a partial list of physics topics, not the standard ones, but topics that might not usually be covered, and that represent how thorough this book is in its coverage of basic physics. Section numbers are given in parentheses.

- use of calculus; variable acceleration (2–8)
- nonuniform circular motion (5–4)
- velocity-dependent forces (5–5)
- gravitational versus inertial mass; principle of equivalence (6–8)
- gravitation as curvature of space; black holes (6–9)
- kinetic energy at very high speed (7–5)
- potential energy diagrams (8–9)
- systems of variable mass (9–10)
- rotational plus translational motion (10–11)
- using $\Sigma\tau_{CM} = I_{CM}\alpha_{CM}$ (10–11)
- derivation of $K = K_{CM} + K_{rot}$ (10–11)
- why does a rolling sphere slow down? (10–12)
- angular momentum and torque for a system (11–4)
- derivation of $d\mathbf{L}_{CM}/dt = \Sigma\boldsymbol{\tau}_{CM}$ (11–4)
- rotational imbalance (11–6)
- the spinning top (11–8)
- rotating reference frames; inertial forces (11–9)
- coriolis effect (11–10)
- trusses (12–7)
- flow in tubes: Poiseuille's equation (13–11)
- surface tension and capillarity (13–12)
- physical pendulum; torsion pendulum (14–6)
- damped harmonic motion: finding the solution (14–7)
- forced vibrations; equation of motion and its solution; Q-value (14–8)
- the wave equation (15–5)
- mathematical representation of waves; pressure wave derivation (16–2)
- intensity of sound related to amplitude (16–3)
- interference in space and in time (16–6)
- atomic theory of expansion (17–4)
- thermal stresses (17–5)
- ideal gas temperature scale (17–10)
- calculations using the Maxwell distribution of molecular speeds (18–2)
- real gases (18–3)
- vapor pressure and humidity (18–4)
- van der Waals equation of state (18–5)
- mean free path (18–6)
- diffusion (18–7)
- equipartition of energy (19–8)
- energy availability; heat death (20–8)
- statistical interpretation of entropy and the second law (20–9)
- thermodynamic temperature scale; absolute zero and the third law (20–10)
- electric dipoles (21–11, 23–6)
- experimental basis of Gauss's and Coulomb's laws (22–4)
- general relation between electric potential and electric field (23–2, 23–8)
- electric fields in dielectrics (24–5)
- molecular description of dielectrics (24–6)
- current density and drift velocity (25–8)
- superconductivity (25–9)
- RC circuits (26–4)
- use of voltmeters and ammeters; effects of meter resistance (26–5)
- transducers (26–6)
- magnetic dipole moment (27–5)
- Hall effect (27–8)
- operational definition of the ampere and coulomb (28–3)
- magnetic materials—ferromagnetism (28–7)
- electromagnets and solenoids (28–8)
- hysteresis (28–9)
- paramagnetism and diamagnetism (28–10)
- counter emf and torque; eddy currents (29–5)
- Faraday's law—general form (29–7)
- force due to changing $\mathbf{B}$ is nonconservative (29–7)
- LC circuits and EM oscillations (30–5)
- AC resonance; oscillators (31–6)
- impedance matching (31–7)
- three phase AC (31–8)
- changing electric fields produce magnetic fields (32–1)
- speed of light from Maxwell's equations (32–5)
- radiation pressure (32–8)
- fiber optics (33–7)
- lens combinations (34–3)
- aberrations of lenses and mirrors (34–10)
- coherence (35–4)
- intensity in double-slit pattern (35–5)
- luminous intensity (35–8)
- intensity for single-slit (36–2)
- diffraction for double–slit (36–3)
- limits of resolution, the λ limit (36–4, 36–5)
- resolution of the human eye and useful magnification (36–6)
- spectroscopy (36–8)
- peak widths and resolving power for a diffraction grating (36–9)
- x-rays and x-ray diffraction (36–10)
- scattering of light by the atmosphere (36–12)
- time–dependent Schrödinger equation (39–6)
- wave packets (39–7)
- tunneling through a barrier (39–9)
- free-electron theory of metals (41–6)
- semiconductor electronics (41–9)
- standard model, symmetry, QCD, GUT (44–9, 44–10)
- astrophysics, cosmology (Ch. 45)

New Applications

Relevant applications to everyday life, to engineering, and to other fields such as geology and medicine, provide students with motivation and offer the instructor the opportunity to show the relevance of physics. Applications are a good response to students who ask "Why study physics?" Many new applications have been added in this edition. Here are some highlights:

- airbags (Chapter 2)
- elevator and counterweight (Chapter 4)
- antilock brakes and skidding (Chapter 5)
- geosynchronous satellites (Chapter 6)
- hard drive and bit speed (Chapter 10)
- star collapse (Chapter 10)
- forces within trusses, bridges, arches, domes (Chapter 12)
- the Titanic (Chapter 12)
- Bernoulli's principle: wings, sailboats, TIA, plumbing traps and bypasses (Chapter 13)
- pumps (Chapter 13)
- car springs, shock absorbers, building dampers for earthquakes (Chapter 14)
- loudspeakers (Chapters 14, 16, 27)
- autofocusing cameras (Chapter 16)
- sonar (Chapter 16)
- ultrasound imaging (Chapter 16)
- thermal stresses (Chapter 17)
- R-values, thermal insulation (Ch. 19)
- engines (Chapter 20)
- heat pumps, refrigerators, AC; coefficient of performance (Chapter 20)
- thermal pollution (Chapter 20)
- electric shielding (Chapters 21, 28)
- photocopier (Chapter 21)
- superconducting cables (Chapter 25)
- jump starting a car (Chapter 26)
- aurora borealis (Chapter 27)
- solenoids and electromagnetics (Ch. 28)
- computer memory and digital information (Chapter 29)
- seismograph (Chapter 29)

- tape recording (Chapter 29)
- loudspeaker cross-over network (Ch. 31)
- antennas, for **E** or **B** (Chapter 32)
- TV and radio; AM and FM (Chapter 32)
- eye and corrective lenses (Chapter 34)
- mirages (Chapter 35)
- liquid crystal displays (Chapter 36)
- CAT scans, PET, MRI (Chapter 43)

Some old favorites retained (and improved):

- pressure gauges (Chapter 13)
- musical instruments (Chapter 16)
- humidity (Chapter 18)
- CRT, TV, computer monitors (Ch. 23, 27)
- electric hazards (Chapter 25)
- power in household circuits (Chapter 25)
- ammeters and voltmeters (Chapter 26)
- microphones (Chapters 26, 29)
- transducers (Chapter 26, and elsewhere)
- electric motors (Chapter 27)
- car alternator (Chapter 29)
- electric power transmission (Chapter 29)
- capacitors as filters (Chapter 31)
- impedance matching (Chapter 31)
- fiber optics (Chapter 33)
- cameras, telescopes, microscopes, other optical instruments (Chapter 34)
- lens coatings (Chapter 35)
- spectroscopy (Chapter 36)
- electron microscopes (Chapter 38)
- lasers, holography, CD players (Ch. 40)
- semiconductor electronics (Chapter 41)
- radioactivity (Chapters 42 and 43)

Deletions

Something had to go, or the book would have been too long. Lots of subjects were shortened—the detail simply isn't necessary at this level. Some topics were dropped entirely: polar coordinates; center-of-momentum reference frame; Reynolds number (now a Problem); object moving in a fluid and sedimentation; derivation of Poiseuille's equation; Stoke's equation; waveguide and transmission line analysis; electric polarization and electric displacement vectors; potentiometer (now a Problem); negative pressure; combinations of two harmonic motions; adiabatic character of sound waves; central forces.

Many topics have been shortened, often a lot, such as: velocity-dependent forces; variable acceleration; instantaneous axis; surface tension and capillarity; optics topics such as some aspects of light polarizarion. Many of the brief historical and philosophical issues have been shortened as well.

General Approach

This book offers an in-depth presentation of physics, and retains the basic approach of the earlier editions. Rather than using the common, dry, dogmatic approach of treating topics formally and abstractly first, and only later relating the material to the students' own experience, my approach is to recognize that physics is a description of reality and thus to start each topic with concrete observations and experiences that students can directly relate to. Then we move on to the generalizations and more formal treatment of the topic. Not only does this make the material more interesting and easier to understand, but it is closer to the way physics is actually practiced.

This new edition, even more than previous editions, aims to explain the physics in a readable and interesting manner that is accessible and clear. It aims to teach students by anticipating their needs and difficulties, but without oversimplifying. Physics is all about us. Indeed, it is the goal of this book to help students "see the world through eyes that know physics."

As mentioned above, this book includes of a wide range of Examples and applications from technology, engineering, architecture, earth sciences, the environment, biology, medicine, and daily life. Some applications serve only as examples of physical principles. Others are treated in depth. But applications do not dominate the text—this is, after all, a physics book. They have been carefully chosen and integrated into the text so as not to interfere with the development of the physics but rather to illuminate it. You won't find essay sidebars here. The applications are integrated right into the physics. To make it easy to spot the applications, a new *Physics Applied* marginal note is placed in the margin (except where diagrams in the margin prevent it).

It is assumed that students have started calculus or are taking it concurrently. Calculus is treated gently at first, usually in an optional Section so as not to burden students taking calculus concurrently. For example, using the integral in kinematics, Chapter 2, is an optional Section. But in Chapter 7, on work, the integral is discussed fully for all readers.

Throughout the text, *Système International* (SI) units are used. Other metric and British units are defined for informational purposes. Careful attention is paid to significant figures. When a certain value is given as, say, 3, with its units, it is meant to be 3, not assumed to be 3.0 or 3.00. When we mean 3.00 we write 3.00. It is important for students to be aware of the uncertainty in any measured value, and not to overestimate the precision of a numerical result.

Rather than start this physics book with a chapter on mathematics, I have instead incorporated many mathematical tools, such as vector addition and multiplication, directly in the text where first needed. In addition, the Appendices contain a review of many mathematical topics such as trigonometric identities, integrals, and the binomial (and other) expansions. One advanced topic is also given an Appendix: integrating to get the gravitational force due to a spherical mass distribution.

It is necessary, I feel, to pay careful attention to detail, especially when deriving an important result. I have aimed at including all steps in a derivation, and have tried to make clear which equations are general, and which are not, by explicitly stating the limitations of important equations in brackets next to the equation, such as

$$x = x_0 + v_0 t + \tfrac{1}{2}at^2. \qquad \text{[constant acceleration]}$$

The more detailed introduction to Newton's laws and their use is of crucial pedagogic importance. The many new worked-out Examples include initially fairly simple ones that provide careful step-by-step analysis of how to proceed in solving dynamics problems. Each succeeding Example adds a new element or a new twist that introduces greater complexity. It is hoped that this strategy will enable even less-well-prepared students to acquire the tools for using Newton's laws correctly. If students don't surmount this crucial hurdle, the rest of physics may remain forever beyond their grasp.

Rotational motion is difficult for most students. As an example of attention to detail (although this is not really a "detail"), I have carefully distinguished the position vector (**r**) of a point and the perpendicular distance of that point from an axis, which is

called R in this book (see Fig. 10–2). This distinction, which enters particularly in connection with torque, moment of inertia, and angular momentum, is often not made clear—it is a disservice to students to use **r** or r for both without distinguishing. Also, I have made clear that it is not always true that $\Sigma\tau = I\alpha$. It depends on the axis chosen (valid if axis is fixed in an inertial reference frame, or through the CM). To not tell this to students can get them into serious trouble. (See pp. 250, 283, 284.) I have treated rotational motion by starting with the simple instance of rotation about an axis (Chapter 10), including the concepts of angular momentum and rotational kinetic energy. Only in Chapter 11 is the more general case of rotation about a point dealt with, and this slightly more advanced material can be omitted if desired (except for Sections 11–1 and 11–2 on the vector product and the torque vector). The end of Chapter 10 has an optional subsection containing three slightly more advanced Examples, using $\Sigma\tau_{CM} = I_{CM}\alpha_{CM}$: car braking distribution, a falling yo-yo, and a sphere rolling with and without slipping.

Among other special treatments is Chapter 28, Sources of Magnetic Field: here, in one chapter, are discussed the magnetic field due to currents (including Ampère's law and the law of Biot-Savart) as well as magnetic materials, ferromagnetism, paramagnetism, and diamagnetism. This presentation is clearer, briefer, and more of a whole, and all the content is there.

Organization

The general outline of this new edition retains a traditional order of topics: mechanics (Chapters 1 to 12); fluids, vibrations, waves, and sound (Chapter 13 to 16); kinetic theory and thermodynamics (Chapters 17 to 20). In the two-volume version of this text, volume I ends here, after Chapter 20. The text continues with electricity and magnetism (Chapters 21 to 32), light (Chapters 33 to 36), and modern physics (Chapters 37 and 38 in the short version, Chapters 37 to 45 in the extended version "with Modern Physics"). Nearly all topics customarily taught in introductory physics courses are included. A number of topics from modern physics are included with the classical physics chapters as discussed earlier.

The tradition of beginning with mechanics is sensible, I believe, because it was developed first, historically, and because so much else in physics depends on it. Within mechanics, there are various ways to order topics, and this book allows for considerable flexibility. I prefer, for example, to cover statics after dynamics, partly because many students have trouble working with forces without motion. Besides, statics is a special case of dynamics—we study statics so that we can prevent structures from becoming dynamic (falling down)—and that sense of being at the limit of dynamics is intuitively helpful. Nonetheless statics (Chapter 12) can be covered earlier, if desired, before dynamics, after a brief introduction to vector addition. Another option is light, which I have placed after electricity and magnetism and EM waves. But light could be treated immediately after the chapters on waves (Chapters 15 and 16). Special relativity is Chapter 37, but could instead be treated along with mechanics—say, after Chapter 9.

Not every chapter need be given equal weight. Whereas Chapter 4 might require $1\frac{1}{2}$ to 2 weeks of coverage, Chapter 16 or 22 may need only $\frac{1}{2}$ week.

Some instructors may find that this book contains more material than can be covered completely in their courses. But the text offers great flexibility in choice of topics. Sections marked with a star (asterisk) are considered optional. These Sections contain slightly more advanced physics material, or material not usually covered in typical courses, and/or interesting applications. They contain no material needed in later chapters (except perhaps in later optional Sections). This does not imply that all nonstarred Sections must be covered: there still remains considerable flexibility in the choice of material. For a brief course, all optional material could be dropped as well as major parts of Chapters 11, 13, 16, 26, 30, 31, and 36 as well as selected parts of Chapters 9, 12, 19, 20, 32, 34, and the modern physics chapters. Topics not covered in class can be a valuable resource for later study; indeed, this text can serve as a useful reference for students for years because of its wide range of coverage.

Thanks

Some 60 physics professors provided input or direct feedback on every aspect of this textbook. The reviewers and contributors to this third edition are listed below. I owe each a debt of gratitude.

Ralph Alexander, University of Missouri at Rolla

Zaven Altounian, McGill University

Charles R. Bacon, Ferris State University

Bruce Birkett, University of California, Berkeley

Art Braundmeier, Southern Illinois University at Edwardsville

Wayne Carr, Stevens Institute of Technology

Edward Chang, University of Massachusetts, Amherst

Charles Chiu, University of Texas at Austin

Lucien Crimaldi, University of Mississippi

Robert Creel, University of Akron

Alexandra Cowley, Community College of Philadelphia

Timir Datta, University of South Carolina

Gary DeLeo, Lehigh University

John Dinardo, Drexel University

Paul Draper, University of Texas, Arlington

Alex Dzierba, Indiana University

William Fickinger, Case Western University

Jerome Finkelstein, San Jose State University

Donald Foster, Wichita State University

Gregory E. Frances, Montana State University

Lothar Frommhold, University of Texas at Austin

Thomas Furtak, Colorado School of Mines

Edward Gibson, California State University, Sacramento

Christopher Gould, University of Southern California

John Gruber, San Jose State University

Martin den Boer, Hunter College

Greg Hassold, General Motors Institute

Joseph Hemsky, Wright State University

Laurent Hodges, Iowa State University

Mark Holtz, Texas Tech University

James P. Jacobs, University of Montana

James Kettler, Ohio University Eastern Campus

Jean Krisch, University of Michigan

Mark Lindsay, University of Louisville

Eugene Livingston, University of Notre Dame

Bryan Long, Columbia State Community College

Daniel Mavlow, Princeton University

Pete Markowitz, Florida International University

John McCullen, University of Arizona, Tucson

Peter Nemeth, New York University

Hon-Kie Ng, Florida State University

Eugene Patroni, Georgia Institute of Technology

Robert Pelcovits, Brown University

William Pollard, Valdosta State University

Joseph Priest, Miami University

Carl Rotter, West Virginia University

Lawrence Rees, Brigham Young University

Peter Riley, University of Texas at Austin

Roy Rubins, University of Texas at Arlington

Mark Semon, Bates College

Robert Simpson, University of New Hampshire

Mano Singham, Case Western University

Harold Slusher, University of Texas at El Paso

Don Sparks, Los Angeles Pierce Community College

Michael Strauss, University of Oklahoma

Joseph Strecker, Wichita State University

William Sturrus, Youngstown State University

Arthur Swift, University of Massachusetts, Amherst

Leo Takahasi, The Pennsylvania State University

Edward Thomas, Georgia Institute of Technology

Som Tyagi, Drexel University

John Wahr, University of Colorado

Robert Webb, Texas A & M University

James Whitmore, The Pennsylvania State University

W. Steve Quon, Ventura College

I owe special thanks to Irv Miller, not only for many helpful physics discussions, but for having worked out all the Problems and managed the team that also worked out the Problems, each checking the other, and finally for producing the Solutions Manual and all the answers to the odd-numbered Problems at the end of this book. He was ably assisted by Zaven Altounian and Anand Batra.

I am particularly grateful to Robert Pelcovits and Peter Riley, as well as to Paul Draper and James Jacobs, who inspired many of the new Examples, Conceptual Examples, and Problems.

Crucial for rooting out errors, as well as providing excellent suggestions, were the perspicacious Edward Gibson and Michael Strauss, both of whom carefully checked all aspects of the physics in page proof.

Special thanks to Bruce Birkett for input of every kind, from illuminating discussions on pedagogy to a careful checking of details in many sections of this book. I wish also to thank Professors Howard Shugart, Joe Cerny, Roger Falcone and Buford Price for helpful discussions, and for hospitality at the University of California, Berkeley. Many thanks also to Prof. Tito Arecchi at the Istituto Nazionale di Ottica, Florence, Italy, and to the staff of the Institute and Museum for the History of Science, Florence, for their hospitality.

Finally, I wish to thank the superb editorial and production work provided by all those with whom I worked directly at Prentice Hall: Susan Fisher, Marilyn Coco, David Chelton, Kathleen Schiaparelli, Trudy Pisciotti, Gus Vibal, Mary Teresa Giancoli, and Jocelyn Phillips.

The biggest thanks of all goes to Paul Corey, whose constant encouragement and astute ability to get things done, provided the single strongest catalyst.

The final responsibility for all errors lies with me, of course. I welcome comments and corrections.

D.C.G.

AVAILABLE SUPPLEMENTS

For the Student

Student Study Guide and Solutions Manual
Douglas Brandt, Eastern Illinois University. (0-13-021475-2)
Contains chapter objectives, summaries with additional examples, self-study quizzes, key mathematical equations, and complete worked-out solutions to alternate odd problems in the text.

Doing Physics with Spreadsheets: A Workbook
Gordon Aubrecht, T. Kenneth Bolland, and Michael Ziegler, all of The Ohio State University.
(0-13-021474-4)
Designed to introduce students to the use of spreadsheets for solving simple and complex physics problems. Students are either provided with spreadsheets or must construct their own, then use the model to most closely approximate natural behavior. The amount of spreadsheet construction and the complexity of the spreadsheet increases as the student gains experience.

Science on the Internet: A Student's Guide, 1999
Andrew Stull and Carl Adler (0-13-021308-X)
The perfect tool to help students take advantage of the *Physics for Scientists and Engineers, Third Edition* Web page. This useful resource gives clear steps to access Prentice Hall's regularly updated physics resources, along with an overview of general World Wide Web navigation strategies. Available FREE for students when packaged with the text.

Prentice Hall/*New York Times* Themes of the Times — Physics
This unique newspaper supplement brings together a collection of the latest physics-related articles from the pages of *The New York Times*. Updated twice per year and available FREE to students when packaged with the text.

For the Instructor

Instructor's Solutions Manual
Irvin A. Miller, Drexel University.
Print version (0-13-021381-0); Electronic (CD-ROM) version (0-13-021481-7)
Contains detailed worked solutions to every problem in the text. Electronic versions are available in CD-ROM (dual platform for both Windows and Macintosh systems) for instructors with Microsoft Word or Word-compatible software.

Test Item File
Robert Pelcovits, Brown University; David Curott, University of North Alabama; and Edward Oberhofer, University of North Carolina at Charlotte (0-13-021482-5)
Contains over 2200 multiple choice questions, about 25% conceptual in nature. All are referenced to the corresponding Section in the text and ranked by difficulty.

Prentice Hall Custom Test Windows (0-13-021477-9); Macintosh (0-13-021476-0)
Based on the powerful testing technology developed by Engineering Software Associates, Inc. (ESA), Prentice Hall Custom Test includes all questions from the Test Item File and allows instructors to create and tailor exams to their own needs. With the Online Testing Program, exams can also be administered on line and data can then be automatically transferred for evaluation. A comprehensive desk reference guide is included along with online assistance.

Transparency Pack (0-13-021470-1)
Includes approximately 400 full color transparencies of images from the text.

Media Supplements

Physics for Scientists and Engineers Web Site www.prenhall.com/giancoli
A FREE innovative online resource that provides students with a wealth of activities and exercises for each text chapter. Features on the site include:

- Practice Questions, Destinations (links to related sites), NetSearch keywords and algorithmically generated numeric Practice Problems by Carl Adler of East Carolina University.
- Physlet Problems (Java-applet simulations) by Wolfgang Christian of Davidson College.
- Warmups and Puzzles essay questions and Applications from Gregor Novak and Andrew Gavrin at Indiana University-Purdue University, Indianapolis.
- Ranking Task Exercises edited by Tom O'Kuma of Lee College, Curtis Hieggelke of Joliet Junior College and David Maloney of Indiana University-Purdue University, Fort Wayne.

Using Prentice Hall CW '99 technology, the website grades and scores all objective questions, and results can be automatically e-mailed directly to the instructors if so desired. Instructors can also create customized syllabi online and link directly to activities on the Giancoli website.

Presentation Manager CD-ROM

Dual Platform (Windows/Macintosh; 0-13-214479-5)
This CD-ROM enables instructors to build custom sequences of Giancoli text images and Prentice Hall digital media for playback in lecture presentations. The CD-ROM contains all text illustrations, digitized segments from the Prentice Hall *Physics You Can See* videotape as well as additional lab and demonstration videos and animations from the Prentice Hall *Interactive Journey Through Physics* CD-ROM. Easy to navigate with Prentice Hall Presentation Manager software, instructors can preview, sequence, and play back images, as well as perform keyword searches, add lecture notes, and incorporate their own digital resources.

Physics You Can See *Video*

(0-205-12393-7)
Contains eleven two- to five-minute demonstrations of classical physics experiments. It includes segments such as "Coin and Feather" (acceleration due to gravity), "Monkey and Gun" (projectile motion), "Swivel Hips" (force pairs), and "Collapse a Can" (atmospheric pressure).

CAPA: A Computer-Assisted Personalized Approach to Assignments, Quizzes, and Exams

CAPA is an on-line homework system developed at Michigan State University that instructors can use to deliver problem sets with randomized variables for each student. The system gives students immediate feedback on their answers to problems, and records their participation and performance. Prentice Hall has arranged to have half of the even-numbered problems of Giancoli, *Physics for Scientists and Engineers, Third Edition*, coded for use with the CAPA system. For additional information about the CAPA system, please visit the web site at http://www.pa.msu.edu/educ/CAPA/.

WebAssign

WebAssign is a web-based homework delivery, collection, grading, and recording service developed and hosted by North Carolina State University. Prentice Hall will arrange for end-of-chapter problems from Giancoli, *Physics for Scientists and Engineers, Third Edition* to be coded for use with the *WebAssign* system for instructors who wish to take advantage of this service. For more information on the *WebAssign* system and its features, please visit http://webassign.net/info or e-mail webassign@ncsu.edu.

NOTES TO STUDENTS AND INSTRUCTORS ON THE FORMAT

1. Sections marked with a star (*) are considered optional. They can be omitted without interrupting the main flow of topics. No later material depends on them except possibly later starred sections. They may be fun to read though.

2. The customary conventions are used: symbols for quantities (such as m for mass) are italicized, whereas units (such as m for meter) are not italicized. Boldface ($\mathbf{F}$) is used for vectors.

3. Few equations are valid in all situations. Where practical, the limitations of important equations are stated in square brackets next to the equation. The equations that represent the great laws of physics are displayed with a tan background, as are a few other equations that are so useful that they are indispensable.

4. The number of significant figures (see Section 1–3) should not be assumed to be greater than given: if a number is stated as (say) 6, with its units, it is meant to be 6 and not 6.0 or 6.00.

5. At the end of each chapter is a set of Questions that students should attempt to answer (to themselves at least). These are followed by Problems which are ranked as level I, II, or III, according to estimated difficulty, with level I Problems being easiest. These Problems are arranged by Section, but Problems for a given Section may depend on earlier material as well. There follows a group of General Problems, which are not arranged by Section nor ranked as to difficulty. Questions and Problems that relate to optional Sections are starred.

6. Being able to solve problems is a crucial part of learning physics, and provides a powerful means for understanding the concepts and principles. This book contains many aids to problem solving: (a) worked-out Examples and their solutions in the text, which are set off with a vertical blue line in the margin, and should be studied as an integral part of the text; (b) special "Problem-solving boxes" placed throughout the text to suggest ways to approach problem solving for a particular topic—but don't get the idea that every topic has its own "techniques," because the basics remain the same; (c) special problem-solving Sections (marked in blue in the Table of Contents); (d) "Problem solving" marginal notes (see point 8 below) which refer to hints for solving problems within the text; (e) some of the worked-out Examples are Estimation Examples, which show how rough or approximate results can be obtained even if the given data are sparse (see Section 1–6); and finally (f) the Problems themselves at the end of each chapter (point 5 above).

7. Conceptual Examples look like ordinary Examples but are conceptual rather than numerical. Each proposes a question or two, which hopefully starts you to think and come up with a response. Give yourself a little time to come up with your own response before reading the Response given.

8. Marginal notes: brief notes in the margin of almost every page are printed in blue and are of four types: (a) ordinary notes (the majority) that serve as a sort of outline of the text and can help you later locate important concepts and equations; (b) notes that refer to the great laws and principles of physics, and these are in capital letters and in a box for emphasis; (c) notes that refer to a problem-solving hint or technique treated in the text, and these say "Problem Solving"; (d) notes that refer to an application of physics, in the text or an Example, and these say "Physics Applied."

9. This book is printed in full color. But not simply to make it more attractive. The color is used above all in the Figures, to give them greater clarity for our analysis, and to provide easier learning of the physical principles involved. The Table on the next page is a summary of how color is used in the Figures, and shows which colors are used for the different kinds of vectors, for field lines, and for other symbols and objects. These colors are used consistently throughout the book.

10. Appendices include useful mathematical formulas (such as derivatives and integrals, trigonometric identities, areas and volumes, expansions), and a table of isotopes with atomic masses and other data. Tables of useful data are located inside the front and back covers.

USE OF COLOR

Vectors

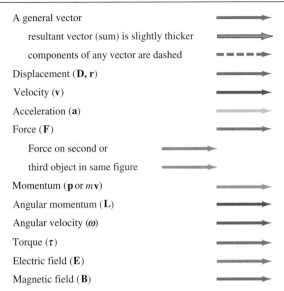

A general vector

    resultant vector (sum) is slightly thicker

    components of any vector are dashed

Displacement (**D**, **r**)

Velocity (**v**)

Acceleration (**a**)

Force (**F**)

    Force on second or

    third object in same figure

Momentum (**p** or m**v**)

Angular momentum (**L**)

Angular velocity (ω)

Torque (τ)

Electric field (**E**)

Magnetic field (**B**)

Electricity and magnetism

Electric field lines

Equipotential lines

Magnetic field lines

Electric charge (+)   + or ● +

Electric charge (−)   − or ● −

Electric circuit symbols

Wire

Resistor

Capacitor

Inductor

Battery

Optics

Light rays

Object

Real image (dashed)

Virtual image (dashed and paler)

Other

Energy level (atom, etc.)

Measurement lines |←—1.0 m—→|

Path of a moving object

Direction of motion or current

Under water, sea creatures and scuba divers experience a buoyant force ($\mathbf{F}_B$) that almost exactly balances their weight $m\mathbf{g}$. The buoyant force is equal to the weight of the volume of fluid displaced (Archimedes' principle) and arises because the pressure increases with depth in the fluid. Sea creatures, and even humans, have a density very close to that of water, so their weight very nearly equals the buoyant force. Actually, humans have a density slightly less than water, so they can float.
When fluids flow, interesting effects occur because the pressure in the fluid is lower where the fluid velocity is higher (Bernoulli's principle): airplanes can fly, sailboats can sail against the wind, smoke goes up a chimney, and many others.

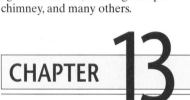

CHAPTER 13

Fluids

Phases of matter

The three common states, or **phases**, of matter are solid, liquid, and gas. We can distinguish these three phases as follows. A **solid** maintains a fixed shape and a fixed size; even if a large force is applied to a solid, it does not readily change its shape or volume. A **liquid** does not maintain a fixed shape—it takes on the shape of its container—but like a solid it is not readily compressible, and its volume can be changed significantly only by a very large force. A **gas** has neither a fixed shape nor a fixed volume—it will expand to fill its container. For example, when air is pumped into an automobile tire, the air does not all run to the bottom of the tire as a liquid would; it spreads out to fill the whole volume of the tire. Since liquids and gases do not maintain a fixed shape, they both have the ability to flow; they are thus often referred to collectively as **fluids**.

13–1 Density and Specific Gravity

It is sometimes said that iron is "heavier" than wood. This cannot really be true since a large log clearly weighs more than an iron nail. What we should say is that iron is more *dense* than wood.

The **density**, ρ, of an object (ρ is the lowercase Greek letter "rho") is defined as its mass per unit volume:

Density

$$\rho = \frac{m}{V}, \tag{13–1}$$

where m is the mass of the object and V its volume. Density is a characteristic property of any pure substance. Objects made of a given pure substance, such as pure gold, can have any size or mass, but the density will be the same for each.

(Sometimes we will find Eq. 13–1 useful for writing the mass of an object as $m = \rho V$, and the weight of an object, mg, as $\rho V g$.)

The SI unit for density is kg/m^3. Since $1\,kg/m^3 = 1000\,g/(100\,cm)^3 = 10^{-3}\,g/cm^3$, then a density given in g/cm^3 must be multiplied by 1000 to give the result in kg/m^3. The densities of a variety of substances are given in Table 13–1. The Table specifies temperature and pressure because they affect the density of substances, although the effect is slight for liquids and solids.

EXAMPLE 13–1 **Mass, given volume and density.** What is the mass of a solid iron wrecking ball of radius 18 cm?

SOLUTION The volume of any sphere is $V = \frac{4}{3}\pi r^3$ so we have

$$V = \tfrac{4}{3}\pi r^3 = \tfrac{4}{3}(3.14)(0.18\,m)^3 = 0.024\,m^3.$$

From Table 13–1, the density of iron is $\rho = 7800\,kg/m^3$, so we have from Eq. 13–1,

$$m = \rho V = (7800\,kg/m^3)(0.024\,m^3) = 190\,kg.$$

The **specific gravity** of a substance is defined as the ratio of the density of that substance to the density of water at 4.0°C. Specific gravity (abbreviated SG) is a number, without dimensions or units. Since the density of water is $1.00\,g/cm^3 = 1.00 \times 10^3\,kg/m^3$, the specific gravity of any substance will be equal numerically to its density specified in g/cm^3, or 10^{-3} times its density specified in kg/m^3. For example (see Table 13–1), the specific gravity of lead is 11.3, and that of alcohol is 0.79.

13–2 Pressure in Fluids

Pressure is defined as force per unit area, where the force F is understood to be acting perpendicular to the surface area A:

$$\text{pressure} = P = \frac{F}{A}. \tag{13–2}$$

The SI unit of pressure is N/m^2. This unit has the official name **pascal** (Pa), in honor of Blaise Pascal (see Section 13–4); that is, $1\,Pa = 1\,N/m^2$. However, for simplicity, we will often use N/m^2. Other units sometimes used are $dynes/cm^2$, $lb/in.^2$ (sometimes abbreviated "psi"). We will meet several other units shortly, and will discuss conversions between them in Section 13–5.

As an example of calculating pressure, a 60-kg person whose two feet cover an area of 500 cm^2 will exert a pressure of

$$F/A = mg/A = (60\,kg)(9.8\,m/s^2)/(0.050\,m^2) = 12 \times 10^3\,N/m^2$$

on the ground. If the person stands on one foot, the force is the same but the area will be half, so the pressure will be twice as much: $24 \times 10^3\,N/m^2$.

The concept of pressure is particularly useful in dealing with fluids. It is an experimental fact that *a fluid exerts a pressure in all directions*. This is well known to swimmers and divers who feel the water pressure on all parts of their bodies. At any point in a fluid at rest, the pressure is the same in all directions. This is illustrated in Fig. 13–1. Consider a tiny cube of the fluid which is so small that we can ignore the force of gravity on it. The pressure on one side of it must equal the pressure on the opposite side. If this weren't true, there would be a net force on the cube and it would start moving. If the fluid is not flowing, then the pressures must be equal.

TABLE 13–1
Densities of Substances†

Substance	Density, $\rho\,(kg/m^3)$
Solids	
Aluminum	2.70×10^3
Iron and steel	7.8×10^3
Copper	8.9×10^3
Lead	11.3×10^3
Gold	19.3×10^3
Concrete	2.3×10^3
Granite	2.7×10^3
Wood (typical)	$0.3–0.9 \times 10^3$
Glass, common	$2.4–2.8 \times 10^3$
Ice	0.917×10^3
Bone	$1.7–2.0 \times 10^3$
Liquids	
Water (4°C)	1.00×10^3
Sea water	1.025×10^3
Blood, plasma	1.03×10^3
Blood, whole	1.05×10^3
Mercury	13.6×10^3
Alcohol, ethyl	0.79×10^3
Gasoline	0.68×10^3
Gases	
Air	1.29
Helium	0.179
Carbon dioxide	1.98
Water (steam) (100°C)	0.598

†Densities are given at 0°C and 1 atm pressure unless otherwise specified.

FIGURE 13–1 Pressure is the same in every direction in a fluid at a given depth; if it weren't the fluid would start to move.

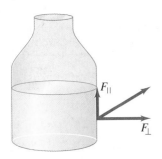

FIGURE 13–2 If there were a component of force parallel to the solid surface, the liquid would move in response to it; for a liquid at rest, $F_{\parallel} = 0$.

FIGURE 13–3 Calculating the pressure at a depth h in a liquid.

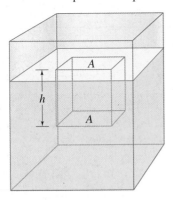

FIGURE 13–4 Forces on a flat, slablike volume of fluid for determining the pressure P at a height y in the fluid.

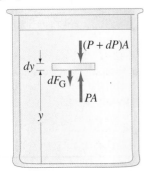

Another important property of a fluid at rest is that the force due to fluid pressure always acts *perpendicular* to any surface it is in contact with. If there were a component of the force parallel to the surface as shown in Fig. 13–2, then according to Newton's third law, the surface would exert a force back on the fluid that also would have a component parallel to the surface. Such a component would cause the fluid to flow, in contradiction to our assumption that the fluid is at rest. Thus the force due to the pressure is perpendicular to the surface.

Let us now calculate quantitatively how the pressure in a liquid of uniform density varies with depth. Consider a point which is at a depth h below the surface of the liquid (that is, the surface is a height h above this point), as shown in Fig. 13–3. The pressure due to the liquid at this depth h is due to the weight of the column of liquid above it. Thus the force due to the weight of liquid acting on the area A is $F = mg = \rho Ahg$, where Ah is the volume of the column, ρ is the density of the liquid (assumed to be constant), and g is the acceleration of gravity. The pressure P due to the weight of liquid is then

$$P = \frac{F}{A} = \frac{\rho Ahg}{A}$$

$$P = \rho gh. \qquad \text{[liquid]} \quad \textbf{(13–3)}$$

Thus the fluid pressure is directly proportional to the density of the liquid, and to the depth within the liquid. In general, the pressure at equal depths within a uniform liquid is the same.

Equation 13–3 tells us what the pressure is at a depth h in the liquid, due to the liquid itself. But what if there is additional pressure exerted at the surface of the liquid, such as the pressure of the atmosphere or a piston pushing down? And what if the density of the fluid is not constant? Gases are quite compressible and hence their density can vary significantly with depth; and liquids, too, can be compressed, although we can often ignore the variation in density. (One exception is in the depths of the ocean where the great weight of water above significantly compresses the water and increases its density.) To cover these, and other cases, we now treat the general case of determining how the pressure in a fluid varies with depth.

Consider any fluid, and let us determine the pressure at any height y above some reference point (such as the ocean floor or the bottom of a tank or swimming pool), as shown in Fig. 13–4.[†] Within this fluid, at the height y, we consider a tiny, flat, slablike volume of the fluid whose area is A and whose (infinitesimal) thickness is dy, as shown. Let the pressure acting upward on its lower surface (at height y) be P. The pressure acting downward on the top surface of our tiny slab (at height $y + dy$) is designated $P + dP$. The fluid pressure acting on our slab thus exerts a force equal to PA upward on our slab and a force equal to $(P + dP)A$ downward on it. The only other force acting vertically on the slab is the (infinitesimal) force of gravity dF_G, which on our slab of mass dm is

$$dF_G = (dm)g = \rho g\, dV = \rho g A\, dy,$$

where ρ is the density of the fluid at the height y. Since the fluid is assumed to be at rest, our slab is in equilibrium so the net force on it must be zero. Therefore we have

$$PA - (P + dP)A - \rho g A\, dy = 0,$$

which when simplified becomes

$$\frac{dP}{dy} = -\rho g. \qquad \textbf{(13–4)}$$

This relation tells us how the pressure varies with height within the fluid. The minus sign indicates that the pressure decreases with an increase in height; or that the pressure increases with depth (reduced height).

[†] Now we are measuring y positive upwards, the reverse of what we did to get Eq. 13–3 where we measured the depth (i.e. downward as positive).

If the pressure at a height y_1 in the fluid is P_1, and at height y_2 it is P_2, then we can integrate Eq. 13–4 to obtain

$$\int_{P_1}^{P_2} dP = -\int_{y_1}^{y_2} \rho g \, dy$$

$$P_2 - P_1 = -\int_{y_1}^{y_2} \rho g \, dy, \qquad\qquad \textbf{(13–5)}$$

where we assume ρ is a function of height y: $\rho = \rho(y)$. This is a general relation, and we apply it now to two special cases: (1) pressure in liquids of uniform density and (2) pressure variations in the Earth's atmosphere.

For liquids in which any variation in density can be ignored, $\rho = $ constant and Eq. 13–5 is readily integrated:

$$P_2 - P_1 = -\rho g(y_2 - y_1). \qquad\qquad \textbf{(13–6a)}$$

For the everyday situation of a liquid in an open container—such as water in a glass, a swimming pool, a lake, or the ocean—there is a free surface at the top. And it is convenient to measure distances from this top surface. That is, we let h be the *depth* in the liquid where $h = y_2 - y_1$ as shown in Fig. 13–5. If we let y_2 be the position of the top surface, then P_2 represents the atmospheric pressure, P_0, at the top surface. Then, from Eq. 13–6a, the pressure $P(= P_1)$ at a depth h in the fluid is

$$P = P_0 + \rho g h. \qquad\qquad \text{[h is depth in liquid]} \quad \textbf{(13–6b)}$$

Note that Eq. 13–6b is simply the liquid pressure (Eq. 13–3) plus the pressure P_0 due to the atmosphere above.

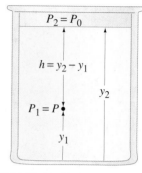

FIGURE 13–5 Pressure at a depth $h = (y_2 - y_1)$ in a liquid of density ρ is $P = P_0 + \rho g h$, where P_0 is the external pressure at the liquid's top surface.

EXAMPLE 13–2 **Pressure at a faucet.** The surface of the water in a storage tank is 30 m above a water faucet in the kitchen of a house, Fig. 13–6. Calculate the water pressure at the faucet.

SOLUTION The same atmospheric pressure acts both at the surface of the water in the storage tank, and on the water leaving the faucet. The pressure difference between the inside and outside of the faucet is

$$\Delta P = \rho g h = \left(1.0 \times 10^3 \, \text{kg/m}^3\right)\left(9.8 \, \text{m/s}^2\right)(30 \, \text{m}) = 2.9 \times 10^5 \, \text{N/m}^2.$$

The height h is sometimes called the **pressure head**. In this Example, the head of the water is 30 m. Note that the very different diameters of the tank and faucet don't affect the result—only pressure does.

➡ **PHYSICS APPLIED**

Water supply

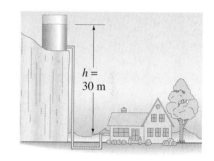

FIGURE 13–6 Example 13–2.

EXAMPLE 13–3 **Force on aquarium window.** Calculate the force due to water pressure exerted on a 1.0 m × 3.0 m aquarium viewing window, Fig. 13–7.

SOLUTION At a depth h, the pressure due to the water is given by Eq. 13–6b. Divide the window up into thin horizontal strips of width $w = 3.0$ m and thickness dy, as shown in Fig. 13–7. We choose a coordinate system with $y = 0$ at the surface of the water and y is positive downward. (With this choice, the minus sign in Eq. 13–6a becomes plus, or we use Eq. 13–6b with $y = h$.) The force due to water pressure on each strip is $dF = P \, dA = \rho g y w \, dy$. The total force on the window is given by the integral:

$$\int_{y_1 = 1.0 \text{ m}}^{y_2 = 2.0 \text{ m}} \rho g y w \, dy = \tfrac{1}{2}\rho g w(y_2^2 - y_1^2)$$

$$= \tfrac{1}{2}\left(1000 \, \text{kg/m}^3\right)\left(9.8 \, \text{m/s}^2\right)(3.0 \, \text{m})\left[(2.0 \, \text{m})^2 - (1.0 \, \text{m})^2\right] = 44{,}000 \, \text{N}.$$

FIGURE 13–7 Example 13–3.

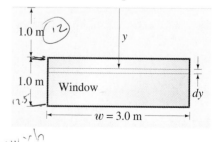

Now let us apply Eq. 13–4 or 13–5 to gases. The density of gases is normally quite small, so the difference in pressure at different heights can usually be ignored if $y_2 - y_1$ is not large (which is why, in Example 13–2, we could ignore the difference in air pressure between the faucet and the top of the storage tank). Indeed, for most ordinary containers of gas, we can assume that the pressure is the same throughout. However, if $y_2 - y_1$, is very large, we cannot make this assumption. An interesting example is the Earth's atmosphere, whose pressure at sea level is about $1.013 \times 10^5 \text{ N/m}^2$ and decreases slowly with altitude.

Air pressure variation with altitude

EXAMPLE 13–4 **Elevation effect on atmospheric pressure.** (a) Determine the variation in pressure in the Earth's atmosphere as a function of height y above sea level, assuming g is constant and that the density of the air is proportional to the pressure. (This last assumption is not terribly accurate, in part because temperature and other weather effects are important.) (b) At what elevation is the air pressure equal to half the pressure at sea level?

SOLUTION (a) We are assuming that ρ is proportional to P, so we can write

$$\frac{\rho}{\rho_0} = \frac{P}{P_0},$$

where $P_0 = 1.013 \times 10^5 \text{ N/m}^2$ is atmospheric pressure at sea level and $\rho_0 = 1.29 \text{ kg/m}^3$ is the density of air at sea level at $0°C$ (Table 13–1). From the differential change in pressure with height, Eq. 13–4, we have

$$\frac{dP}{dy} = -\rho g = -P\left(\frac{\rho_0}{P_0}\right)g,$$

so

$$\frac{dP}{P} = -\frac{\rho_0}{P_0} g \, dy.$$

We integrate this from $y = 0$ (Earth's surface) and $P = P_0$, to the height y where the pressure is P:

$$\int_{P_0}^{P} \frac{dP}{P} = -\frac{\rho_0}{P_0} g \int_0^y dy$$

$$\ln \frac{P}{P_0} = -\frac{\rho_0}{P_0} gy$$

since $\ln P - \ln P_0 = \ln(P/P_0)$. Thus

$$P = P_0 e^{-(\rho_0 g/P_0)y}.$$

Exponential decrease in pressure with altitude

So, based on our assumptions, we find that the air pressure in our atmosphere decreases approximately exponentially with height. (Note that the atmosphere does not have a distinct top surface, so there is no natural point from which to measure depth in the atmosphere, as we can do for a liquid.)

(b) The constant $(\rho_0 g/P_0)$ has the value

$$\frac{\rho_0 g}{P_0} = \frac{(1.29 \text{ kg/m}^3)(9.80 \text{ m/s}^2)}{(1.013 \times 10^5 \text{ N/m}^2)}$$

$$= 1.25 \times 10^{-4} \text{ m}^{-1}.$$

Then, when we set $P = \frac{1}{2} P_0$, we have

$$\frac{1}{2} = e^{-(1.25 \times 10^{-4} \text{ m}^{-1})y}$$

or

$$y = (\ln 2.00)/(1.25 \times 10^{-4} \text{ m}^{-1}) = 5550 \text{ m},$$

where $\ln 2 = 0.693$. Thus, at an elevation of about 5500 m (about 18,000 ft), atmospheric pressure drops to half what it is at sea level. It is not surprising that mountain climbers often use oxygen tanks at very high altitudes.

13–3 | Atmospheric Pressure and Gauge Pressure

The pressure of the Earth's atmosphere varies with altitude, as we have seen. But even at a given altitude, it varies slightly according to the weather. As already mentioned, at sea level the pressure of the atmosphere on the average is $1.013 \times 10^5\,\text{N/m}^2$ (or $14.7\,\text{lb/in}^2$). This value is used to define a commonly used unit of pressure, the **atmosphere** (abbreviated atm):

$$1\,\text{atm} = 1.013 \times 10^5\,\text{N/m}^2 = 101.3\,\text{kPa}.$$

One atmosphere (unit)

Another unit of pressure sometimes used (in meteorology and on weather maps) is the **bar**, which is defined as $1\,\text{bar} = 1.00 \times 10^5\,\text{N/m}^2$. Thus standard atmospheric pressure is slightly more than 1 bar.

The bar (unit)

The pressure due to the weight of the atmosphere is exerted on all objects immersed in this great sea of air, including our bodies. How does a human body withstand the enormous pressure on its surface? The answer is that living cells maintain an internal pressure that closely equals the external pressure, just as the pressure inside a balloon only slightly exceeds the outside pressure of the atmosphere. An automobile tire, because of its rigidity, can maintain internal pressures much greater than the external pressure.

It is important to note that tire gauges, and most other pressure gauges, register the pressure over and above atmospheric pressure. This is called **gauge pressure**. Thus, to get the absolute pressure, P, one must add the atmospheric pressure, P_A, to the gauge pressure, P_G:

Gauge pressure

$$P = P_A + P_G.$$

If a tire gauge registers $220\,\text{kPa}$, the absolute pressure within the tire is $220\,\text{kPa} + 101\,\text{kPa} = 321\,\text{kPa}$. This is equivalent to about 3.2 atm (2.2 atm gauge pressure).

Absolute pressure = atmospheric pressure + gauge pressure

CONCEPTUAL EXAMPLE 13–5 | **Finger holds water in a straw.** You insert a straw of length L into a tall glass of your favorite beverage. You place your finger over the top of the straw so that no air can get in or out, and then lift the straw from the liquid. You find that the straw retains the liquid such that the distance from the bottom of your finger to the top of the liquid is h. (See Fig. 13–8.) Does the air in the space between your finger and the top of the liquid have a pressure P that is greater than, equal to, or less than, the atmospheric pressure P_A outside the straw?

RESPONSE Consider the forces on the column of liquid. Atmospheric pressure on the outside of the straw pushes upward on the liquid at the bottom of the straw, gravity pulls the liquid downward, and the air pressure inside the top of the straw pushes downward on the liquid. Since the liquid is in equilibrium, the upward force due to atmospheric pressure must balance the two downward forces. The only way this is possible is for the air pressure inside the straw to be rather less than the atmosphere pressure outside the straw.

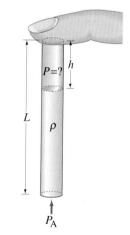

FIGURE 13–8 Example 13–5.

13–4 | Pascal's Principle

The Earth's atmosphere exerts a pressure on all objects with which it is in contact, including other fluids. External pressure acting on a fluid is transmitted throughout that fluid. For instance, according to Eq. 13–3, the pressure due to the water at a depth of 100 m below the surface of a lake is $P = \rho g h = (1000\,\text{kg/m}^3)(9.8\,\text{m/s}^2)(100\,\text{m}) = 9.8 \times 10^5\,\text{N/m}^2$, or 9.7 atm. However, the total pressure at this point is due to the pressure of water plus the pressure of the air above it (Eq. 13–6b). Hence the total pressure (if the lake is near sea level) is $9.7\,\text{atm} + 1.0\,\text{atm} = 10.7\,\text{atm}$. This is just one example of a general principle attributed to the French philosopher and scientist Blaise Pascal (1623–1662). **Pascal's principle** states that *pressure applied to a confined fluid increases the pressure throughout by the same amount.*

Pascal's principle

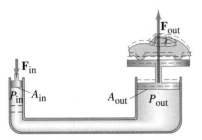

FIGURE 13–9 Application of Pascal's principle: hydraulic lift.

A number of practical devices make use of Pascal's principle. One example is the hydraulic lift, illustrated in Fig. 13–9, in which a small force is used to exert a large force by making the area of the output piston larger than the area of the input piston. To see how this works, we assume the input and output pistons are at the same height (at least approximately). Then the external input force F_{in}, by Pascal's principle, increases the pressure equally throughout so that at the same level (see Fig. 13–9):

$$P_{out} = P_{in}$$

where the input quantities are represented by the subscript "in" and the output by "out." Thus

$$\frac{F_{out}}{A_{out}} = \frac{F_{in}}{A_{in}},$$

or

$$\frac{F_{out}}{F_{in}} = \frac{A_{out}}{A_{in}}.$$

The quantity F_{out}/F_{in} is called the "mechanical advantage" of the hydraulic lift, and is equal to the ratio of the areas. For example, if the area of the output piston is 20 times that of the input cylinder, the force is multiplied by a factor of 20: thus a force of 200 lb could lift a 4000-lb car.

13–5 Measurement of Pressure; Gauges and the Barometer

Many devices have been invented to measure pressure, some of which are shown in Fig. 13–10. The simplest is the open-tube *manometer* (Fig. 13–10a) which is a U-shaped tube partially filled with a liquid, usually mercury or water. The pressure P being measured is related to the difference in height h of the two levels of the liquid by the relation

$$P = P_0 + \rho g h,$$

where P_0 is atmospheric pressure (acting on the top of the fluid in the left-hand tube), and ρ is the density of the liquid. Note that the quantity $\rho g h$ is the "gauge pressure"—the amount by which P exceeds atmospheric pressure. If the liquid in the left-hand column were lower than that in the right-hand column, this would indicate that P was less than atmospheric pressure (and h would be negative).

FIGURE 13–10 Pressure gauges: (a) open-tube manometer, (b) aneroid gauge, and (c) common tire pressure gauge.

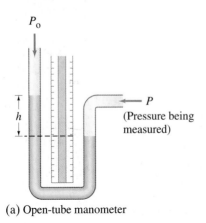

(a) Open-tube manometer

(b) Aneroid gauge (used mainly for air pressure and then called an aneroid barometer)

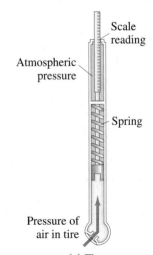

(c) Tire gauge

Instead of calculating the product $\rho g h$, it is common to simply specify the height h. In fact, pressures are sometimes specified as so many "millimeters of mercury" (mm-Hg), and sometimes as so many "mm of water" (mm-H_2O). The unit mm-Hg is equivalent to a pressure of $133\,N/m^2$, since $\rho g h$ for $1\,mm = 1.0 \times 10^{-3}\,m$ of mercury gives

$$\rho g h = (13.6 \times 10^3\,kg/m^3)(9.80\,m/s^2)(1.00 \times 10^{-3}\,m) = 1.33 \times 10^2\,N/m^2.$$

The unit mm-Hg is also called the **torr** in honor of Evangelista Torricelli (1608–1647), who invented the barometer (see below). Conversion factors among the various units of pressure (an incredible nuisance!) are given in Table 13–2. It is important that only $N/m^2 = Pa$, the proper SI unit, be used in calculations involving other quantities specified in SI units.

The torr (unit)

➡ **PROBLEM SOLVING**

Use SI unit in calculations:
$1\,Pa = 1\,N/m^2$

TABLE 13–2 Conversion Factors Between Different Units of Pressure

In Terms of $1\,Pa = 1\,N/m^2$	Related to 1 atm
$1\,atm = 1.013 \times 10^5\,N/m^2$	$1\,atm = 1.013 \times 10^5\,N/m^2$
$= 1.013 \times 10^5\,Pa = 101.3\,kPa$	
$1\,bar = 1.000 \times 10^5\,N/m^2$	$1\,atm = 1.013\,bar$
$1\,dyne/cm^2 = 0.1\,N/m^2$	$1\,atm = 1.013 \times 10^6\,dyne/cm^2$
$1\,lb/in.^2 = 6.90 \times 10^3\,N/m^2$	$1\,atm = 14.7\,lb/in.^2$
$1\,lb/ft^2 = 47.9\,N/m^2$	$1\,atm = 2.12 \times 10^3\,lb/ft^2$
$1\,cm\text{-}Hg = 1.33 \times 10^3\,N/m^2$	$1\,atm = 76\,cm\text{-}Hg$
$1\,mm\text{-}Hg = 133\,N/m^2$	$1\,atm = 760\,mm\text{-}Hg$
$1\,torr = 133\,N/m^2$	$1\,atm = 760\,torr$
$1\,mm\text{-}H_2O\ (4°C) = 9.81\,N/m^2$	$1\,atm = 1.03 \times 10^4\,mm\text{-}H_2O\ (4°C)$

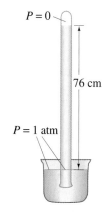

FIGURE 13–11 Diagram of a mercury barometer, when the air pressure is 76 cm-Hg.

FIGURE 13–12 A water barometer: water was added at the top, and the spigot at the top then closed; the water level dropped, leaving a vacuum between its upper surface and the spigot. Why? Because air pressure could not support a column of water more than 10 m high.

Another type of pressure gauge is the aneroid gauge (Fig. 13–10b) in which the pointer is linked to the flexible ends of an evacuated thin metal chamber. In an electronic gauge, the pressure may be applied to a thin metal diaphragm whose resulting distortion is translated into an electrical signal by a transducer. How a common tire gauge is constructed is shown in Fig. 13–10c.

Atmospheric pressure is often measured by a modified kind of mercury manometer with one end closed, called a mercury **barometer** (Fig. 13–11). The glass tube is completely filled with mercury and then inverted into the bowl of mercury. If the tube is long enough, the level of the mercury will drop, leaving a vacuum at the top of the tube, since atmospheric pressure can support a column of mercury only about 76 cm high (exactly 76.0 cm at standard atmospheric pressure). That is, a column of mercury 76 cm high exerts the same pressure as the atmosphere:

$$P = \rho g h = (13.6 \times 10^3\,kg/m^3)(9.80\,m/s^2)(0.760\,m) = 1.013 \times 10^5\,N/m^2 = 1.00\,atm.$$

Household barometers are usually of the aneroid type, either mechanical (Fig. 3–10b), or electronic.

A calculation similar to that above will show that atmospheric pressure can maintain a column of water 10.3 m high in a tube whose top is under vacuum (Fig. 13–12). A few centuries ago, it was a source of wonder and frustration that no matter how good a vacuum pump was, it could not lift water more than about 10 m. The only way to pump water out of deep mine shafts, for example, was to use multiple stages for depths greater than 10 m. Galileo studied this problem, and his student Torricelli was the first to explain it. The point is that a pump does not really suck water up a tube—it merely reduces the pressure at the top of the tube. Atmospheric air pressure *pushes* the water up the tube if the top end is at low pressure (under a vacuum), just as it is air pressure that pushes (or maintains) the mercury 76 cm high in a barometer.

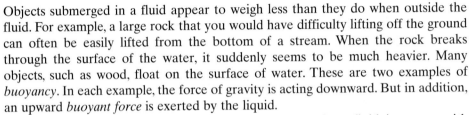

CONCEPTUAL EXAMPLE 13–6 **Suction.** You sit in a meeting where a novice NASA engineer proposes suction cup shoes for Space Shuttle astronauts working on the exterior of the spacecraft. Having just studied this chapter, you gently remind him of the fallacy of this plan. What is it?

RESPONSE Suction cups work by pushing out the air underneath the cup. What holds the cup in place is the air pressure outside the cup. (This can be a substantial force on Earth. For example, a 10 cm diameter cup has an area of $7.8 \times 10^{-3} \, \text{m}^2$. The force of the atmosphere on it is $(7.8 \times 10^{-3} \, \text{m}^2)(1.0 \times 10^5 \, \text{N/m}^2) \approx 800 \, \text{N}$, about 180 lbs!). But in outer space, there is no air pressure to hold the suction cup on the spacecraft.

We sometimes mistakenly think of suction as something we actively do. For example, we intuitively think that we pull the soda up through a straw. Instead, all we do is lower the pressure at the top of the straw, and the atmosphere *pushes* the soda up the straw.

13–6 | Buoyancy and Archimedes' Principle

Objects submerged in a fluid appear to weigh less than they do when outside the fluid. For example, a large rock that you would have difficulty lifting off the ground can often be easily lifted from the bottom of a stream. When the rock breaks through the surface of the water, it suddenly seems to be much heavier. Many objects, such as wood, float on the surface of water. These are two examples of *buoyancy*. In each example, the force of gravity is acting downward. But in addition, an upward *buoyant force* is exerted by the liquid.

The buoyant force occurs because the pressure in a fluid increases with depth. Thus the upward pressure on the bottom surface of a submerged object is greater than the downward pressure on its top surface. To see the effect of this, consider a cylinder of height h whose top and bottom ends have an area A and which is completely submerged in a fluid of density ρ_F, as shown in Fig. 13–13. The fluid exerts a pressure $P_1 = \rho_F g h_1$ at the top surface of the cylinder. The force due to this pressure on top of the cylinder is $F_1 = P_1 A = \rho_F g h_1 A$, and it is directed downward. Similarly, the fluid exerts an upward force on the bottom of the cylinder equal to $F_2 = P_2 A = \rho_F g h_2 A$. The net force due to the fluid pressure, which is the **buoyant force**, $\mathbf{F_B}$, acts upward and has the magnitude

$$F_B = F_2 - F_1 = \rho_F g A (h_2 - h_1)$$
$$= \rho_F g A h = \rho_F g V,$$

where $V = Ah$ is the volume of the cylinder. Since ρ_F is the density of the fluid, the product $\rho_F g V = m_F g$ is the weight of fluid which takes up a volume equal to the volume of the cylinder. Thus the buoyant force on the cylinder is equal to the weight of fluid displaced[†] by the cylinder. This result is valid no matter what the shape of the object. Its discovery is credited to Archimedes (287?–212 B.C.), and it is called **Archimedes' principle**: *the buoyant force on a body immersed in a fluid is equal to the weight of the fluid displaced by that object.*

We can derive Archimedes' principle in general by the following simple but elegant argument. The irregularly shaped object D shown in Fig. 13–14a is acted on by the force of gravity (its weight, $m\mathbf{g}$, downward) and the buoyant force, $\mathbf{F_B}$, upward. We wish to determine F_B. To do so, we next consider a body, this time made of our same fluid (D' in Fig. 13–14b) with the same shape and size as the original object, and located at the same depth. You might think of this body of fluid as being separated from the rest of the fluid by an imaginary membrane. The buoyant force F_B on this body of fluid will be exactly the same as that on the

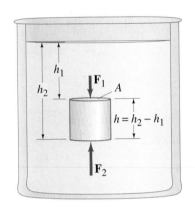

FIGURE 13–13 Determination of the buoyant force.

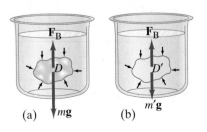

FIGURE 13–14 Archimedes' principle.

[†] By "fluid displaced," we mean a volume of fluid equal to the volume of the object, or that part of the object submerged if it floats or is only partly submerged (the fluid that used to be where the object is). If the object is placed in a glass or tub initially filled to the brim with water, the water that flows over the top represents the water displaced by the object.

original object since the surrounding fluid, which exerts F_B, is in exactly the same configuration. Now the body of fluid D' is in equilibrium (the fluid as a whole is at rest). Therefore, $F_B = m'g$, where $m'g$ is the weight of the body of fluid. Hence the buoyant force F_B is equal to the weight of the body of fluid whose volume equals the volume of the original submerged object, which is Archimedes' principle.

CONCEPTUAL EXAMPLE 13–7 | **Two pails of water.** Consider two identical pails of water filled to the brim. One pail contains only water while the other has a piece of wood floating in it. Which one has the greater weight?

RESPONSE Both pails weigh the same. Recall Archimedes' principle: the wood displaces a volume of water with weight equal to the weight of the wood. Some water will overflow the pail, but the spilled water has weight equal to that of the wood; so the pails have the same weight.

EXAMPLE 13–8 | **Recovering a submerged statue.** A 70-kg ancient statue lies at the bottom of the sea. Its volume is $3.0 \times 10^4 \, cm^3$. How much force is needed to lift it?

SOLUTION The buoyant force on the statue due to the water is equal to the weight of $3.0 \times 10^4 \, cm^3 = 3.0 \times 10^{-2} \, m^3$ of water (for seawater $\rho = 1.025 \times 10^3 \, kg/m^3$):

$$F_B = m_{H_2O} \, g = \rho_{H_2O} \, gV$$
$$= (1.025 \times 10^3 \, kg/m^3)(9.8 \, m/s^2)(3.0 \times 10^{-2} \, m^3) = 3.0 \times 10^2 \, N.$$

The weight of the statue is $mg = (70 \, kg)(9.8 \, m/s^2) = 6.9 \times 10^2 \, N$. Hence the force needed to lift it is $690 \, N - 300 \, N = 390 \, N$. It is as if the statue had a mass of only $(390 \, N)/(9.8 \, m/s^2) = 40 \, kg$.

Archimedes is said to have discovered his principle in his bath while thinking how he might determine whether the king's new crown was pure gold or a fake. Gold has a specific gravity of 19.3, somewhat higher than that of most metals, but a determination of specific gravity or density is not readily done directly because, even if the mass is easily known, the volume of an irregularly shaped object is not easily calculated. However, if the object is weighed in air $(= w)$ and also "weighed" while it is under water $(= w')$, the density can be determined using Archimedes' principle, as the following Example shows. The quantity w' is called the *apparent weight* in water, and is what a scale reads when the object is submerged in water (see Fig. 13–15); w' equals the true weight $(w = mg)$ minus the buoyant force.

EXAMPLE 13–9 | **Archimedes: Is the crown gold?** When a crown of mass 14.7 kg is submerged in water, an accurate scale reads only 13.4 kg. Is the crown made of gold?

SOLUTION See analysis in Fig. 13–15. The apparent weight of the submerged object, $w' \, (= F_T'$ in Fig. 13–15b), equals its actual weight $w(= mg)$ minus the buoyant force F_B as shown:

$$w' = F_T' = w - F_B = \rho_O gV - \rho_F gV,$$

where V is the volume of the object, ρ_O the object's density, and ρ_F the density of the fluid (water in this case). From this relation, we can see that $F_B = w - w' = \rho_F gV$. Then we can write

$$\frac{w}{w - w'} = \frac{\rho_O gV}{\rho_F gV} = \frac{\rho_O}{\rho_F}.$$

Thus $w/(w - w')$ is equal to the specific gravity of the object if the fluid in which it is submerged is water. For the crown we have

$$\frac{\rho_O}{\rho_{H_2O}} = \frac{w}{w - w'} = \frac{(14.7 \, kg)g}{(14.7 \, kg - 13.4 \, kg)g} = \frac{14.7 \, kg}{1.3 \, kg} = 11.3.$$

This corresponds to a density of $11,300 \, kg/m^3$. The crown seems to be made of lead (see Table 13–1)!

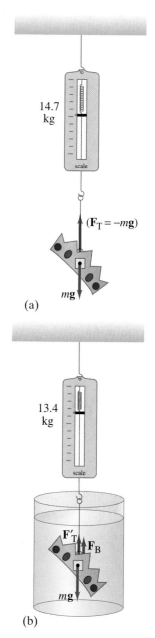

FIGURE 13–15 (a) A scale reads the mass of an object in air—in this case the crown of Example 13–9. All objects are at rest, so the tension F_T in the connecting cord equals the weight w of the object: $F_T = mg$. Note that we show the free-body diagram of the crown, and that F_T is what causes the scale reading (it's equal to the net downward force on the scale). (b) The object submerged has an additional force on it, the buoyant force F_B. The net force is zero, so $F_T' + F_B = mg(= w)$. The scale now reads $m' = 13.4 \, kg$, where m' is related to the effective weight by $w' = m'g$, where $F_T' = w' = w - F_B$.

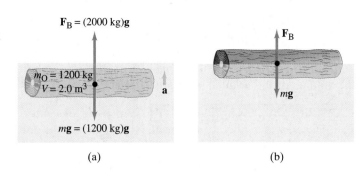

FIGURE 13–16 (a) The fully submerged log accelerates upward because $F_B > mg$. It comes to equilibrium (b) when $\Sigma F = 0$, so $F_B = mg = (1200\,\text{kg})g$. Thus 1200 kg, or 1.2 m³, of water is displaced.

(a)

(b)

Floating

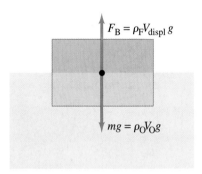

FIGURE 13–17 An object floating in equilibrium: $F_B = mg$.

FIGURE 13–18 A hydrometer. Example 13–10.

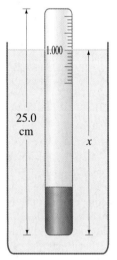

Archimedes' principle applies equally well to objects that float, such as wood. In general, *an object floats on a fluid if its density is less than that of the fluid.* This is readily seen from Fig. 13–16a, where a submerged object will experience a net upward force and float to the surface if $F_B > mg$; that is, if $\rho_F V g > \rho_O V g$ or $\rho_F > \rho_O$. At equilibrium—that is, when floating—the buoyant force on an object has magnitude equal to the weight of the object. For example, a log whose specific gravity is 0.60 and whose volume is 2.0 m³ has a mass $m = \rho_O V = (0.60 \times 10^3\,\text{kg/m}^3)(2.0\,\text{m}^3) = 1200\,\text{kg}$. If the log is fully submerged, it will displace a mass of water $m_F = \rho_F V = (1000\,\text{kg/m}^3)(2.0\,\text{m}^3) = 2000\,\text{kg}$. Hence the buoyant force on the log will be greater than its weight, and it will float upward to the surface (Fig. 13–16). It will come to equilibrium when it displaces 1200 kg of water, which means that 1.2 m³ of its volume will be submerged. This 1.2 m³ corresponds to 60 percent of the volume of the log $(1.2/2.0 = 0.60)$, so 60 percent of the log is submerged. In general when an object floats, we have $F_B = mg$, which we can write as (see Fig. 13–17)

$$\rho_F V_{\text{displ}}\, g = \rho_O V_O\, g$$

where V_O is the full volume of the object and V_{displ} is the volume of fluid it displaces (= volume submerged). Thus

$$\frac{V_{\text{displ}}}{V_O} = \frac{\rho_O}{\rho_F}.$$

That is, the fraction of the object submerged is given by the ratio of the object's density to that of the fluid.

EXAMPLE 13–10 **Hydrometer calibration.** A **hydrometer** is a simple instrument used to indicate specific gravity of a liquid by measuring how deeply it sinks in the liquid. A particular hydrometer (Fig. 13–18) consists of a glass tube, weighted at the bottom, which is 25.0 cm long, 2.00 cm² in cross-sectional area, and has a mass of 45.0 g. How far from the end should the 1.000 mark be placed?

SOLUTION The hydrometer has an overall density

$$\rho = \frac{m}{V} = \frac{45.0\,\text{g}}{(2.00\,\text{cm}^2)(25.0\,\text{cm})} = 0.900\,\text{g/cm}^3.$$

Thus, when placed in water, it will come to equilibrium when 0.900 of its volume is submerged. Since it is of uniform cross section, $(0.900)(25.0\,\text{cm}) = 22.5\,\text{cm}$ of its length will be submerged. Since the specific gravity of water is defined to be 1.000, the mark should be placed 22.5 cm from the end.

➡ **PHYSICS APPLIED**

Continental drift—plate tectonics

Weight affected by buoyancy of air

Archimedes' principle is also useful in geology. According to the theory of plate tectonics and continental drift, the continents can be considered to be floating on a fluid "sea" of slightly deformable rock (mantle rock). Some interesting calculations can be done using very simple models, which we consider in the problems at the end of the chapter.

Air is a fluid and it too exerts a buoyant force. Ordinary objects weigh less in air than they do if weighed in a vacuum. Because the density of air is so small, the effect for ordinary solids is slight. There are objects, however, that float in air—helium-filled balloons, for example, because the density of helium is less than the density of air.

EXAMPLE 13–11 **Helium balloon.** What volume V of helium is needed if a balloon is to lift a load of 180 kg (including the weight of the empty balloon)?

SOLUTION The buoyant force on the helium balloon, F_B, which is equal to the weight of displaced air, must be at least equal to the weight of the helium plus the load (Fig. 13–19).

$$F_B = (m_{He} + 180 \text{ kg})g.$$

This equation can be written in terms of density:

$$\rho_{air} Vg = (\rho_{He}V + 180 \text{ kg})g.$$

Solving now for V, we find

$$V = \frac{180 \text{ kg}}{\rho_{air} - \rho_{He}} = \frac{180 \text{ kg}}{(1.29 \text{ kg/m}^3 - 0.18 \text{ kg/m}^3)} = 160 \text{ m}^3.$$

This is the volume needed near the Earth's surface, where $\rho_{air} = 1.29 \text{ kg/m}^3$. To reach a high altitude, a greater volume would be needed since the density of air decreases with altitude.

FIGURE 13–19 Example 13–11.

13–7 | Fluids in Motion; Flow Rate and the Equation of Continuity

We now turn from the study of fluids at rest to the more complex subject of fluids in motion, which is called **f**   **l**   **u**   **i**
n   **a**   **m**   **i**   **c**
bulence as a manifestation of chaos is a "hot" topic today). Nonetheless, with certain simplifying assumptions, a good understanding of this subject can be obtained.

To begin with, we can distinguish two main types of fluid flow. If the flow is smooth, such that neighboring layers of the fluid slide by each other smoothly, the flow is said to be **s**   **t**   **r**   **e**   **a**   **m**
the fluid follows a smooth path, called a **s**   **t**   **r**   **e**   **a**   **m**
over one another (Fig. 13–20a). Above a certain speed, the flow becomes turbulent. **T**   **u**   **r**   **b**   **u**   **l**   **e**   **n**   **t**
eddy currents or eddies (Fig. 13–20b). Eddies absorb a great deal of energy, and

† The word laminar means "in layers."

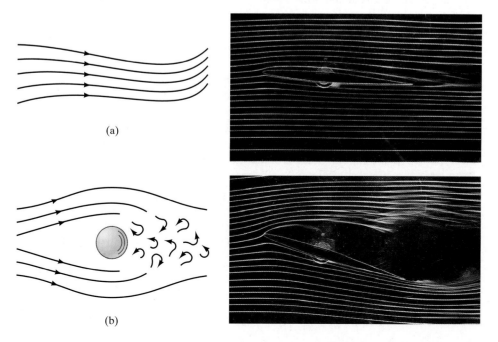

(a)

(b)

FIGURE 13–20 (a) Streamline or laminar flow; (b) turbulent flow.

although a certain amount of internal friction, called **viscosity**, is present even during streamline flow, it is much greater when the flow is turbulent. A few tiny drops of ink or food coloring dropped into a moving liquid can quickly reveal whether the flow is streamline or turbulent.

We will assume in this chapter that the fluid is essentially incompressible (no significant variations in density) and that the flow past any point is steady.

Let us consider the steady laminar flow of a fluid through an enclosed tube or pipe as shown in Fig. 13–21. First we determine how the speed of the fluid changes when the size of the tube changes. The mass **flow rate** is defined as the mass Δm of fluid that passes a given point per unit time Δt: mass flow rate $= \Delta m/\Delta t$. In Fig. 13–21, the volume of fluid passing point 1 (that is, through area A_1) in a time Δt is just $A_1 \Delta l_1$, where Δl_1 is the distance the fluid moves in time Δt. Since the velocity[†] of fluid passing point 1 is $v_1 = \Delta l_1/\Delta t$, the mass flow rate $\Delta m_1/\Delta t$ through area A_1 is

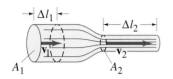

FIGURE 13–21 Fluid flow through a pipe of varying diameter.

$$\frac{\Delta m_1}{\Delta t} = \frac{\rho_1 \Delta V_1}{\Delta t} = \frac{\rho_1 A_1 \Delta l_1}{\Delta t} = \rho_1 A_1 v_1,$$

where $\Delta V_1 = A_1 \Delta l_1$ is the volume of mass Δm_1, and ρ_1 is the fluid density. Similarly, at point 2 (through area A_2), the flow rate is $\rho_2 A_2 v_2$. Since no fluid flows in or out the sides, the flow rates through A_1 and A_2 must be equal. Thus, since:

$$\frac{\Delta m_1}{\Delta t} = \frac{\Delta m_2}{\Delta t},$$

then

$$\rho_1 A_1 v_1 = \rho_2 A_2 v_2.$$

This is called the **equation of continuity**. If the fluid is incompressible (ρ doesn't change with pressure), which is an excellent approximation for liquids under most circumstances (and sometimes for gases as well), then $\rho_1 = \rho_2$, and the equation of continuity becomes

$$A_1 v_1 = A_2 v_2. \qquad [\rho = \text{constant}] \quad \textbf{(13–7)}$$

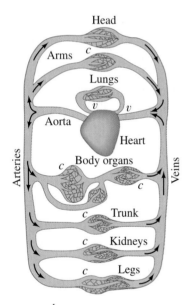

Head

Arms c

Lungs

Aorta v v

Heart

c Body organs

Arteries — Veins

c

c Trunk

c Kidneys

c Legs

v = valves
c = capillaries

FIGURE 13–22
Human circulatory system.

➡ **PHYSICS APPLIED**

Blood flow

Notice that the product Av represents the *volume rate of flow* (volume of fluid passing a given point per second), since $\Delta V/\Delta t = A \, \Delta l/\Delta t = Av$, which in SI units is m^3/s. Equation 13–7 tells us that where the cross-sectional area is large the velocity is small, and where the area is small the velocity is large. That this is reasonable can be seen by looking at a river. A river flows slowly through a meadow where it is broad, but speeds up to torrential speed when passing through a narrow gorge.

EXAMPLE 13–12 **ESTIMATE** **Blood flow.** In humans, blood flows from the heart into the aorta, from which it passes into the major arteries. These branch into the small arteries (arterioles), which in turn branch into myriads of tiny capillaries, Fig. 13–22. The blood returns to the heart via the veins. The radius of the aorta is about 1.0 cm and the blood passing through it has a speed of about 30 cm/s. A typical capillary has a radius of about 4×10^{-4} cm, and blood flows through it at a speed of about 5×10^{-4} m/s. Estimate how many capillaries there are in the body.

[†] If there were no viscosity, the velocity would be the same across a cross section of the tube. Real fluids have viscosity, and this internal friction causes different layers of the fluid to flow at different speeds. In this case v_1 and v_2 represent the average speeds at each cross section.

SOLUTION Let A_1 be the area of the aorta, and A_2 be the area of *all* the capillaries through which blood flows. Then $A_2 = N \pi r_{cap}^2$ where N is the number of capillaries and $r_{cap} \approx 4 \times 10^{-4}$ cm is the estimated average radius of one capillary. From the equation of continuity (Eq. 13–7), we have

$$v_2 A_2 = v_1 A_1$$

$$v_2 N \pi r_{cap}^2 = v_1 \pi r_{aorta}^2$$

so

$$N = \frac{v_1}{v_2} \frac{r_{aorta}^2}{r_{cap}^2} = \left(\frac{0.30 \, \text{m/s}}{5 \times 10^{-4} \, \text{m/s}} \right) \left(\frac{1.0 \times 10^{-2} \, \text{m}}{4 \times 10^{-6} \, \text{m}} \right)^2 \approx 4 \times 10^9,$$

or about 4 billion capillaries.

Another Example that makes use of the equation of continuity and the argument leading up to it is the following.

EXAMPLE 13–13 **Heating duct to a room.** How large must a heating duct be if air moving 3.0 m/s along it can replenish the air every 15 minutes in a room of 300-m³ volume? Assume the air's density remains constant.

SOLUTION We can apply the equation of continuity (Eq. 13–7) if we consider the room (call it point 2) as a large section of the duct, Fig. 13–23. Reasoning in the same way we did to obtain Eq. 13–7 (changing Δt to t), we see that $A_2 v_2 = A_2 l_2 / t = V_2 / t$ where V_2 is the volume of the room. Then $A_1 v_1 = A_2 v_2 = V_2 / t$ and

$$A_1 = \frac{V_2}{v_1 t} = \frac{300 \, \text{m}^3}{(3.0 \, \text{m/s})(900 \, \text{s})} = 0.11 \, \text{m}^2.$$

If the duct is square, then each side has length $l = \sqrt{A} = 0.33$ m or 33 cm. A rectangular duct 20 cm × 55 cm will also do.

➡ **PHYSICS APPLIED**

Heating duct

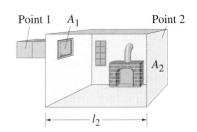

Point 1 A_1 Point 2

A_2

l_2

FIGURE 13–23 Example 13–13.

13–8 Bernoulli's Equation

Have you ever wondered why smoke goes up a chimney, why a car's convertible top bulges upward at high speeds, or how a sailboat can move against the wind? These are examples of a principle worked out by Daniel Bernoulli (1700–1782) in the early eighteenth century. In essence, **Bernoulli's principle** states that *where the velocity of a fluid is high, the pressure is low, and where the velocity is low, the pressure is high.* For example, if the pressures at points 1 and 2 in Fig. 13–21 are measured, it will be found that the pressure is lower at point 2, where the velocity is greater, than it is at point 1, where the velocity is smaller. At first glance, this might seem strange; you might expect that the greater speed at point 2 would imply a higher pressure. But this cannot be the case. For if the pressure at point 2 were higher than at 1, this higher pressure would slow the fluid down, whereas in fact it has speeded up in going from point 1 to point 2. Thus the pressure at point 2 must be less than at point 1, to be consistent with the fact that the fluid accelerates.

Bernoulli's principle

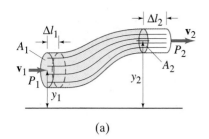

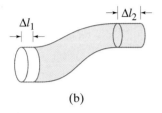

FIGURE 13–24 Fluid flow: for derivation of Bernoulli's equation.

Bernoulli developed an equation that expresses this principle quantitatively. To derive Bernoulli's equation, we assume the flow is steady and laminar, the fluid is incompressible, and the viscosity is small enough to be ignored. To be general, we assume the fluid is flowing in a tube of nonuniform cross section that varies in height above some reference level, Fig. 13–24. We will consider the amount of fluid shown in color and calculate the work done to move it from the position shown in (a) to that shown in (b). In this process, fluid at point 1 flows a distance Δl_1 and forces the fluid at point 2 to move a distance Δl_2. The fluid to the left of point 1 exerts a pressure P_1 on our section of fluid and does an amount of work

$$W_1 = F_1 \Delta l_1 = P_1 A_1 \Delta l_1.$$

At point 2, the work done on our section of fluid is

$$W_2 = -P_2 A_2 \Delta l_2;$$

the negative sign is present because the force exerted on the fluid is opposite to the motion (thus the fluid shown in color does work on the fluid to the right of point 2). Work is also done on the fluid by the force of gravity. Since the net effect of the process shown in Fig. 13–24 is to move a mass m of volume $A_1 \Delta l_1 (= A_2 \Delta l_2$, since the fluid is incompressible) from point 1 to point 2, the work done by gravity is

$$W_3 = -mg(y_2 - y_1),$$

where y_1 and y_2 are heights of the center of the tube above some (arbitrary) reference level. Notice that in the case shown in Fig. 13–24, this term is negative since the motion is uphill against the force of gravity. The net work W done on the fluid is thus:

$$W = W_1 + W_2 + W_3$$

$$W = P_1 A_1 \Delta l_1 - P_2 A_2 \Delta l_2 - mgy_2 + mgy_1.$$

According to the work-energy principle (Section 7–4), the net work done on a system is equal to its change in kinetic energy. Thus

$$\tfrac{1}{2}mv_2^2 - \tfrac{1}{2}mv_1^2 = P_1 A_1 \Delta l_1 - P_2 A_2 \Delta l_2 - mgy_2 + mgy_1.$$

The mass m has volume $A_1 \Delta l_1 = A_2 \Delta l_2$. Thus we can substitute $m = \rho A_1 \Delta l_1 = \rho A_2 \Delta l_2$, and also divide through by $A_1 \Delta l_1 = A_2 \Delta l_2$, to obtain:

$$\tfrac{1}{2}\rho v_2^2 - \tfrac{1}{2}\rho v_1^2 = P_1 - P_2 - \rho gy_2 + \rho gy_1$$

which we rearrange to get

Bernoulli's equation

$$P_1 + \tfrac{1}{2}\rho v_1^2 + \rho gy_1 = P_2 + \tfrac{1}{2}\rho v_2^2 + \rho gy_2. \tag{13–8}$$

This is **Bernoulli's equation**. Since points 1 and 2 can be any two points along a tube of flow, Bernoulli's equation can be written:

$$P + \tfrac{1}{2}\rho v^2 + \rho gy = \text{constant}$$

at every point in the fluid, where y is the height of the center of the tube above a fixed reference level. [Note that if there is no flow ($v_1 = v_2 = 0$), then Eq. 13–8 reduces to the hydrostatic equation, Eq. 13–6a: $P_2 - P_1 = -\rho g(y_2 - y_1)$.]

Bernoulli's equation is an expression of the law of energy conservation, since we derived it from the work-energy principle.

EXAMPLE 13–14 **Flow and pressure in hot-water heating systems.** Water circulates throughout a house in a hot-water heating system. If the water is pumped at a speed of 0.50 m/s through a 4.0-cm-diameter pipe in the basement under a pressure of 3.0 atm, what will be the flow speed and pressure in a 2.6-cm-diameter pipe on the second floor 5.0 m above? Assume the pipes do not divide into branches.

➡ PHYSICS APPLIED

Hot-water heating system

SOLUTION We first calculate the flow speed on the second floor, calling it v_2, using the equation of continuity, Eq. 13–7. Noting that the areas are proportional to the radii squared $(A = \pi r^2)$, we call the basement point 1 and obtain

$$v_2 = \frac{v_1 A_1}{A_2} = \frac{v_1 \pi r_1^2}{\pi r_2^2} = (0.50 \text{ m/s}) \frac{(0.020 \text{ m})^2}{(0.013 \text{ m})^2} = 1.2 \text{ m/s}.$$

To find the pressure, we use Bernoulli's equation:

$$P_2 = P_1 + \rho g(y_1 - y_2) + \tfrac{1}{2}\rho(v_1^2 - v_2^2)$$
$$= (3.0 \times 10^5 \text{ N/m}^2) + (1.0 \times 10^3 \text{ kg/m}^3)(9.8 \text{ m/s}^2)(-5.0 \text{ m})$$
$$+ \tfrac{1}{2}(1.0 \times 10^3 \text{ kg/m}^3)[(0.50 \text{ m/s})^2 - (1.2 \text{ m/s})^2]$$
$$= 3.0 \times 10^5 \text{ N/m}^2 - 4.9 \times 10^4 \text{ N/m}^2 - 6.0 \times 10^2 \text{ N/m}^2$$
$$= 2.5 \times 10^5 \text{ N/m}^2,$$

or 2.5 atm. Notice that the velocity term contributes very little in this case.

13–9 Applications of Bernoulli's Principle: From Torricelli to Sailboats, Airfoils, and TIA

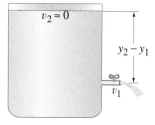

FIGURE 13–25 Torricelli's theorem: $v_1 = \sqrt{2g(y_2 - y_1)}$.

Bernoulli's equation can be applied to many situations. One example is to calculate the velocity, v_1, of a liquid flowing out of a spigot at the bottom of a reservoir, Fig. 13–25. We choose point 2 in Eq. 13–8 to be the top surface of the liquid. Assuming the diameter of the reservoir is large compared to that of the spigot, v_2 will be almost zero. Points 1 (the spigot) and 2 (top surface) are open to the atmosphere so the pressure at both points is equal to atmospheric pressure: $P_1 = P_2$. Then Bernoulli's equation becomes

$$\tfrac{1}{2}\rho v_1^2 + \rho g y_1 = \rho g y_2$$

or

$$v_1 = \sqrt{2g(y_2 - y_1)}. \tag{13–9}$$

Torricelli's theorem

This result is called **Torricelli's theorem.** Although it is seen to be a special case of Bernoulli's equation, it was discovered a century before Bernoulli by Evangelista Torricelli, a student of Galileo, hence its name. Equation 13–9 tells us that the liquid leaves the spigot with the same speed that a freely falling object would attain falling the same height. This should not be too surprising since the derivation of Bernoulli's equation relies on the conservation of energy.

Another special case of Bernoulli's equation arises when a fluid is flowing horizontally with no appreciable change in height; that is, $y_1 = y_2$. Then Eq. 13–8 becomes

$$P_1 + \tfrac{1}{2}\rho v_1^2 = P_2 + \tfrac{1}{2}\rho v_2^2, \tag{13–10}$$

which tells us quantitatively that where the speed is high the pressure is low, and vice versa. It explains many common phenomena, some of which are illustrated in

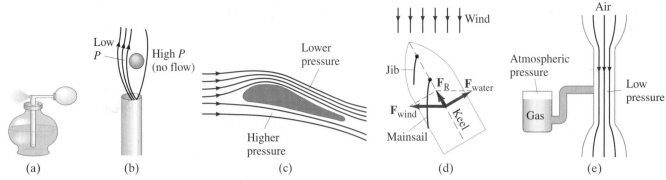

FIGURE 13–26 Examples of Bernoulli's principle: (a) atomizer, (b) Ping-Pong ball in jet of air, (c) airplane wing, (d) sailboat, (e) carburetor barrel.

➡ **PHYSICS APPLIED**

Perfume atomizer

➡ **PHYSICS APPLIED**

Airplanes and dynamic lift

➡ **PHYSICS APPLIED**

Sailing against the wind

FIGURE 13–27 Venturi meter.

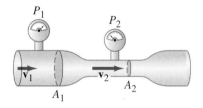

➡ **PHYSICS APPLIED**

Smoke up a chimney

Fig. 13–26. The pressure in the air blown at high speed across the top of the vertical tube of a perfume atomizer (Fig. 13–26a) is less than the normal air pressure acting on the surface of the liquid in the bowl. Thus perfume is pushed up the tube because of the reduced pressure at the top. A Ping-Pong ball can be made to float above a blowing jet of air (some vacuum cleaners can blow air), Fig. 13–26b; if the ball begins to leave the jet of air, the higher pressure in the still air outside the jet (Bernoulli's principle) pushes the ball back in.

Airplane wings and other airfoils moving rapidly relative to the air are designed to deflect the air so that, although streamline flow is largely maintained, the streamlines are crowded together above the wing, Fig. 13–26c. Just as the flow lines are crowded together in a pipe constriction where the velocity is high (see Fig. 13–21), so the crowded streamlines above the wing indicate that the air speed is greater above the wing than below it. Hence the air pressure above the wing is less than that below and there is thus a net upward force, called **dynamic lift**. Bernoulli's principle is only one aspect of the lift on a wing. Wings are usually tilted slightly upward so that air striking the bottom surface is deflected downward; the change in momentum of the rebounding air molecules results in an additional upward force on the wing. Turbulence also plays an important role.

A sailboat can move against the wind, Fig. 13–26d and the Bernoulli effect aids in this considerably if the sails are arranged so that the air velocity increases in the narrow constriction between the two sails. The normal atmospheric pressure behind the mainsail is larger than the reduced pressure in front of it (due to the fast moving air in the narrow slot between the sails), and this pushes the boat forward. When going against the wind, the mainsail is set at an angle, as shown in Fig. 13–26d, such that the net force on the sail (wind and Bernoulli) acts nearly perpendicular to the sail ($\mathbf{F}_{wind}$). This would tend to make the boat move sideways if it weren't for the keel that extends vertically downward beneath the water—for the water exerts a force ($\mathbf{F}_{water}$) on the keel nearly perpendicular to the keel. The resultant of these two forces ($\mathbf{F}_R$) is almost directly forward as shown.

A **venturi tube** is essentially a pipe with a narrow constriction (the throat). One example of a venturi tube is the barrel of a carburetor in a car, Fig. 13–26e. The flowing air speeds up as it passes through this constriction, and so the pressure is lower. Because of the reduced pressure, gasoline under atmospheric pressure in the carburetor reservoir is forced into the air stream in the throat and mixes with the air before entering the cylinders.

The venturi tube is also the basis of the *venturi meter*, which is used to measure the flow speed of fluids, Fig. 13–27. Venturi meters can be used to measure the flow velocities of gases and liquids, including blood velocity in arteries.

Why does smoke go up a chimney? It's partly because hot air rises (it's less dense and therefore buoyant). But Bernoulli's principle also plays a role. When wind blows across the top of a chimney, the pressure is less there than inside the

house. Hence, air and smoke are pushed up the chimney. Even on an apparently still night there is usually enough ambient air flow at the top of a chimney to assist upward flow of smoke.

In medicine, one of many applications of Bernoulli's principle is to explain a TIA, a transient ischemic attack (meaning a temporary lack of blood supply to the brain), caused by the so-called "subclavian steal syndrome." A person suffering a TIA may experience symptoms such as dizziness, double vision, headache, and weakness of the limbs. A TIA can occur as follows. Blood normally flows up to the brain at the back of the head via the two vertebral arteries—one going up each side of the neck—which meet to form the basilar artery just below the brain, as shown in Fig. 13–28. The vertebral arteries issue from the subclavian arteries, as shown, before the latter pass to the arms. When an arm is exercised vigorously, blood flow increases to meet the needs of the arm's muscles. If the subclavian artery on one side of the body is partially blocked, however, say by arteriosclerosis, the blood velocity will have to be higher on that side to supply the needed blood. (Recall the equation of continuity: smaller area means larger velocity for the same flow rate, Eq. 13–7.) The increased blood velocity past the opening to the vertebral artery results in lower pressure (Bernoulli's principle). Thus, blood rising in the vertebral artery on the "good" side at normal pressure can be *diverted down* into the other vertebral artery because of the low pressure on that side (like the Venturi effect), instead of passing upward into the basilar artery and the brain. Hence the blood supply to the brain can be reduced due to "subclavian steal syndrome": the fast-moving blood in the subclavian artery "steals" the blood away from the brain. The resulting dizziness or weakness usually causes the person to stop the exertions, followed by a return to normal.

The sink where you brush your teeth contains a *trap*—a U-shaped pipe in the drain designed to retain water as a barrier against disagreeable odors (Fig. 13–29a). The air pressure on the two sides of the trap is the same, because of the *main vent*, which usually opens through the roof. However, the water in the trap could be pushed out of the trap, via the Bernoulli effect, if flush water from above passes at high speed—and low pressure—as shown in Fig. 13–29b. Building codes specify that drains of this type must have a separate vent (Fig. 13–29c) that maintains the atmospheric pressure on both sides of the trap, even when flush water passes through the system.

Bernoulli's equation ignores the effects of friction (viscosity) and the compressibility of the fluid. The energy that is transformed to internal (or potential) energy due to compression and to thermal energy by friction can be taken into account by adding terms to the right side of Eq. 13–8. These terms are difficult to calculate theoretically and are normally determined empirically. They do not significantly alter the explanations for the phenomena described above.

➡ P H Y S I C S   A P P L I E D
Medicine—TIA

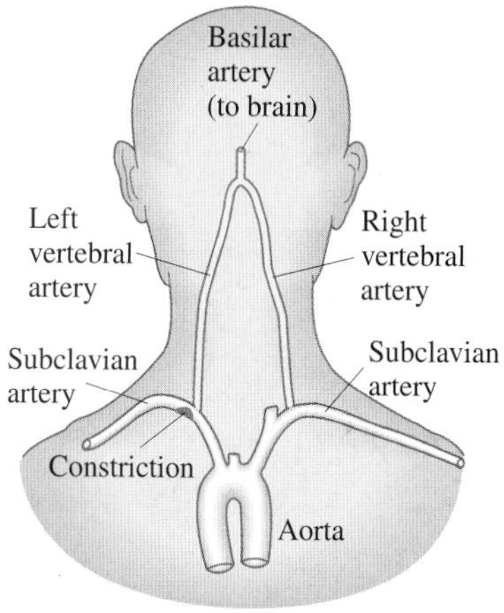

FIGURE 13–28 Rear of the head and shoulders showing arteries leading to the brain and to the arms. High blood velocity past the constriction in the left subclavian artery causes low pressure in the left vertebral artery, in which a reverse (downward) blood flow can then result: so-called "subclavian steal syndrome," resulting in a TIA (see text).

➡ P H Y S I C S   A P P L I E D
Sink trap

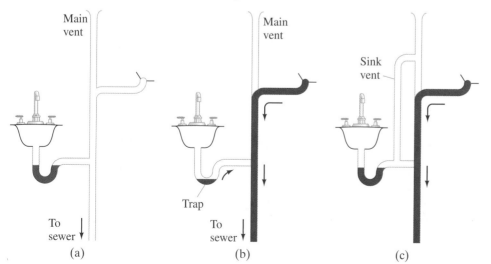

FIGURE 13–29
Making sure a trap works.

As already mentioned, real fluids have a certain amount of internal friction which is called **viscosity**. Viscosity exists in both liquids and gases, and is essentially a frictional force between adjacent layers of fluid as the layers move past one another. In liquids, viscosity is due to the cohesive forces between the molecules. In gases, it arises from collisions between the molecules.

Different fluids possess different amounts of viscosity: syrup is more viscous than water; grease is more viscous than engine oil; liquids in general are much more viscous than gases. The viscosity of different fluids can be expressed quantitatively by a *coefficient of viscosity*, η (the Greek lowercase letter eta), which is defined in the following way. A thin layer of fluid is placed between two flat plates. One plate is stationary and the other is made to move, Fig. 13–30. The fluid directly in contact with each plate is held to the surface by the adhesive force between the molecules of the liquid and those of the plate. Thus the upper surface of the fluid moves with the same speed v as the upper plate, whereas the fluid in contact with the stationary plate remains stationary. The stationary layer of fluid retards the flow of the layer just above it, which in turn retards the flow of the next layer, and so on. Thus the velocity varies continuously from 0 to v, as shown. The increase in velocity divided by the distance over which this change is made—equal to v/l—is called the *velocity gradient*. To move the upper plate requires a force, which you can verify by moving a flat plate across a puddle of syrup on a table. For a given fluid, it is found that the force required, F, is proportional to the area of fluid in contact with each plate, A, and to the speed, v, and is inversely proportional to the separation, l, of the plates: $F \propto vA/l$. For different fluids, the more viscous the fluid, the greater is the required force. Hence the proportionality constant for this equation is defined as the *coefficient of viscosity*, η:

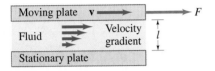

FIGURE 13–30
Determination of viscosity.

$$F = \eta A \frac{v}{l}. \qquad \textbf{(13–11)}$$

Solving for η, we find $\eta = Fl/vA$. The SI unit for η is $N \cdot s/m^2 = Pa \cdot s$ (pascal·second). In the cgs system, the unit is $dyne \cdot s/cm^2$ and this unit is called a *poise* (P). Viscosities are often given in centipoise $(1 \, cP = 10^{-2} \, P)$. Table 13–3 lists the coefficient of viscosity for various fluids. The temperature is also specified, since it has a strong effect; the viscosity of liquids such as motor oil, for example, decreases rapidly as temperature increases.[†]

[†]The Society of Automotive Engineers assigns numbers to represent the viscosity of oils: 30-weight (SAE 30) is more viscous than 10-weight. Multigrade oils, such as 20–50, are designed to maintain viscosity as temperature increases; 20–50 means the oil is 20-wt when cool but is like a 50-wt pure oil when it is hot (engine running temperature).

TABLE 13–3 Coefficient of Viscosity for Various Fluids

Fluid	Temperature (°C)	Coefficient of Viscosity, $\eta (Pa \cdot s)$[†]
Water	0	1.8×10^{-3}
	20	1.0×10^{-3}
	100	0.3×10^{-3}
Whole blood	37	$\approx 4 \times 10^{-3}$
Blood plasma	37	$\approx 1.5 \times 10^{-3}$
Ethyl alcohol	20	1.2×10^{-3}
Engine oil (SAE 10)	30	200×10^{-3}
Glycerine	20	1500×10^{-3}
Air	20	0.018×10^{-3}
Hydrogen	0	0.009×10^{-3}
Water vapor	100	0.013×10^{-3}

[†]$1 \, Pa \cdot s = 10 \, P = 1000 \, cP$

*13–11 | Flow in Tubes: Poiseuille's Equation

If a fluid had no viscosity, it could flow through a level tube or pipe without a force being applied. Because of viscosity, a pressure difference between the ends of a tube is necessary for the steady flow of any real fluid, be it water or oil in a pipe, or blood in the circulatory system of a human, even when the tube is level.

The rate of flow of a fluid in a round tube depends on the viscosity of the fluid, the pressure difference, and the dimensions of the tube. The French scientist J. L. Poiseuille (1799–1869), who was interested in the physics of blood circulation (and after whom the "poise" is named), determined how the variables affect the flow rate of an incompressible fluid undergoing laminar flow in a cylindrical tube. His result, known as *Poiseuille's equation*, is:

$$Q = \frac{\pi R^4 (P_1 - P_2)}{8 \eta L}, \qquad \textbf{(13–12)}$$

where R is the inside radius of the tube, L is its length, $P_1 - P_2$ is the pressure difference between the ends, η is the coefficient of viscosity, and Q is the volume rate of flow (volume of fluid flowing past a given point per unit time which in SI has units of m^3/s). Equation 13–12 applies only to laminar flow.

Poiseuille's equation tells us that the flow rate Q is directly proportional to the "pressure gradient," $(P_1 - P_2)/L$, and it is inversely proportional to the viscosity of the fluid. This is just what we might expect. It may be surprising, however, that Q also depends on the *fourth* power of the tube's radius. This means that for the same pressure gradient, if the tube radius is halved, the flow rate is decreased by a factor of 16! Thus the rate of flow, or alternately the pressure required to maintain a given flow rate, is greatly affected by only a small change in tube radius. Applying this to blood flow—although only approximately because of the presence of corpuscles and turbulence—we can see how reduction of artery radius by buildup of cholesterol and/or arteriosclerosis requires the heart to work much harder to maintain proper flow rates.

*13–12 | Surface Tension and Capillarity

Up to now in this chapter, we have mainly been studying what happens to fluids as a whole. But the *surface* of a liquid at rest also behaves in an interesting way. A number of common observations suggest that the surface of a liquid acts like a stretched membrane under tension. For example, a drop of water on the end of a dripping faucet, or hanging from a thin branch in the early morning dew (Fig. 13–31), forms into a nearly spherical shape as if it were a tiny balloon filled with water. A steel needle can be made to float on the surface of water even though it is denser than the water. The surface of a liquid acts as if it is under tension, and this tension, acting parallel to the surface, arises from the attractive forces between the molecules. This effect is called *surface tension*. More specifically, a quantity called the *surface tension*, γ (the Greek letter gamma), is defined as the force F per unit length L that acts across any line in a surface, tending to pull the surface closed:

$$\gamma = \frac{F}{L}. \qquad \textbf{(13–13)}$$

To understand this, consider the U-shaped apparatus shown in Fig. 13–32, which encloses a thin film of liquid. Because of surface tension, a force F is required to pull the movable wire and thus increase the surface area of the liquid. The liquid contained by the wire apparatus is a thin film having both a top and a bottom surface. Hence the length of the surface being increased is $2l$, and the surface tension is $\gamma = F/2l$. A delicate apparatus of this type can be used to measure the surface tension of various liquids. The surface tension of water is $0.072\ N/m$ at $20°C$. Table 13–4 gives the values for other liquids. Note that temperature has a considerable effect on the surface tension.

Poiseuille's equation for flow rate in a tube

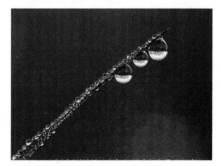

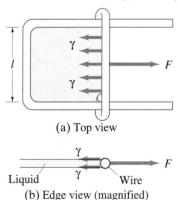

FIGURE 13–31 Spherical water droplets, dew on a grass blade.

FIGURE 13–32 U-shaped wire apparatus holding a film of liquid to measure surface tension ($\gamma = F/2l$).

(a) Top view

(b) Edge view (magnified)

TABLE 13–4 Surface Tension of Some Substances

Substance	Surface Tension (N/m)
Mercury (20°C)	0.44
Blood, whole (37°C)	0.058
Blood, plasma (37°C)	0.073
Alcohol, ethyl (20°C)	0.023
Water (0°C)	0.076
(20°C)	0.072
(100°C)	0.059
Benzene (20°C)	0.029
Soap solution (20°C)	≈ 0.025
Oxygen (−193°C)	0.016

FIGURE 13–33 A water strider.

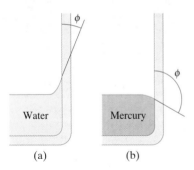

(a)   (b)

FIGURE 13–35 Water (a) "wets" the surface of glass, whereas (b) mercury does not "wet" the glass.

FIGURE 13–36 Capillarity.

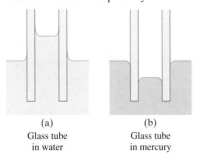

(a)   (b)
Glass tube   Glass tube
in water   in mercury

Because of surface tension, insects (Fig. 13–33) can walk on water; and objects more dense than water, such as a steel needle, can actually float on the surface. Figure 13–34a shows how the surface tension can support the weight w of an object. Actually, the object sinks slightly into the fluid, so w is the "effective weight" of that object—its true weight less the buoyant force. If the object is spherical in shape, the surface tension acts at all points around a horizontal circle of approximately radius r (Fig. 13–34a). Only the vertical component, $\gamma \cos \theta$, acts to balance w. We set the length L equal to the circumference of the circle, $L \approx 2\pi r$, so the net upward force due to surface tension is $F \approx (\gamma \cos \theta)L \approx 2\pi r \gamma \cos \theta$.

EXAMPLE 13–15 ESTIMATE Insect walks on water. The base of an insect's leg is approximately spherical in shape, with a radius of about 2.0×10^{-5} m. The 0.0030-g mass of the insect is supported equally by the six legs. Estimate the angle θ (see Fig. 13–34) for an insect on the surface of water. Assume the water temperature is 20°C.

SOLUTION Since the insect is in equilibrium, the upward surface tension force is equal to the effective pull of gravity downward on each leg:

$$2\pi r \gamma \cos \theta \approx w,$$

where w is one-sixth the weight of the insect (since it has six legs). Then

$$(6.28)(2.0 \times 10^{-5}\,\text{m})(0.072\,\text{N/m}) \cos \theta \approx \tfrac{1}{6}(3.0 \times 10^{-6}\,\text{kg})(9.8\,\text{m/s}^2)$$

$$\cos \theta \approx \frac{0.49}{0.90} = 0.54.$$

So $\theta \approx 57°$. Notice that if $\cos \theta$ were greater than 1, this would indicate that the surface tension would not be great enough to support the weight.

FIGURE 13–34 Surface tension acting on (a) a sphere, and (b) an insect leg. Example 13–15.

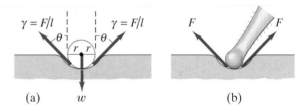

(a)   (b)

Soaps and detergents have the effect of lowering the surface tension of water. This is desirable for washing and cleaning since the high surface tension of pure water prevents it from penetrating easily between the fibers of material and into tiny crevices. Substances that reduce the surface tension of a liquid are called *surfactants*.

Surface tension plays a role in another interesting phenomenon, capillarity. It is a common observation that water in a glass container rises up slightly where it touches the glass, Fig. 13–35a. The water is said to "wet" the glass. Mercury, on the other hand, is depressed when it touches the glass, Fig. 13–35b; the mercury does not wet the glass. Whether or not a liquid wets a solid surface is determined by the relative strength of the cohesive forces between the molecules of the liquid compared to the adhesive forces between the molecules of the liquid and those of the container. (*Cohesion* refers to the force between molecules of the same type and *adhesion* to the force between molecules of different types.) Water wets glass because the water molecules are more strongly attracted to the glass molecules than they are to other water molecules. The opposite is true for mercury: the cohesive forces are stronger than the adhesive forces.

In tubes having very small diameters, liquids are observed to rise or fall relative to the level of the surrounding liquid. This phenomenon is called **capillarity**, and such thin tubes are called **capillaries**. Whether the liquid rises or falls (Fig. 13–36) depends on the relative strengths of the adhesive and cohesive forces. Thus water rises in a glass tube whereas mercury falls. The actual amount of rise (or fall) depends on the surface tension—which is what keeps the liquid surface from breaking apart.

*13–13 Pumps, and the Heart

We conclude this chapter with a brief discussion of pumps of various types, including the heart. Pumps can be classified into categories according to their function. A *vacuum pump* is designed to reduce the pressure (usually of air) in a given vessel. A *force pump*, on the other hand, is a pump that is intended to increase the pressure—for example, to lift a liquid (such as water from a well) or to push a fluid through a pipe. Figure 13–37 illustrates the principle behind a simple reciprocating pump. It could be a vacuum pump, in which case the intake is connected to the vessel to be evacuated. A similar mechanism is used in some force pumps, and in this case the fluid is forced under increased pressure through the outlet. Other kinds of pumps are illustrated in Fig. 13–38. The centrifugal pump, or any force pump, can be used as a *circulating pump*—that is, to circulate a fluid around a closed path, such as the cooling water or lubricating oil in an automobile.

The heart of a human (and of other animals as well) is essentially a circulating pump. The action of a human heart is shown in Fig. 13–39. There are actually two separate paths for blood flow. The longer path takes blood to the parts of the body, via the arteries, bringing oxygen to body tissues and picking up carbon dioxide which it carries back to the heart via veins. This blood is then pumped into the lungs (the second path), where the carbon dioxide is released and oxygen is taken up. The oxygen-laden blood is returned to the heart, where it is again pumped to the tissues of the body.

➡ PHYSICS APPLIED

Pumps and the heart

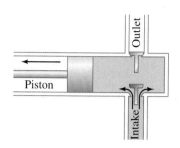

FIGURE 13–37 An example of one kind of pump. As the diagram indicates, the intake valve opens and air (or fluid that is being pumped) fills the empty space when the piston moves to the left. When the piston moves to the right (not shown), the outlet valve opens and fluid is forced out.

FIGURE 13–38 (a) Centrifugal pump: the rotating blades force fluid through the outlet pipe; this kind of pump is used in vacuum cleaners and as a water pump in automobiles. (b) Rotary oil-seal pump, used to obtain vacuums as low as 10^{-4} mm-Hg: gas (usually air) from the vessel to be evacuated diffuses into the space G via the intake pipe I; the rotating off-center cylinder C traps the gas in G and pushes it out the exhaust valve E, in the meantime allowing more gas to diffuse into G for the next cycle. The sliding valve V is kept in contact with C by a spring S, and this prevents the exhaust gas from returning to G. (c) Diffusion pump, used to obtain vacuums as low as 10^{-8} mm-Hg: air molecules from the vessel to be evacuated diffuse into the jet, where the rapidly moving jet of oil sweeps the molecules away. A "forepump" is needed, which is a mechanical pump, such as the rotary type (b), and acts as a first stage in reducing the pressure.

FIGURE 13–39 (a) In the diastole phase, the heart relaxes between beats. Blood moves into the heart; both atria are filled rapidly. (b) When the atria contract, the systole, or pumping, phase begins. The contraction pushes the blood through the mitral and tricuspid valves into the ventricles. (c) The contraction of the ventricles forces the blood through the semilunar valves into the pulmonary artery which leads to the lungs, and to the aorta (the body's largest artery) which leads to the arteries serving all the body. (d) When the heart relaxes, the semilunar valves close; blood fills the atria, beginning the cycle again.

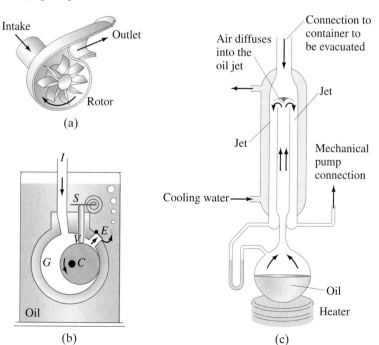

(a)

(b)

(c)

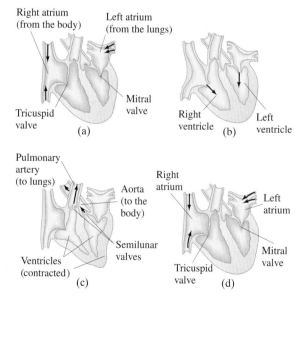

Summary

The three common phases of matter are **solid, liquid**, and **gas**. Liquids and gases are collectively called **fluids**, meaning they have the ability to flow. The **density** of a material is defined as its mass per unit volume. **Specific gravity** is the ratio of the density of the material to the density of water (at 4°C).

Pressure is defined as force per unit area. The pressure at a depth h in a liquid is given by

$$P = \rho g h,$$

where ρ is the density of the liquid and g is the acceleration due to gravity. If the density of a fluid is not uniform, the pressure P varies with height y as

$$\frac{dP}{dy} = -\rho g.$$

Pascal's principle says that an external pressure applied to a confined fluid is transmitted throughout the fluid.

Pressure is measured using a manometer or other type of gauge. A **barometer** is used to measure atmospheric pressure. Standard atmospheric pressure (average at sea level) is $1.013 \times 10^5 \text{ N/m}^2$. **Gauge pressure** is the total pressure less atmospheric pressure.

Archimedes' principle states that an object submerged wholly or partially in a fluid is buoyed up by a force equal to the weight of fluid it displaces.

Fluid flow can be characterized either as **streamline** (sometimes called **laminar**), in which the layers of fluid move smoothly and regularly along paths called streamlines, or as **turbulent**, in which case the flow is not smooth and regular but is characterized by irregularly shaped whirlpools.

Fluid flow rate is the mass or volume of fluid that passes a given point per unit time. The **equation of continuity** states that for an incompressible fluid flowing in an enclosed tube, the product of the velocity of flow and the cross-sectional area of the tube remains constant:

$$Av = \text{constant}.$$

Bernoulli's principle tells us that where the velocity of a fluid is high, the pressure in it is low, and where the velocity is low, the pressure is high. **Bernoulli's equation** for steady laminar flow of an incompressible and nonviscous fluid is

$$P_1 + \tfrac{1}{2}\rho v_1^2 + \rho g y_1 = P_2 + \tfrac{1}{2}\rho v_2^2 + \rho g y_2.$$

for two points along the flow.

Viscosity refers to friction within a fluid that prevents the fluid from flowing freely and is essentially a frictional force between adjacent layers of fluid as they move past one another.

Questions

1. If one material has a higher density than another, does this mean the molecules of the first must be heavier than those of the second? Explain.

2. Airplane travelers often note that their cosmetics bottles and other containers have leaked after a trip. What might cause this?

3. The three containers in Fig. 13–40 are filled with water to the same height and have the same surface area at the base; hence the water pressure, and the total force on the base of each, is the same. Yet the total weight of water is different for each. Explain this "hydrostatic paradox."

FIGURE 13–40   Question 3.

4. Consider what happens when you push both a pin and the blunt end of a pen against your skin with the same force. Decide what determines whether your skin suffers a cut— the net force applied to it or the pressure.

5. It is often said that water seeks its own level. Explain.

6. A small amount of water is boiled in a one-gallon gasoline can. The can is removed from the heat and the lid put on. Shortly thereafter the can collapses. Explain.

7. Explain how the tube in Fig. 13–41, known as a **siphon**, can transfer liquid from one container to a lower one even though the liquid must flow uphill for part of its journey. (Note that the tube must be filled with liquid to start with.)

FIGURE 13–41
A siphon. Question 7.

8. An ice cube floats in a glass of water filled to the brim. What can you say about the density of ice? As the ice melts, will the glass overflow?

9. Will an ice cube float in a glass of alcohol? Why or why not?

10. A barge filled high with sand approaches a low bridge over the river and cannot quite pass under it. Should sand be added to, or removed from, the barge?

11. Will an empty balloon have precisely the same apparent weight on a scale as one that is filled with air? Explain.

12. Does the buoyant force on a diving bell deep beneath the ocean have precisely the same value as when the bell is just beneath the surface? Explain.

13. A small wooden boat floats in a swimming pool, and the level of the water at the edge of the pool is marked. Consider the following situations and determine whether the level of the water will rise, fall, or stay the same. (*a*) The boat is removed from the water. (*b*) The boat in the water holds an iron anchor which is removed from the boat and placed on the shore. (*c*) The iron anchor is removed from the boat and dropped in the pool.

14. Explain why helium weather balloons, which are used to measure atmospheric conditions at high altitude, are normally released while filled to only 10%–20% of their maximum volume.

15. Why do you float more easily in salt water than in fresh?

16. Roofs of houses are sometimes "blown" off (or are they pushed off?) during a tornado or hurricane. Explain, using Bernoulli's principle.

17. If you dangle two pieces of paper vertically, a few inches apart (Fig. 13–42), and blow between them, how do you think the papers will move? Try it and see. Explain.

FIGURE 13–42 Question 17.

18. Why does the canvas top of a convertible bulge out when the car is traveling at high speed?

19. Children are told to avoid standing too close to a rapidly moving train because they might get sucked under it. Is this possible? Explain.

20. Why does a sailboat need a keel? In a small sailboat, the keel (a vertical "board" that extends below the boat into the water) is removed when the boat is anchored. Why?

21. A tall Styrofoam cup is filled with water. Two holes are punched in the cup near the bottom, and water begins rushing out. If the cup is dropped so it falls freely, will the water continue to flow from the holes? Explain.

22. With a little effort, you can blow across a dime on a table and make it land in a cup as shown in Fig. 13–43 without touching either cup or dime. Explain (and try it).

FIGURE 13–43 Question 22.

23. Why do airplanes normally take off into the wind?

24. Why does the stream of water from a faucet become narrower as it falls (Fig. 13–44)?

FIGURE 13–44 Water coming from a faucet. Question 24 and Problem 93.

25. A baseball pitcher puts spin on the ball when throwing a curve. Use Bernoulli's principle to explain in detail why the ball curves. Explain why a spinning ball with a very smooth surface curves in the opposite direction to one with a rough surface (such as a baseball or tennis ball). [This is a challenging question. See "The Physics of Baseball," *Physics Today*, May 1995, p. 29.]

26. Two ships moving in parallel paths close to one another risk colliding. Why?

Problems

Section 13–1

1. (I) The approximate volume of the granite monolith known as El Capitan in Yosemite National Park (Fig. 13–45) is about $10^8 \, m^3$. What is its approximate mass?

FIGURE 13–45
Problem 1.

2. (I) What is the approximate mass of air in a living room $4.8 \, m \times 3.8 \, m \times 2.8 \, m$?

3. (I) If you tried to nonchalantly smuggle gold bricks by filling your backpack, whose dimensions are $60 \, cm \times 25 \, cm \times 15 \, cm$, what would its mass be?

4. (I) Estimate your volume. [*Hint*: Because you can swim on or just under the surface of the water in a swimming pool, you have a pretty good idea of your density.]

5. (II) A bottle has a mass of 35.00 g when empty and 98.44 g when filled with water. When filled with another fluid, the mass is 88.78 g. What is the specific gravity of this other fluid?

6. (II) If 5.0 L of antifreeze solution (specific gravity = 0.80) is added to 4.0 L of water to make a 9.0-L mixture what is the specific gravity of the mixture?

Sections 13–2 to 13–5

7. (I) Estimate the pressure exerted on a floor by (*a*) a pointed loudspeaker leg (60 kg on four legs) of area $= 0.05 \, cm^2$, and compare it (*b*) to the pressure exerted by a 1500-kg elephant standing on one foot $\left(\text{area} = 800 \, cm^2\right)$.

8. (I) (*a*) Calculate the total force of the atmosphere acting on the top of a table that measures $1.6 \, m \times 2.9 \, m$. (*b*) What is the total force acting upward on the underside of the table?

9. (II) In a movie, Tarzan is shown evading his captors by hiding underwater for many minutes while breathing through a long thin reed. Assuming the maximum pressure difference lungs can manage and still breathe is -80 mm-Hg, calculate the deepest he could have been.

10. (II) The gauge pressure in each of the four tires of an automobile is 240 kPa. If each tire has a "footprint" of $200 \, cm^2$, estimate the mass of the car.

11. (II) The maximum gauge pressure in a hydraulic lift is 17.0 atm. What is the largest size vehicle (kg) it can lift if the diameter of the output line is 24.5 cm?

12. (II) How high would the level be in an alcohol barometer at normal atmospheric pressure?

13. (II) What is the total force and the absolute pressure on the bottom of a swimming pool 22.0 m by 8.5 m whose uniform depth is 2.0 m? What will be the pressure against the *side* of the pool near the bottom?

14. (II) How high would the atmosphere extend if it were of uniform density, equal to that at sea level throughout?

15. (II) Water and then oil (which don't mix) are poured into a U-shaped tube, open at both ends. They come to equilibrium as shown in Fig. 13–46. What is the density of the oil? [*Hint*: Pressures at points a and b are equal. Why?]

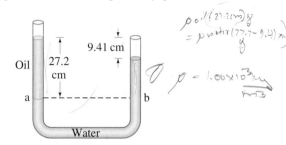

FIGURE 13–46   Problem 15.

16. (II) Determine the water gauge pressure at a house at the bottom of a hill fed by a full tank of water 5.0 m deep and connected to the house by a pipe that is 100 m long at an angle of 60° from the horizontal (Fig. 13–47). Neglect turbulence, and frictional and viscous effects. How high would the water shoot if it came vertically out of a broken pipe in front of the house?

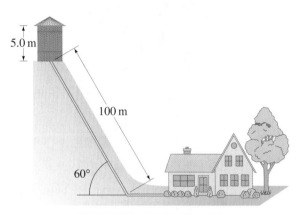

FIGURE 13–47   Problem 16.

17. (II) Estimate the air pressure on the summit of Mt. Everest (8850 m above sea level).

18. (II) Determine the minimum gauge pressure needed in the water pipe leading into a building if water is to come out of a faucet on the twelfth floor, 36.5 m above.

19. (II) An open-tube mercury manometer is used to measure the pressure in an oxygen tank. On a day when the atmospheric pressure is 1040 mbar, what is the absolute pressure (in Pa) in the tank if the height of the mercury in the open tube is (*a*) 28.0 cm higher, (*b*) 4.2 cm lower, than the mercury in the tube connected to the tank?

20. (II) A hydraulic press for compacting powdered samples has a large cylinder which is 10.0 cm in diameter, and a small cylinder with a diameter of 2.0 cm (Fig. 13–48). A lever is attached to the small cylinder as shown. The sample, which is placed on the large cylinder, has an area of 4.0 cm². What is the pressure on the sample if 300 N is applied to the lever?

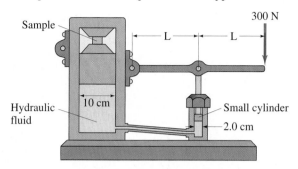

FIGURE 13–48   Problem 20.

21. (II) In working out his principle, Pascal showed dramatically how force can be multiplied with fluid pressure. He placed a long thin tube of 0.30-cm radius vertically into a 20-cm-radius wine barrel, Fig. 13–49. He found that when the barrel was filled with water and the tube filled to a height of 12 m, the barrel burst. Calculate (a) the mass of fluid in the tube, and (b) the net force on the lid of the barrel.

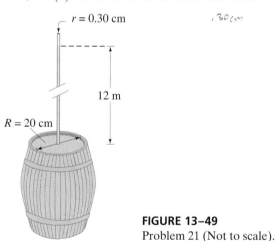

FIGURE 13–49
Problem 21 (Not to scale).

22. (III) A beaker of liquid accelerates from rest, on a horizontal surface, with acceleration a to the right. (a) Show that the surface of the liquid makes an angle $\theta = \tan^{-1} a/g$ with the horizontal. (b) Which edge of the water surface is higher? (c) How does the pressure vary with depth below the surface?

23. (III) Water stands at a height h behind a vertical dam of uniform width b. (a) Use integration to show that the total force of the water on the dam is $F = \frac{1}{2}\rho g h^2 b$. (b) Show that the torque about the base of the dam due to this force can be considered to act with a lever arm equal to $h/3$. (c) For a freestanding concrete dam of uniform thickness t and height h, what minimum thickness is needed to prevent overturning? Do you need to add in atmospheric pressure for this last part? Explain.

24. (III) Estimate the density of the water 6.0 km deep in the sea. (See Section 12–5 and Table 12–1.) By what fraction does it differ from the density at the surface?

25. (III) A cylindrical bucket of liquid (density ρ) is rotated about its symmetry axis, which is vertical. If the angular velocity is ω, show that the pressure at a distance r from the rotation axis is

$$P = P_0 + \tfrac{1}{2}\rho\omega^2 r^2,$$

where P_0 is the pressure at $r = 0$.

Section 13–6

26. (I) The hydrometer of Example 13–10 sinks to a depth of 22.9 cm when placed in a fermenting vat. What is the density of the brewing liquid?

27. (I) A geologist finds that a moon rock whose mass is 7.85 kg has an apparent mass of 6.18 kg when submerged in water. What is the density of the rock?

28. (I) What fraction of a piece of aluminum will be submerged when it floats in mercury?

29. (II) A spherically shaped balloon has a radius of 7.35 m, and is filled with helium. How large a cargo can it lift, assuming that the skin and structure of the balloon have a mass of 1000 kg? Neglect the buoyant force on the cargo volume itself.

30. (II) A 78-kg person has an apparent mass of 54 kg (because of buoyancy) when standing in water that comes up to the hips. Estimate the mass of each leg. Assume the body has SG = 1.00.

31. (II) What is the likely identity of a metal (see Table 13–1) if a sample has a mass of 63.5 g when measured in air and an apparent mass of 56.4 g when submerged in water?

32. (II) Calculate the true mass (in vacuum) of a piece of aluminum whose apparent mass is 2.0000 kg when weighed in air.

33. (II) An undersea research chamber for aquanauts is spherical with an external diameter of 6.0 m. The mass of the chamber, when occupied, is 75,000 kg. It is anchored to the sea bottom by a cable. What is (a) the buoyant force on the chamber, and (b) the tension in the cable?

34. (II) A scuba diver is diving off the shores of the Cayman Islands. The diver and her gear displace a volume of 65.0 L and have a total mass of 63.0 kg. (a) What is the buoyant force on the diver? (b) Will the diver sink or float?

35. (II) (a) Show that the buoyant force F_B on a partially submerged object such as a ship acts at the center of gravity of the fluid before it is displaced. This point is called the **center of buoyancy**. (b) For a ship to be stable, should its center of buoyancy be above, below, or at the same point as, its center of gravity? Explain (See Fig. 13–50.)

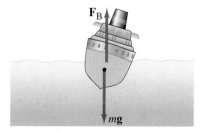

FIGURE 13–50   Problem 35.

36. (II) Archimedes' principle can be used not only to determine the specific gravity of a solid using a known liquid (Example 13–9). The reverse can be done as well. (a) As an example, a 3.40-kg aluminum ball has an apparent mass of 2.10 kg when submerged in a particular liquid: calculate the density of the liquid. (b) Derive a simple formula for determining the density of a liquid using this procedure.

37. (II) A 0.48-kg piece of wood floats in water but is found to sink in alcohol (specific gravity = 0.79) in which it has an apparent mass of 0.047 kg. What is the SG of the wood?

38. (II) The specific gravity of ice is 0.917, whereas that for seawater is 1.025. What fraction of an iceberg is above the surface of the water?

39. (III) A polar bear partially supports herself by pulling part of her body out of the water onto a rectangular slab of ice. The ice sinks down so that only half of what was once exposed now is exposed, and the bear has 70 percent of her volume (and weight) out of the water. Estimate the bear's mass, assuming that the total volume of the ice is 10 m³, and the bear's specific gravity is 1.0.

40. (III) A 3.15-kg piece of wood (SG = 0.50) floats on water. What minimum mass of lead, hung from it by a string, will cause it to sink?

41. (III) If an object floats in water, its density can be determined by tying a sinker on it so that both the object and the sinker are submerged. Show that the specific gravity is given by $w/(w_1 - w_2)$, where w is the weight of the object alone in air, w_1 is the apparent weight when a sinker is tied to it and the sinker only is submerged, and w_2 is the apparent weight when both the object and the sinker are submerged.

Sections 13–7 to 13–9

42. (I) Using the data of Example 13–12, calculate the average speed of blood flow in the major arteries of the body which have a total cross-sectional area of about 2.0 cm².

43. (I) A 15-cm-radius air duct is used to replenish the air of a room 9.2 m × 5.0 m × 4.5 m every 12 min. How fast does the air flow in the duct?

44. (I) Show that Bernoulli's equation reduces to the hydrostatic variation of pressure with depth (Eq. 13–6) when there is no flow $(v_1 = v_2 = 0)$.

45. (I) How fast does water flow from a hole at the bottom of a very wide, 4.6-m-deep storage tank filled with water? Ignore viscosity.

46. (II) A $\frac{5}{8}$-inch (inside) diameter garden hose is used to fill a round swimming pool 6.1 m in diameter. How long will it take to fill the pool to a depth of 1.2 m if water issues from the hose at a speed of 0.33 m/s?

47. (II) What gauge pressure in the water mains is necessary if a firehose is to spray water to a height of 15 m?

48. (II) A 6.0-cm-diameter pipe gradually narrows to 4.0 cm. When water flows through this pipe at a certain rate, the gauge pressure in these two sections is 32 kPa and 24 kPa. What is the volume rate of flow?

49. (II) What is the volume rate of flow of water from a 1.85-cm-diameter faucet if the pressure head is 12.0 m?

50. (II) If wind blows at 25 m/s over your house, what is the net force on the roof if its area is 240 m²?

51. (II) What is the lift (in newtons) due to Bernoulli's principle on a wing of area 86 m² if the air passes over the top and bottom surfaces at speeds of 340 m/s and 290 m/s, respectively?

52. (II) Estimate the air pressure inside a category 5 hurricane where the wind speed is 300 km/h. (See Fig. 13–51.)

FIGURE 13–51 Problem 52.

53. (II) Show that the power needed to drive a fluid through a pipe is equal to the volume rate of flow, Q, times the pressure difference, $P_1 - P_2$.

54. (II) Water at a gauge pressure of 3.8 atm at street level flows into an office building at a speed of 0.60 m/s through a pipe 5.0 cm in diameter. The pipes taper down to 2.6 cm in diameter by the top floor, 20 m above (Fig. 13–52). Calculate the flow velocity and the gauge pressure in such a pipe on the top floor. Assume no branch pipes and ignore viscosity.

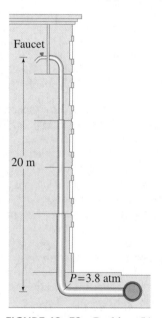

FIGURE 13–52 Problem 54.

55. (II) In Fig. 13–53, take into account the speed of the top surface of the tank and show that the speed of fluid leaving the opening at the bottom is

$$v_1 = \sqrt{2gh/(1 - A_1^2/A_2^2)},$$

where $h = y_2 - y_1$, and A_1 and A_2 are the areas of the opening and the top surface, respectively. Assume $A_1 \ll A_2$ so that the flow remains nearly steady and laminar.

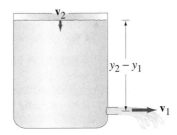

FIGURE 13–53 Problems 55, 56, 59, and 60.

56. (II) Suppose the top surface of the vessel in Fig. 13–53 is subjected to an external gauge pressure P_2. (a) Derive a formula for the speed, v_1, at which the liquid flows from the opening at the bottom into atmospheric pressure, P_A. Assume the velocity of the liquid surface, v_2, is approximately zero. (b) If $P_2 = 0.85$ atm and $y_2 - y_1 = 2.1$ m, determine v_1 for water.

57. (III) *Thrust of a rocket.* (a) Use Bernoulli's equation and the equation of continuity to show that the emission speed of the propelling gases of a rocket is

$$v = \sqrt{2(P - P_0)/\rho},$$

where ρ is the density of the gas, P is the pressure of the gas inside the rocket, and P_0 is atmospheric pressure just outside the exit orifice. Assume that the gas density stays approximately constant, and that the area of the exit orifice, A_0, is much smaller than the cross-sectional area, A, of the inside of the rocket (take it to be a large cylinder). Assume also that the gas speed is not so high that significant turbulence or non-steady flow sets in. (b) Show that the thrust force on the rocket due to the emitted gases is

$$F = 2A_0(P - P_0).$$

58. (III) (a) Show that the flow velocity measured by a venturi meter is given by the relation

$$v_1 = A_2 \sqrt{2(P_1 - P_2)/\rho(A_1^2 - A_2^2)}.$$

See Fig. 13–27. (b) A venturi tube is measuring the flow of water; it has a main diameter of 3.0 cm tapering down to a throat diameter of 1.0 cm; if the pressure difference is measured to be 18 mm-Hg, what is the velocity of the water?

59. (III) Suppose the opening in the tank of Fig. 13–53 is a height h_1 above the base and the liquid surface is a height h_2 above the base. The tank rests on level ground. (a) At what horizontal distance from the base of the tank will the fluid strike the ground? (b) At what other height, h_1', can a hole be placed so that the emerging liquid will have the same "range"?

60. (III) (a) In Fig. 13–53, show that Bernoulli's principle predicts that the level of the liquid, $h = y_2 - y_1$, drops at a rate

$$\frac{dh}{dt} = -\sqrt{\frac{2ghA_1^2}{A_2^2 - A_1^2}},$$

where A_1 and A_2 are the areas of the opening and the top surface, respectively, assuming $A_1 \ll A_2$, and viscosity is ignored. (b) Determine h as a function of time by integrating. Let $h = h_0$ at $t = 0$. (c) How long would it take to empty a 9.4-cm-tall cylinder filled with 1.0 L of water if the opening is at the bottom and has a 0.50-cm diameter?

* Section 13–10

*61. (II) A viscometer consists of two concentric cylinders, 10.20 cm and 10.60 cm in diameter. A particular liquid fills the space between them to a depth of 12.0 cm. The outer cylinder is fixed, and a torque of 0.024 m·N keeps the inner cylinder turning at a steady rotational speed of 62 rev/min. What is the viscosity of the liquid?

* Section 13–11

*62. (I) A gardener feels it is taking him too long to water a garden with a $\frac{3}{8}$-in-diameter hose. By what factor will his time be cut if he uses a $\frac{5}{8}$-in-diameter hose? Assume nothing else is changed.

*63. (I) Engine oil (assume SAE 10, Table 13–3) passes through a fine 1.80-mm-diameter tube in a prototype engine. The tube is 5.5 cm long. What pressure difference is needed to maintain a flow rate of 5.6 mL/min?

*64. (II) What must be the pressure difference between the two ends of a 1.9-km section of pipe, 29 cm in diameter, if it is to transport oil $(\rho = 950 \text{ kg/m}^3,\ \eta = 0.20 \text{ Pa·s})$ at a rate of 450 cm^3/s?

*65. (II) What diameter must a 17.5-m-long air duct have if the ventilation and heating system is to replenish the air in a room 9.0 m × 12.0 m × 4.0 m every 10 min? Assume the pump can exert a gauge pressure of 0.71×10^{-3} atm.

*66. (II) Assuming a constant pressure gradient, by what factor does a blood vessel decrease in radius if the blood flow is reduced by 75 percent?

*67. (II) Poiseuille's equation does not hold if the flow velocity is high enough that turbulence sets in. The onset of turbulence occurs when the so-called **Reynolds number**, Re, exceeds approximately 2000. Re is defined as

$$Re = \frac{2\bar{v}r\rho}{\eta},$$

where $\bar{v}$ is the average speed of the fluid, ρ is its density, η is its viscosity, and r is the radius of the tube in which the fluid is flowing. (a) Determine if blood flow through the aorta is laminar or turbulent, when the average speed of blood in the aorta ($r = 1.0$ cm) during the resting part of the heart's cycle is about 30 cm/s. (b) During exercise, the blood-flow speed approximately doubles. Calculate the Reynolds number in this case and determine if the flow is laminar or turbulent.

*68. (II) Water is to be pumped through a 10.0-cm-diameter pipe over a distance of 300 m. The far end of the pipe is 20 m above the pump and is at normal atmospheric pressure. What gauge pressure must the pump develop for there to be any flow at all?

*69. (III) A patient is to be given a blood transfusion. The blood is to flow through a tube from a raised bottle to a needle inserted in the vein (Fig. 13–54). The inside diameter of the 4.0-cm-long needle is 0.40 mm and the required flow rate is 4.0 cm^3 of blood per minute. How high should the bottle be placed above the needle? Obtain ρ and η from the Tables. Assume the blood pressure is 18 torr above atmospheric pressure.

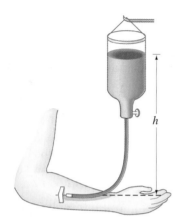

FIGURE 13–54 Problems 69 and 75.

* Section 13–12

*70. (I) If the force F needed to move the wire in Fig. 13–32 is 5.1×10^{-3} N, calculate the surface tension γ of the enclosed fluid. Assume $l = 0.070$ m.

*71. (I) Calculate the force needed to move the wire in Fig. 13–32 if it is immersed in a soapy solution and the wire is 18.2 cm long.

*72. (II) If the base of an insect's leg has a radius of about 3.0×10^{-5} m and its mass is 0.016 g, would you expect the six-legged insect to remain on top of the water?

*73. (II) The surface tension of a liquid can be determined by measuring the force F needed to just lift a circular platinum ring of radius r from the surface of the liquid. (a) Find a formula for γ in terms of F and r. (b) At 30°C, if $F = 8.40 \times 10^{-3}$ N and $r = 2.8$ cm, calculate γ for the tested liquid.

*74. (III) Show that inside a soap bubble, there must be a pressure ΔP in excess of that outside equal to $\Delta P = 4\gamma/r$, where r is the radius of the bubble and γ is the surface tension. [Hint: Think of the bubble as two hemispheres in contact with each other; and remember that there are two surfaces to the bubble. Note that this result applies to any kind of membrane, where 2γ is the tension per unit length in that membrane.]

General Problems

75. Intravenous infusions are often made under gravity, as shown in Fig. 13–54. Assuming the fluid has a density of 1.00 g/cm^3, at what height h should the bottle be placed so the liquid pressure is (a) 65 mm-Hg, (b) 550 mm-H$_2$O? (c) If the blood pressure is 18 mm-Hg above atmospheric pressure, how high should the bottle be placed so that the fluid just barely enters the vein?

76. A 2.4-N force is applied to the plunger of a hypodermic needle. If the diameter of the plunger is 1.3 cm and that of the needle 0.20 mm, (a) with what force does the fluid leave the needle? (b) What force on the plunger would be needed to push fluid into a vein where the gauge pressure is 18 mm-Hg? Answer for the instant just before the fluid starts too move.

77. A bicycle pump is used to inflate a tire. The initial tire pressure is 210 kPa (30 psi). At the end of the pumping process, the final pressure is 310 kPa (45 psi). If the diameter of the plunger in the cylinder of the pump is 3.0 cm, what is the range of the force that needs to be applied to the pump handle from beginning to end?

78. Estimate the pressure on the mountains underneath the Antarctic ice pack, which is typically 4 km thick.

79. What is the approximate difference in air pressure between the top and the bottom of the World Trade Center buildings in New York City? They are 410 m tall and are located at sea level. Express as a fraction of atmospheric pressure at sea level.

80. Giraffes are a wonder of cardiovascular engineering. Calculate the difference in pressure (in atmospheres) that the blood vessels in a giraffe's head have to accommodate as it lowers its head from a full upright position to ground level for a drink. The height of an average giraffe is about 6 meters.

81. When you drive up into the mountains, or descend rapidly from the mountains, your ears "pop," which means that the pressure behind the eardrum is being equalized to that outside. If this did not happen, what would be the approximate force on an eardrum of area 0.50 cm^2 if a change in altitude of 1000 m takes place?

82. One arm of a U-shaped tube (open at both ends) contains water and the other alcohol. If the two fluids meet exactly at the bottom of the U, and the alcohol is at a height of 18.0 cm, at what height will the water be?

83. A simple model (Fig. 13–55) considers a continent as a block (density $= 2800 \text{ kg/m}^3$) floating in the mantle rock around it (density $= 3300 \text{ kg/m}^3$). Assuming the continent is 35 km thick (the average thickness of the Earth's crust), estimate the height of the continent above the surrounding rock.

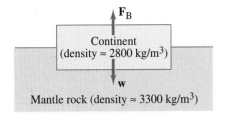

FIGURE 13–55
Problem 83.

84. The contraction of the left ventricle (chamber) of the heart pumps blood to the body. Assuming that the inner surface of the left ventricle has an area of 82 cm² and the maximum pressure in the blood is 120 mm-Hg, estimate the force exerted by the ventricle at maximum pressure.

85. Estimate the total mass of the Earth's atmosphere using the known value of atmospheric pressure at sea level.

86. Suppose a person can reduce the pressure in the lungs to −80 mm-Hg gauge pressure. How high can water then be sucked up a straw?

87. How high should the pressure head be if water is to come from a faucet at a speed of 7.2 m/s? Ignore viscosity.

88. A ship, carrying freshwater to a desert island in the Caribbean, has a horizontal cross-sectional area of 2650 m² at the waterline. When unloaded, the ship rises 8.50 m higher in the sea. How much water was delivered?

89. A raft is made of 10 logs lashed together. Each is 38 cm in diameter and has a length of 6.1 m. How many people can the raft hold before they start getting their feet wet, assuming the average person has a mass of 70 kg? Do *not* neglect the weight of the logs. Assume the specific gravity of wood is 0.60.

90. During each heartbeat, approximately 70 cm³ of blood is pushed from the heart at an average pressure of 105 mm-Hg. Calculate the power output of the heart, in watts, assuming 70 beats per minute.

91. A bucket of water is accelerated upward at 2.4 *g*. What is the buoyant force on a 3.0-kg granite rock (SG = 2.7) submerged in the water? Will the rock float? Why or why not?

92. The drinking fountain outside your classroom shoots water about 16 cm up in the air from a nozzle of diameter 0.60 cm. The pump at the base of the unit (1.1 m below the nozzle) pushes water into a 1.2-cm-diameter supply pipe that goes up to the nozzle. What gauge pressure does the pump have to provide? Ignore the viscosity; your answer will therefore be an underestimate.

93. The stream of water from a faucet decreases in diameter as it falls (Fig. 13–44). Derive an equation for the diameter of the stream as a function of the distance *y* below the faucet, given that the water has speed v_0 when it leaves the faucet, whose diameter is *D*.

94. Four lawn sprinkler heads are fed by a 1.9-cm-diameter pipe. The water comes out of the heads at an angle of 30° to the horizontal and covers a radius of 8.0 m. (*a*) What is the velocity of the water coming out of the sprinkler head? (Assume zero air resistance.) (*b*) If the output diameter of each head is 3.0 mm, how many liters of water do the four heads deliver per second? (*c*) How fast is the water flowing inside the 1.9-cm-diameter pipe?

95. You need to siphon water from a clogged sink. The sink has an area of 0.48 m² and is filled to a height of 4.0 cm. Your siphon tube rises 50 cm above the bottom of the sink and then descends 100 cm to a pail as shown in Fig. 13–56. The siphon tube has a diameter of 2.0 cm. (*a*) Assuming that the water enters the siphon tube with almost zero velocity, calculate its velocity when it enters the pail. (*b*) Estimate how long it will take to empty the sink.

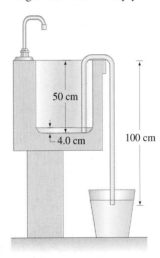

FIGURE 13–56
Problems 95 and 96.

96. Consider a siphon which transfers water (20°C) from one vessel to a second (lower) one, as in Fig. 13–56. Determine the rate of flow if the hose has a 1.2-cm-diameter and the difference in water levels of the two containers is 64 cm.

97. An airplane has a mass of 2.0×10^6 kg, and the air flows past the lower surface of the wings at 100 m/s. If the wings have a surface area of 1200 m², how fast must the air flow over the upper surface of the wing if the plane is to stay in the air? Consider only the Bernoulli effect.

98. A hydraulic lift is used to jack a 1000 kg car 10 cm off the floor. The diameter of the output piston is 15 cm and the input force is 250 N. (*a*) What is the area of the input piston? (*b*) What is the work done in lifting the car 10 cm? (*c*) If the travel for each stroke of the input piston is 12 cm, how high does the car move up for each stroke? (*d*) How many strokes are required to jack the car up 10 cm? (*e*) Show that energy is conserved.

The pendulum of a clock is an example of oscillatory motion. Many kinds of oscillatory motion are sinusoidal, or nearly so, and are referred to as being simple harmonic motion. Real systems generally have at least some friction, and the motion is damped. When an external sinusoidal force is exerted on a system able to oscillate, resonance occurs if the driving force is at or near the natural frequency of vibration.

CHAPTER 14

Oscillations

Many objects vibrate or oscillate—an object on the end of a spring, a tuning fork, the balance wheel of an old watch, a pendulum, a plastic ruler held firmly over the edge of a table and gently struck, the strings of a guitar or piano. Spiders detect prey by the vibrations of their webs, cars oscillate up and down when they hit a bump, buildings and bridges vibrate when heavy trucks pass or the wind is fierce. Indeed, because most solids are elastic (see Chapter 12), most material objects vibrate (at least briefly) when given an impulse. Electrical oscillations occur in radio and television sets. At the atomic level, atoms vibrate within a molecule, and the atoms of a solid vibrate about their relatively fixed positions. Because it is so common in everyday life and occurs in so many areas of physics, oscillatory (or vibrational) motion is of great importance. Vibrational motion is not really a "new" phenomenon because vibrations of mechanical systems are fully described on the basis of Newtonian mechanics.

14–1 Oscillations of a Spring

When a **vibration** or an **oscillation** repeats itself, back and forth, over the same path, the motion is **periodic**. The simplest form of periodic motion is represented by an object oscillating on the end of a coil spring. Because many other types of vibrational motion closely resemble this system, we will look at it in detail. We assume that the mass of the spring can be ignored, and that the spring is mounted horizontally (Fig. 14–1a), so that the object of mass m slides without friction on the horizontal surface. Any spring has a natural length at which the net force on the mass m is zero; the position of the mass at this point is called the **equilibrium position**. If the mass is moved either to the left, which compresses the spring, or to the right, which stretches it, the spring exerts a force on the mass that acts in the direction of returning the mass to the equilibrium position; hence it is called a "restoring force." The magnitude of the restoring force F is found to be directly proportional to the displacement x the spring has been stretched or compressed from the equilibrium position (Fig. 14–1b and c):

$$F = -kx. \tag{14-1}$$

Note that the equilibrium position is at $x = 0$. Equation 14–1, which is often referred to as Hooke's law (see Sections 8–2 and 12–5), is accurate as long as the spring is not compressed to the point where the coils come close to touching, or stretched beyond the elastic region (see Fig. 12–19). The minus sign in Eq. 14–1 indicates that the restoring force is always in the direction opposite to the displacement x. For example, if we choose the positive direction to the right in Fig. 14–1, x is positive when the spring is stretched, but the direction of the restoring force is to the left (negative direction). If the spring is compressed, x is negative (to the left) but the force F acts toward the right (Fig. 14–1c).

The proportionality constant k in Eq. 14–1 is called the "spring constant." In order to stretch the spring a distance x, one has to exert an (external) force on the spring at least equal to $F = +kx$. The greater the value of k, the greater the force needed to stretch a spring a given distance.

Note that the force F in Eq. 14–1 is *not* a constant, but varies with position. Therefore the acceleration of the mass m is not constant, so we *cannot* use the equations for constant acceleration developed in Chapter 2.

Let us examine what happens when the spring is initially stretched a distance $x = A$, as shown in Fig. 14–2a, and then released. The spring exerts a force on the mass that pulls it toward the equilibrium position. But because the mass has been accelerated by the force, it passes the equilibrium position with considerable speed. Indeed, as the mass reaches the equilibrium position, the force on it decreases to zero, but its speed at this point is a maximum, Fig. 14–2b. As it moves farther to the left, the force on it acts to slow it down, and it stops momentarily at $x = -A$, Fig. 14–2c. It then begins moving back in the opposite direction, Fig. 14–2d, until it reaches the original starting point, $x = A$, Fig. 14–2e. It then repeats the motion, moving back and forth symmetrically between $x = A$ and $x = -A$.

To discuss vibrational motion, we need to define a few terms. The distance x of the mass from the equilibrium point at any moment is called the **displacement**. The maximum displacement—the greatest distance from the equilibrium point—is called the **amplitude**, A. One **cycle** refers to the complete to-and-fro motion from some initial point back to that same point, say from $x = A$ to $x = -A$ back to $x = A$. The **period**, T, is defined as the time required for one complete cycle. Finally, the **frequency**, f, is the number of complete cycles per second. Frequency is generally specified in hertz (Hz), where $1 \, \text{Hz} = 1$ cycle per second (s^{-1}). It is easy to see, from their definitions, that frequency and period are inversely related:

$$f = \frac{1}{T} \quad \text{and} \quad T = \frac{1}{f}; \tag{14-2}$$

for example, if the frequency is 5 cycles per second, then each cycle takes $\frac{1}{5}$ s.

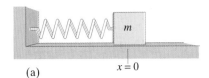
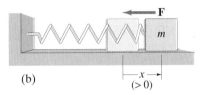
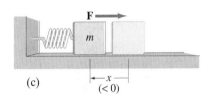

FIGURE 14–1 Mass vibrating at the end of a spring.

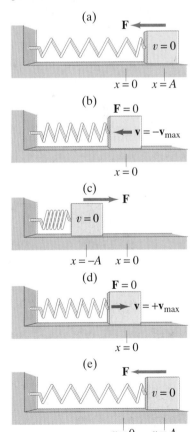

FIGURE 14–2 Force on, and velocity of, mass at different positions of its oscillation.

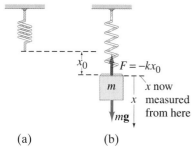

(a)   **(b)**

FIGURE 14–3 (a) Free spring, hung vertically. (b) Mass m attached to spring in new equilibrium position, which occurs when $\Sigma F = 0 = kx_0 - mg$.

FIGURE 14–4 Photo of a car's spring. (Also visible is the shock absorber—see Fig. 14–21.)

The oscillation of a spring hung vertically is essentially the same as that of a horizontal spring. Because of the force of gravity, the length of the vertical spring at equilibrium will be longer than when it is horizontal, as shown in Fig. 14–3. The spring is in equilibrium when $\Sigma F = 0 = kx_0 - mg$, so the spring stretches an extra amount $x_0 = mg/k$ to be in equilibrium. If x is measured from this new equilibrium position, Eq. 14–1 can be used directly with the same value of k.

EXAMPLE 14–1 **Car springs.** When a family of four people with a total mass of 200 kg step into their 1200-kg car, the car's springs compress 3.0 cm. (*a*) What is the spring constant of the car's springs (Fig. 14–4), assuming they act as a single spring? (*b*) How far will the car lower if loaded with 300 kg?

SOLUTION (*a*) The added force of $(200\,\text{kg})(9.8\,\text{m/s}^2) = 1960\,\text{N}$ causes the springs to compress $3.0 \times 10^{-2}\,\text{m}$. Therefore, by Eq. 14–1, the spring constant is

$$k = \frac{F}{x} = \frac{1960\,\text{N}}{3.0 \times 10^{-2}\,\text{m}} = 6.5 \times 10^4\,\text{N/m}.$$

(*b*) If the car is loaded with 300 kg,

$$x = \frac{F}{k} = \frac{(300\,\text{kg})(9.8\,\text{m/s}^2)}{(6.5 \times 10^4\,\text{N/m})} = 4.5 \times 10^{-2}\,\text{m},$$

or 4.5 cm. We could have obtained this answer without solving for k: since x is proportional to F, if 200 kg compresses the spring 3.0 cm, then 1.5 times the force will compress the spring 1.5 times as much, or 4.5 cm.

14–2 | Simple Harmonic Motion

Any vibrating system for which the net restoring force is directly proportional to the negative of the displacement (as in Eq. 14–1, $F = -kx$) is said to exhibit **simple harmonic motion** (SHM). Such a system is often called a **simple harmonic oscillator** (SHO). We saw in Chapter 12 (Section 12–5) that most solid materials stretch or compress according to Eq. 14–1 as long as the displacement is not too great. Because of this, many natural vibrations are simple harmonic or close to it.

Let us now determine the position x as a function of time for a mass attached to the end of a simple spring with spring constant k. To do so, we make use of Newton's second law, $F = ma$. Since the acceleration $a = d^2x/dt^2$, we have

$$ma = \Sigma F$$
$$m\frac{d^2x}{dt^2} = -kx,$$

where m is the mass[†] which is oscillating. We rearrange this to obtain

Equation of motion (SHM)

$$\frac{d^2x}{dt^2} + \frac{k}{m}x = 0, \tag{14–3}$$

which is known as the **equation of motion** for the simple harmonic oscillator. Mathematically it is called a *differential equation*, since it involves derivatives. We want to determine what function of time, $x(t)$, satisfies this equation. We might guess the form of the solution by noting that if a pen were attached to a vibrating mass (Fig. 14–5) and a sheet of paper moved at a steady rate beneath it, the pen would trace the curve shown. The shape of this curve looks a lot like it might be

---

[†] In the case of a mass, m', on the end of a spring, the spring itself also oscillates and at least a part of its mass must be included. It can be shown—see the Problems—that approximately one-third the mass of the spring, m_s, must be included, so $m = m' + \frac{1}{3}m_s$ in our equation. Often m_s is small enough to be ignored.

sinusoidal (such as cosine or sine) as a function of time, and its height is the amplitude A. Let us then guess that the general solution to Eq. 14–3 can be written in a form such as

$$x = A \cos(\omega t + \phi),\qquad\text{(14–4)}$$

General solution: position as functions of time

where we include the constant ϕ in the argument to be general.[†] Let us now put this trial solution into Eq. 14–3 and see if it really works. We need to differentiate the $x = x(t)$ twice:

$$\frac{dx}{dt} = \frac{d}{dt}[A\cos(\omega t + \phi)] = -\omega A \sin(\omega t + \phi)$$

$$\frac{d^2 x}{dt^2} = -\omega^2 A \cos(\omega t + \phi).$$

We now put the latter into Eq. 14–3, along with Eq. 14–4 for x:

$$\frac{d^2 x}{dt^2} + \frac{k}{m}x = 0$$

$$-\omega^2 A \cos(\omega t + \phi) + \frac{k}{m}A\cos(\omega t + \phi) = 0$$

or

$$\left(\frac{k}{m} - \omega^2\right)A\cos(\omega t + \phi) = 0.$$

Our solution, Eq. 14–4, does indeed satisfy the equation of motion (Eq. 14–3) for any time t. But it does so only if $(k/m - \omega^2) = 0$; hence

$$\omega^2 = \frac{k}{m}.\qquad\text{(14–5)}$$

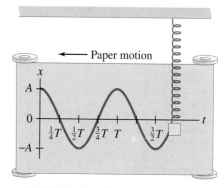

FIGURE 14–5 Sinusoidal nature of SHM as a function of time. In this case, $x = A\cos(2\pi t/T)$.

Equation 14–4 is the general solution and it contains two arbitrary constants A and ϕ, which we should expect because the second derivative in Eq. 14–3 implies that two integrations are needed, each yielding a constant. They are "arbitrary" only in a calculus sense, in that they can be anything and still satisfy the differential equation, Eq. 14–3. In real physical situations, however, A and ϕ are determined by the **initial conditions**. Suppose, for example, the mass is started at its maximum displacement and is released from rest. This is, in fact, what is shown in Fig. 14–5, and for this case $x = A \cos \omega t$. Let us confirm it: we are given $v = 0$ at $t = 0$, where

Initial conditions

$$v = \frac{dx}{dt} = \frac{d}{dt}[A\cos(\omega t + \phi)] = -\omega A \sin(\omega t + \phi) = 0. \quad [\text{at } t = 0]$$

For v to be zero at $t = 0$, then $\sin(\omega t + \phi) = \sin(0 + \phi)$ is zero if $\phi = 0$ (ϕ could also be π, 2π, etc.), and when $\phi = 0$, then

$$x = A \cos \omega t,$$

as we expected. We see immediately that A is the amplitude of the motion, and it is determined initially by how far you pulled the mass m from equilibrium before releasing it.

[†]Another possible way to write the solution is the combination $x = a \cos \omega t + b \sin \omega t$, where a and b are constants. This is equivalent to Eq. 14–4 as can be seen using the trigonometric identity $\cos(A \pm B) = \cos A \cos B \mp \sin A \sin B$.

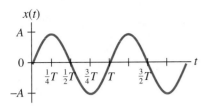

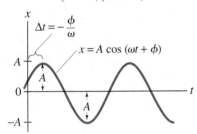

FIGURE 14–6 Special case of SHM where the mass m starts, at $t = 0$, at the equilibrium position $x = 0$ and has initial velocity toward positive values of x ($v > 0$ at $t = 0$).

FIGURE 14–7 A plot of $x = A \cos(\omega t + \phi)$ when $\phi < 0$.

Position as function of time

f and T are independent of amplitude

Consider another interesting case: at $t = 0$, the mass m is at $x = 0$ and is struck, giving it an initial velocity toward increasing values of x. Then at $t = 0$, $x = 0$, so we can write $x = A \cos(\omega t + \phi) = A \cos \phi = 0$, which can happen only if $\phi = \pm \pi/2$ (or $\pm 90°$). Whether $\phi = +\pi/2$ or $-\pi/2$ depends on $v = dx/dt = -\omega A \sin(\omega t + \phi) = -\omega A \sin \phi$, which we are given as positive ($v > 0$ at $t = 0$); hence $\phi = -\pi/2 [\sin(-90°) = -1]$. Thus our solution for this case is

$$x = A \cos\left(\omega t - \frac{\pi}{2}\right)$$
$$= A \sin \omega t,$$

where we used $\cos(\theta - \pi/2) = \sin \theta$. The solution in this case is a pure sine wave, Fig. 14–6, where A is still the amplitude.

Lots of other situations are possible, such as that shown in Fig. 14–7. The constant ϕ is called the **phase angle**, and it tells us how long after (or before) $t = 0$ the peak at $x = A$ is reached. Notice that the value of ϕ does not affect the shape of the $x(t)$ curve, but only affects the displacement at some arbitrary time, $t = 0$. Simple harmonic motion is thus always *sinusoidal*. Indeed, simple harmonic motion is *defined* as motion that is purely sinusoidal.

Since our oscillating mass repeats its motion after a time equal to its period T, it must be at the same position and moving in the same direction at $t = T$ as it was at $t = 0$. And since a sine or cosine function repeats itself after every 2π radians, then from Eq. 14–4, we must have

$$\omega T = 2\pi.$$

Hence

$$\omega = \frac{2\pi}{T} = 2\pi f,$$

where f is the frequency of the motion. We call ω the **angular frequency** (units are rad/s) to distinguish it from the frequency f (units are $s^{-1} = Hz$). Thus we can write Eq. 14–4 as

$$x = A \cos\left(\frac{2\pi t}{T} + \phi\right) \tag{14–6a}$$

or

$$x = A \cos(2\pi f t + \phi), \tag{14–6b}$$

where, because of Eq. 14–5

$$f = \frac{1}{2\pi}\sqrt{\frac{k}{m}}, \tag{14–7a}$$

$$T = 2\pi\sqrt{\frac{m}{k}}. \tag{14–7b}$$

Note that the *frequency and period do not depend on the amplitude*. Changing the amplitude of a simple harmonic oscillator does not affect its frequency. Equation 14–7a tells us that the greater the mass, the lower the frequency; and the stiffer the spring, the higher the frequency. This makes sense since a greater mass means more inertia and therefore slower response (or acceleration); and larger k means greater force and therefore quicker response. The frequency f (Eq. 14–7a) at which a SHO oscillates naturally is called its **natural frequency** (to distinguish it from a frequency at which it might be forced to oscillate by an outside force, as discussed in Section 14–8).

The simple harmonic oscillator is important in physics because whenever we have a net restoring force proportional to the displacement ($F = -kx$), which is at least a good approximation for a variety of systems, then the motion is simple harmonic—that is, sinusoidal.

EXAMPLE 14–2 **Car springs again.** What are the period and frequency of the car in Example 14–1 after hitting a bump? Assume the shock absorbers are poor, so the car really oscillates up and down.

➡ **PHYSICS APPLIED**
Car springs

SOLUTION From Eq. 14–7b,

$$T = 2\pi \sqrt{\frac{m}{k}} = 6.28 \sqrt{\frac{1400\,\text{kg}}{6.5 \times 10^4\,\text{N/m}}} = 0.92\,\text{s},$$

or slightly less than a second. The frequency $f = 1/T = 1.09$ Hz.

Let us continue our analysis of a simple harmonic oscillator. The velocity and acceleration of the oscillating mass can be obtained by differentiation of Eq. 14–4

$$v = \frac{dx}{dt} = -\omega A \sin(\omega t + \phi) \tag{14–8}$$

$$a = \frac{d^2x}{dt^2} = \frac{dv}{dt} = -\omega^2 A \cos(\omega t + \phi). \tag{14–9}$$

The velocity and acceleration of a SHO also vary sinusoidally. In Fig. 14–8 we plot the displacement, velocity, and acceleration of a SHO as a function of time for the case when $\phi = 0$. As can be seen, the speed reaches its maximum

$$v_{max} = \omega A = \sqrt{\frac{k}{m}}\, A$$

when the oscillating object is passing through its equilibrium point, $x = 0$. And the speed is zero at points of maximum displacement, $x = \pm A$. This is in accord with our discussion of Fig. 14–2. Similarly, the acceleration has its maximum value

$$a_{max} = \omega^2 A = \frac{k}{m}\, A$$

which occurs where $x = \pm A$, and a is zero at $x = 0$, as we expect, since $ma = F = -kx$.

For the general case when $\phi \neq 0$, we can relate the constants A and ϕ to the initial values of x, v, and a by setting $t = 0$ in Eqs. 14–4, 14–8, and 14–9:

$$x_0 = x(0) = A \cos\phi$$
$$v_0 = v(0) = -\omega A \sin\phi = -v_{max} \sin\phi$$
$$a_0 = a(0) = -\omega^2 A \cos\phi = -a_{max} \cos\phi.$$

FIGURE 14–8 Displacement, x, velocity, dx/dt, and acceleration, d^2x/dt^2, of a simple harmonic oscillator when $\phi = 0$.

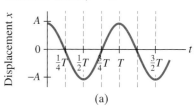

(a)

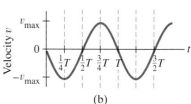

(b)

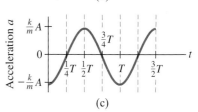

(c)

EXAMPLE 14–3 **Loudspeaker cone.** The cone of a loudspeaker vibrates in SHM at a frequency of 262 Hz ("middle C"). The amplitude at the center of the cone is $A = 1.5 \times 10^{-4}$ m, and at $t = 0$, $x = A$. (a) What is the equation describing the motion of the center of the cone? (b) What is its maximum velocity and maximum acceleration? (c) What is the position of the cone at $t = 1.00$ ms?

➡ **PHYSICS APPLIED**
Loudspeaker oscillations

SOLUTION (a) The amplitude $A = 1.5 \times 10^{-4}$ m and $\omega = 2\pi f = (6.28\,\text{rad})(262\,\text{s}^{-1}) = 1650$ rad/s. The motion begins ($t = 0$) with the cone at its maximum displacement ($x = A$ at $t = 0$), so we use the cosine function with $\phi = 0$:

$$x = A \cos\omega t = A \cos 2\pi f t = (1.5 \times 10^{-4}\,\text{m}) \cos(1650t).$$

(b) From Eq. 14–8,

$$v_{max} = \omega A = 2\pi A f = 2\pi(1.5 \times 10^{-4}\,\text{m})(262\,\text{s}^{-1}) = 0.25\,\text{m/s}.$$

From Eqs. 14–9 and 14–5:

$$a_{max} = \omega^2 A = (2\pi f)^2 A$$
$$= 4\pi^2(262\,\text{s}^{-1})^2(1.5 \times 10^{-4}\,\text{m}) = 410\,\text{m/s}^2,$$

which is more than 40 g's.

(c) At $t = 1.00 \times 10^{-3}$ s,

$$x = (1.5 \times 10^{-4}\,\text{m}) \cos[(1650\,\text{rad/s})(1.00 \times 10^{-3}\,\text{s})]$$
$$= (1.5 \times 10^{-4}\,\text{m}) \cos(1.65\,\text{rad}) = -1.2 \times 10^{-5}\,\text{m}.$$

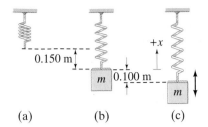

(a)   (b)   (c)

FIGURE 14-9 Example 14-4.

EXAMPLE 14-4 **Spring calculations.** In a testing device, a spring stretches 0.150 m when a 0.300-kg mass is hung from it (Fig. 14–9). The spring is then stretched an additional 0.100 m from this equilibrium point and released. Determine: (a) the values of the spring constant k and the angular frequency ω; (b) the amplitude of the oscillation A; (c) the maximum velocity, v_{max}; (d) the magnitude of the maximum acceleration of the mass; (e) the period T and frequency f; (f) the displacement x as a function of time; and (g) the velocity at $t = 0.150$ s.

SOLUTION (a) Since the spring stretches 0.150 m when 0.300 kg is hung from it, we find k from Eq. 14–1 to be

$$k = \frac{F}{x} = \frac{mg}{x} = \frac{(0.300 \text{ kg})(9.80 \text{ m/s}^2)}{0.150 \text{ m}} = 19.6 \text{ N/m}.$$

Also,

$$\omega = \sqrt{\frac{k}{m}} = \sqrt{\frac{19.6 \text{ N/m}}{0.300 \text{ kg}}} = 8.08 \text{ s}^{-1}.$$

(b) Since the spring is stretched 0.100 m from equilibrium (Fig. 14–9c) and is given no initial speed, $A = 0.100$ m.
(c) From Eq. 14–8, the maximum velocity is

$$v_{max} = \omega A = (8.08 \text{ s}^{-1})(0.100 \text{ m}) = 0.808 \text{ m/s}.$$

(d) Since $F = ma$, the maximum acceleration occurs where the force is greatest—that is, when $x = A = 0.100$ m. Thus

$$a_{max} = \frac{kA}{m} = \frac{(19.6 \text{ N/m})(0.100 \text{ m})}{0.300 \text{ kg}} = 6.53 \text{ m/s}^2.$$

(e) Equations 14–7a and b give

$$T = 2\pi \sqrt{\frac{m}{k}} = 6.28 \sqrt{\frac{0.300 \text{ kg}}{19.6 \text{ N/m}}} = 0.777 \text{ s}$$

$$f = \frac{1}{T} = 1.29 \text{ Hz}.$$

(f) The motion begins at a point of maximum displacement downward. If we take x positive upward, then at $t = 0$, $x = x_0 = -A = -0.100$ m. So we need a sinusoidal curve that has its maximum negative valve at $t = 0$; this is just a negative cosine:

$$x = -A \cos \omega t.$$

To write this in the form of Eq. 14–4 (no minus sign), recall that $\cos \theta = -\cos(\theta - \pi)$; then, putting in numbers,

$$x = -(0.100 \text{ m}) \cos 8.08t$$
$$= (0.100 \text{ m}) \cos(8.08t - \pi),$$

where t is in seconds and x is in meters. Note that the phase angle (Eq. 14–4) is $\phi = \pi$ or $180°$.
(g) The velocity at any time t is (see part c)

$$v = \frac{dx}{dt} = A\omega \sin \omega t = (0.808 \text{ m/s}) \sin 8.08t.$$

At $t = 0.150$ s, $v = (0.808 \text{ m/s}) \sin(1.21 \text{ rad}) = 0.756 \text{ m/s}$, and is upward (+).

EXAMPLE 14–5 **Spring is started with a push.** Suppose the spring of Example 14–4 is stretched 0.100 m from equilibrium ($x_0 = -0.100$ m) but is given an upward shove of $v_0 = 0.400$ m/s. Determine (a) the phase angle ϕ, (b) the amplitude A, and (c) the displacement x as a function of time, $x(t)$.

SOLUTION (a) From Eq. 14–8, at $t = 0$, $v_0 = -\omega A \sin\phi$, and from Eq. 14–4, $x_0 = A \cos\phi$. Combining these, we get

$$\tan\phi = \frac{\sin\phi}{\cos\phi} = \frac{(v_0/-\omega A)}{(x_0/A)} = -\frac{v_0}{\omega x_0} = -\frac{0.400 \text{ m/s}}{(8.08 \text{ s}^{-1})(-0.100 \text{ m})} = 0.495.$$

A calculator gives the angle as 26.3°, but we note from this equation that both the sine and cosine are negative, so our angle is in the third quadrant. Hence

$$\phi = 26.3° + 180° = 206.3° = 3.60 \text{ rad}.$$

(b) Again using Eq. 14–4 at $t = 0$,

$$A = x_0/\cos\phi = (-0.100 \text{ m})/\cos(3.60 \text{ rad}) = 0.112 \text{ m}.$$

(c) $x = A\cos(\omega t + \phi) = 0.112 \cos(8.08t + 3.60)$.

14–3 Energy in the Simple Harmonic Oscillator

When dealing with forces that are not constant, such as here with simple harmonic motion, it is often convenient and useful to use the energy approach.

For a simple harmonic oscillator, such as a mass m oscillating on the end of a massless spring, the restoring force is given by

$$F = -kx.$$

The potential energy function, as we have already seen in Chapter 8, is given by

$$U = -\int F \, dx = \tfrac{1}{2}kx^2,$$

where we set the constant of integration equal to zero so $U = 0$ at $x = 0$ (the equilibrium position).

The total mechanical energy is the sum of the kinetic and potential energies,

$$E = \tfrac{1}{2}mv^2 + \tfrac{1}{2}kx^2,$$

where v is the velocity of the mass m when it is a distance x from the equilibrium position. SHM can occur only if there is no friction, so the total mechanical energy E remains constant. As the mass oscillates back and forth, the energy continuously changes from potential energy to kinetic energy, and back again (Fig. 14–10). At the extreme points, $x = A$ and $x = -A$, all the energy is stored in the spring as potential energy (and is the same whether the spring is compressed or stretched to the full amplitude). At these extreme points, the mass stops momentarily as it changes direction, so $v = 0$ and:

$$E = \tfrac{1}{2}m(0)^2 + \tfrac{1}{2}kA^2 = \tfrac{1}{2}kA^2. \tag{14–10a}$$

Thus, the **total mechanical energy of a simple harmonic oscillator is proportional to the square of the amplitude**. At the equilibrium point, $x = 0$, all the energy is kinetic:

$$E = \tfrac{1}{2}mv^2 + \tfrac{1}{2}k(0)^2 = \tfrac{1}{2}mv_{max}^2, \tag{14–10b}$$

where v_{max} is the maximum velocity during the motion. At intermediate points the energy is part kinetic and part potential, and because energy is conserved

$$E = \tfrac{1}{2}mv^2 + \tfrac{1}{2}kx^2 = \tfrac{1}{2}kA^2 = \tfrac{1}{2}mv_{max}^2. \tag{14–10c}$$

We can confirm Eqs. 14–10a and b explicitly by inserting Eqs. 14–4 and 14–8 into this last relation:

$$E = \tfrac{1}{2}m\omega^2 A^2 \sin^2(\omega t + \phi) + \tfrac{1}{2}kA^2 \cos^2(\omega t + \phi).$$

Substituting with $\omega^2 = k/m$, or $kA^2 = m\omega^2 A^2 = mv_{max}^2$, and noting that $\sin^2(\omega t + \phi) + \cos^2(\omega t + \phi) = 1$, we obtain Eqs. 14–10a and b:

$$E = \tfrac{1}{2}kA^2 = \tfrac{1}{2}mv_{max}^2.$$

FIGURE 14–10 Energy changes from kinetic energy to potential energy and back again as the spring oscillates.

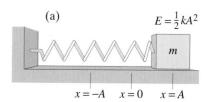

(a) $E = \tfrac{1}{2}kA^2$

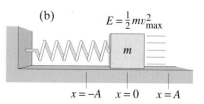

(b) $E = \tfrac{1}{2}mv_{max}^2$

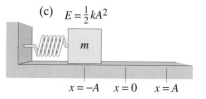

(c) $E = \tfrac{1}{2}kA^2$

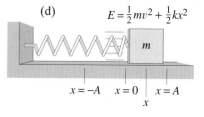

(d) $E = \tfrac{1}{2}mv^2 + \tfrac{1}{2}kx^2$

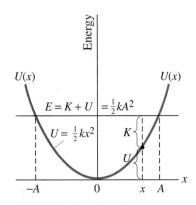

FIGURE 14–11 Graph of potential energy, $U = \frac{1}{2}kx^2$. $K + U = E = $ constant for any point x where $-A \leq x \leq A$. Values of K and U are indicated for an arbitrary point x.

We can now obtain an equation for the velocity v as a function of x by solving for v^2 in Eq. 14–10c:

$$v = \pm \sqrt{\frac{k}{m}\left(A^2 - x^2\right)} \qquad \textbf{(14–11a)}$$

or, since $v_{\max} = A\sqrt{k/m}$,

$$v = \pm v_{\max}\sqrt{1 - \frac{x^2}{A^2}}. \qquad \textbf{(14–11b)}$$

Again we see that v is a maximum at $x = 0$, and is zero at $x = \pm A$.

The potential energy, $U = \frac{1}{2}kx^2$, is plotted in Fig. 14–11. The upper horizontal line represents a particular value of the total energy $E = \frac{1}{2}kA^2$. The distance between the E line and the U curve represents the kinetic energy, K, and the motion[†] is restricted to x values between $-A$ and $+A$. These results are, of course, consistent with our full solution of the previous Section.

Energy conservation is a convenient way to obtain v, for example, if x is given (or vice versa), without having to deal with time t.

EXAMPLE 14–6 **Energy calculations.** For the simple harmonic oscillation of Example 14–4, determine (a) the total energy, (b) the kinetic and potential energies as a function of time, (c) the velocity when the mass is 0.050 m from equilibrium, (d) the kinetic and potential energies at half amplitude ($x = \pm A/2$).

SOLUTION (a) Since $k = 19.6\,\text{N/m}$ and $A = 0.100\,\text{m}$, the total energy E from Eq. 14–10a is

$$E = \tfrac{1}{2}kA^2 = \tfrac{1}{2}(19.6\,\text{N/m})(0.100\,\text{m})^2 = 9.80 \times 10^{-2}\,\text{J}.$$

(b) We have, from parts (f) and (g) of Example 14–4, $x = -(0.100\,\text{m})\cos 8.08t$ and $v = (0.808\,\text{m/s})\sin 8.08t$, so

$$U = \tfrac{1}{2}kx^2 = \left(9.80 \times 10^{-2}\,\text{J}\right)\cos^2 8.08t$$
$$K = \tfrac{1}{2}mv^2 = \left(9.80 \times 10^{-2}\,\text{J}\right)\sin^2 8.08t.$$

(c) We use Eq. 14–11b and find

$$v = v_{\max}\sqrt{1 - x^2/A^2} = 0.70\,\text{m/s}.$$

(d) At $x = A/2 = 0.050\,\text{m}$, we have

$$U = \tfrac{1}{2}kx^2 = 2.5 \times 10^{-2}\,\text{J}$$
$$K = E - U = 7.3 \times 10^{-2}\,\text{J}.$$

CONCEPTUAL EXAMPLE 14–7 **Doubling the amplitude.** Suppose the spring in Fig. 14–10 is stretched twice as far (to $x = 2A$). What happens to (a) the energy of the system, (b) the maximum velocity, (c) the maximum acceleration?

RESPONSE (a) From Eq. 14–10a, the energy is related to the square of the amplitude, so stretching it twice as far quadruples the energy [You may protest: "I did work stretching the spring from $x = 0$ to $x = A$. Don't I do the same work stretching it from A to $2A$?" No. The force you have to exert for the second leg is more than for the first leg (because $F = -kx$), so the work done is more too.]
(b) From Eq. 14–10b, we can see that since the energy is quadrupled, the maximum velocity must be double what it was before.
(c) Since the force is twice as great when we stretch it twice as far, the acceleration here is also twice as great.

[†] See Section 8–9 for more discussion.

14–4 | Simple Harmonic Motion Related to Uniform Circular Motion

Simple harmonic motion has an interesting simple relationship to a particle rotating in a circle with uniform speed. Consider a mass m rotating in a circle of radius A with speed v_M on top of a table as shown in Fig. 14–12. As viewed from above, the motion is a circle. But a person who looks at the motion from the edge of the table, sees an oscillatory motion back and forth, and this corresponds precisely to SHM as we shall now see. What the person sees, and what we are interested in, is the projection of the circular motion onto the x axis, Fig. 14–12. To see that this motion is analogous to SHM, let us calculate the x component of the velocity v_M which is labeled v in Fig. 14–12. The two right triangles indicated in Fig. 14–12 are similar, so

$$\frac{v}{v_M} = \frac{\sqrt{A^2 - x^2}}{A}$$

or

$$v = v_M\sqrt{1 - \frac{x^2}{A^2}}.$$

This is exactly the equation for the speed of a mass oscillating with SHM, Eq. 14–11b, where $v_M = v_{max}$. Furthermore, we can see from Fig. 14–12 that if the angular displacement at $t = 0$ is ϕ, then after a time t the particle will have rotated through an angle $\theta = \omega t$, and so

$$x = A\cos(\theta + \phi) = A\cos(\omega t + \phi).$$

But what is ω here? The linear velocity v_M of our particle undergoing rotational motion is related to ω by $v_M = \omega A$ where A is the radius of the circle (see Eq. 10–4). To make one revolution requires a time T, so we also have $v_M = 2\pi A/T$ where $2\pi A$ is the circle's circumference. Hence

$$\omega = \frac{v_M}{A} = \frac{2\pi A/T}{A} = 2\pi/T = 2\pi f$$

where T is the time required for one rotation and f is the frequency. This corresponds precisely to the back-and-forth motion of a simple harmonic oscillator. Thus, the projection on the x axis of a particle rotating in a circle has the same motion as a mass undergoing SHM. Indeed, we can say that the projection of circular motion onto a straight line is SHM.

The projection of uniform circular motion onto the y axis is also simple harmonic. Thus uniform circular motion can be thought of as two simple harmonic motions operating at right angles.

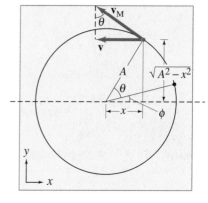

(a)

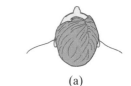

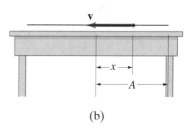

(b)

FIGURE 14–12 Analysis of simple harmonic motion as a side view (b) of circular motion (a).

FIGURE 14–13 Strobe-light photo of an oscillating pendulum.

14–5 | The Simple Pendulum

A **simple pendulum** consists of a small object (the pendulum bob) suspended from the end of a lightweight cord, Fig. 14–13. We assume that the cord doesn't stretch and that its mass can be ignored relative to that of the bob. The motion of a simple pendulum moving back and forth (Fig. 14–13) with negligible friction resembles simple harmonic motion: the pendulum oscillates along the arc of a circle with equal amplitude on either side of its equilibrium point (where it hangs vertically) and as it passes through the equilibrium point it has its maximum speed. But is it really undergoing SHM? That is, is the restoring force proportional to its displacement? Let us find out.

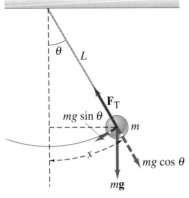

FIGURE 14–14 Simple pendulum.

The displacement of the pendulum along the arc is given by $x = L\theta$, where θ is the angle the cord makes with the vertical and L is the length of the cord, as shown in Fig. 14–14. If the restoring force is proportional to x or to θ, the motion will be simple harmonic. The restoring force is the component of the weight, mg, tangent to the arc:

$$F = -mg\sin\theta,$$

where the minus sign, as in Eq. 14–1, means the force is in the direction opposite to the angular displacement θ. Since F is proportional to the sine of θ and not to θ itself, the motion is *not* SHM. However, if θ is small, then $\sin\theta$ is very nearly equal to θ when the latter is specified in radians. This can be seen by looking at the series expansion[†] of $\sin\theta$ (or by looking at the trigonometry table inside the back cover), or by noting in Fig. 14–14 that the arc length x ($= L\theta$) is nearly the same length as the chord ($= L\sin\theta$) indicated by the straight dashed line, *if θ is small*. For angles less than 15°, the difference between θ (in radians) and $\sin\theta$ is less than 1 percent. Thus, to a very good approximation for small angles,

$$F = -mg\sin\theta \approx -mg\theta.$$

Using $x = L\theta$, we have

$$F \approx -\frac{mg}{L}x.$$

Thus, for small displacements, the motion is essentially simple harmonic, since this equation fits Hooke's law, $F = -kx$ where the effective force constant is $k = mg/L$. Thus we can write

$$\theta = \theta_{max}\cos(\omega t + \phi)$$

where θ_{max} is the maximum angular displacement and $\omega = 2\pi f = 2\pi/T$. To obtain ω we use Eq. 14–5, where for k we substitute mg/L: that is,[‡] $\omega = \sqrt{(mg/L)/m}$, or

$$\omega = \sqrt{\frac{g}{L}}. \qquad [\theta \text{ small}] \quad \textbf{(14–12a)}$$

Then the frequency f is

$$f = \frac{\omega}{2\pi} = \frac{1}{2\pi}\sqrt{\frac{g}{L}}, \qquad [\theta \text{ small}] \quad \textbf{(14–12b)}$$

and the period T is

$$T = \frac{1}{f} = 2\pi\sqrt{\frac{L}{g}}. \qquad [\theta \text{ small}] \quad \textbf{(14–12c)}$$

A surprising result is that the period does not depend on the mass of the pendulum bob! You may have noticed this if you pushed a small child and a large one on the same swing. The period *does* depend on the length L.

We saw in Section 14–2 that the period of an object undergoing SHM, including a simple pendulum, does not depend on the amplitude. Galileo is said to have first noted this fact while watching a swinging lamp in the cathedral at Pisa (Fig. 14–15). This discovery led to the pendulum clock, the first really precise time-piece, which became the standard for centuries.

Because a pendulum does not undergo *precisely* SHM, the period does depend slightly on the amplitude, the more so for large amplitudes. The accuracy

FIGURE 14–15 The swinging motion of this lamp, hanging by a very long cord from the ceiling of the cathedral at Pisa, is said to have been observed by Galileo and to have inspired him to the conclusion that the period of a pendulum does not depend on amplitude.

[†] $\sin\theta = \theta - \dfrac{\theta^3}{3!} + \dfrac{\theta^5}{5!} - \dfrac{\theta^7}{7!} + \cdots.$

[‡] Be careful not to think that $\omega = d\theta/dt$ as in rotational motion. Here θ is the angle of the pendulum at any instant (Fig. 14–14), but we use ω now to represent *not* the rate this angle θ changes, but rather as a constant related to the period, $\omega = 2\pi f = \sqrt{g/L}$.

of a pendulum clock would be affected, after many swings, by the decrease in amplitude due to friction; but the mainspring in a pendulum clock (or the falling weight in a grandfather clock) supplies energy to compensate for the friction and to maintain the amplitude constant, so that the timing remains accurate.

EXAMPLE 14–8 **Measuring g.** For calibrating instruments that measure the acceleration of gravity, a geologist uses a simple pendulum whose length is 37.10 cm, and has a frequency of 0.8190 Hz at a particular location on the Earth. What is the acceleration of gravity at this location?

SOLUTION From Eq. 14–12b, we have
$$f = \frac{1}{2\pi}\sqrt{\frac{g}{L}}.$$
Solving for g, we obtain
$$g = (2\pi f)^2 L = (6.283 \times 0.8190\,\text{s}^{-1})^2(0.3710\,\text{m}) = 9.824\,\text{m/s}^2.$$

*14–6 The Physical Pendulum and the Torsion Pendulum

Physical Pendulum

The term physical pendulum refers to any real extended body which oscillates back and forth, in contrast to the rather idealized simple pendulum where all the mass is assumed concentrated in the tiny pendulum bob. An example of a physical pendulum is a baseball bat suspended from the point O, as shown in Fig. 14–16. The force of gravity acts at the center of gravity (CG) of the body located a distance h from the pivot point O. The physical pendulum is best analyzed using the equations of rotational motion. The torque on a physical pendulum, calculated about point O, is
$$\tau = -mgh\sin\theta.$$
Newton's second law for rotational motion, Eq. 10–14, states that
$$\Sigma\tau = I\alpha = I\frac{d^2\theta}{dt^2},$$
where I is the moment of inertia of the body about the pivot point and $\alpha = d^2\theta/dt^2$ is the angular acceleration. Thus we have
$$I\frac{d^2\theta}{dt^2} = -mgh\sin\theta$$
or
$$\frac{d^2\theta}{dt^2} + \frac{mgh}{I}\sin\theta = 0,$$
where I is calculated about an axis through point O. For small angular amplitude, $\sin\theta \approx \theta$, so we have
$$\frac{d^2\theta}{dt^2} + \left(\frac{mgh}{I}\right)\theta = 0. \qquad \text{[small angular displacement]} \quad \textbf{(14–13)}$$

This is just the equation for SHM, Eq. 14–3, except that θ replaces x and mgh/I replaces k/m. Thus, for small angular displacements, a physical pendulum undergoes SHM, given by
$$\theta = \theta_{\text{max}}\cos(\omega t + \phi),$$
where θ_{max} is the maximum angular displacement and $\omega = 2\pi/T$. The period, T, is (see Eq. 14–7b, replacing m/k with I/mgh):
$$T = 2\pi\sqrt{\frac{I}{mgh}}. \qquad \text{[small angular displacement]} \quad \textbf{(14–14)}$$

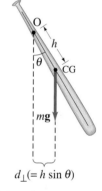

FIGURE 14–16 A physical pendulum suspended from point O.

Period of physical pendulum

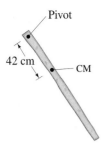

FIGURE 14–17 Example 14–9.

EXAMPLE 14–9 **Moment of inertia measurement.** An easy way to measure the moment of inertia of an object about any axis is to measure the period of oscillation about that axis. (*a*) Suppose a nonuniform 1.0-kg stick can be balanced at a point 42 cm from one end. If it is pivoted about that end (Fig. 14–17), it oscillates with a period of 3.0 s. What is its moment of inertia about this end? (*b*) What is its moment of inertia about an axis perpendicular to the stick through its center of mass?

SOLUTION (*a*) Given $T = 3.0$ s, and $h = 0.42$ m, we solve Eq. 14–14 for I:

$$I = mghT^2/4\pi^2 = 0.94 \text{ kg·m}^2.$$

Since $I = \frac{1}{3}ML^2$ for a uniform stick of length L pivoted about one end (Fig. 10–21), do you think our stick is longer or shorter than 84 cm?

(*b*) We use the parallel-axis theorem (Section 10–8). The CM is where the stick balanced, 42 cm from the end, so from Eq. 10–17,

$$I_{CM} = I - Mh^2 = 0.94 \text{ kg·m}^2 - (1.0 \text{ kg})(0.42 \text{ m})^2 = 0.76 \text{ kg·m}^2.$$

Since an object does not oscillate about its CM, we can't measure I_{CM} directly, so the parallel-axis theorem provides a convenient method to determine I_{CM}.

Torsion Pendulum

FIGURE 14–18 A torsion pendulum. The disc oscillates in SHM between θ_{max} and $-\theta_{max}$.

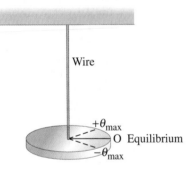

Another type of oscillatory motion is a **torsion pendulum**, in which a disc (Fig. 14–18) or a bar (as in Cavendish's apparatus, Fig. 6–3) is suspended from a wire. The twisting (torsion) of the wire serves as the elastic force. The motion here will be SHM since the restoring torque is very closely proportional to the negative of the angular displacement,

$$\tau = -K\theta,$$

where K is a constant that depends of the properties of the system. Then

$$\omega = \sqrt{\frac{K}{I}}.$$

There is no small angle restriction here, as there is for the physical pendulum (where gravity acts), as long as the wire responds linearly in accordance with Hooke's law.

14–7 Damped Harmonic Motion

The amplitude of any real oscillating spring or swinging pendulum slowly decreases in time until the oscillations stop altogether. Figure 14–19 shows a typical graph of the displacement as a function of time. This is called **damped harmonic motion**. The damping[†] is generally due to the resistance of air and to internal friction within the oscillating system. The energy that is dissipated to thermal energy is reflected in a decreased amplitude of oscillation.

[†] To "damp" means to diminish, restrain, or extinguish, as to "dampen one's spirits."

FIGURE 14–19 Damped harmonic motion. The solid red curve represents a cosine times a decreasing exponential (the dashed curves).

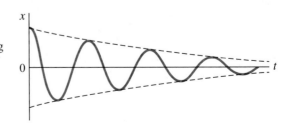

Since natural oscillating systems are damped in general, why do we even talk about (undamped) simple harmonic motion? The answer is that SHM is much easier to deal with mathematically. And if the damping is not large, the oscillations can be thought of as simple harmonic motion on which the damping is superposed, as represented by the dashed curves in Fig. 14–19. Although frictional damping does alter the frequency of vibration, the effect is usually small if the damping is small. Let us look at this in more detail.

The damping force depends on the speed of the oscillating object, and opposes the motion. In some simple cases the damping force can be approximated as being directly proportional to the speed:

$$F_{\text{damping}} = -bv,$$

where b is a constant.[†] For a mass oscillating on the end of a spring, the restoring force of the spring is $F = -kx$; so Newton's second law ($ma = \Sigma F$) becomes

$$ma = -kx - bv.$$

We bring all terms to the left side of the equation and substitute $v = dx/dt$ and $a = d^2x/dt^2$ and obtain

$$m\frac{d^2x}{dt^2} + b\frac{dx}{dt} + kx = 0, \qquad (14\text{--}15)$$

Equation of motion for damped harmonic motion

which is the equation of motion. To solve this equation, we guess at a solution and then check to see if it works. If the damping constant b is small, x as a function of t is as plotted in Fig. 14–19, which looks like a cosine function times a factor (represented by the dashed lines) that decreases in time. A simple function that does this is the exponential, $e^{-\alpha t}$, and the solution that satisfies Eq. 14–15 is

$$x = Ae^{-\alpha t}\cos\omega' t. \qquad (14\text{--}16)$$

General solution: x as function of t

where A, α, and ω' are assumed to be constants, and $x = A$ at $t = 0$. We have called the angular frequency ω' (and **not** ω) because it is not the same as the ω for SHM without damping $\left(\omega = \sqrt{k/m}\right)$.

If we substitute Eq. 14–16 into Eq. 14–15 (we do this in the optional subsection below) we find that Eq. 14–15 is indeed a solution if α and ω' have the values

$$\alpha = \frac{b}{2m} \qquad (14\text{--}17)$$

$$\omega' = \sqrt{\frac{k}{m} - \frac{b^2}{4m^2}}. \qquad (14\text{--}18)$$

Thus x as a function of time t for a (lightly) damped harmonic oscillator is

$$x = Ae^{-(b/2m)t}\cos\omega' t. \qquad (14\text{--}19)$$

Of course a phase constant, ϕ, can be added to the argument of the cosine in Eq. 14–19. As it stands with $\phi = 0$, it is clear that the constant A in Eq. 14–19 is simply the initial displacement, $x = A$ at $t = 0$. The frequency f is

$$f = \frac{\omega'}{2\pi} = \frac{1}{2\pi}\sqrt{\frac{k}{m} - \frac{b^2}{4m^2}}. \qquad (14\text{--}20)$$

The frequency is lower, and the period longer, than for undamped SHM. (In many practical cases of light damping, however, ω' differs only slightly from $\omega = \sqrt{k/m}$.) This makes sense since we expect friction to slow down the motion. Equation 14–20 reduces to Eq. 14–7a, as it should, when there is no friction ($b = 0$). The constant $\alpha = b/2m$ is a measure of how quickly the oscillations decrease toward zero (Fig. 14–19). The time $t_L = 2m/b$ is the time taken for the oscillations to drop to $1/e$ of the original amplitude; t_L is called the "mean lifetime" of the oscillations. Note that the larger b is, the more quickly the oscillations die away.

[†] Such velocity-dependent forces were discussed in Section 5–5.

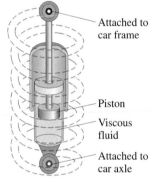

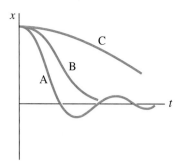

FIGURE 14–20 Underdamped (A), critically damped (B), and overdamped (C) motion.

Shock absorbers and building dampers

FIGURE 14–21 Automobile spring and shock absorber to provide damping so that car won't bounce up and down endlessly.

Attached to car frame
Piston
Viscous fluid
Attached to car axle

FIGURE 14–22 These huge dampers placed in a building structure look a lot like huge automobile shock absorbers, and they serve a similar purpose—to reduce the amplitude and the acceleration of movement when the shock of an earthquake hits.

The solution, Eq. 14–19, is not valid if b is so large that

$$b^2 > 4mk$$

since then ω' (Eq. 14–18) would become imaginary. In this case the system does not oscillate at all but returns directly to its equilibrium position, as we now discuss. Three common cases of *heavily damped* systems are shown in Fig. 14–20. Curve C represents the situation when the damping is so large $(b^2 \gg 4mk)$ that it takes a long time to reach equilibrium; the system is **overdamped**. Curve A represents an **underdamped** situation in which the system makes several swings before coming to rest $(b^2 < 4mk)$ and corresponds to a more heavily damped version of Eq. 14–19. Curve B represents **critical damping**: $b^2 = 4mk$; in this case equilibrium is reached in the shortest time. These terms all derive from the use of practical damped systems such as door closing mechanisms and shock absorbers in a car (Fig. 14–21) which are usually designed to give critical damping. But as they wear out, underdamping occurs: a door slams and a car bounces up and down several times whenever it hits a bump.

In many systems, the oscillatory motion is what counts, as in clocks and watches, and damping needs to be minimized. In other systems, oscillations are the problem, such as a car's springs, so a proper amount of damping (i.e., critical) is desired. Well-designed damping is needed for all kinds of applications. Large buildings, especially in California, are now built (or retrofitted) with huge dampers to reduce earthquake damage (Fig. 14–22).

EXAMPLE 14–10 **Simple pendulum with damping.** A simple pendulum has a length of 1.0 m (Fig. 14–23). It is set swinging with small-amplitude oscillations. After 5.0 minutes, the amplitude is only 50% of what it was initially. (*a*) What is the value of α for the motion? (*b*) By what factor does the frequency, ω', differ from ω, the undamped frequency?

SOLUTION (*a*) The equation of motion for damped harmonic motion is

$$x = Ae^{-\alpha t}\cos\omega' t, \quad \text{where} \quad \alpha = \frac{b}{2m} \quad \text{and} \quad \omega' = \sqrt{\frac{k}{m} - \frac{b^2}{4m^2}},$$

for motion of a mass on the end of a spring. For the simple pendulum without damping, we saw in Section 14–5 that

$$F = -mg\theta.$$

Since $F = ma = mL\dfrac{d^2\theta}{dt^2}$, then

$$L\frac{d^2\theta}{dt^2} + g\theta = 0.$$

Introducing a damping term, $b(d\theta/dt)$, we have

$$L\frac{d^2\theta}{dt^2} + b\frac{d\theta}{dt} + g\theta = 0,$$

which is the same as Eq. 14–15 with θ replacing x, and L and g replacing m and k. Thus

$$\alpha = \frac{b}{2L} \quad \text{and} \quad \omega' = \sqrt{\frac{g}{L} - \frac{b^2}{4L^2}}.$$

At $t = 0$, Eq. 14–16 with θ replacing x is

$$\theta_0 = Ae^{-\alpha \cdot 0}\cos\omega' \cdot 0 = A.$$

376 CHAPTER 14 Oscillations

Then at $t = 5.0\,\text{min} = 300\,\text{s}$, the amplitude has fallen to $0.50\,A$, so

$$Ae^{-\alpha(300\,\text{s})} = 0.50\,A.$$

We solve this for α and obtain $\alpha = \ln 2.0/(300\,\text{s}) = 2.3 \times 10^{-3}\,\text{s}^{-1}$.
(b) We have $L = 1.0\,\text{m}$, so $b = 2\alpha L = 4.6 \times 10^{-3}\,\text{m/s}$. Thus $(b^2/4L^2)$ is very much less than $g/L(= 9.8\,\text{s}^{-2})$, and the angular frequency of the motion remains almost the same as that of the undamped motion. Specifically,

$$\omega' = \sqrt{\frac{g}{L}}\left[1 - \frac{L}{g}\left(\frac{b^2}{4L^2}\right)\right]^{1/2} \approx \sqrt{\frac{g}{L}}\left[1 - \frac{1}{2}\frac{L}{g}\left(\frac{b^2}{4L^2}\right)\right]$$

where we used the binomial expansion. Then, with $\omega = \sqrt{g/L}$ (Eq. 14–12a),

$$\frac{\omega - \omega'}{\omega} \approx \frac{1}{2}\frac{L}{g}\left(\frac{b^2}{4L^2}\right) = 2.7 \times 10^{-7}.$$

So ω' differs from ω by less than one part in a million.

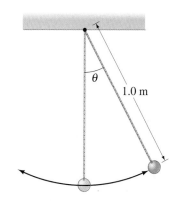

FIGURE 14–23 Example 14–10.

*Showing $x = Ae^{-\alpha t}\cos\omega't$ is a Solution

We start with Eq. 14–16, to see if it is a solution to Eq. 14–15. First we take the first and second derivatives

$$\frac{dx}{dt} = -\alpha Ae^{-\alpha t}\cos\omega't - \omega'Ae^{-\alpha t}\sin\omega't$$

$$\frac{d^2x}{dt^2} = \alpha^2 Ae^{-\alpha t}\cos\omega't + \alpha A\omega'e^{-\alpha t}\sin\omega't + \omega'\alpha Ae^{-\alpha t}\sin\omega't - \omega'^2 Ae^{-\alpha t}\cos\omega't.$$

We next substitute these relations back into Eq. 14–15 and reorganize to obtain

$$Ae^{-\alpha t}\left[(m\alpha^2 - m\omega'^2 - b\alpha + k)\cos\omega't + (2\omega'\alpha m - b\omega')\sin\omega't\right] = 0. \quad \textbf{(i)}$$

The left side of this equation must equal zero for all times t, but this can only be so for certain values of α and ω'. To determine α and ω', we choose two values of t that will make their evaluation easy. At $t = 0$, $\sin\omega't = 0$, so the above relation reduces to $A(m\alpha^2 - m\omega'^2 - b\alpha + k) = 0$, which means[†] that

$$m\alpha^2 - m\omega'^2 - b\alpha + k = 0. \quad \textbf{(ii)}$$

And at $t = \pi/2\omega'$, $\cos\omega't = 0$ so Eq. (i) can be valid only if

$$2\alpha m - b = 0. \quad \textbf{(iii)}$$

From Eq. (iii) we have

$$\alpha = \frac{b}{2m}$$

and from Eq. (ii)

$$\omega' = \sqrt{\alpha^2 - \frac{b\alpha}{m} + \frac{k}{m}} = \sqrt{\frac{k}{m} - \frac{b^2}{4m^2}}.$$

Thus we see that Eq. 14–16 is a solution to the equation of motion for the damped harmonic oscillator as long as α and ω' have these specific values (already given in Eqs. 14–17 and 14–18).

[†]It would also be satisfied by $A = 0$, but this gives the trivial and uninteresting solution $x = 0$ for all t—that is, no oscillation.

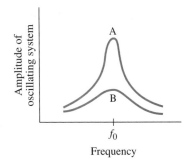

FIGURE 14–24 Resonance for lightly damped (A) and heavily damped (B) systems. (See Fig. 14–27 for a more detailed graph.)

FIGURE 14–25 This goblet breaks as it vibrates in resonance to a trumpet call.

FIGURE 14–26 (a) Large-amplitude oscillations of the Tacoma Narrows Bridge, due to gusty winds, led to its collapse (1940). (b) Collapse of freeway in California, due to the 1989 earthquake.

(a)

(b)

14–8 | Forced Vibrations; Resonance

When a vibrating system is set into motion, it vibrates at its natural frequency (Eqs. 14–7a and 14–12b). However, a system may have an external force applied to it that has its own particular frequency and then we have a **forced vibration**.

For example, we might pull the mass on the spring of Fig. 14–1 back and forth at a frequency f. The mass then vibrates at the frequency f of the external force, even if this frequency is different from the **natural frequency** of the spring, which we will now denote by f_0 where (see Eqs. 14–5 and 14–7a)

$$\omega_0 = 2\pi f_0 = \sqrt{\frac{k}{m}}.$$

In a forced vibration, the amplitude of vibration, and hence the energy transferred to the vibrating system, is found to depend on the difference between ω and ω_0 as well as on the amount of damping, reaching a maximum when the frequency of the external force equals the natural frequency of the system—that is, when $f = f_0$. The amplitude is plotted in Fig. 14–24 as a function of the external frequency f. Curve A represents light damping and curve B heavy damping. The amplitude can become large when the driving frequency f is near the natural frequency, $f \approx f_0$, as long as the damping is not too large. When the damping is small, the increase in amplitude near $f = f_0$ is very large (and often dramatic). This effect is known as **resonance**. The natural vibrating frequency f_0 of a system is called its **resonant frequency**.

A simple illustration of resonance is pushing a child on a swing. A swing, like any pendulum, has a natural frequency of oscillation that depends on its length L. If you push on the swing at a random frequency, the swing bounces around and reaches no great amplitude. But if you push with a frequency equal to the natural frequency of the swing, the amplitude increases greatly. The swing clearly illustrates that at resonance, relatively little effort is required to obtain a large amplitude.

The great tenor Enrico Caruso was said to be able to shatter a crystal goblet by singing a note of just the right frequency at full voice. This is an example of resonance, for the sound waves emitted by the voice act as a forced vibration on the glass. At resonance, the resulting vibration of the goblet may be large enough in amplitude that the glass exceeds its elastic limit and breaks (Fig. 14–25).

Since material objects are, in general, elastic, resonance is an important phenomenon in a variety of situations. It is particularly important in structural engineering, although the effects are not always foreseen. For example, it has been reported that a railway bridge collapsed because a nick in one of the wheels of a crossing train set up a resonant vibration in the bridge. Indeed, marching soldiers break step when crossing a bridge to avoid the possibility that their normal rhythmic march might match a resonant frequency of the bridge. The famous collapse of the Tacoma Narrows Bridge (Fig. 14–26a) in 1940 occurred as a result of strong gusting winds driving the span into large-amplitude oscillatory motion. The Oakland freeway collapse in the 1989 California earthquake (Fig. 14–26b) involved resonant oscillation which reached large amplitude on a section that was built on mudfill that readily transmitted that frequency.

We will meet important examples of resonance later. We will also see that vibrating objects often have not one, but many resonant frequencies.

* Equation of Motion and Its Solution

We now look at the equation of motion for a forced vibration, and its solution. Suppose the external force is sinusoidal and can be represented by

$$F_{ext} = F_0 \cos \omega t,$$

where $\omega = 2\pi f$ is the angular frequency applied externally to the oscillator. Then the equation of motion (with damping) is

$$ma = -kx - bv + F_0 \cos \omega t.$$

This can be written as

$$m \frac{d^2 x}{dt^2} + b \frac{dx}{dt} + kx = F_0 \cos \omega t. \qquad \text{(14–21)}$$

Equation of motion for forced vibration

The external force, on the right of the equation, is the only term that does not involve x or one of its derivatives. It is left as an exercise (Problem 63) to show that

$$x = A_0 \sin(\omega t + \phi_0) \qquad \text{(14–22)}$$

General solution: x as function of t

is a solution to Eq. 14–21, by direct substitution, where

$$A_0 = \frac{F_0}{m \sqrt{(\omega^2 - \omega_0^2)^2 + b^2 \omega^2 / m^2}} \qquad \text{(14–23)}$$

Amplitude

and

$$\phi_0 = \tan^{-1} \frac{\omega_0^2 - \omega^2}{\omega(b/m)}. \qquad \text{(14–24)}$$

Phase angle

Actually, the general solution to Eq. 14–21 is Eq. 14–22 plus another term of the form of Eq. 14–19 for the natural damped motion of the oscillator; this second term approaches zero in time, so in many cases we need to be concerned only with Eq. 14–22.

The amplitude of forced harmonic motion, A_0, depends strongly on the difference between the applied and the natural frequency. A plot of A_0 (Eq. 14–23) as a function of the applied frequency, ω, is shown in Fig. 14–27 (a more detailed version of Fig. 14–24) for three specific values of the damping constant b. Curve A $\left(b = \frac{1}{6} m\omega_0\right)$ represents light damping, curve B $\left(b = \frac{1}{2} m\omega_0\right)$ fairly heavy damping, and curve C $\left(b = \sqrt{2} m\omega_0\right)$ overdamped motion. The amplitude can become large when the driving frequency ω is near the natural frequency, $\omega \approx \omega_0$, as long as the damping is not too large. When the damping is small, the increase in amplitude near $\omega = \omega_0$ is very large and, as we saw, is known as *resonance*. The natural vibrating frequency ω_0 of a system is its *resonant frequency*.[†] If $b = 0$, resonance occurs at $\omega = \omega_0$ and the resonant peak (of A_0) becomes infinite; in such a case, energy is being continuously transferred into the system and none is dissipated. For real systems, b is never precisely zero, and the resonant peak is finite. The peak does not occur precisely at $\omega = \omega_0$ (because of the term $b^2 \omega^2 / m^2$ in the denominator of Eq. 14–23), although it is quite close to ω_0 unless the damping is very large. If the damping is large, there is little or no peak (curve C in Fig. 14–27).

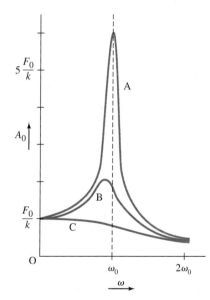

FIGURE 14–27 Amplitude of a forced harmonic oscillator as a function of ω. Curves A, B, and C correspond to light, heavy, and overdamped systems, respectively $(Q = m\omega_0/b = 6, 2, 0.71)$.

[†]Sometimes the resonant frequency is defined as the actual value of ω at which the amplitude has its maximum value, and this depends somewhat on the damping constant. Except for very heavy damping, this value is quite close to ω_0.

The height and narrowness of a resonant peak is often specified by its **quality factor** or **Q value**, defined as

*Quality factor
or Q-value*

$$Q = \frac{m\omega_0}{b}.$$

(14–25)

In Fig. 14–27, curve A has $Q = 6$, curve B has $Q = 2$, and curve C has $Q = 1/\sqrt{2}$. The smaller the damping constant b, the larger the Q value becomes, and the higher the resonance peak. The Q value is also a measure of the width of the peak. To see why, let ω_1 and ω_2 be the frequencies where the square of the amplitude A_0 has half its maximum value (we use the square because the power transferred to the system is proportional A_0^2—see the problems); then $\Delta\omega = \omega_1 - \omega_2$, which is called the *width* of the resonance peak, is related to Q by

$$\frac{\Delta\omega}{\omega_0} = \frac{1}{Q}.$$

(14–26)

(The proof of this relation, which is accurate only for weak damping, is the subject of Problem 67.) The larger the Q value, the narrower will be the resonance peak relative to its height. Thus a large Q value, representing a system of high quality, has a high, narrow resonance peak.

Summary

A vibrating object undergoes **simple harmonic motion** (SHM) if the restoring force is proportional to the displacement,

$$F = -kx.$$

The maximum displacement is called the **amplitude**.

The **period**, T, is the time required for one complete cycle (back and forth), and the **frequency**, f, is the number of cycles per second; they are related by

$$f = \frac{1}{T}.$$

The period of vibration for a mass m on the end of an ideal massless spring is given by

$$T = 2\pi \sqrt{m/k}.$$

SHM is **sinusoidal**, which means that the displacement as a function of time follows a sine or cosine curve. The general solution can be written

$$x = A \cos(\omega t + \phi)$$

where A is the amplitude, ϕ is the **phase angle**, and

$$\omega = 2\pi f = \sqrt{\frac{k}{m}}.$$

The values of A and ϕ depend on the **initial** conditions (x and v at $t = 0$).

During SHM, the total energy $E = \frac{1}{2}mv^2 + \frac{1}{2}kx^2$ is continually changing from potential to kinetic and back again.

A swinging **simple pendulum** of length L approximates SHM if its amplitude is small and friction can be ignored. Its period is then given by (for small amplitudes)

$$T = 2\pi \sqrt{L/g},$$

where g is the acceleration of gravity.

When friction is present (for all real springs and pendulums), the motion is said to be **damped**. The maximum displacement decreases in time, and the mechanical energy is eventually all transformed to thermal energy. If the friction is very large, so no oscillations occur, the system is said to be **overdamped**. If the friction is small enough that oscillations occur, the system is **underdamped**, and the displacement is given by

$$x = Ae^{-\alpha t} \cos \omega' t,$$

where α and ω' are constants. For a **critically damped** system, no oscillations occur and equilibrium is reached in the shortest time.

If an oscillating force is applied to a system capable of vibrating, the amplitude of vibration can be very large if the frequency of the applied force is near the **natural** (or **resonant**) **frequency** of the oscillator; this is called **resonance**.

Questions

1. Give some everyday examples of vibrating objects. Which follow SHM, at least approximately?
2. Contrast the equations for x, v, and a for uniformly accelerated linear motion ($a = $ constant) with those for simple harmonic motion. Discuss their similarities and differences.
3. If a particle undergoes SHM with amplitude A, what is the total distance it travels in one period?
4. Real springs have mass. How will the true period and frequency differ from those given by the equations for a mass oscillating on the end of an idealized massless spring?
5. For a simple harmonic oscillator, when (if ever) are the displacement and velocity vectors in the same direction? When are the displacement and acceleration vectors in the same direction?
6. How could you double the maximum speed of a SHO?
7. A mass m hangs from a spring with stiffness constant k. The spring is cut in half and the same mass hung from it. Will the new arrangement have a higher or a lower stiffness constant than the original spring?
8. Two equal masses are attached to separate identical springs next to one another. One mass is pulled so its spring stretches 20 cm and the other pulled so its spring stretches only 10 cm. The masses are released simultaneously. Which mass reaches the equilibrium point first?
9. A 10-kg fish is attached to the hook of a vertical spring scale, and is then released. Describe the scale reading as a function of time.
10. Is the motion of a piston in an automobile engine simple harmonic? Explain.
11. If a pendulum clock is accurate at sea level, will it gain or lose time when taken to high altitude?
12. A tire swing hangs from a branch nearly to the ground. How could you estimate the height of the branch using only a stopwatch?
13. Does a car bounce on its springs faster when it is empty or when it is fully loaded?
14. What happens to the period of a playground swing if you rise up from sitting to a standing position?
15. Describe the possible motion of a solid object that is suspended so it is free to rotate about its center of gravity. Is it a physical pendulum?
16. A thin uniform rod of mass m is suspended from one end and oscillates with a frequency f. If a small sphere of mass $2m$ is attached to the other end, does the frequency increase or decrease? Explain.
17. Is the acceleration of a simple harmonic oscillator ever zero? If so, when? What about a damped harmonic oscillator?
18. A tuning fork of natural frequency 264 Hz sits on a table at the front of a room. At the back of the room, two tuning forks, one of natural frequency 260 Hz and one of 420 Hz are initially silent, but when the tuning fork at the front of the room is set into vibration, the 260-Hz fork spontaneously begins to vibrate but the 420-Hz fork does not. Explain.
19. Give several everyday examples of resonance.
20. Is a rattle in a car ever a resonance phenomenon? Explain.
21. Over the years, buildings have been able to be built out of lighter and lighter materials. How has this affected the natural vibration frequencies of buildings and the problems of resonance due to passing trucks, airplanes, or by wind and other natural sources of vibration?

Problems

Sections 14–1 and 14–2

1. (I) If a particle undergoes SHM with amplitude 0.15 m, what is the total distance it travels in one period?
2. (I) A fisherman's scale stretches 2.8 cm when a 3.7-kg fish hangs from it. (a) What is the spring constant? (b) What will be the amplitude and frequency of vibration if the fish is pulled down 2.5 cm more and released so that it vibrates up and down?
3. (I) When an 80-kg person climbs into a 1000-kg car, the car's springs compress vertically by 1.40 cm. What will be the frequency of vibration when the car hits a bump? (Ignore damping.)
4. (I) (a) What is the equation describing the motion of a spring that is stretched 8.8 cm from equilibrium and then released, and whose period is 0.75 s? (b) What will be its displacement after 1.8 s?
5. (II) A small fly of mass 0.60 g is caught in a spider's web. The web vibrates predominantly with a frequency of 10 Hz. (a) What is the value of the effective spring constant k for the web? (b) At what frequency would you expect the web to vibrate if an insect of mass 0.40 g were trapped?
6. (II) Determine the phase constant ϕ in Eq. 14–4 if, at $t = 0$, the oscillating mass is at (a) $x = -A$, (b) $x = 0$, (c) $x = A$, (d) $x = \frac{1}{2}A$, (e) $x = -\frac{1}{2}A$, (f) $x = A/\sqrt{2}$.
7. (II) A mass on the end of a spring is stretched a distance x_0 from equilibrium and released. At what distance from equilibrium will it have (a) velocity equal to half its maximum velocity and (b) acceleration equal to half its maximum acceleration?
8. (II) A balsa wood block of mass 50 g floats on a lake, bobbing up and down at a frequency of 2.5 Hz. (a) What is the value of the effective spring constant of the water? (b) A partially filled water bottle of mass 0.25 kg and almost the same size and shape of the balsa block is tossed into the water. At what frequency would you expect the bottle to bob up and down? Assume SHM.
9. (II) At what displacement from equilibrium is the speed of a SHO half the maximum value?

10. (II) A mass m at the end of a spring vibrates with a frequency of 0.88 Hz; when an additional 1.25 kg mass is added to m, the frequency is 0.48 Hz. What is the value of m?

11. (II) A block of mass m is supported by two parallel vertical springs, with spring constant k_1 and k_2 (Fig. 14–28). What will be the frequency of vibration?

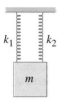

FIGURE 14–28 Problem 11.

12. (II) The graph of displacement versus time for a small mass at the end of spring is shown in Fig. 14–29. At $t = 0$, $x = 0.43$ cm. (a) If $m = 14.3$ g, find the spring constant, k. (b) Write the equation for displacement x as a function of time.

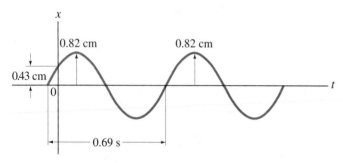

FIGURE 14–29 Problem 12.

13. (II) The position of a SHO as a function of time is given by $x = 3.8 \cos (7\pi t/4 + \pi/6)$ where t is in seconds and x in meters. Find (a) the period and frequency, (b) the position and velocity at $t = 0$, and (c) the velocity and acceleration at $t = 2.0$ s.

14. (II) A tuning fork vibrates at a frequency of 264 Hz and the tip of each prong moves 1.5 mm to either side of center. Calculate (a) the maximum speed and (b) the maximum acceleration of the tip of a prong.

15. (II) A spring vibrates with a frequency of 3.0 Hz when a weight of 0.50 kg is hung from it. What will its frequency be if only 0.35 kg hangs from it?

16. (II) (a) Show that

$$x = a \sin \omega t + b \cos \omega t$$

is a general solution of Eq. 14–3, and (b) determine the constants a and b in terms of the A and ϕ of Eq. 14–4.

17. (II) If a mass m hangs from a vertical spring, as shown in Fig. 14–3, show that $F = -kx$ holds for stretching or compression of the spring, where x is the displacement from the (vertical position) equilibrium point.

18. (II) A mass of 1.62 kg stretches a vertical spring 0.315 m. If the spring is stretched an additional 0.130 m and released, how long does it take to reach the (new) equilibrium position again?

19. (II) A spring of force constant 345 N/m vibrates with an amplitude of 22.0 cm when 0.250 kg hangs from it. (a) What is the equation describing this motion as a function of time? Assume that the mass passes downwards through the equilibrium point at $t = 0$. (b) At what times will the spring have its maximum and minimum extensions? Take y positive upwards.

20. (II) A 450-g object oscillates from a vertically hanging light spring once every 0.55 s. (a) Write down the equation giving its position y (+ upward) as a function of time t, assuming it started by being compressed 10 cm from the equilibrium position (where $y = 0$), and released. (b) How long will it take to get to the equilibrium position for the first time? (c) What will be its maximum speed? (d) What will be its maximum acceleration, and where will it first be attained?

21. (II) A uniform meter stick of mass M is pivoted on a hinge at one end and held horizontal by a spring with spring constant k attached at the other end (Fig. 14–30). If the stick oscillates up and down slightly, what is its frequency? [Hint: Write a torque equation about the hinge.]

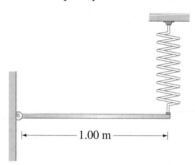

FIGURE 14–30
Problem 21.

22. (III) A mass m is at rest on the end of a spring of spring constant k. At $t = 0$ it is given an impulse J by a hammer. Write the formula for the subsequent motion in terms of $m, k, J,$ and t.

23. (III) A mass m is connected to two springs, with spring constants k_1 and k_2, in two different ways as shown in Fig. 14–31a and b. Show that the period for the configuration shown in part (a) is given by

$$T = 2\pi \sqrt{m\left(\frac{1}{k_1} + \frac{1}{k_2}\right)}$$

and for that in part (b) is given by

$$T = 2\pi \sqrt{\frac{m}{k_1 + k_2}}.$$

Ignore friction.

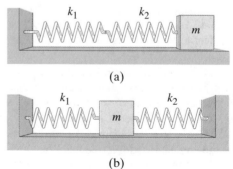

(a)

(b)

FIGURE 14–31
Problem 23.

24. (III) Two equal masses, m_1 and m_2, are connected by three identical springs of spring constant k as shown in Fig. 14–32. (*a*) Apply $\Sigma F = ma$ to each mass and obtain two differential equations for the displacements x_1 and x_2. (*b*) Determine the possible frequencies of vibration by assuming a solution of the form $x_1 = A_1 \cos \omega t$, $x_2 = A_2 \cos \omega t$.

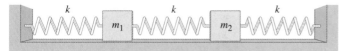

FIGURE 14–32 Problem 24.

25. (III) A spring with spring constant 250 N/m vibrates with an amplitude of 12.0 cm when 0.380 kg hangs from it. (*a*) What is the equation describing this motion as a function of time? Assume the mass passes through the equilibrium point, toward positive x (upward), at $t = 0.110$ s. (*b*) At what times will the spring have its maximum and minimum lengths? (*c*) What is the displacement at $t = 0$? (*d*) What is the force exerted by the spring at $t = 0$? (*e*) What is the maximum speed and when is it first reached after $t = 0$?

Section 14–3

26. (I) (*a*) At what displacement of a SHO is the energy half kinetic and half potential? (*b*) What fraction of the total energy of a SHO is kinetic and what fraction potential when the displacement is half the amplitude?

27. (I) A 2.00-kg mass vibrates according to the equation $x = 0.650 \cos 8.40t$ where x is in meters and t in seconds. Determine (*a*) the amplitude, (*b*) the frequency, (*c*) the total energy, and (*d*) the kinetic energy and potential energy when $x = 0.260$ m.

28. (II) A 0.35-kg mass at the end of a spring vibrates 3.0 times per second with an amplitude of 0.15 m. Determine (*a*) the velocity when it passes the equilibrium point, (*b*) the velocity when it is 0.10 m from equilibrium, (*c*) the total energy of the system, and (*d*) the equation describing the motion of the mass, assuming that at $t = 0$, x was a maximum.

29. (II) It takes a force of 95.0 N to compress the spring of a popgun 0.185 m to "load" a 0.200-kg ball. With what speed will the ball leave the gun?

30. (II) A 0.0125 kg bullet strikes a 0.300-kg block attached to a fixed horizontal spring whose spring constant is 2.25×10^3 N/m and sets it into vibration with an amplitude of 12.4 cm. What was the speed of the bullet if the two objects move together after impact?

31. (II) If one vibration has 10 times the energy of a second one of equal frequency, but the first's spring constant k is twice as large as the second's, how do their amplitudes compare?

32. (II) A mass of 240 g oscillates on a horizontal frictionless surface at a frequency of 3.5 Hz and with amplitude of 4.5 cm. (*a*) What is the effective spring constant for this motion? (*b*) How much energy is involved in this motion?

33. (II) A mass sitting on a horizontal, frictionless surface is attached to one end of a spring; the other end is fixed to a wall. 3.0 J of work is required to compress the spring by 0.12 m. If the mass is released from rest with the spring compressed, it experiences a maximum acceleration of 15 m/s^2. Find the value of (*a*) the spring constant and (*b*) the mass.

34. (II) An object with mass 2.1 kg is executing simple harmonic motion, attached to a spring with spring constant $k = 280$ N/m. When the object is 0.020 m from its equilibrium position, it is moving with a speed of 0.55 m/s. (*a*) Calculate the amplitude of the motion. (*b*) Calculate the maximum velocity attained by the object.

35. (II) Nikita devised the following method of measuring the muzzle velocity of a rifle (Fig. 14–33). She fires a bullet into a 6.023-kg wooden block resting on a smooth surface, and attached to a spring of spring constant $k = 142.7$ N/m. The bullet, whose mass is 7.870 g, remains embedded in the wooden block. She measures the distance that the block recoils and compresses the spring to be 9.460 cm. What is the speed v of the bullet?

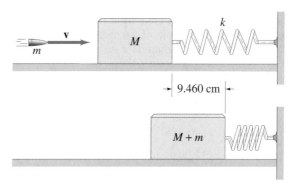

FIGURE 14–33 Problem 35.

36. (II) At $t = 0$, a 650-g mass at rest on the end of a horizontal spring ($k = 184$ N/m) is struck by a hammer which gives it an initial speed of 2.26 m/s. Determine (*a*) the period and frequency of the motion, (*b*) the amplitude, (*c*) the maximum acceleration, (*d*) the position as a function of time, (*e*) the total energy, and (*f*) the kinetic energy when $x = 0.40A$ where A is the amplitude.

37. (II) Obtain the displacement x as a function of time for the simple harmonic oscillator using the conservation of energy, Eq. 14–10. [*Hint*: Integrate Eq. 14–11a with $v = dx/dt$.]

Section 14–5

38. (I) A pendulum makes 42 vibrations in 50 s. What is its (*a*) period, and (*b*) frequency?

39. (I) How long must a simple pendulum be if it is to make exactly one swing per second? That is, one complete vibration takes exactly two seconds.

40. (I) What is the period of a simple pendulum on Mars, where the acceleration of gravity is about 0.37 that on Earth, if the pendulum has a period of 0.80 s on Earth?

41. (II) (*a*) Determine the length of a simple pendulum whose period is 1.00 s. (*b*) What would be the period of a 1.00-m-long simple pendulum?

42. (II) What is the period of a simple pendulum 73 cm long (*a*) on the Earth, and (*b*) when it is in a freely falling elevator?

43. (II) A simple pendulum is 0.30 m long. At $t = 0$ it is released starting at an angle of 14°. Ignoring friction, what will be the angular position of the pendulum at (*a*) $t = 0.65$ s, (*b*) $t = 1.95$ s, and (*c*) $t = 5.00$ s?

44. (II) Derive a formula for the maximum speed v_0 of a simple pendulum bob in terms of g, the length L, and the maximum angle of swing θ.

45. (II) A simple pendulum vibrates with an amplitude of 10.0°. What fraction of the time does it spend between +5.0 and −5.0°? Assume SHM.

46. (II) The length of a simple pendulum is 0.68 m and it is released at an angle of 12° to the vertical. (*a*) With what frequency does it vibrate? (*b*) What is the pendulum bob's speed when it passes through the lowest point of the swing?

* Section 14–6

* 47. (II) A plywood disk of radius 20.0 cm and mass 3.00 kg has a small hole drilled through it, 2.00 cm from its edge (Fig. 14–34). The disk is hung from the wall by means of a metal pin through the hole, and is used as a pendulum. What is the period of this pendulum for small oscillations?

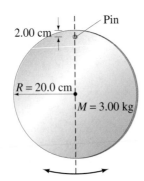

FIGURE 14–34   Problem 47.

* 48. (II) A pendulum consists of a tiny bob of mass M and a uniform cord of mass m and length L. (*a*) Determine a formula for the period. (*b*) What would be the fractional error if one used the formula for a simple pendulum, Eq. 14–12a?

* 49. (II) The balance wheel of a watch is a thin ring of radius 0.95 cm and oscillates with a frequency of 3.10 Hz. If a torque of $1.1 \times 10^{-5}\ \mathrm{m \cdot N}$ causes the wheel to rotate 60°, calculate the mass of the balance wheel.

* 50. (II) (*a*) Determine the equation of motion (for θ as a function of time) for a torsion pendulum, Fig. 14–18, and show that the motion is simple harmonic. (*b*) Show that the period T is $T = 2\pi \sqrt{I/K}$. [The balance wheel of a mechanical watch is an example of a torsion pendulum in which the restoring torque is applied by a coil spring.]

* 51. (II) The human leg can be compared to a physical pendulum, with a "natural" swinging period at which walking is easiest. Consider the leg as two rods joined rigidly together at the knee; the axis for the leg is the hip joint. The length of each rod is about the same, 50 cm. The upper rod has a mass of 7.0 kg and the lower rod has a mass of 4.0 kg. (*a*) Calculate the natural swinging period of the system. (*b*) Check your answer by standing on a chair and measuring the time for one or more complete back-and-forth swings. The effect of a shorter leg is, of course, a shorter swinging period, enabling a faster, although a shorter, "natural" stride.

* 52. (II) A student wants to use a meter stick as a pendulum. She plans to drill a small hole through the meter stick and suspend it from a smooth pin attached to the wall (Fig. 14–35). Where in the meter stick should she drill the hole to obtain the shortest possible period? How short an oscillation period can she obtain with a meter stick in this way?

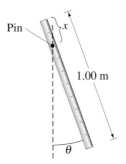

FIGURE 14–35   Problem 52.

* 53. (II) A meter stick is hung at its center from a thin wire (Fig. 14–36a). It is twisted and oscillates with a period of 6.0 s. The meter stick is sawed off to a length of 70.0 cm. This piece is again balanced at its center and set in oscillation (Fig. 14–36b). With what period does it oscillate?

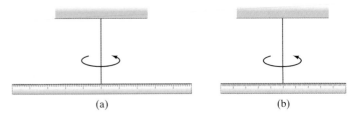

(a)   (b)

FIGURE 14–36   Problem 53.

* 54. (II) An aluminum disk, 12.5 cm in diameter and 500 g in mass, is mounted on a vertical shaft with an air bearing (Fig. 14–37). The disk also floats on an air bearing. One end of a flat coil spring is attached to the disk, the other end to the base of the apparatus. The disk is set into rotational oscillation and the frequency is 0.331 Hz. What is the torsional spring constant K ($\tau = -K\theta$)?

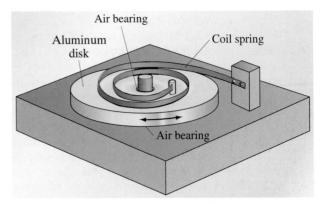

FIGURE 14–37   Problem 54.

Section 14–7

55. (I) A 750-g block oscillates on the end of a spring whose force constant is $k = 56.0 \text{ N/m}$. The mass moves in a fluid which offers a resistive force $F = -bv$, where $b = 0.162 \text{ N·s/m}$. (a) What is the period of the motion? (b) What is the fractional decrease in amplitude per cycle? (c) Write the displacement as a function of time if at $t = 0$, $x = 0$, and at $t = 1.00 \text{ s}$, $x = 0.120 \text{ m}$.

56. (II) A physical pendulum consists of an 80-cm-long, 300-g-mass, uniform wooden rod hung from a nail near one end (Fig. 14–38). The motion is damped because of friction in the pivot; the damping force is approximately proportional to $d\theta/dt$. The rod is set in oscillation by displacing it 15° from its equilibrium position and releasing it. After 8.0 seconds, the amplitude of the oscillation has been reduced to 5.5°. If the angular displacement can be written as $\theta = Ae^{-\alpha t} \cos \omega' t$, find (a) α, (b) the approximate period of the motion, and (c) how long it takes for the amplitude to be reduced to $\frac{1}{2}$ of its original value.

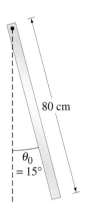

FIGURE 14–38 Problem 56.

57. (II) A damped harmonic oscillator loses 5.0 percent of its mechanical energy per cycle. (a) By what percentage does its frequency differ from the natural frequency $\omega_0 = \sqrt{k/m}$? (b) After how many periods will the amplitude have decreased to $1/e$ of its original value?

58. (III) (a) Show that the total mechanical energy, $E = \frac{1}{2}mv^2 + \frac{1}{2}kx^2$, as a function of time for a lightly damped harmonic oscillator is

$$E = \tfrac{1}{2}kA^2 e^{-(b/m)t} = E_0 e^{-(b/m)t},$$

where E_0 is the total mechanical energy at $t = 0$. (Assume $\omega' \gg b/2m$.) (b) Show that the fractional energy lost per period is

$$\frac{\Delta E}{E} = \frac{2\pi b}{m\omega_0} = \frac{2\pi}{Q},$$

where $\omega_0 = \sqrt{k/m}$ and $Q = m\omega_0/b$ is called the *quality factor* or *Q value* of the system. A larger Q value means the system can undergo oscillations for a longer time.

59. (III) A glider on an air track is connected by springs to either end of the track (Fig. 14–39). Both springs have the same spring constant, k, and the glider has mass M. (a) Determine the frequency of the oscillation, assuming no damping, if $k = 100 \text{ N/m}$ and $M = 200$ grams. (b) It is observed that after 55 oscillations, the amplitude of the oscillation has dropped to one-half of its initial value. Estimate the value of α, using Eq. 14–16. (c) How long does it take the amplitude to decrease to one-quarter of its initial value?

FIGURE 14–39 Problem 59.

Section 14–8

*60. (II) (a) At resonance $(\omega = \omega_0)$, what is the value of the phase angle ϕ_0? (b) What, then, is the displacement at a time when the driving force F_{ext} is a maximum, and at a time when $F_{\text{ext}} = 0$? (c) What is the phase difference (in degrees) between the driving force and the displacement in this case?

*61. (II) Construct an accurate resonance curve, from $\omega = 0$ to $\omega = 2\omega_0$, for $Q = 4.0$.

*62. (II) The amplitude of a driven harmonic oscillator reaches a value of $28.6\ F_0/m$ at a resonant frequency of 382 Hz. What is the Q value of this system?

*63. (II) By direct substitution, show that Eq. 14–22, with Eqs. 14–23 and 14–24, is a solution of the equation of motion (Eq. 14–21) for the forced oscillator.

*64. (II) Differentiate Eq. 14–23 to show that the resonant amplitude peaks at

$$\omega = \sqrt{\omega_0^2 - b^2/2m^2}.$$

*65. (III) Consider a simple pendulum (point mass bob) 0.50 m long with a Q of 400. (a) How long does it take for the amplitude (assumed small) to decrease by two-thirds? (b) If the amplitude is 2.0 cm and the bob has mass 0.20 kg, what is the initial energy loss rate of the pendulum in watts? (c) If we are to stimulate resonance with a sinusoidal driving force, how close must the driving frequency be to the natural frequency of the pendulum?

*66. (III) *Power transferred to driven oscillator.* (a) Show that the power input to a forced oscillator due to the external force F_{ext} is

$$P = F_{\text{ext}} v = \frac{F_0^2 \omega \cos\phi_0 \cos^2 \omega t - \frac{1}{2}F_0^2 \omega \sin\phi_0 \sin 2\omega t}{m\sqrt{(\omega^2 - \omega_0^2)^2 + \omega^2 b^2/m^2}}.$$

(b) Show that the average power input, averaged over one (or many) full cycles, is

$$\overline{P} = \frac{\omega F_0^2 \cos\phi_0}{2m\sqrt{(\omega^2 - \omega_0^2)^2 + \omega^2 b^2/m^2}} = \tfrac{1}{2}\omega A_0 F_0 \cos\phi_0$$

or

$$\overline{P} = \tfrac{1}{2}F_0 v_{\text{max}} \cos\phi_0,$$

where v_{max} is the maximum value of dx/dt. (c) Plot $\overline{P}$ versus ω from $\omega = 0$ to $\omega = 2\omega_0$ for $Q = 6.0$. Note that although the amplitude is not zero at $\omega = 0$, $\overline{P}$ is zero at $\omega = 0$.

*67. (III) Derive Eq. 14–26.

68. A 52-kg person jumps from a window to a fire net 20.0 m below, which stretches the net 1.1 m. Assume that the net behaves like a simple spring, and calculate how much it would stretch if the same person were lying in it. How much would it stretch if the person jumped from 35 m?

69. The length of a simple pendulum is 0.63 m, the pendulum bob has a mass of 365 grams, and it is released at an angle of 15° to the vertical. (*a*) With what frequency does it vibrate? (*b*) What is the pendulum bob's speed when it passes through the lowest point of the swing? Assume SHM. (*c*) What is the total energy stored in this oscillation assuming no losses?

70. A 0.650-kg mass vibrates according to the equation $x = 0.25 \sin (5.50t)$ where x is in meters and t is in seconds. Determine (*a*) the amplitude, (*b*) the frequency, (*c*) the period, (*d*) the total energy, and (*e*) the kinetic energy and potential energy when x is 10 cm.

71. A bungee jumper with mass 72.0 kg jumps from a high bridge. After reaching his lowest point, he oscillates up and down, hitting a low point eight more times in 34.7 s. He finally comes to rest 25.0 m below the level of the bridge. Calculate the spring constant and the unstretched length of the bungee cord.

72. (*a*) A crane has hoisted a 1200-kg car at the junkyard. The steel crane cable is 20.0 m long and has a diameter of 6.4 mm. A breeze starts the car bouncing at the end of the cable. What is the period of the bouncing? [*Hint*: Refer to Table 12–1]. (*b*) What amplitude of bouncing will likely cause the cable to snap? (See Table 12–2, and assume Hooke's law holds all the way up to the breaking point.)

73. An energy-absorbing car bumper has a spring constant of 500 kN/m. Find the maximum compression of the bumper if the car, with mass 1500 kg, collides with a wall at a speed of 2.0 m/s (approximately 5 mi/h).

74. A "seconds" pendulum has a period of exactly 2.0 seconds—each one-way swing takes 1.0 s. What is the length of a seconds pendulum in Austin, Texas, where $g = 9.793 \text{ m/s}^2$? If the pendulum is moved to Paris, where $g = 9.809 \text{ m/s}^2$, by how many millimeters must we lengthen the pendulum? What is the length of a "seconds" pendulum on the Moon, where $g = 1.62 \text{ m/s}^2$?

75. A block of jello rests on a cafeteria plate as shown in Fig. 14–40 (which also gives the dimensions of the block). You push it sideways as shown, and then you let go. The jello springs back and begins to vibrate. In analogy to a mass vibrating on a spring, estimate the frequency of this vibration, given that the shear modulus of jello is 520 N/m² and its density is 1300 kg/m³.

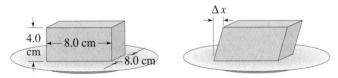

FIGURE 14–40   Problem 75.

76. A simple pendulum oscillates with frequency f. What is its frequency if it accelerates at $\frac{1}{2}g$ (*a*) upward, and (*b*) downward?

77. A 420-kg wooden raft floats on a lake. When a 75-kg man stands on the raft, it sinks 3.5 cm deeper into the water. When he steps off, the raft vibrates for a while. (*a*) What is the frequency of vibration? (*b*) What is the total energy of vibration (ignoring damping)?

78. A 5.0-kg box slides into a spring of spring constant 310 N/m (Fig. 14–41), compressing it 24 cm. (*a*) What is the incoming speed of the block? (*b*) How long is the box in contact with the spring before it bounces off in the opposite direction?

FIGURE 14–41   Problem 78.

79. A 1.60-kg table is supported on four springs. A 0.55-kg chunk of modeling clay is held above the table and dropped so that it hits the table with a speed of 1.65 m/s (Fig. 14–42). The clay makes an inelastic collision with the table and the table and clay oscillate up and down. After a long time the table comes to rest 6.0 cm below its original position. (*a*) What is the effective spring constant of all four springs taken together? (*b*) With what maximum amplitude does the platform oscillate?

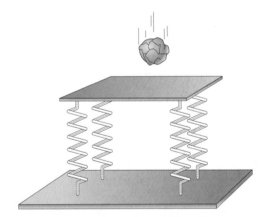

FIGURE 14–42   Problem 79.

80. A diving board oscillates with simple harmonic motion of frequency 5.0 cycles per second. What is the maximum amplitude with which the end of the board can vibrate in order that a pebble placed there (Fig. 14–43) does not lose contact with the board during the oscillation?

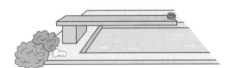

FIGURE 14–43   Problem 80.

81. A rectangular block of wood floats in a calm lake. Show that, if friction is ignored, when the block is pushed gently down into the water, it will then vibrate with SHM. Also, determine an equation for the force constant.

82. A mass m is gently placed on the end of a freely hanging spring. The mass then falls 22.0 cm before it stops and begins to rise. What is the frequency of the oscillation?

83. The water in a U-shaped tube is displaced an amount Δx from equilibrium. (The level in one side is $2\Delta x$ above the level in the other side.) If friction is neglected, will the water oscillate harmonically? Determine a formula for the equivalent spring constant k. Does k depend on the density of the liquid, the cross-section of the tube, or the length of the water column?

84. In some diatomic molecules, the force each atom exerts on the other can be approximated by $F = -C/r^2 + D/r^3$, where C and D are positive constants. (*a*) Graph F versus r from $r = 0$ to $r = 2D/C$. (*b*) Show that equilibrium occurs at $r = r_0 = D/C$. (*c*) Let $\Delta r = r - r_0$ be a small displacement from equilibrium, where $\Delta r \ll r_0$. Show that for such small displacements, the motion is approximately simple harmonic, and (*d*) determine the force constant. (*e*) What is the period of such motion? [*Hint*: Assume one atom is kept at rest.]

85. A mass m is connected to two springs with equal spring constants k (Fig. 14–44). In the horizontal position shown, each spring is stretched by an amount Δa. The mass is raised vertically and begins to oscillate up and down. Assuming that the displacement is small, and ignoring gravity, show that the motion is simple harmonic and find the period. [This kind of motion is called transverse oscillation. If we hooked many masses together like this, their transverse oscillations would be a model for a vibrating string in a musical instrument.]

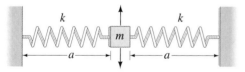

FIGURE 14–44   Problem 85.

86. Carbon dioxide is a linear molecule. The carbon–oxygen bonds in this molecule act very much like springs. Figure 14–45 shows one possible way the oxygen atoms in this molecule can vibrate: the oxygen atoms vibrate symmetrically in and out, while the central carbon atom remains at rest. Hence each oxygen atom acts like a simple harmonic oscillator with a mass equal to the mass of an oxygen atom. It is observed that this oscillation occurs with a frequency of $f = 2.83 \times 10^{13}$ Hz. What is the spring constant of the C–O bond?

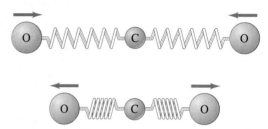

FIGURE 14–45   Problem 86, the CO_2 molecule.

87. A thin, straight, uniform rod of length $l = 1.00$ m and mass $m = 160$ g hangs from a pivot at one end. (*a*) What is its period for small-amplitude oscillations? (*b*) What is the length of a simple pendulum that will have the same period?

88. Imagine that a 10-cm-diameter circular hole were drilled all the way through the center of the Earth (Fig. 14–46). At one end of the hole, you drop an apple into the hole. Show that, if you assume that the Earth has a constant density, the subsequent motion of the apple is simple harmonic. How long will the apple take to return? Assume that we can ignore all frictional effects.

FIGURE 14–46   Problem 88.

Waves—such as water waves or waves traveling along a cord or slinky—travel away from their source. The waves on a stretched cord shown in these four photographs, however, are "standing waves." They seem not to be traveling; but each can be thought of as the sum of a wave traveling to the right interfering with its reflection traveling back toward the left. We can also think of these standing waves as oscillations of the stretched cord at resonance. Each of these standing waves occurs at a particular frequency. Can you guess how the frequency of each is related to the others?

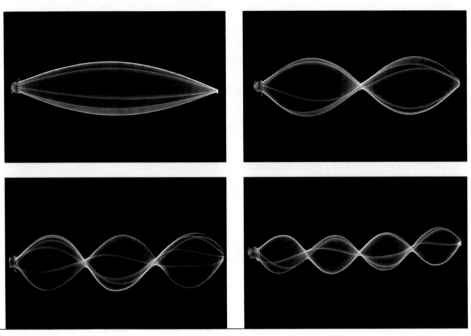

Wave Motion

When you throw a stone into a lake or pool of water, circular waves form and move outward, Fig. 15–1. Waves will also travel along a cord (or a "slinky") that is stretched out straight on a table if you vibrate one end back and forth as shown in Fig. 15–2. Water waves and waves on a cord are two common examples of wave motion.

Vibrations and wave motion are intimately related subjects. Waves—whether ocean waves, waves on a string, earthquake waves, or sound waves in air—have as their source a vibration. In the case of sound, not only is the source a vibrating object, but so is the detector—the eardrum or the membrane of a microphone.

FIGURE 15–1 A stone thrown into a lake: water waves spread outward from the source.

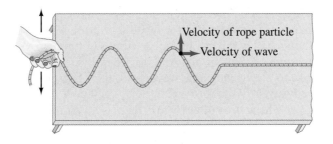

Velocity of rope particle

Velocity of wave

FIGURE 15-2 Wave traveling on a rope or cord. The wave travels to the right along the cord. Particles of the cord oscillate back and forth on the tabletop.

Indeed, the medium through which such "mechanical waves" travel itself vibrates (such as air for sound waves). In this chapter we will mainly be concerned with **mechanical waves** such as waves on water or on a cord. In Chapter 16 we will study sound waves, and in later chapters we will encounter other forms of wave motion, including electromagnetic waves and light.

If you have ever watched ocean waves moving toward shore (before they break), you may have wondered if the waves were carrying water from out at sea into the beach. This is, in fact, not the case.[†] Water waves move with a recognizable velocity. But each particle (or molecule) of the water itself merely oscillates about an equilibrium point. This is clearly demonstrated by observing leaves on a pond as waves move by. The leaves (or a cork) are not carried forward by the waves, but simply oscillate about an equilibrium point because this is the motion of the water itself. Similarly, the wave on the rope or cord of Fig. 15-2 moves to the right, but each piece of the rope only vibrates to and fro. These are general features of mechanical waves: (1) a wave can move over large distances with a particular speed; and (2) each particle of the medium in which the wave travels (the water or the rope) oscillates about an equilibrium point which is simple harmonic if the wave is sinusoidal. Thus, although a wave is not matter, the wave pattern can travel in matter.

Waves carry energy from one place to another. Energy is given to a water wave, for example, by a rock thrown into the water, or by wind far out at sea. The energy is transported by waves to the shore. The oscillating hand in Fig. 15-2 transfers energy to the rope, which is then transported down the rope and can be transferred to an object at the other end. All forms of traveling waves transport energy.

FIGURE 15-3 Motion of a wave pulse. Arrows indicate velocity of rope particles. Note that the wave velocity is to the right.

15-1 | Characteristics of Wave Motion

Let us look at how a wave is formed and how it comes to "travel." We first look at a single wave bump or **pulse**. A single pulse can be formed on a rope by a quick up and down motion of the hand, Fig. 15-3. The hand pulls up on one end of the rope, and because the end piece is attached to adjacent pieces, these also feel an upward force and they too begin to move upward. As each succeeding piece of rope moves upward, the wave crest moves outward along the rope. Meanwhile, the end piece of rope has been returned to its original position by the hand, and as each succeeding piece of rope reaches its peak position, it too is pulled back down again by the adjacent section of rope. Thus, the source of a traveling wave pulse is a disturbance, and cohesive forces between adjacent pieces of rope cause the pulse to travel outward. Waves in other media are created and propagate outward in a similar fashion.

A **continuous** or **periodic wave**, such as that shown starting in Fig. 15-2, has as its source a disturbance that is continuous and oscillating. That is, the source is a *vibration* or *oscillation*. In Fig. 15-2, a hand oscillates one end of the rope. Water

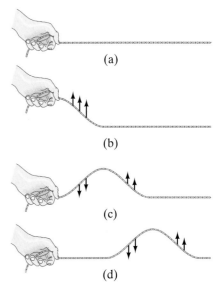

(a)

(b)

(c)

(d)

[†] Do not be confused by the "breaking" of ocean waves, which occurs when a wave interacts with the ground in shallow water and hence is no longer a simple wave.

waves may be produced by any vibrating object placed at the surface such as your hand; or the water itself is made to vibrate when wind blows across it or a rock or tennis ball is thrown into it. A vibrating tuning fork or drum membrane gives rise to sound waves in air. And, as we will see later, oscillating electric charges give rise to electromagnetic waves and light. Indeed, almost any vibrating object sends out waves.

The source of any wave, then, is a vibration. And it is the *vibration* that propagates outward and thus constitutes the wave. If the source vibrates sinusoidally in simple harmonic motion (SHM), then the wave itself—if the medium is perfectly elastic—will have a sinusoidal shape both in space and in time. If you were to take a picture of the wave spread throughout space at a given instant of time, the wave would have the shape of a sine or cosine function. On the other hand, if you look at the motion of the medium at one place over a long period of time—for example, if you look between two closely spaced posts of a pier or out of a ship's porthole as water waves pass by—the up and down motion of that small segment of water will be simple harmonic motion—the water moves up and down sinusoidally in time.

Sinusoidal wave:

(a) in space

(b) in time,

Some of the important quantities used to describe a periodic sinusoidal wave are shown in Fig. 15–4. The high points on a wave are called crests, the low points troughs. The **amplitude** is the maximum height of a crest, or depth of a trough, relative to the normal (or equilibrium) level. The total swing from a crest to a trough is twice the amplitude. The distance between two successive crests is called the **wavelength**, λ (the Greek letter lambda). The wavelength is also equal to the distance between *any* two successive identical points on the wave. The **frequency**, f (sometimes called ν, the Greek letter nu) is the number of crests—or complete cycles—that pass a given point per unit time. The **period**, T, is the time required for one complete oscillation or one complete cycle of the wave to pass a given point along the line of travel. As before (Eq. 14–2), $T = 1/f$.

Amplitude

Wavelength, λ

Frequency, f

FIGURE 15–4 Characteristics of a continuous wave.

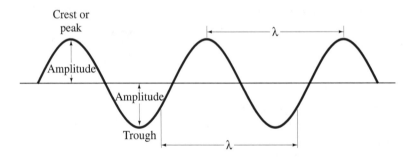

The **wave velocity**, v, is the velocity at which wave crests (or any other part of the waveform) move. The wave velocity is often referred to as the **phase** velocity, and it must be distinguished from the velocity of a particle of the medium itself. For example, for a wave traveling along a cord in Fig. 15–2 the wave velocity is to the right, along the cord, whereas the velocity of a particle of the cord is up or down.

Wave velocity (don't confuse with the velocity of a particle)

A wave crest travels a distance of one wavelength, λ, in one period, T. Thus the wave velocity v is equal to λ/T, or (since $1/T = f$):

$v = \lambda f$ (sinusoidal waves)

$$v = \lambda f. \tag{15–1}$$

For example, suppose a wave has a wavelength of 5 m and a frequency of 3 Hz. Since three crests pass a given point per second, and the crests are 5 m apart, the first crest (or any other part of the wave) must travel a distance of 15 m during the 1 s. So its speed is 15 m/s.

Transverse and Longitudinal Waves

For a wave that travels down a cord, say from the left to right as in Fig. 15–2, the particles of the cord vibrate up and down in a direction transverse (or perpendicular) to the motion of the wave itself. Such a wave is called a **transverse wave**. A second type of wave is a **longitudinal wave**, in which the vibration of the particles of the medium is along the *same* direction as the motion of the wave. Longitudinal waves are readily formed on a stretched spring or "slinky" by alternately compressing and expanding one end, as shown in Fig. 15–5b (a transverse wave is shown in Fig. 15–5a for comparison). A series of compressions and expansions propagate along the spring. The *compressions* are those areas where the coils are momentarily close together. *Expansions* (sometimes called *rarefactions*) are regions where the coils are momentarily far apart. Compressions and expansions correspond to the crests and troughs of a transverse wave.

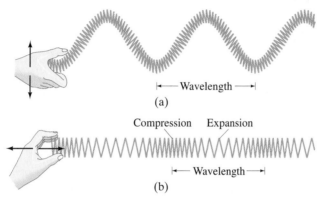

(a)

(b)

FIGURE 15–5 (a) Transverse wave; (b) longitudinal wave.

An important example of a longitudinal wave is a sound wave in air. A vibrating drum head, for example, alternately compresses and rarefies the air just in front of it, producing a longitudinal wave that travels outward in the air, as shown in Fig. 15–6.

As in the case of transverse waves, each particle of the medium in which a longitudinal wave passes oscillates over a very small distance, whereas the wave itself can travel large distances. What are the wavelength, frequency, and wave velocity for a longitudinal wave? The wavelength is the distance between successive compressions (or between successive expansions), and frequency is the number of compressions that pass a given point per second. The wave velocity is the velocity with which each compression appears to move; and it is equal to the product of wavelength and frequency, $v = \lambda f$ (Eq. 15–1).

A longitudinal wave can be represented graphically by plotting the density of air molecules (or coils of a slinky) versus position, as shown in Fig. 15–7b. We will often use such a graphical representation because it is much easier to illustrate what is happening. Note that the graph looks much like a transverse wave.

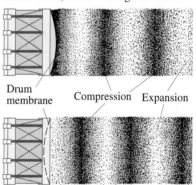

FIGURE 15–6 Production of a sound wave, which is longitudinal.

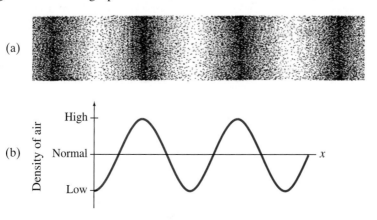

(a)

(b)

FIGURE 15–7 (a) A longitudinal wave with (b) its graphical representation.

Velocity of Transverse Waves

The velocity of a wave depends on the properties of the medium in which it travels. The velocity of a transverse wave on a stretched string or cord, for example, depends on the tension in the cord, F_T, and on the mass per unit length of the cord, μ (the Greek letter mu). For waves of small amplitude, the relationship is

Velocity of wave on a cord

$$v = \sqrt{\frac{F_T}{\mu}}. \qquad \left[\begin{array}{c}\text{transverse} \\ \text{wave on a cord}\end{array}\right] \quad (15\text{--}2)$$

Before giving a derivation of this formula, it is worth noting that at least qualitatively it makes sense on the basis of Newtonian mechanics. That is, we do expect the tension to be in the numerator and the mass per unit length in the denominator. Why? Because when the tension is greater, we expect the velocity to be greater since each segment of cord is in tighter contact with its neighbor. And, the greater the mass per unit length, the more inertia the cord has and the more slowly the wave would be expected to propagate.

We can make a simple derivation of Eq. 15–2 using a simple model of a cord under a tension F_T as shown in Fig. 15–8a. The cord is pulled upward at a speed v' by the force F_y. As shown in Fig. 15–8b all points of the cord to the left of point A move upward at the speed v', and those to the right are still at rest. The speed of propagation, v, of this wave pulse is the speed of point A, the leading edge of the pulse. Point A moves to the right a distance vt in a time t, whereas the end of the cord moves upward a distance $v't$. By similar triangles we have the approximate relation

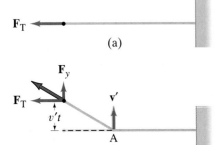

FIGURE 15–8 Diagram of simple wave pulse on a cord for derivation of Eq. 15–2. The vector shown in (b) as the resultant of $\mathbf{F}_T + \mathbf{F}_y$ has to be directed along the rope because the rope is flexible.

$$\frac{F_T}{F_y} = \frac{vt}{v't} = \frac{v}{v'},$$

which is accurate for small displacements ($v't \ll vt$) so that F_T does not change appreciably. As we saw in Chapter 9, the impulse given to an object is equal to its change in momentum. During the time t the total upward impulse is $F_y t = (v'/v)F_T t$. The change in momentum of the cord, Δp, is the mass of cord moving upward times its velocity. Since the upward moving segment of cord has mass equal to the mass per unit length μ times its length vt we have

$$F_y t = \Delta p$$

$$\frac{v'}{v} F_T t = (\mu vt)v'.$$

Solving for v we find $v = \sqrt{F_T/\mu}$ which is Eq. 15–2. Although it was derived for a special case, it is valid for any wave shape since other shapes can be considered to be made up of many tiny such lengths. But it is valid only for small displacements (as was our derivation). Experiment is in accord with this result derived from Newtonian mechanics.

EXAMPLE 15–1 **Wave on a wire.** A wave whose wavelength is 0.30 m is traveling down a 300-m-long wire whose total mass is 15 kg. If the wire is under a tension of 1000 N, what are the velocity and frequency of this wave?

SOLUTION From Eq. 15–2, the velocity is

$$v = \sqrt{\frac{1000\,\text{N}}{(15\,\text{kg})/(300\,\text{m})}} = 140\,\text{m/s}.$$

The frequency then is (Eq. 15–1)

$$f = \frac{v}{\lambda} = \frac{140\,\text{m/s}}{0.30\,\text{m}} = 470\,\text{Hz}.$$

Note that a higher tension would increase both v and f, whereas a thicker, denser wire would reduce v and f.

Velocity of Longitudinal Waves

The velocity of a longitudinal wave has a form similar to that for a transverse wave on a cord (Eq. 15–2); that is,

$$v = \sqrt{\frac{\text{elastic force factor}}{\text{inertia factor}}}.$$

In particular, for a longitudinal wave traveling down a long solid rod,

$$v = \sqrt{\frac{E}{\rho}}, \qquad \begin{bmatrix} \text{longitudinal} \\ \text{wave in a long rod} \end{bmatrix} \quad \textbf{(15–3)}$$

where E is the elastic modulus (Section 12–5) of the material and ρ is its density. For a longitudinal wave traveling in a liquid or gas,

$$v = \sqrt{\frac{B}{\rho}}, \qquad \begin{bmatrix} \text{longitudinal wave} \\ \text{in a fluid} \end{bmatrix} \quad \textbf{(15–4)}$$

where B is the bulk modulus (Section 12–5) and ρ again is the density.

We now derive Eq. 15–4. Consider a wave pulse traveling in a fluid in a long tube, so that the wave motion is one dimensional. The tube is fitted with a piston at the end and is filled with a fluid which, at $t = 0$, is of uniform density ρ and at uniform pressure P_0, Fig. 15–9a. At this moment the piston is abruptly made to start moving to the right with speed v', compressing the fluid in front of it. In the (short) time t the piston moves a distance $v't$. The compressed fluid itself also moves with speed v', but the leading edge of the compressed region moves to the right at the characteristic speed v of compression waves in that fluid; we assume the wave speed v is much larger than the piston speed v'. The leading edge of the compression in the fluid (which at $t = 0$ was at the piston face) thus moves a distance vt in time t as shown in Fig. 15–9b. Let the pressure in the compression be $P_0 + \Delta P$, which is ΔP higher than in the uncompressed fluid. To move the piston to the right requires an external force $(P_0 + \Delta P)A$ acting to the right, where A is the area of the tube. The *net* force on the compressed region of the fluid is

$$F_{\text{net}} = (P_0 + \Delta P)A - P_0 A = A \Delta P$$

since the uncompressed fluid exerts a force $P_0 A$ to the left at the leading edge. Hence the impulse given to the compressed fluid, which equals its change in momentum, is

$$F_{\text{net}} t = \Delta m v'$$
$$A \Delta P t = (\rho A v t) v',$$

where $(\rho A v t)$ represents the mass of fluid which is given the speed v' (the compressed fluid of area A moves a distance vt, Fig. 15–9, so the volume moved is Avt). Hence we have

$$\Delta P = \rho v v'.$$

From the definition of the bulk modulus, B (Eq. 12–7a):

$$B = -\frac{\Delta P}{\Delta V / V_0} = -\frac{\rho v v'}{\Delta V / V_0},$$

where $\Delta V / V_0$ is the fractional change in volume due to compression. The original volume of the compressed fluid is $V_0 = Avt$ (see Fig. 15–9), and it has been compressed by an amount $\Delta V = -Av't$ (Fig. 15–9b). Thus

$$B = -\frac{\rho v v'}{\Delta V / V_0} = -\rho v v' \left(\frac{Avt}{-Av't} \right) = \rho v^2$$

and so

$$v = \sqrt{\frac{B}{\rho}},$$

which is Eq. 15–4.

The derivation of Eq. 15–3 follows similar lines, but takes into account the expansion of the sides of a rod when the end of the rod is compressed.

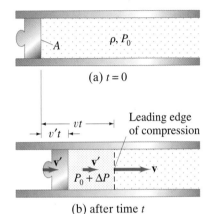

FIGURE 15–9 Determining the speed of a one-dimensional longitudinal wave in a fluid contained in a long narrow tube.

EXAMPLE 15–2 **Echolocation.** Echolocation is a form of sensory perception used by animals such as bats, toothed whales and porpoises. The animal emits a pulse of sound (a longitudinal wave) which is reflected from objects; the reflected pulse is detected by the animal. Echolocation waves emitted by whales have frequencies of about 200,000 Hz. (*a*) What is the wavelength of the whale's echolocation wave? (*b*) If an obstacle is 100 m from the whale, how long after the whale emits a wave will the reflected wave return to him?

SOLUTION (*a*) To find the wavelength given the frequency we first need to compute the velocity of longitudinal (sound) waves in sea water, using Eq. 15–4 and Tables 12–1 and 13–1. We find,

$$v = \sqrt{\frac{B}{\rho}} = \sqrt{\frac{2.0 \times 10^9 \,\text{N/m}^2}{1.025 \times 10^3 \,\text{kg/m}^3}} = 1.40 \times 10^3 \,\text{m/s}.$$

Then using Eq. 15–1 we find,

$$\lambda = v/f = (1.40 \times 10^3 \,\text{m/s})/(2.0 \times 10^5 \,\text{Hz}) = 7.0 \,\text{mm}.$$

As we shall see later, waves can be used to "resolve" objects whose sizes are comparable to or larger than the wavelength. Thus, a whale can resolve tiny objects with echolocation.

(*b*) The time required for the round-trip between the whale and the object is

$$t = \frac{\text{distance}}{\text{speed}} = \frac{2(100 \,\text{m})}{1.40 \times 10^3 \,\text{m/s}} = 0.14 \,\text{s}.$$

Other Waves

Both transverse and longitudinal waves are produced when an earthquake occurs. The transverse waves that travel through the body of the Earth are called *S* waves (*S* for shear) and the longitudinal waves are called *P* waves (*P* for pressure). Both longitudinal and transverse waves can travel through a solid since the atoms or molecules can vibrate about their relatively fixed positions in any direction. But in a fluid, only longitudinal waves can propagate, because any transverse motion would experience no restoring force since a fluid is readily deformable. This fact was used by geophysicists to infer that a portion of the Earth's core must be liquid: longitudinal waves are detected diametrically across the Earth, but not transverse waves.

Besides these two types of waves that can pass through the body of the Earth (or other substance), there can also be *surface waves* that travel along the boundary between two materials. A wave on water is actually a surface wave that moves on the boundary between water and air. The motion of each particle of water at the surface is circular or elliptical (Fig. 15–10), so it is a combination of transverse and longitudinal motions. Below the surface, there is also transverse plus longitudinal wave motion, as shown. At the bottom, the motion is only longitudinal. (When a wave approaches shore, the water drags at the bottom and is slowed down, while the crests move ahead at higher speed (Fig. 15–11) and "spill" over the top.)

Surface waves are also set up on the Earth when an earthquake occurs. The waves that travel along the surface are mainly responsible for the damage caused by earthquakes.

Waves which travel along a line in one dimension, such as transverse waves on a stretched string or longitudinal waves in a rod or fluid-filled tube, are *linear* or *one-dimensional waves*. Surface waves, such as the water waves of Fig. 15–1, are *two-dimensional waves*. Finally, waves that move out from a source in all directions, such as sound from a speaker or earthquake waves through the Earth, are *three-dimensional waves*. We will be concerned with all three types, but especially linear waves, since they are simpler and clearer to understand.

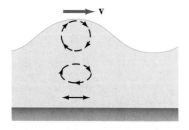

FIGURE 15–10 A water wave is an example of a *surface wave*, which is a combination of transverse and longitudinal wave motions.

FIGURE 15–11 How a wave breaks. The green arrows represent the local velocity of water molecules.

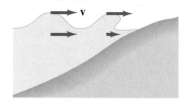

15–3 | Energy Transported by Waves

Waves transport energy from one place to another. As waves travel through a medium, the energy is transferred as vibrational energy from particle to particle of the medium. For a sinusoidal wave of frequency f, the particles move in SHM as a wave passes, and each particle has an energy $E = \frac{1}{2}kD_M^2$ where D_M is the maximum displacement (amplitude) of its motion, either transversely or longitudinally (see Eq. 14–10a, in which we have replaced A by D_M). Using Eq. 14–7a we can write $k = 4\pi^2 mf^2$, where m is the mass of a particle (or small volume) of the medium. Then in terms of the frequency

$$E = \frac{1}{2}kD_M^2 = 2\pi^2 mf^2 D_M^2.$$

For three-dimensional waves traveling in an elastic medium, the mass $m = \rho V$, where ρ is the density of the medium and V is the volume of a small slice of the medium. The volume $V = Al$ where A is the cross-sectional area through which the wave travels (Fig. 15–12), and we can write l as the distance the wave travels in a time t as $l = vt$, where v is the speed of the wave. Thus $m = \rho V = \rho Al = \rho Avt$ and

$$E = 2\pi^2 \rho Avtf^2 D_M^2. \tag{15–5}$$

From this equation we have the important result that the **energy transported by a wave is proportional to the square of the amplitude**, **and to the square of the frequency**. The average *rate* of energy transferred is the average power P:

$$\overline{P} = \frac{E}{t} = 2\pi^2 \rho Avf^2 D_M^2. \tag{15–6}$$

Finally, the **intensity**, I, of a wave is defined as the average power transferred across unit area perpendicular to the direction of energy flow:

$$I = \frac{\overline{P}}{A} = 2\pi^2 v\rho f^2 D_M^2. \tag{15–7}$$

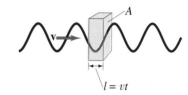

FIGURE 15–12 Calculating the energy carried by a wave moving with velocity v.

$l = vt$

Wave energy $\propto$ (amplitude)^2
Wave energy $\propto (f)^2$

Intensity

If a wave flows out from the source in all directions, it is a three-dimensional wave. Examples are sound traveling in the open air, earthquake waves, and light waves. If the medium is isotropic (same in all directions), the wave is said to be a *spherical wave* (Fig. 15–13). As the wave moves outward, the energy it carries is spread over a larger and larger area since the surface area of a sphere of radius r is $4\pi r^2$. Thus the intensity of a wave is

$$I = \frac{\overline{P}}{A} = \frac{\overline{P}}{4\pi r^2}.$$

If the power output $\overline{P}$ is constant, then the intensity decreases as the inverse square of the distance from the source:

$$I \propto \frac{1}{r^2}. \qquad \text{[spherical wave]} \quad \textbf{(15–8a)}$$

If we consider two points at distances r_1 and r_2 from the source, as in Fig. 15–13, then $I_1 = \overline{P}/4\pi r_1^2$ and $I_2 = \overline{P}/4\pi r_2^2$, so

$$\frac{I_2}{I_1} = \frac{r_1^2}{r_2^2}. \tag{15–8b}$$

Thus, for example, when the distance doubles $(r_2/r_1 = 2)$, then the intensity is reduced to $\frac{1}{4}$ its earlier value: $I_2/I_1 = \left(\frac{1}{2}\right)^2 = \frac{1}{4}$.

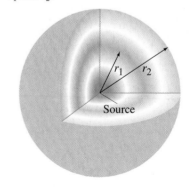

FIGURE 15–13 Wave traveling outward from source has spherical shape. Two different crests (or compressions) are shown, of radius r_1 and r_2.

r_1 r_2

Source

Intensity $I \propto \dfrac{1}{r^2}$

The amplitude of a wave also decreases with distance. Since the intensity is proportional to the square of the amplitude $(I \propto D_M^2$, Eq. 15–7), then the amplitude must decrease as $1/r$, so that $I \propto D_M^2$ will be proportional to $1/r^2$, Eq. 15–8a; that is,

Amplitude $\propto \dfrac{1}{r}$

$$D_M \propto \frac{1}{r}.$$

To see this directly from Eq. 15–6, we consider again two different distances from the source, r_1 and r_2. For constant power output, $A_1 D_{M1}^2 = A_2 D_{M2}^2$ where D_{M1} and D_{M2} are the amplitudes of the wave at r_1 and r_2, respectively. Since $A_1 = 4\pi r_1^2$ and $A_2 = 4\pi r_2^2$, we have $(D_{M1}^2 r_1^2) = (D_{M2}^2 r_2^2)$, or

$$\frac{D_{M2}}{D_{M1}} = \frac{r_1}{r_2}.$$

When the wave is twice as far from the source, the amplitude is half as large, as long as we can ignore damping due to friction.

EXAMPLE 15–3 **Earthquake intensity.** If the intensity of an earthquake P wave 100 km from the source is $1.0 \times 10^6 \, \text{W/m}^2$, what is the intensity 400 km from the source?

SOLUTION The intensity decreases as the square of the distance from the source. Therefore, at 400 km, the intensity will be $\left(\frac{1}{4}\right)^2 = \frac{1}{16}$ of its value at 100 km, or $6.2 \times 10^4 \, \text{W/m}^2$. Alternatively, Eq. 15–8b could be used: $I_2 = I_1 r_1^2 / r_2^2 = (1.0 \times 10^6 \, \text{W/m}^2)(100 \, \text{km})^2 / (400 \, \text{km})^2 = 6.2 \times 10^4 \, \text{W/m}^2$.

The situation is different for a one-dimensional wave, such as a transverse wave on a string or a longitudinal wave pulse traveling down a thin uniform metal rod. The area remains constant, so the amplitude D_M also remains constant (ignoring friction). Thus the amplitude and the intensity do not decrease with distance.

In practice, frictional damping is generally present, and some of the energy is transformed into thermal energy. Thus the amplitude and intensity of a one-dimensional wave decrease with distance from the source, and for a three-dimensional wave the decrease will be greater than that discussed above, although the effect may often be small.

15–4 | Mathematical Representation of a Traveling Wave

Let us now consider a one-dimensional wave traveling along the x axis. It could be, for example, a transverse wave on a cord or a longitudinal wave traveling in a rod or in a fluid-filled tube. Let us assume the wave shape is sinusoidal and has a particular wavelength λ and frequency f. At $t = 0$, suppose the wave shape is given by

$$D(x) = D_M \sin \frac{2\pi}{\lambda} x, \tag{15–9}$$

as shown by the solid curve in Fig. 15–14. $D(x)$ is the **displacement**[†] of the wave (be it a longitudinal or transverse wave) at position x, and D_M is the **amplitude**

[†] Some books use $y(x)$ in place of $D(x)$. To avoid confusion, we reserve y (and z) for the coordinate positions of waves in two or three dimensions. Our $D(x)$ can stand for pressure (in longitudinal waves), position displacement (transverse mechanical waves) or—as we will see later—electric or magnetic fields (for electromagnetic waves).

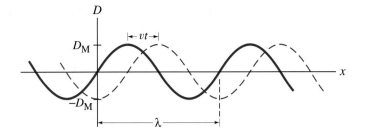

(maximum displacement) of the wave. This relation gives a shape that repeats itself every wavelength, which is what we want so that the displacement is the same at, for example, $x = 0$, $x = \lambda$, $x = 2\lambda$, and so on (since $\sin 4\pi = \sin 2\pi = \sin 0$).

Now suppose the wave is moving to the right with velocity v. Then, after a time t, each part of the wave (indeed, the whole wave "shape") has moved to the right a distance vt; see the dashed curve in Fig. 15–14. Consider any point on the wave at $t = 0$: say, a crest which is at some position x. After a time t, that crest will have traveled a distance vt so its new position is a distance vt greater than its old position. To describe this same point on the wave shape, the argument of the sine function must be the same, so we replace x in Eq. 15–9 by $(x - vt)$:

$$D(x, t) = D_{\text{M}} \sin\left[\frac{2\pi}{\lambda}(x - vt)\right]. \qquad \textbf{(15–10a)}$$

1-D wave, moving in positive x direction

Said another way, if you are riding on a crest, the argument of the sine function, $(2\pi/\lambda)(x - vt)$, remains the same $(= \pi/2, 5\pi/2$, and so on); as t increases, x must increase at the same rate so that $(x - vt)$ remains constant.

Equation 15–10a is the mathematical representation of a sinusoidal wave traveling along the x axis to the right (increasing x). It gives the displacement $D(x, t)$ of the wave at any chosen point x at any time t. The function $D(x, t)$ describes a curve that represents the actual shape of the wave in space at time t. Since $v = \lambda f$ (Eq. 15–1) we can write Eq. 15–10a in other ways that are often convenient:

$$D(x, t) = D_{\text{M}} \sin\left(\frac{2\pi x}{\lambda} - \frac{2\pi t}{T}\right), \qquad \textbf{(15–10b)}$$

1-D wave, moving in positive x direction

where $T = 1/f = \lambda/v$ is the period; and

$$D(x, t) = D_{\text{M}} \sin(kx - \omega t), \qquad \textbf{(15–10c)}$$

1-D wave, moving in positive x direction

where $\omega = 2\pi f = 2\pi/T$ is the angular frequency and

$$k = \frac{2\pi}{\lambda} \qquad \textbf{(15–11)}$$

is called the **wave number**. (Do not confuse the wave number k with the spring constant k; they are very different quantities.) All three forms, Eq. 15–10a, b, and c, are equivalent; Eq. 15–10c is the simplest to write and is perhaps the most common. The quantity $(kx - \omega t)$, and its equivalent in the other two equations, is called the **phase** of the wave. The velocity v of the wave is often called the **phase velocity**, since it describes the velocity of the phase (or shape) of the wave and it can be written in terms of ω and k:

$$v = \lambda f = \left(\frac{2\pi}{k}\right)\left(\frac{\omega}{2\pi}\right) = \frac{\omega}{k}. \qquad \textbf{(15–12)}$$

For a wave traveling along the x axis to the left (decreasing values of x), we start again with Eq. 15–9 and note that a particular point on the wave changes position by $-vt$ in a time t. So x in Eq. 15–9 must be replaced by $(x + vt)$. Thus, for a wave traveling to the left with velocity v,

1-D wave

moving in

negative x

direction

$$D(x, t) = D_M \sin\left[\frac{2\pi}{\lambda}(x + vt)\right] \qquad \textbf{(15–13a)}$$

$$= D_M \sin\left(\frac{2\pi x}{\lambda} + \frac{2\pi t}{T}\right) \qquad \textbf{(15–13b)}$$

$$= D_M \sin(kx + \omega t). \qquad \textbf{(15–13c)}$$

In other words, we simply replace v in Eqs. 15–10 by $-v$.

Let us look for a moment at Eq. 15–13c (or, just as well, at Eq. 15–10c). At $t = 0$ we have

$$D(x, 0) = D_M \sin kx,$$

which is what we started with, a sinusoidal wave shape. If we look at the wave shape in space at a particular later time t_1, then we have

$$D(x, t_1) = D_M \sin(kx + \omega t_1).$$

That is, if we took a picture of the wave at $t = t_1$, we would see a sine wave with a phase constant ωt_1. Thus, for fixed $t = t_1$, the wave has a sinusoidal shape in space. On the other hand, if we consider a fixed point in space, say $x = 0$, we can see how the wave varies in time:

$$D(0, t) = D_M \sin \omega t$$

where we used Eq. 15–13c. This is just the equation for simple harmonic motion (see Section 14–2, Eq. 14–4). For any other fixed value of x, say $x = x_1$, $D = D_M \sin(\omega t + kx_1)$ which differs only by a phase constant kx_1. Thus, at any fixed point in space, the displacement undergoes the oscillations of simple harmonic motion in time. Equations 15–10 and 15–13 combine both these aspects to give us the representation for a **traveling sinusoidal wave** (also called a **harmonic wave**).

The argument of the sine in Eqs. 15–10 and 15–13 can in general contain a phase angle ϕ,

$$D(x, t) = D_M \sin(kx \pm \omega t + \phi),$$

to adjust for the position of the wave at $t = 0$, $x = 0$, just as in Section 14–2 (see Fig. 14–7). If the displacement is zero at $t = 0$, $x = 0$, as in Fig. 14–6, then $\phi = 0$.

Now let us consider a general wave (or wave pulse) of any shape. If frictional losses are small, experiment shows that the wave maintains its shape as it travels. Thus we can make the same arguments as we did right after Eq. 15–9. Suppose our wave has some shape at $t = 0$, given by

$$D(x, 0) = D(x)$$

where $D(x)$ is the displacement of the wave at x and is not necessarily sinusoidal. Then at some later time, if the wave is traveling to the right along the x axis, the wave will have the same shape but all parts will have moved a distance vt where v is the phase velocity of the wave. Hence we must replace x by $x - vt$ to obtain the amplitude at time t:

$$D(x, t) = D(x - vt). \qquad \textbf{(15–14)}$$

Similarly, if the wave moves to the left, we must replace x by $x + vt$, so

$$D(x, t) = D(x + vt). \qquad \textbf{(15–15)}$$

Thus, any wave traveling along the x axis must have the form of Eq. 15–14 or 15–15.

EXAMPLE 15–4 **A traveling wave.** The left-hand end of a long horizontal stretched cord oscillates transversely in SHM with frequency $f = 250\,\text{Hz}$ and amplitude 2.6 cm. The cord is under a tension of 140 N and has a linear density $\mu = 0.12\,\text{kg/m}$. At $t = 0$, the end of the cord has an upward displacement of 1.6 cm and is falling (Fig. 15–15). Determine (a) the wavelength of waves produced and (b) the equation for the traveling wave.

SOLUTION (a) The wave velocity is

$$v = \sqrt{\frac{F_T}{\mu}} = \sqrt{\frac{140\,\text{N}}{0.12\,\text{kg/m}}} = 34\,\text{m/s}.$$

Then

$$\lambda = \frac{v}{f} = \frac{34\,\text{m/s}}{250\,\text{Hz}} = 0.14\,\text{m} \quad \text{or} \quad 14\,\text{cm}.$$

(b) Let $x = 0$ at the left-hand end of the cord. The phase of the wave at $t = 0$ is not zero in general as was assumed in Eqs. 15–9, 10, and 13. The general form for a wave traveling to the right is then

$$D(x, t) = D_M \sin(kx - \omega t + \phi),$$

where ϕ is the phase angle. In our case, the amplitude $D_M = 2.6\,\text{cm}$; and at $t = 0$, $x = 0$, we are given $D = 1.6\,\text{cm}$. Thus

$$1.6 = 2.6 \sin\phi,$$

so $\phi = 38° = 0.66\,\text{rad}$. We also have $\omega = 2\pi f = 1570\,\text{s}^{-1}$ and $k = 2\pi/\lambda = 45\,\text{m}^{-1}$. Hence

$$D = 0.026 \sin(45x - 1570t + 0.66),$$

where D and x are in meters and t in seconds.

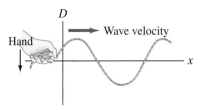

FIGURE 15–15 Example 15–4: The wave at $t = 0$ (the hand is falling). Not to scale.

15–5 The Wave Equation

Many types of waves satisfy an important general equation that is the equivalent of Newton's second law of motion for particles. This "equation of motion for a wave" is called the **wave equation**, and we derive it now for waves traveling on a stretched horizontal string.

We assume the amplitude of the wave is small compared to the wavelength so that each point on the string can be assumed to move only vertically and the tension in the string, F_T, does not vary during a vibration. We apply Newton's second law, $\Sigma F = ma$, to the vertical motion of a tiny section of the string as shown in Fig. 15–16. The amplitude of the wave is small, so the angles θ_1 and θ_2 that the string makes with the horizontal are small. The length of this section is then approximately Δx, and its mass is $\mu\,\Delta x$, where μ is the mass per unit length of the string. The net vertical force on this section of string is $F_T \sin\theta_2 - F_T \sin\theta_1$. So Newton's second law applied to the vertical (y) direction gives

$$\Sigma F_y = ma_y$$

$$F_T \sin\theta_2 - F_T \sin\theta_1 = (\mu\,\Delta x)\frac{\partial^2 D}{\partial t^2}.$$

We have written the acceleration as $a_y = \partial^2 D/\partial t^2$ since the motion is only vertical, and we use the partial derivative notation because the displacement D is a function of both x and t.

FIGURE 15–16 Deriving the wave equation from Newton's second law: a segment of string under tension F_T.

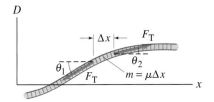

Because the angles θ_1 and θ_2 are assumed small, $\sin \theta \approx \tan \theta$ and $\tan \theta$ is equal to the slope s of the string at each point:

$$\sin \theta \approx \tan \theta = \frac{\partial D}{\partial x} = s.$$

Thus our equation at the bottom of the previous page becomes

$$F_T(s_2 - s_1) = \mu \, \Delta x \, \frac{\partial^2 D}{\partial t^2}$$

or

$$F_T \frac{\Delta s}{\Delta x} = \mu \, \frac{\partial^2 D}{\partial t^2}, \tag{a}$$

where $\Delta s = s_2 - s_1$ is the difference in the slope between the two ends of our tiny section. Now we take the limit of $\Delta x \to 0$, so that

$$F_T \lim_{\Delta x \to 0} \frac{\Delta s}{\Delta x} = F_T \frac{\partial s}{\partial x}$$

$$= F_T \frac{\partial}{\partial x} \left(\frac{\partial D}{\partial x} \right) = F_T \frac{\partial^2 D}{\partial x^2}$$

since the slope $s = \partial D/\partial x$, as we saw above. Substituting this into the equation labelled (a) above gives

$$F_T \frac{\partial^2 D}{\partial x^2} = \mu \, \frac{\partial^2 D}{\partial t^2}$$

or

$$\frac{\partial^2 D}{\partial t^2} = \frac{F_T}{\mu} \frac{\partial^2 D}{\partial t^2}.$$

We saw earlier in this chapter (Eq. 15–2) that the velocity of waves on a string is given by $v = \sqrt{F_T/\mu}$, so we can write this last equation as

Wave equation (1–D)
$$\frac{\partial^2 D}{\partial x^2} = \frac{1}{v^2} \frac{\partial^2 D}{\partial t^2}. \tag{15–16}$$

This is the **one-dimensional wave equation**, and it can describe not only small amplitude waves on a stretched string, but also small amplitude longitudinal waves (such as sound waves) in gases, liquids, and elastic solids, in which case D can refer to the pressure variations. In this case, the wave equation is a direct consequence of Newton's second law applied to a continuous elastic medium. The wave equation also describes electromagnetic waves for which D refers to the electric or magnetic field, as we shall see in Chapter 32. Equation 15–16 applies to waves traveling in one dimension only. For waves spreading out in three dimensions, the wave equation is the same, with the addition of $\partial^2 D/\partial y^2$ and $\partial^2 D/\partial z^2$ to the left side of Eq. 15–16.

The wave equation is a *linear* equation: the displacement D appears singly in each term. There are no terms that contain D^2, or $D(\partial D/\partial x)$, or the like in which D appears more than once. Thus, if $D_1(x, t)$ and $D_2(x, t)$ are two different solutions of a linear equation, such as the wave equation, then the linear combination

Superposition principle
$$D_3(x, t) = aD_1(x, t) + bD_2(x, t),$$

where a and b are constants, is also a solution. This is readily seen by direct substitution into the wave equation. This is the essence of the *superposition principle*, which we discuss in the next Section. Basically it says that if two waves pass through the same region of space at the same time, the actual displacement is the sum of the separate displacements. For waves on a string, or for sound waves, this is valid only for small-amplitude waves. If the amplitude is not small enough, the equations for wave propagation may become nonlinear and the principle of superposition would not hold and other new effects may occur.

EXAMPLE 15–5 **Wave equation solution.** Verify that the sinusoidal wave of Eq. 15–10c, $D(x, t) = D_M \sin(kx - \omega t)$, satisfies the wave equation.

SOLUTION We take the derivative of Eq. 15–10c twice with respect to t:

$$\frac{\partial D}{\partial t} = -\omega D_M \cos(kx - \omega t)$$

$$\frac{\partial^2 D}{\partial t^2} = -\omega^2 D_M \sin(kx - \omega t).$$

With respect to x, the derivatives are

$$\frac{\partial D}{\partial x} = k D_M \cos(kx - \omega t)$$

$$\frac{\partial^2 D}{\partial x^2} = -k^2 D_M \sin(kx - \omega t).$$

If we now divide the second derivatives we get

$$\frac{\partial^2 D/\partial t^2}{\partial^2 D/\partial x^2} = \frac{-\omega^2 D_M \sin(kx - \omega t)}{-k^2 D_M \sin(kx - \omega t)} = \frac{\omega^2}{k^2}.$$

From Eq. 15–12 we have $\omega^2/k^2 = v^2$, so we see that Eq. 15–10 does satisfy the wave equation (Eq. 15–16).

15–6 The Principle of Superposition

When two or more waves pass through the same region of space at the same time, it is found that for many waves *the actual displacement is the vector* (or *algebraic*) *sum of the separate displacements*. This is called the **principle of superposition**. It is valid for mechanical waves as long as the displacements are not too large and there is a linear relationship between the displacement and the restoring force of the oscillating medium.[†] If the amplitude of a mechanical wave, for example, is so large that it goes beyond the elastic region of the medium, and Hooke's law is no longer operative, the superposition principle is no longer accurate.[‡] For the most part, we will consider systems for which the superposition principle can be assumed to hold.

One result of the superposition principle is that if two waves pass through the same region of space, they continue to move independently of one another. You may have noticed, for example, that the ripples on the surface of water (two-dimensional waves) that form from two rocks striking the water at different places will pass through each other.

Figure 15–17, shows an example of the superposition principle. In this case there are three waves present, on a stretched string, each of different amplitude and frequency. At any time, such as at the instant shown, the actual amplitude at any position x is the algebraic sum of the amplitude of the three waves at that position. The actual wave is not a simple sinusoidal wave and is called a *composite* (or *complex*) *wave*. (Amplitudes are exaggerated in Fig. 15–17.)

It can be shown that any complex wave can be considered as being composed of many simple sinusoidal waves of different amplitudes, wavelengths, and frequencies. This is known as *Fourier's theorem*. A complex periodic wave of period T can be represented as a sum of pure sinusoidal terms whose frequencies are integral multiples of $f = 1/T$. If the wave is not periodic, the sum becomes an integral (called a *Fourier integral*). Although we will not go into the details here, we see the importance of considering sinusoidal waves (and simple harmonic motion): because any other wave shape can be considered a sum of such pure sinusoidal waves.

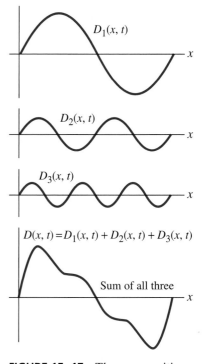

FIGURE 15–17 The superposition principle for one-dimensional waves. Composite wave formed from three sinusoidal waves of different amplitudes and frequencies $(f_0, 2f_0, 3f_0)$ at a certain instant in time. The amplitude of the composite wave at each point in space, at any time, is the algebraic sum of the amplitudes of the component waves. Amplitudes are shown exaggerated; for the superposition principle to hold, they must be small compared to the wavelengths.

[†] For electromagnetic waves in vacuum, Chapter 32, the superposition principle always holds.

[‡] Intermodulation distortion in high-fidelity equipment is an example of the superposition principle not holding when two frequencies do not combine linearly in the electronics.

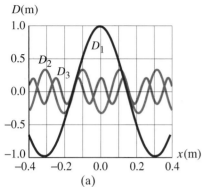

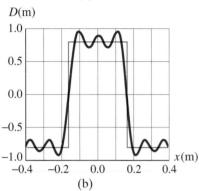

FIGURE 15–18 Example 15–6. Making a square wave.

FIGURE 15–19 Reflection of a wave pulse on a rope when the end of the rope is (a) fixed and (b) free.

EXAMPLE 15–6 **Making a square wave.** At $t = 0$, three waves are given by $D_1 = D_M \cos kx$, $D_2 = \frac{1}{3} D_M \cos 3kx$ and $D_3 = \frac{1}{5} D_M \cos 5kx$ where $D_M = 1.0$ m and $k = 10$ m^{-1}. Plot the sum of the three waves from $x = -0.4$ to $+0.4$ m. (These three waves are the first three Fourier components of a "square wave.")

SOLUTION The first wave, D_1 has amplitude of 1.0 m and wavelength $\lambda = 2\pi/k = 2\pi/10$ m $= 0.628$ m. The second wave, D_2 has amplitude of 0.33 m and wavelength $\lambda = 2\pi/3k = 2\pi/30$ m $= 0.209$ m. The third wave, D_3 has amplitude of 0.20 m and wavelength $\lambda = 2\pi/5k = 2\pi/50$ m $= 0.126$ m. Each wave is plotted in Fig. 15–18a. The sum of the three waves is shown in Fig. 15–18b. The sum begins to resemble a "square wave," shown in blue in Fig. 15–18b.

When the restoring force is not precisely proportional to the displacement for mechanical waves in some continuous medium, the speed of sinusoidal waves depends on the frequency. The variation of speed with frequency is called **dispersion**. The different sinusoidal waves that compose a complex wave will travel with slightly different speeds in such a case. Consequently, a complex wave will change shape as it travels if the medium is "dispersive." A pure sine wave will not change shape under these conditions, however, except by the influence of friction or dissipative forces. If there is no dispersion (or friction), even a complex linear wave doesn't change shape.

15–7 Reflection and Transmission

When a wave strikes an obstacle, or comes to the end of the medium it is traveling in, at least a part of the wave is reflected. You have probably seen water waves reflect off of a rock or the side of a swimming pool. And you may have heard a shout reflected from a distant cliff—which we call an "echo."

A wave pulse traveling down a rope is reflected as shown in Fig. 15–19. You can observe this for yourself (try it with a rope lying on a table) and see that the reflected pulse is inverted as in Fig. 15–19a if the end of the rope is fixed; and returns right side up if the end is free as in Fig. 15–19b. When the end is fixed to a

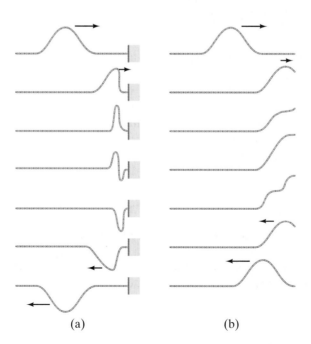

(a)   (b)

support, as in Fig. 15–19a, the pulse reaching that fixed end exerts a force (upward) on the support. The support exerts an equal but opposite force (Newton's third law) downward on the rope. This downward force on the rope is what "generates" the inverted reflected pulse. The inversion of the reflected pulse in part (a) is said to be a phase change of 180°. It is as if the phase shifted by $\frac{1}{2}\lambda$, or 180°, from a crest to a trough. In Fig. 15–19b, the free end is constrained by neither a support nor by additional rope. It therefore tends to overshoot—its displacement being momentarily greater than that of the traveling pulse. The overshooting end exerts an upward pull on the rope, and this is what generates the reflected pulse, which is not inverted (no phase change).

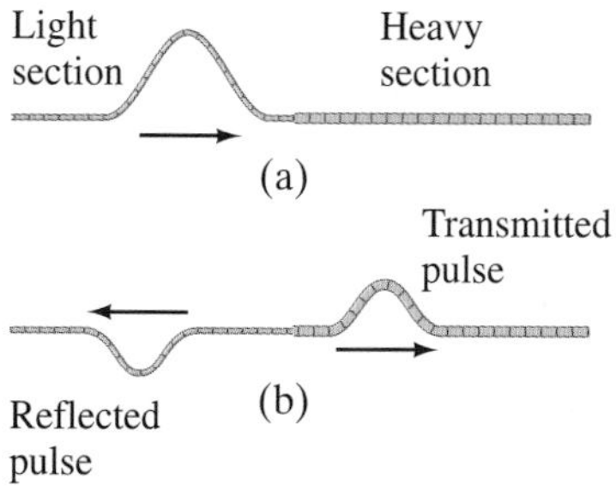

When the wave pulse on the rope in Fig. 15–19a reaches the wall, not all of the energy is reflected. Some of it is absorbed by the wall. Part of the absorbed energy is transformed into thermal energy, and part continues to propagate through the material of the wall. This is more clearly illustrated by considering a pulse that travels down a rope which consists of a light section and a heavy section, as shown in Fig. 15–20. When the wave reaches the boundary between the two sections, part of the pulse is reflected and part is transmitted, as shown. The heavier the second section, the less that is transmitted; and when the second section is a wall or rigid support, very little is transmitted. For a periodic wave, the frequency does not change across the boundary since the boundary point oscillates at that frequency. Thus if the transmitted wave has a lower speed, its wavelength is also less ($\lambda = v/f$).

For a two- or three-dimensional wave, such as a water wave, we are concerned with **wave fronts**, by which we mean all the points along the wave forming the wave crest (what we usually refer to simply as a "wave" when we are at the seashore). A line drawn in the direction of motion, perpendicular to the wave front, is called a **ray**, as shown in Fig. 15–21. Note in Fig. 15–21b that wave fronts far from the source have lost almost all their curvature and are nearly straight, as ocean waves often are; they are then called **plane waves**.

For reflection of a two- or three-dimensional plane wave, as shown in Fig. 15–22, the angle that the incoming or *incident wave* makes with the reflecting surface is equal to the angle made by the reflected wave. This is the **law of reflection: the angle of reflection equals the angle of incidence**. The "angle of incidence" is defined as the angle the incident ray makes with the perpendicular to the reflecting surface (or the wave front makes with a tangent to the surface), and the "angle of reflection" is the corresponding angle for the reflected wave.

Wave fronts

Plane waves

Law of reflection

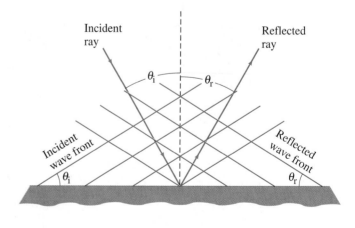

FIGURE 15–20 When a wave pulse traveling to the right (a) reaches a discontinuity, then (b) part is reflected and part is transmitted.

FIGURE 15–21 Rays, signifying the direction of motion, are always perpendicular to the wave fronts (wave crests). (a) Circular or spherical waves near the source. (b) Far from the source, the wave fronts are nearly straight or flat, and are called plane waves.

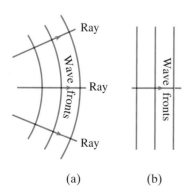

FIGURE 15–22 Law of reflection.

15–8 | Interference

Interference refers to what happens when two waves pass through the same region of space at the same time. It is an example of the superposition principle. Consider for example the two wave pulses on a string traveling toward each other as shown in Fig. 15–23. In part (a) the two pulses have the same amplitude but one is a crest and the other a trough; in part (b) they are both crests. In both cases, the waves meet and pass right by each other. However, when they overlap, the resultant displacement is the *algebraic sum of their separate displacements* (the principle of superposition). In Fig. 15–23a, the two wave amplitudes are opposite one another as they pass by and the result is called **destructive interference**. In Fig. 15–23b, the resultant displacement is greater than that of either pulse and the result is called **constructive interference**.

Destructive interference

Constructive interference

FIGURE 15–23 Two waves pulses pass each other. Where they overlap, interference occurs: (a) destructive; (b) constructive.

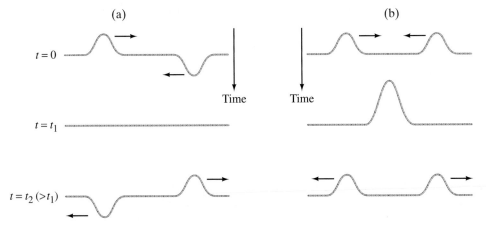

When two rocks are thrown into a pond simultaneously, the two sets of circular waves interfere with one another as shown in Fig. 15–24. In some areas of overlap, crests of one wave meet crests of the other (and troughs meet troughs); this is constructive interference and the water oscillates up and down with greater amplitude than either wave separately. In other areas, destructive interference occurs

FIGURE 15–24 Interference of water waves. Constructive interference occurs where one wave's maximum (a crest) meets the other's maximum. Destructive interference ("flat water") occurs where one wave's maximum (a crest) meets the other's miminum (a trough).

(a)

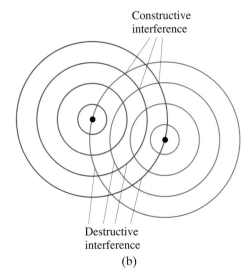

(b)

where the water actually does not move at all—this is where crests of one wave meet troughs of the other, and vice versa. In the first case, of constructive interference, the two waves are **in phase**, whereas in destructive interference the two waves are **out of phase** by one-half wavelength or 180°. Of course the relative phase of the two waves in most areas is intermediate between these two extremes, resulting in partially destructive interference. All three of these situations are shown in Fig. 15–25, where the amplitudes are plotted versus time at a given point in space. We will deal with interference in more detail when we discuss sound and light.

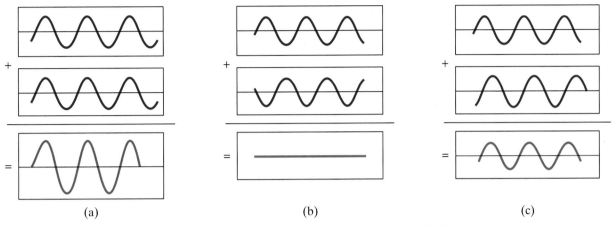

(a)   (b)   (c)

FIGURE 15–25 Two waves interfere: (a) constructively, (b) destructively, (c) partially destructively.

15–9 | Standing Waves; Resonance

If you shake one end of a cord (or slinky) and the other end is kept fixed, a continuous wave will travel down to the fixed end and be reflected back, inverted (Fig. 15–19a). As you continue to vibrate the cord, there will be waves traveling in both directions, and the wave traveling down the cord will interfere with the reflected wave coming back. Usually there will be quite a jumble. But if you vibrate the cord at just the right frequency, the two traveling waves will interfere in such a way that a large-amplitude **standing wave** will be produced, Fig. 15–26. It is called a "standing wave" because it doesn't appear to be traveling. The cord simply appears to have segments that oscillate up and down in a fixed pattern. The points of destructive interference, where the cord remains still at all times, are called **nodes**. Points of constructive interference, where the end oscillates with maximum amplitude, are called **antinodes**. The nodes and antinodes remain in fixed positions for a given frequency.

Standing waves can occur at more than one frequency. The lowest frequency of vibration that produces a standing wave gives rise to the pattern shown in Fig. 15–26a. The standing waves shown in parts (b) and (c) are produced at precisely twice and three times the lowest frequency, respectively, assuming the tension in the cord is the same. The cord can also vibrate with four loops at four times the lowest frequency, and so on.

The frequencies at which standing waves are produced are the **natural frequencies** or **resonant frequencies** of the cord, and the different standing wave patterns shown in Fig. 15–26 are different "resonant modes of vibration." Although a standing wave on a cord is the result of the interference of two waves traveling in opposite directions, it is also a vibrating object at resonance. Standing waves then represent the same phenomenon as the resonance of a vibrating spring or pendulum, which we discussed in Chapter 14. The only difference is that a spring or pendulum has only one resonant frequency, whereas the cord has an infinite number of resonant frequencies, each of which is a whole-number multiple of the lowest resonant frequency.

FIGURE 15–26 Standing waves corresponding to three resonant frequencies.

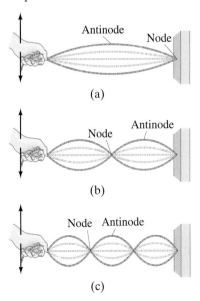

(a)

(b)

(c)

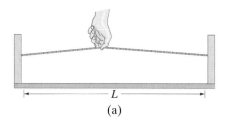

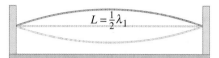

$L = \frac{1}{2}\lambda_1$

Fundamental or first harmonic, f_1

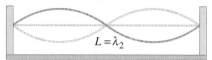

$L = \lambda_2$

First overtone or second harmonic, $f_2 = 2f_1$

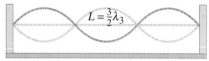

$L = \frac{3}{2}\lambda_3$

Second overtone or third harmonic, $f_3 = 3f_1$

(b)

FIGURE 15–27 (a) A string is plucked. (b) Only standing waves corresponding to resonant frequencies persist for long.

Now let us consider a cord stretched between two supports that is plucked like a guitar or violin string, Fig. 15–27a. Waves of a great variety of frequencies will travel in both directions along the string, will be reflected at the ends, and will travel back in the opposite direction. Most of these waves interfere in a random way with each other and quickly die away. However, those waves that correspond to the resonant frequencies of the string will persist. The ends of the string, since they are fixed, will be nodes. There may be other nodes as well. Some of the possible resonant modes of vibration (standing waves) are shown in Fig. 15–27b. Generally, the motion will be a combination of these different resonant modes; but only those frequencies that correspond to a resonant frequency will be present.

To determine the resonant frequencies, we first note that the wavelengths of the standing waves bear a simple relationship to the length L of the string. The lowest frequency, called the **fundamental frequency**, corresponds to one antinode (or loop). And as can be seen in Fig. 15–27b, the whole length corresponds to one-half wavelength. Thus $L = \frac{1}{2}\lambda_1$, where λ_1 stands for the wavelength of the fundamental. The other natural frequencies are called **overtones**; when they are integral multiples of the fundamental (as they are for a simple string), they are also called **harmonics**, with the fundamental being referred to as the **first harmonic**.[†] The next mode after the fundamental has two loops and is called the **second harmonic** (or first overtone); the length of the string L at the second harmonic corresponds to one complete wavelength: $L = \lambda_2$. For the third and fourth harmonics, $L = \frac{3}{2}\lambda_3$, and $L = 2\lambda_4$, respectively, and so on. In general, we can write

$$L = \frac{n\lambda_n}{2}, \qquad \text{where } n = 1, 2, 3, \cdots.$$

The integer n labels the number of the harmonic: $n = 1$ for the fundamental, $n = 2$ for the second harmonic, and so on. We solve for λ_n and find

$$\lambda_n = \frac{2L}{n}, \qquad n = 1, 2, 3, \cdots. \tag{15–17}$$

In order to find the frequency f of each vibration we use Eq. 15–1, $f = v/\lambda$, and we see that

$$f_n = \frac{v}{\lambda_n} = \frac{nv}{2L} = nf_1,$$

where $f_1 = v/\lambda_1 = v/2L$ is the fundamental frequency. We see that each resonant frequency is an integer multiple of the fundamental frequency.

Because a standing wave is equivalent to two traveling waves moving in opposite directions, the concept of wave velocity still makes sense and is given by Eq. 15–2 in terms of the tension F_T in the string and its mass per unit length μ; that is, $v = \sqrt{F_T/\mu}$ for both traveling waves.

EXAMPLE 15–7 **Piano string.** A piano string is 1.10 m long and has a mass of 9.00 g. (a) How much tension must the string be under if it is to vibrate at a fundamental frequency of 131 Hz? (b) What are the frequencies of the first four harmonics?

SOLUTION (a) The wavelength of the fundamental is $\lambda = 2L = 2.20$ m (Eq. 15–17). The velocity is then $v = \lambda f = (2.20 \text{ m})(131 \text{ s}^{-1}) = 288$ m/s. Then, from Eq. 15–2, we have

$$F_T = \mu v^2$$
$$= \left(\frac{9.00 \times 10^{-3} \text{ kg}}{1.10 \text{ m}}\right)(288 \text{ m/s})^2 = 679 \text{ N}.$$

(b) The frequencies of the second, third, and fourth harmonics are two, three, and four times the fundamental frequency: 262, 393, and 524 Hz.

[†] The term "harmonic" comes from music, because such integral multiples of frequencies "harmonize."

A standing wave does appear to be standing in place (and a traveling wave appears to move). The term "standing" wave is also meaningful from the point of view of energy. Since the string is at rest at the nodes, no energy flows past these points. Hence the energy is not transmitted down the string but "stands" in place in the string.

Standing waves are produced not only on strings, but on any object that is set into vibration. Even when a rock or a piece of wood is struck with a hammer, standing waves are set up that correspond to the natural resonant frequencies of that object. In general, the resonant frequencies depend on the dimensions of the object, just as for a string they depend on its length. For example a small object does not have as low resonant frequencies as does a large object. All musical instruments depend on standing waves to produce their musical sounds, from stringed instruments to wind instruments (in which a column of air vibrates as a standing wave) to drums and other percussion instruments, as we will discuss in the next chapter.

Mathematical Representation of a Standing Wave

In Section 15–4 we saw how to write an equation for the displacement D of a linear traveling wave as a function of position x and time t. We can do the same for a standing wave on a string. We already discussed that a standing wave can be considered to consist of two traveling waves that move in opposite directions. These can be written (see Eqs. 15–10c and 15–13c)

$$D_1(x, t) = D_M \sin(kx - \omega t)$$
$$D_2(x, t) = D_M \sin(kx + \omega t)$$

since, assuming no damping, the amplitudes are equal as are the frequencies and wavelengths. The sum of these two traveling waves produces a standing wave which can be written mathematically as

$$D = D_1 + D_2 = D_M[\sin(kx - \omega t) + \sin(kx + \omega t)].$$

From the trigonometric identity $\sin\theta_1 + \sin\theta_2 = 2\sin\frac{1}{2}(\theta_1 + \theta_2)\cos\frac{1}{2}(\theta_1 - \theta_2)$, we can rewrite this as

$$D = 2D_M \sin kx \cos \omega t. \tag{15-18}$$

Standing wave

If we let $x = 0$ at the left-hand end of the string, then the right-hand end is at $x = L$ where L is the length of the string. Since the string is fixed at its two ends (Fig. 15–27), $D(x, t)$ must be zero at $x = 0$ and at $x = L$. Equation 15–18 already satisfies the first condition ($D = 0$ at $x = 0$) and satisfies the second condition if $\sin kL = 0$ which means

$$kL = \pi, 2\pi, 3\pi, \ldots, n\pi, \ldots$$

where $n =$ integer, or, since $k = 2\pi/\lambda$,

$$\lambda = \frac{2L}{n}. \quad (n = \text{integer})$$

This is just Eq. 15–17.

Equation 15–18, with the condition $\lambda = 2L/n$, is the mathematical representation of a standing wave. We see that a particle at any position x vibrates in simple harmonic motion (because of the factor $\cos \omega t$). All particles of the string vibrate with the same frequency $f = \omega/2\pi$, but the amplitude depends on x and equals $2D_M \sin kx$. (Compare this to a traveling wave for which all particles vibrate with the same amplitude.) The amplitude has a maximum, equal to $2D_M$, when $kx = \pi/2, 3\pi/2, 5\pi/2$, and so on—that is, at

$$x = \frac{\lambda}{4}, \frac{3\lambda}{4}, \frac{5\lambda}{4}, \ldots.$$

These are, of course, the positions of the antinodes (see Fig. 15–27).

EXAMPLE 15–8 **Wave forms.** Two waves traveling in opposite directions on a string fixed at $x = 0$ are described by the functions

$$D_1 = (0.20\,\text{m})\sin(2.0x - 4.0t)$$

$$D_2 = (0.20\,\text{m})\sin(2.0x + 4.0t)$$

and they produce a standing wave pattern (x is in m, t is in s). (a) Determine the function for the standing wave. (b) What is the maximum amplitude at $x = 0.45\,\text{m}$? (c) Where is the other end fixed ($x > 0$)? (d) What is the maximum amplitude, and where does it occur?

SOLUTION (a) The two waves are of the form $D = D_M \sin(kx \pm \omega t)$, so

$$k = 2.0\,\text{m}^{-1} \quad \text{and} \quad \omega = 4.0\,\text{s}^{-1}.$$

These combine to form a standing wave of the form of Eq. 15–18:

$$D = 2D_M \sin kx \cos \omega t = (0.40\,\text{m})\sin(2.0x)\cos(4.0t).$$

(b) At $x = 0.45\,\text{m}$,

$$D = (0.40\,\text{m})\sin(0.90)\cos(4.0t) = (0.31\,\text{m})\cos(4.0t).$$

The maximum amplitude at this point is $D = 0.31\,\text{m}$ and occurs when the cosine = 1.

(c) These waves make a standing wave pattern, so both ends of the string must be nodes. Nodes occur every half wavelength, which for our string is

$$\frac{\lambda}{2} = \frac{1}{2}\frac{2\pi}{k} = \frac{\pi}{2.0}\,\text{m} = 1.57\,\text{m}.$$

If the string includes only one loop, its length is $L = 1.57\,\text{m}$. But since we aren't given more information, it could be twice as long, including two loops, $L = 3.14\,\text{m}$, or any integral number times 1.57 m, and still provide a standing wave pattern for these waves—see Fig. 15–28.

(d) The nodes occur at $x = 0$, $x = 1.57\,\text{m}$, and, if the string is longer than $L = 1.57\,\text{m}$, at $x = 3.14\,\text{m}$, 4.71 m, and so on. The maximum amplitude (antinode) is 0.40 m [from part (b) above] and occurs midway between the nodes. For $L = 1.57\,\text{m}$, there is only one antinode, at $x = 0.79\,\text{m}$.

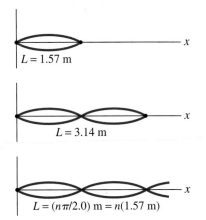

$L = 1.57\,\text{m}$

$L = 3.14\,\text{m}$

$L = (n\pi/2.0)\,\text{m} = n(1.57\,\text{m})$

FIGURE 15–28 Example 15–8: possible lengths for the string.

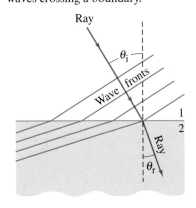

FIGURE 15–29 Refraction of waves crossing a boundary.

Ray

θ_i

Wave fronts

1

2

Ray

θ_r

* | 15–10 | Refraction†

When any wave strikes a boundary, some of the energy is reflected and some is transmitted or absorbed. When a two- or three-dimensional wave traveling in one medium crosses a boundary into a medium where its velocity is different, the transmitted wave may move in a different direction than the incident wave, as shown in Fig. 15–29. This phenomenon is known as **refraction**. One example is a water wave; the velocity decreases in shallow water and the waves refract, Fig. 15–30. [When the wave velocity changes gradually, as in Fig. 15–30, without a sharp boundary, the waves change direction (refract) gradually.] In Fig. 15–29, the velocity of the wave in medium 2 is less than in medium 1. In this case, the direction of the wave bends so it travels more nearly perpendicular to the boundary. That is, the *angle of refraction*, θ_r, is less than the *angle of incidence*, θ_i. To see why this is so, and to help us get a quantitative relation between θ_r and θ_i, let us think of each wave front as a row of soldiers. The soldiers are marching from firm

---

†This Section and the next are covered in more detail in Chapters 33 to 36, on optics.

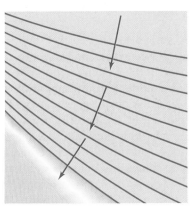
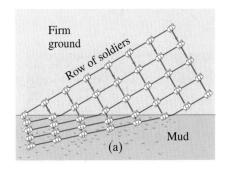

FIGURE 15–30 Water waves refracting as they approach the shore, where their velocity is less. There is no distinct boundary, as in Fig. 15–29, and the wave velocity changes gradually.

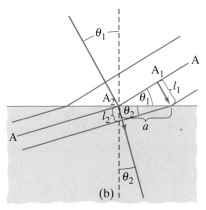

FIGURE 15–31 (a) Soldier analogy, to (b) derive law of refraction for waves.

ground (medium 1) into mud (medium 2) and hence are slowed down. The soldiers that reach the mud first are slowed down first and the row bends as shown in Fig. 15–31a. Let us consider the wave front (or row of soldiers) labeled A in Fig. 15–31b. In the same time t that A_1 moves a distance $l_1 = v_1 t$, we see that A_2 moves a distance $l_2 = v_2 t$. The two triangles shown (one includes θ_1 and l_1, the other θ_2 and l_2) have the side labeled a in common. Thus

$$\sin \theta_1 = \frac{l_1}{a} = \frac{v_1 t}{a}$$

and

$$\sin \theta_2 = \frac{l_2}{a} = \frac{v_2 t}{a}.$$

Dividing these two equations, we find that

$$\frac{\sin \theta_2}{\sin \theta_1} = \frac{v_2}{v_1}. \qquad\qquad \textbf{(15–19)} \qquad \textit{Law of refraction}$$

Since θ_1 is the angle of incidence (θ_i), and θ_2 is the angle of refraction (θ_r), Eq. 15–19 gives the quantitative relation between the two. Of course, if the wave were going in the opposite direction, the argument would not be changed. Only θ_1 and θ_2 would change roles: θ_1 would be the angle of refraction and θ_2 the angle of incidence. Clearly then, if the wave travels into a medium where it can move faster, it will bend in the opposite way, $\theta_r > \theta_i$. We see from Eq. 15–19 that if the velocity increases, the angle increases, and vice versa.

Earthquake waves refract within the Earth as they travel through rock of different densities (and therefore the velocity is different) just as water waves do. Light waves refract as well, and when we discuss light we shall find Eq. 15–19 very useful.

EXAMPLE 15–9 Refraction of earthquake wave. An earthquake P wave passes across a boundary in rock where its velocity increases from 6.5 km/s to 8.0 km/s. If it strikes this boundary at 30°, what is the angle of refraction?

Earthquake wave refraction

SOLUTION Since $\sin 30° = 0.50$, Eq. 15–19 yields

$$\sin \theta_2 = \frac{(8.0 \text{ m/s})}{(6.5 \text{ m/s})} (0.50) = 0.62$$

so $\theta_2 = 38°$.

FIGURE 15–32 Wave diffraction. The waves come from the upper left. Note how the waves, as they pass the obstacle, bend around behind it, into the "shadow region."

Waves spread as they travel, and when they encounter an obstacle they bend around it somewhat and pass into the region behind as shown in Fig. 15–32 for water waves. This phenomenon is called **diffraction**.

The amount of diffraction depends on the wavelength of the wave and on the size of the obstacle, as shown in Fig. 15–33. If the wavelength is much larger than the object, as with the grass blades of Fig. 15–33a, the wave bends around them almost as if they were not there. For larger objects, parts (b) and (c), there is more of a "shadow" region behind the obstacle where we might not expect the waves to penetrate—but they do, at least a little. But notice in part (d), where the obstacle is the same as in part (c) but the wavelength is longer, that there is more diffraction into the shadow region. As a rule of thumb, *only if the wavelength is smaller than the size of the object will there be a significant shadow region.* It is worth noting that this rule applies to *reflection* from an obstacle as well. Very little of a wave is reflected unless the wavelength is smaller than the size of the obstacle.

A rough guide to the amount of diffraction is

$$\theta(\text{radians}) \approx \frac{\lambda}{L}$$

where θ is roughly the angular spread of waves after passing through an opening of width L or around an obstacle of width L.

That waves can bend around obstacles, and thus can carry energy to areas behind obstacles, is clearly different from energy carried by material particles. A clear example is the following: if you are standing around a corner on one side of a building, you can't be hit by a baseball thrown from the other side, but you can hear a shout or other sound because the sound waves diffract around the edges.

FIGURE 15–33 Water waves passing objects of various sizes. Note that the larger the wavelength compared to the size of the object, the more diffraction there is into the "shadow region."

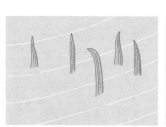   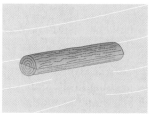

| (a) Water waves passing blades of grass | (b) Stick in water | (c) Short-wavelength waves passing log | (d) Long-wavelength waves passing log |

Summary

Vibrating objects act as sources of **waves** that travel outward from the source. Waves on water and on a string are examples. The wave may be a **pulse** (a single crest) or it may be continuous (many crests and troughs).

The **wavelength** of a continuous wave is the distance between two successive crests (or any two identical points on the wave shape).

The **frequency** is the number of full wavelengths (or crests) that pass a given point per unit time.

The **wave velocity** (how fast a crest moves) is equal to

the product of wavelength and frequency,

$$v = \lambda f.$$

The **amplitude** of a wave is the maximum height of a crest, or depth of a trough, relative to the normal (or equilibrium) level.

In a **transverse wave**, the oscillations are perpendicular to the direction in which the wave travels. An example is a wave on a string.

In a **longitudinal wave**, the oscillations are along (parallel to) the line of travel; sound is an example.

The velocity of both longitudinal and transverse waves in matter is proportional to the square root of an elastic force factor divided by an inertia factor (or density).

Waves carry energy from place to place without matter being carried. The energy transported by a wave, the power (energy transported per unit time), and the **intensity** of a wave (energy transported across unit area per unit time) are all proportional to the square of the amplitude of the wave.

For a wave traveling outward in three dimensions from a point source, the intensity (ignoring damping) decreases with the square of the distance from the source,

$$I \propto \frac{1}{r^2}.$$

The amplitude decreases linearly with distance from the source.

A one-dimensional transverse wave traveling to the right along the x axis (x increasing) can be represented by a formula for its amplitude as a function of position and time as

$$D(x, t) = D_M \sin\left[\left(\frac{2\pi}{\lambda}\right)(x - vt)\right] = D_M \sin(kx - \omega t)$$

where

$$k = \frac{2\pi}{\lambda} \quad \text{and} \quad \omega = 2\pi f.$$

If a wave is traveling toward decreasing values of x,

$$D(x, t) = D_M \sin(kx + \omega t).$$

When two or more waves pass through the same region of space at the same time, the displacement at any given point will be the vector sum of the displacements of the separate waves. This is the **principle of superposition**. It is valid for mechanical waves if the amplitudes are small enough that the restoring force of the medium is proportional to displacement.

Waves reflect off objects in their path. When a wave strikes a boundary between two materials in which it can travel, part of the wave is reflected and part is transmitted.

When the **wave front** of a two or three dimensional wave strikes an object, the angle of reflection equals the angle of incidence.

When two waves pass through the same region of space at the same time, they **interfere**. From the superposition principle, the resultant displacement at any point and time is the sum of their separate displacements. This can result in **constructive interference, destructive interference,** or something in between depending on the amplitudes and relative phases of the waves.

Waves traveling on a cord (or other medium) of fixed length interfere with waves that have reflected off the end and are traveling back in the opposite direction. At certain frequencies, **standing waves** can be produced in which the waves seem to be standing still rather than traveling. The cord (or other medium) is vibrating as a whole. This is a resonance phenomenon and the frequencies at which standing waves occur are called **resonant frequencies**. The points of destructive interference (no vibration) are called **nodes**. Points of constructive interference (maximum amplitude of vibration) are called **antinodes**.

The wavelengths of standing waves are given by $\lambda_n = 2L/n$ where n is an integer.

Questions

1. Is the frequency of a simple periodic wave equal to the frequency of its source? Why or why not?

2. Explain the difference between the speed of a transverse wave traveling down a rope and the speed of a tiny piece of the rope.

3. Why do the strings used for the lowest-frequency notes on a piano normally have wire wrapped around them?

4. What kind of waves do you think will travel down a horizontal metal rod if you strike its end (a) vertically from above and (b) horizontally parallel to its length?

5. Since the density of air decreases with an increase in temperature, but the bulk modulus B is nearly independent of temperature, how would you expect the speed of sound waves in air to vary with temperature?

6. Give examples, other than those already mentioned, of one-, two-, and three-dimensional waves.

7. The speed of sound in most solids is somewhat greater than in air, yet the density of solids is much greater $(10^3–10^4$ times). Explain.

8. Give two reasons why circular water waves decrease in amplitude as they travel away from the source.

9. Two linear waves have the same amplitude and otherwise are identical, except one has half the wavelength of the other. Which transmits more energy? By what factor?

10. The intensity of a sound in real-life situations does not decrease precisely with the square of the distance from the source as we might expect from Eq. 15–8. Why not?

11. Will any function of $(x - vt)$—see Eq. 15–14—represent a wave motion? Why or why not? If not, give an example.

12. When a sinusoidal wave crosses the boundary between two sections of rope as in Fig. 15–20, the frequency does not change (although the wavelength and velocity do change). Explain why.

13. If a sinusoidal wave on a two-section string (Fig. 15–20) is inverted upon reflection, does the transmitted wave have a longer or shorter wavelength?

14. Is energy always conserved when two waves interfere? Explain.

15. If a string is vibrating in three segments, are there any places one can touch it with a knife blade without disturbing the motion?

16. When a standing wave exists on a string, the vibrations of incident and reflected waves cancel at the nodes. Does this mean that energy was destroyed? Explain.

17. Why can you make water slosh back and forth in a pan only if you shake the pan at a certain frequency?

18. Can the amplitude of the standing waves in Fig. 15–26 be greater than the amplitude of the vibrations that cause them (up and down motion of the hand)?

19. When a cord is vibrated as in Fig. 15–26 by hand or by a mechanical vibrator, the "nodes" are not quite true nodes (at rest). Explain. [*Hint*: Consider damping and energy flow from hand or vibrator.]

* 20. AM radio signals can usually be heard behind a hill, but FM often cannot. That is, AM signals bend more than FM. Explain. (Radio signals, as we shall see, are carried by electromagnetic waves whose wavelength for AM is typically 200 to 600 m and for FM about 3 m.)

Problems

Sections 15–1 and 15–2

1. (I) A fisherman notices that wave crests pass the bow of his anchored boat every 4.0 s. He measures the distance between two crests to be 9.0 m. How fast are the waves traveling?

2. (I) AM radio signals have frequencies between 550 kHz and 1600 kHz (kilohertz) and travel with a speed of 3.0×10^8 m/s. What are the wavelengths of these signals? On FM the frequencies range from 88 MHz to 108 MHz (megahertz) and travel at the same speed. What are their wavelengths?

3. (I) A sound wave in air has a frequency of 262 Hz and travels with a speed of 330 m/s. How far apart are the wave crests (compressions)?

4. (I) Calculate the speed of longitudinal waves in (*a*) water and (*b*) granite.

5. (I) Determine the wavelength of a 5000-Hz sound wave traveling along an iron rod.

6. (II) A rope of mass 0.65 kg is stretched between two supports 30 m apart. If the tension in the rope is 120 N, how long will it take a pulse to travel from one support to the other?

7. (II) A 0.40-kg rope is stretched between two supports, 4.8 m apart. When one support is struck by a hammer, a transverse wave travels down the rope and reaches the other support in 0.85 s. What is the tension in the rope?

8. (II) A sailor strikes the side of his ship just below the surface of the sea. He hears the echo of the wave reflected from the ocean floor directly 3.5 s later. How deep is the ocean at this point?

9. (II) A gondola is connected to the top of a hill by a steel cable of length 600 m and diameter 1.5 cm. As the gondola comes to the end of its run, it bumps into the side and sends a wave pulse along the cable. It is observed that it took 16 s for the pulse to return. (*a*) What is the speed of the pulse? (*b*) What is the tension in the cable?

10. (II) The wave on a string shown in Fig. 15–34 is moving to the right with a speed of 1.80 m/s. (*a*) Draw the shape of the string 1.00 s later and indicate which parts of the string are moving up and which down at that instant. (*b*) Estimate the vertical speed of point A on the string at the instant shown in the figure.

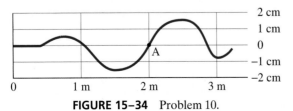

FIGURE 15–34 Problem 10.

11. (II) *S* and *P* waves from an earthquake travel at different speeds and this difference helps in the determination of the earthquake "epicenter" (where the disturbance took place). (*a*) Assuming typical speeds of 9.0 km/s and 5.5 km/s for *P* and *S* waves, respectively, how far away did the earthquake occur if a particular seismic station detects the arrival of these two types of waves exactly 94 s apart? (*b*) Is one seismic station sufficient to determine the position of the focus? Explain.

Section 15–3

12. (I) Two earthquake waves have the same frequency as they travel through the same portion of the Earth, but one is carrying twice the energy. What is the ratio of the amplitudes of the two waves?

13. (I) Compare (*a*) the intensities and (*b*) the amplitudes of an earthquake *P* wave as it passes two points 10 km and 20 km from the source.

14. (II) The intensity of a particular earthquake wave is measured to be 2.2×10^6 W/m² at a distance of 100 km from the source. (*a*) What was the intensity when it passed a point only 4.0 km from the source? (*b*) What was the total power passing through an area of 5.0 m² at a distance of 4.0 km?

15. (II) Show that if damping is ignored the amplitude D_M of circular water waves decreases as the square root of the distance r from the source: $D_M \propto 1/\sqrt{r}$.

16. (II) (*a*) Show that the intensity of a wave is equal to the energy density (energy per unit volume) in the wave times the wave speed. (*b*) What is the energy density 2.0 m from a 100-W light bulb? Light travels 3.0×10^8 m/s.

17. (II) A small steel wire of diameter 1.0 mm is connected to a vibrator and is under a tension of 4.5 N. The frequency of the vibrator is 60.0 Hz and it is observed that the amplitude of the wave on the steel wire is 0.50 cm. (*a*) What is the power output of the vibrator, assuming that the wave is not reflected back? (*b*) If the power output stays constant but the frequency is doubled, what is the amplitude of the wave?

18. (II) (a) Show that the average rate with which energy is transported along a cord by a mechanical wave of frequency f and amplitude D_M is

$$\bar{P} = 2\pi^2 \mu v f^2 D_M^2,$$

where v is the speed of the wave and μ is the mass per unit length of the cord. (b) If the cord is under a tension $F_T = 100\,N$ and has mass per unit length $0.10\,kg/m$, what power is required to transmit 120-Hz transverse waves of amplitude $2.0\,cm$?

Section 15–4

19. (I) Suppose at $t = 0$, a wave shape is represented by $D = D_M \sin(2\pi x/\lambda + \phi)$; that is, it differs from Eq. 15–9 by a constant phase factor ϕ. What then will be the equation for a wave traveling to the left along the x axis as a function of x and t?

20. (II) A transverse traveling wave on a cord is represented by $D = 0.48 \sin(5.6x + 84t)$ where D and x are in meters and t in seconds. For this wave determine (a) the wavelength, (b) frequency, (c) velocity (magnitude and direction), (d) amplitude, and (e) maximum and minimum speeds of particles of the cord.

21. (II) Consider a point on the string of Example 15–4 that is $1.00\,m$ from the left-hand end. Determine (a) the maximum velocity of this point, and (b) its maximum acceleration. (c) What is its velocity and acceleration at $t = 2.0\,s$?

22. (II) Show, for a sinusoidal transverse wave traveling on a string, that the slope of the string at any point x is equal to the ratio of the transverse speed of the particle to the speed of the wave at that point.

23. (II) A transverse wave pulse travels to the right along a string with a speed $v = 2.0\,m/s$. At $t = 0$ the shape of the pulse is given by the function

$$D = 0.45 \cos(3.0x + 1.2),$$

where D and x are in meters. (a) Plot D versus x at $t = 0$. (b) Determine a formula for the wave pulse at any time t assuming there are no frictional losses. (c) Plot $D(x, t)$ versus x at $t = 1.0\,s$. (d) Repeat parts (b) and (c) assuming the pulse is traveling to the left.

24. (II) A 440-Hz longitudinal wave in air has a speed of $345\,m/s$. (a) What is the wavelength? (b) How much time is required for the phase to change by $90°$ at a given point in space? (c) At a particular instant, what is the phase difference (in degrees) between two points $4.4\,cm$ apart?

25. (II) Write down the equation for the wave in Problem 24 if its amplitude is $0.020\,cm$ and at $t = 0$, $x = 0$, and $D = -0.020\,cm$.

26. (II) A sinusoidal wave traveling on a string in the negative x direction has amplitude $1.00\,cm$, wavelength $3.00\,cm$, and frequency $200\,Hz$. At $t = 0$, the particle of string at $x = 0$ is displaced a distance $D = 0.80\,cm$ above the origin and is moving upward. (a) Sketch the shape of the wave at $t = 0$ and (b) determine the function of x and t that describes the wave.

* Section 15–5

*27. (II) Determine if the function $D = D_M \sin kx \cos \omega t$ is a solution of the wave equation.

*28. (II) Show by direct substitution that the following functions satisfy the wave equation: (a) $D(x, t) = D_M \ln(x + vt)$; (b) $D(x, t) = (x - vt)^4$.

*29. (II) Show that the wave forms of Eqs. 15–13 and 15–15 satisfy the wave equation, Eq. 15–16.

*30. (II) Let two linear waves be represented by $D_1 = f_1(x, t)$ and $D_2 = f_2(x, t)$. If both these waves satisfy the wave equation (Eq. 15–16), show that any combination $D = C_1 D_1 + C_2 D_2$ does as well, where C_1 and C_2 are constants.

Section 15–7

31. (II) Consider a sine wave traveling down the stretched two-part cord of Fig. 15–20. Determine a formula (a) for the ratio of the speeds of the wave in the two sections, v_2/v_1, and (b) for the ratio of the wavelengths in the two sections. (The frequency is the same in both sections. Why?) (c) Is the wavelength larger in the heavier cord or the lighter?

32. (II) A cord has two sections with linear densities of $0.10\,kg/m$ and $0.20\,kg/m$, Fig. 15–35. An incident wave, given by $D = (0.050\,m) \sin(6.0x - 12.0t)$, where x is in meters and t in seconds, travels along the lighter cord. (a) What is the wavelength on the lighter section of the cord? (b) What is the tension in the cord? (c) What is the wavelength when the wave travels on the heavier section?

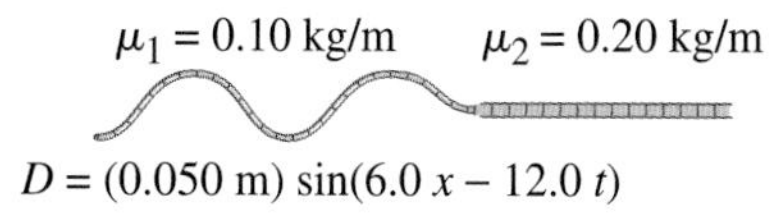

$\mu_1 = 0.10\,kg/m \qquad \mu_2 = 0.20\,kg/m$

$D = (0.050\,m) \sin(6.0\,x - 12.0\,t)$

FIGURE 15–35 Problem 32.

33. (III) A cord stretched to a tension F_T, consists of two sections (as in Fig. 15–21) whose linear densities are μ_1 and μ_2. Take $x = 0$ to be the point (a knot) where they are joined, with μ_1 referring to that section of cord to the left and μ_2 that to the right. A sinusoidal wave, $D = A \sin[k_1(x - v_1 t)]$, starts at the left end of the cord. When it reaches the knot, part of it is reflected and part is transmitted. Let the equation of the reflected wave be $D = A_R \sin[k_1(x + v_1 t)]$ and that for the transmitted wave be $D = A_T \sin[k_2(x - v_2 t)]$. Since the frequency must be the same in both sections, we have $\omega_1 = \omega_2$ or $k_1 v_1 = k_2 v_2$. (a) Using the fact that the rope is continuous—so that a point an infinitesimal distance to the left of the knot has the same displacement at any moment (due to incident plus reflected waves) as a point just to the right of the knot (due to the transmitted wave)— show that $A = A_T + A_R$. (b) Assuming that the slope $(\partial D/\partial x)$ of the string just to the left of the knot is the same as the slope just to the right of the knot, show that the amplitude of the reflected wave is given by

$$A_R = \left(\frac{v_1 - v_2}{v_1 + v_2}\right) A = \left(\frac{k_2 - k_1}{k_2 + k_1}\right) A.$$

(c) What is A_T in terms of A?

Section 15–8

34. (I) The two pulses shown in Fig. 15–36 are moving toward each other. (*a*) Sketch the shape of the string at the moment they directly overlap. (*b*) Sketch the shape of the string a few moments later. (*c*) In Fig. 15–23a, at the moment the pulses pass each other, the string is straight. What has happened to the energy at this moment?

FIGURE 15–36 Problem 34.

35. (II) Suppose two linear waves of equal amplitude and frequency have a phase difference ϕ as they travel in the same medium. They can be represented by

$$D_1 = D_M \sin(kx - \omega t)$$

$$D_2 = D_M \sin(kx - \omega t + \phi).$$

(*a*) Use the trigonometric identity $\sin\theta_1 + \sin\theta_2 = 2\sin\frac{1}{2}(\theta_1 + \theta_2)\cos\frac{1}{2}(\theta_1 - \theta_2)$ to show that the resultant wave is given by

$$D = \left(2D_M \cos\frac{\phi}{2}\right)\sin\left(kx - \omega t + \frac{\phi}{2}\right).$$

(*b*) What is the amplitude of this resultant wave? Is the wave purely sinusoidal, or not? (*c*) Show that constructive interference occurs if $\phi = 0, 2\pi, 4\pi$, and so on, and destructive interference occurs if $\phi = \pi, 3\pi, 5\pi$, etc. (*d*) Describe the resultant wave, by equation and in words, if $\phi = \pi/2$.

Section 15–9

36. (I) A violin string vibrates at 294 Hz when unfingered. At what frequency will it vibrate if it is fingered one-fourth of the way down from the end?

37. (I) If a violin string vibrates at 440 Hz as its fundamental frequency, what are the frequencies of the first four harmonics?

38. (I) A particular string resonates in four loops at a frequency of 264 Hz. Give at least three other frequencies at which it will resonate.

39. (I) In an earthquake, it is noted that a footbridge oscillated up and down in a one loop (fundamental standing wave) pattern once every 2.0 s. What other possible resonant periods of motion are there for this bridge? What frequencies do they correspond to?

40. (II) The velocity of waves on a string is 270 m/s. If the frequency of standing waves is 131 Hz, how far apart are the nodes?

41. (II) If two successive harmonics of a vibrating string are 280 Hz and 350 Hz, what is the frequency of the fundamental?

42. (II) A guitar string is 90.0 cm long and has a mass of 3.6 g. From the bridge to the support post ($= L$) is 60.0 cm and the string is under a tension of 520 N. What are the frequencies of the fundamental and first two overtones?

43. (II) Show that the frequency of standing waves on a string of length L and linear density μ, which is stretched to a tension F_T, is given by

$$f = \frac{n}{2L}\sqrt{\frac{F_T}{\mu}}$$

where n is an integer.

44. (II) One end of a horizontal string of linear density 4.8×10^{-4} kg/m is attached to a small amplitude mechanical 60-Hz vibrator. The string passes over a pulley, a distance $L = 1.40$ m away, and weights are hung from this end. What mass must be hung from this end of the string to produce (*a*) one loop, (*b*) two loops, and (*c*) five loops of a standing wave? Assume the string at the vibrator is a node, which is nearly true. Why can the amplitude of the standing wave be much greater than the vibrator amplitude?

45. (II) In Problem 44, the length of the string may be adjusted by moving the pulley. If the hanging mass is fixed at 0.080 kg, how many different standing wave patterns may be achieved by varying L between 10 cm and 1.5 m?

46. (II) The displacement of a standing wave is given by $D = 8.6 \sin(0.60x)\cos(58t)$, where x and D are in centimeters and t is in seconds. (*a*) What is the distance (cm) between nodes? (*b*) Give the amplitude, frequency, and speed of each of the component waves. (*c*) Find the speed of a particle of the string at $x = 3.20$ cm when $t = 2.5$ s.

47. (II) The displacement of a transverse wave traveling on a string is represented by $D = 4.2 \sin(0.71x - 47t + 2.1)$, where D and x are in cm and t in s. (*a*) Find an equation that represents a wave which, when traveling in the opposite direction, will produce a standing wave when added to this one. (*b*) What is the equation describing the standing wave?

48. (II) When you slosh the water back and forth in a tub at just the right frequency, the water alternately rises and falls at each end, remaining relatively calm at the center. Suppose the frequency to produce such a standing wave in a 60-cm-wide tub is 0.85 Hz. What is the speed of the water wave?

49. (II) A particular violin string plays at a frequency of 294 Hz. If the tension is increased 10 percent what will the new frequency be?

50. (II) Two traveling waves are described by the functions

$$D_1 = D_M \sin(kx - \omega t)$$

$$D_2 = D_M \sin(kx + \omega t),$$

where $D_M = 0.15$ m, $k = 3.5$ m^{-1}, and $\omega = 1.2$ s^{-1}. (*a*) Plot these two waves, from $x = 0$ to a point $x(> 0)$ that includes one full wavelength. Choose $t = 1.0$ s. Plot the sum of the two waves and identify the nodes and antinodes in the plot, and compare to the analytic (mathematical) representation.

51. (II) Plot the two waves given in Problem 50, and their sum, as a function of time from $t = 0$ to $t = T$ (one period). Choose (*a*) $x = 0$ and (*b*) $x = \lambda/4$. Interpret your results.

52. (II) A standing wave on a 1.80-m-long horizontal string displays three loops when the string vibrates at 120 Hz. The maximum swing of the string (top to bottom) at the center of each loop is 12.0 cm. (*a*) What is the function describing the standing wave? (*b*) What are the functions describing the two equal-amplitude waves traveling in opposite directions that make up the standing wave?

* 53. (I) An earthquake P wave traveling 8.0 km/s strikes a boundary within the Earth between two kinds of material. If it approaches the boundary at an incident angle of 50° and the angle of refraction is 31°, what is the speed in the second medium?

* 54. (I) Water waves approach an underwater "shelf" where the velocity changes from 2.8 m/s to 2.5 m/s. If the incident wave crests make a 40° angle with the shelf, what will be the angle of refraction?

* 55. (II) A longitudinal earthquake wave strikes a boundary between two types of rock at a 25° angle. As it crosses the boundary, the specific gravity of the rock changes from 3.7 to 2.8. Assuming that the elastic modulus is the same for both types of rock, determine the angle of refraction.

* 56. (II) It is found for any type of wave, say an earthquake wave, that if it reaches a boundary beyond which its speed is increased, there is a maximum incident angle if there is to be a transmitted refracted wave. This maximum incident angle θ_{iM} corresponds to an angle of refraction equal to 90°. If $\theta_i > \theta_{iM}$, all the wave is reflected at the boundary and none is refracted (because this would correspond to $\sin \theta_r > 1$, where θ_r is the angle of refraction, which is impossible). This phenomenon is referred to as *total internal reflection*. (a) Find a formula for θ_{iM} using Eq. 15–19. (b) At what angles of incidence will there be only reflection and no transmission for an earthquake P wave traveling 7.5 km/s where it reaches a different kind of rock where its speed is 9.3 km/s?

* 57. (II) A sound wave is traveling in warm air when it hits a layer of cold dense air. If the sound wave hits the cold air interface at an angle of 25°, what is the angle of refraction? Assume that the cold air temperature is −10°C and the warm air temperature is +10°C. The speed of sound as a function of temperature can be approximated by $v = (331 + 0.60\,T)$ m/s, where T is in °C.

General Problems

58. When you walk with a cup of coffee (diameter 8 cm) at just the right pace of about 1 step per second, the coffee builds up its "sloshing" until eventually, after a few steps, it starts to spill over the top. What is the speed of the waves in the coffee?

59. Two solid rods have the same bulk modulus but one is twice as dense as the other. In which rod will the speed of longitudinal waves be greater, and by what factor?

60. Two waves traveling along a stretched string have the same frequency, but one transports three times the power of the other. What is the ratio of the amplitudes of the two waves?

61. A bug on the surface of a pond is observed to move up and down a total vertical distance of 0.10 m, lowest to highest point, as a wave passes. (a) What is the amplitude of the wave? (b) If the ripples increase to 0.15 m, by what factor does the bug's maximum kinetic energy change?

62. A particular guitar string is supposed to vibrate at 200 Hz, but it is measured to actually vibrate at 205 Hz. By what percentage should the tension in the string be changed to get the frequency to the correct value?

63. An earthquake-produced surface wave can be approximated by a sinusoidal transverse wave. Assuming a frequency of 0.50 Hz (typical of earthquakes, which actually include a mixture of frequencies), what minimum amplitude will cause objects to leave contact with the ground?

64. A uniform cord of length L and mass m is hung vertically from a support. (a) Show that the speed of transverse waves in this cord is $\sqrt{gh}$, where h is the height above the lower end. (b) How long does it take for a pulse to travel upward from one end to the other?

65. A transverse wave pulse travels to the right along a string with a speed $v = 3.0$ m/s. At $t = 0$ the shape of the pulse is given by the function

$$D = \frac{4.0}{x^2 - 2.0},$$

where D and x are in meters. (a) Plot D versus x at $t = 0$. (b) Determine a formula for the wave pulse at any time t assuming there are no frictional losses. (c) Plot $D(x, t)$ versus x at $t = 0.50$ s. (d) Repeat parts (b) and (c) assuming the pulse is traveling to the left.

66. (a) Show that if the tension in a stretched string is changed by a small amount ΔF_T, the frequency of the fundamental is changed by an amount $\Delta f = \frac{1}{2}(\Delta F_T / F_T)f$. (b) By what percent must the tension in a piano string be increased or decreased to raise the frequency from 438 Hz to 442 Hz. (c) Does the formula in part (a) apply to the overtones as well?

67. Two strings on a musical instrument are tuned to play at 392 Hz (G) and 440 Hz (A). (a) What are the first two overtones for each string? (b) If the two strings have the same length and are under the same tension, what must be the ratio of their masses (M_G / M_A)? (c) If the strings, instead, have the same mass per unit length and are under the same tension, what is the ratio of their lengths (L_G / L_A)? (d) If their masses and lengths are the same, what must be the ratio of the tensions in the two strings?

68. A highway overpass was observed to resonate as one full loop $(\frac{1}{2}\lambda)$ when a small earthquake shook the ground vertically at 4.0 Hz. The highway department put a support at the center of the overpass, anchoring it to the ground as shown in Fig. 15–37. What resonant frequency would you now expect for the overpass? It is noted that earthquakes rarely do significant shaking above 5 or 6 Hz. Did the modifications do any good? Explain.

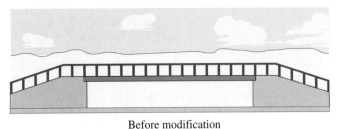

Before modification

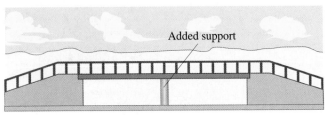

Added support

After modification

FIGURE 15–37   Problem 68.

69. A string can have a "free" end if that end is attached to a ring that can slide without friction on a vertical pole (Fig. 15–38). Determine the wavelengths of the resonant vibrations of such a string with one end fixed and the other free.

Fixed end

Free end

FIGURE 15–38
Problem 69.

70. A string fixed at two ends is pulled up at the center into the triangular shape shown in Fig. 15–39. Assuming that the tension F_T remains constant, calculate the energy of the vibrations of the string when it is released. [*Hint*: What work does it take to stretch the string up?]

FIGURE 15–39
Problem 70.

71. Figure 15–40 shows the wave shape of a sinusoidal wave traveling to the right at two instants of time. What is the mathematical representation of this wave?

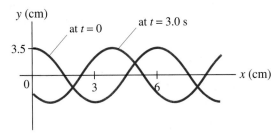

FIGURE 15–40   Problem 71.

72. Estimate the average power of a water wave when it hits the chest of an adult standing in the water at the seashore. Assume that the amplitude of the wave is 0.50 m, the wavelength is 2.5 m, and the period is 5.0 s.

73. Two wave pulses are traveling in opposite directions with the same speed of 5.0 cm/s as shown in Fig. 15–41. At $t = 0$, the leading edges of the two pulses are 15 cm apart. Sketch the wave pulses at $t = 1.0, 2.0$ and 3.0 s.

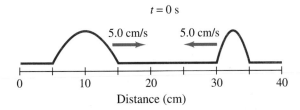

FIGURE 15–41   Problem 73.

74. For a spherical wave traveling uniformly away from a point source, show that the displacement can be represented by

$$D = \left(\frac{A}{r}\right)\sin(kr - \omega t),$$

where r is the radial distance from the source and A is a constant.

If music be the food of physics, play on.

This orchestra contains stringed instruments whose sound depends on transverse standing waves on strings, and wind instruments whose sound comes from longitudinal standing waves in an air column. Percussive instruments create more complicated standing waves. The human voice utilizes vocal chords (almost like strings) and the throat cavity (vibrating air) to produce its particular timbre.

We also study the decibel intensity scale, the ear's response, sound interference and the Doppler effect.

CHAPTER 16

Sound

S ound is associated with our sense of hearing and, therefore, with the physiology of our ears and the psychology of our brain which interprets the sensations that reach our ears. The term *sound* also refers to the physical sensation that stimulates our ears: namely, longitudinal waves.

We can distinguish three aspects of any sound. First, there must be a *source* for a sound; and as with any wave, the source of a sound wave is a vibrating object. Second, the energy is transferred from the source in the form of longitudinal sound *waves*. And third, the sound is *detected* by an ear or an instrument. Sound is an important factor in the design of buildings, especially theaters and auditoriums, but also in factories and other work places. We start this Chapter by looking at some aspects of sound waves themselves.

16–1 | Characteristics of Sound

We already saw in Chapter 15, Fig. 15–6, how a vibrating drumhead produces a sound wave in air. Indeed, we usually think of sound waves traveling in the air, for normally it is the vibrations of the air that force our eardrums to vibrate. But sound waves can also travel in other materials.

Two stones struck together under water can be heard by a swimmer beneath the surface, for the vibrations are carried to the ear by the water. When you put your ear flat against the ground, you can hear an approaching train or truck. In this case the ground does not actually touch your eardrum, but the longitudinal wave transmitted by the ground is called a sound wave just the same, for its vibrations cause the outer ear and the air within it to vibrate. Clearly, sound cannot travel in

TABLE 16–1   Speed of
Sound in Various Materials,
at 20°C and 1 atm

Material	Speed (m/s)
Air	343
Air (0°C)	331
Helium	1005
Hydrogen	1300
Water	1440
Seawater	1560
Iron and steel	≈ 5000
Glass	≈ 4500
Aluminum	≈ 5100
Hardwood	≈ 4000

➡ PHYSICS APPLIED

How far away is the lightning?

Loudness

Pitch

Audible frequency range

➡ PHYSICS APPLIED

Autofocusing camera

the absence of matter. For example, a bell ringing inside an evacuated jar cannot be heard, nor does sound travel through the empty reaches of outer space.

The **speed of sound** is different in different materials. In air at 0°C and 1 atm, sound travels at a speed of 331 m/s. We saw in Eq. 15–4 $(v = \sqrt{B/\rho})$ that the speed depends on the elastic modulus, B, and the density, ρ, of the material. Thus for helium, whose density is much less than that of air but whose elastic modulus is not greatly different, the speed is about three times as great as in air. In liquids and solids, which are much less compressible and therefore have much greater elastic moduli, the speed is larger still. The speed of sound in various materials is given in Table 16–1. The values depend somewhat on temperature, but this is significant mainly for gases. For example, in air near room temperature, the speed increases approximately 0.60 m/s for each Celsius degree increase in temperature:

$$v \approx (331 + 0.60\,T)\ \text{m/s}, \qquad \text{[Speed of sound in air]}$$

where T is the temperature in °C. Unless stated otherwise, we will assume in this chapter that $T = 20°C$, so that $v = [331 + (0.60)(20)]\ \text{m/s} = 343\ \text{m/s}$.

CONCEPTUAL EXAMPLE 16–1 **Distance from a lightning strike.** A rule of thumb that tells how close lightning has hit is: "one mile for every five seconds before the thunder is heard." Justify, noting that the speed of light is so high $(3 \times 10^8\ \text{m/s})$ that the time for light to travel is negligible compared to the time for sound.

RESPONSE The speed of sound in air is about 340 m/s, so to travel 1 km = 1000 m takes about 3 seconds. One mile is about 1.6 kilometers, so the time for the thunder to travel a mile is about $(1.6)(3) \approx 5$ seconds.

Two aspects of any sound are immediately evident to a human listener. These are "loudness" and "pitch," and each refers to a sensation in the consciousness of the listener. But to each of these subjective sensations there corresponds a physically measurable quantity. **Loudness** is related to the energy in the sound wave, and we shall discuss it in Section 16–3.

The **pitch** of a sound refers to whether it is high, like the sound of a piccolo or violin, or low, like the sound of a bass drum or string bass. The physical quantity that determines pitch is the frequency, as was first noted by Galileo. The lower the frequency, the lower the pitch, and the higher the frequency, the higher the pitch.[†] The human ear responds to frequencies in the range from about 20 Hz to about 20,000 Hz. (Recall that 1 Hz is 1 cycle per second.) This is called the **audible range**. These limits vary somewhat from one individual to another. One general trend is that as people age, they are less able to hear the high frequencies, so that the high-frequency limit may be 10,000 Hz or less.

Sound waves whose frequencies are outside the audible range may reach the ear, but we are not generally aware of them. Frequencies above 20,000 Hz are called **ultrasonic** (do not confuse with *supersonic*, which is used for an object moving with a speed faster than the speed of sound). Many animals can hear ultrasonic frequencies; dogs, for example, can hear sounds as high as 50,000 Hz, and bats can detect frequencies as high as 100,000 Hz.

EXAMPLE 16–2 **Autofocusing with sound waves.** Autofocusing cameras emit a pulse of very high frequency (ultrasonic) sound that travels to the object being photographed, and include a sensor that detects the returning reflected sound, as shown in Fig. 16–1. To get an idea of the time sensitivity of the detector, calculate the travel time of the pulse for an object (*a*) 1.0 m away, (*b*) 20 m away.

[†] Although pitch is determined mainly by frequency, it also depends to a slight extent on loudness. For example, a very loud sound may seem slightly lower in pitch than a quiet sound of the same frequency.

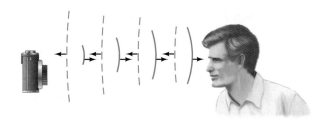

SOLUTION We assume the temperature is about 20°C, so the speed of sound, as calculated above, is 343 m/s. (*a*) The pulse travels 1.0 m to the object and 1.0 m back, for a total of 2.0 m. Since speed = distance/time, we have

$$t = \frac{\text{distance}}{\text{speed}} = \frac{2.0 \text{ m}}{343 \text{ m/s}} = 0.0059 \text{ s} = 5.9 \text{ ms}.$$

(*b*) The total distance now is $2 \times 20 \text{ m} = 40 \text{ m}$, so

$$t = \frac{40 \text{ m}}{343 \text{ m/s}} = 0.12 \text{ s} = 120 \text{ ms}.$$

Sound waves whose frequencies are below the audible range (that is, less than 20 Hz) are called **infrasonic**. Sources of infrasonic waves include earthquakes, thunder, volcanoes, and waves produced by vibrating heavy machinery. This last source can be particularly troublesome to workers, for infrasonic waves—even though inaudible—can cause damage to the human body. These low-frequency waves act in a resonant fashion, causing considerable motion and irritation of internal organs of the body.

16–2 Mathematical Representation of Longitudinal Waves

In Section 15–4, we saw that a one-dimensional sinusoidal wave traveling along the *x* axis can be represented by the relation (Eq. 15–10c)

$$D = D_M \sin(kx - \omega t). \tag{16–1}$$

Here the wave number *k* is related to the wavelength λ by $k = 2\pi/\lambda$, and $\omega = 2\pi f$ where *f* is the frequency; *D* is the displacement at position *x* and time *t*, and D_M is its maximum value, the *amplitude*. For a transverse wave—such as a wave on a string—the displacement *D* is perpendicular to the direction of wave propagation along the *x* axis. But for a longitudinal wave the displacement *D* is *along the direction of wave propagation*. That is, *D* is parallel to *x* and represents the displacement of a tiny volume element of the medium from its equilibrium position.

Longitudinal (sound) waves can also be considered from the point of view of variations in pressure rather than displacement. Indeed, longitudinal waves are often called **pressure waves**. The pressure variation is usually easier to measure than the displacement (see Example 16–4). As can be seen in Fig. 16–2, in a wave "compression" (where molecules are closest together), the pressure is higher than normal, whereas in an expansion (or rarefaction) the pressure is less than normal. Figure 16–3 shows a graphical representation of a sound wave in air in terms of (a) displacement and (b) pressure. Note that the displacement wave is a quarter wavelength, or 90° ($\pi/2$ rad), out of phase with the pressure wave: where the pressure is a maximum or minimum, the molecules are momentarily at rest in equilibrium so the displacement from equilibrium is zero; and where the pressure variation is zero, the displacement is a maximum or minimum.

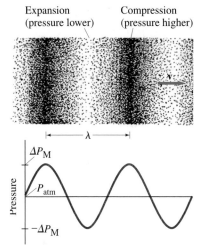

FIGURE 16–2 Longitudinal sound wave traveling to the right, and its graphical representation in terms of pressure.

FIGURE 16–3 Representation of a sound wave in terms of (a) displacement and (b) pressure.

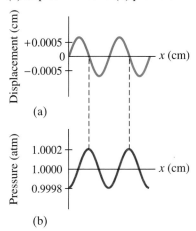

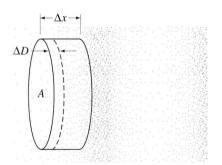

FIGURE 16–4 Longitudinal wave in a fluid moves to the right. A thin layer of fluid, in a thin cylinder of area A and thickness Δx, changes in volume as a result of pressure variation as the wave passes. At the moment shown, the pressure will increase as the wave moves to the right, so the thickness of our layer will decrease, by an amount ΔD.

Pressure Wave Derivation

Let us now derive the mathematical representation of the pressure variation in a traveling longitudinal wave. From the definition of the bulk modulus, B (Eq. 12–7a),

$$\Delta P = -B(\Delta V / V),$$

where $\Delta V / V$ is the fractional change in volume of the medium due to a pressure change ΔP, and ΔP represents the pressure difference from the normal pressure P_0 (no wave present). The negative sign reflects the fact that the volume decreases ($\Delta V < 0$) if the pressure is increased. Consider now a layer of fluid through which the longitudinal wave is passing (Fig. 16–4). If this layer has thickness Δx and area A, then its volume is $V = A\,\Delta x$. As a result of pressure variation in the wave, the volume will change by an amount $\Delta V = A\,\Delta D$ where ΔD is the change in thickness of this layer as it compresses or expands. (Remember that D represents the displacement of the medium.) Thus we have

$$\Delta P = -B\,\frac{A\,\Delta D}{A\,\Delta x}.$$

To be precise, we take the limit of $\Delta x \to 0$, so we obtain

$$\Delta P = -B\,\frac{\partial D}{\partial x}, \tag{16–2}$$

where we use the partial derivative notation since D is a function of both x and t. If the displacement D is sinusoidal as given by Eq. 16–1, then we have from Eq. 16–2 that

$$\Delta P = -(B D_M k)\cos(kx - \omega t). \tag{16–3}$$

Thus the pressure varies sinusoidally as well, but is out of phase from the displacement by 90° or a quarter wavelength; see Fig. 16–3. The quantity $B D_M k$ is called the **pressure amplitude**, ΔP_M. It represents the maximum and minimum amounts by which the pressure varies from the normal ambient pressure. We can thus write

$$\Delta P = -\Delta P_M \cos(kx - \omega t), \tag{16–4}$$

where, using $v = \sqrt{B/\rho}$ (Eq. 15–4), and $k = \omega/v = 2\pi f/v$ (Eq. 15–12), then

$$\begin{aligned}
\Delta P_M &= B D_M k \\
&= \rho v^2 D_M k \\
&= 2\pi \rho v D_M f. \tag{16–5}
\end{aligned}$$

16–3 Intensity of Sound; Decibels

Intensity

Like pitch, **loudness** is a sensation in the consciousness of a human being. It too is related to a physically measurable quantity, the **intensity** of the wave. Intensity is defined as the energy transported by a wave per unit time across unit area perpendicular to the energy flow, and, as we saw in the previous chapter, is proportional to the square of the wave amplitude. Intensity has units of power per unit area, or watts/meter² (W/m^2).

The human ear can detect sounds with an intensity as low as $10^{-12}\,\text{W/m}^2$ and as high as $1\,\text{W/m}^2$ (and even higher, although above this it is painful). This is an incredibly wide range of intensity, spanning a factor of 10^{12} from lowest to highest. Presumably because of this wide range, what we perceive as loudness is not directly proportional to the intensity. To produce a sound that seems twice as loud requires a sound wave that has about ten times the intensity. This is roughly true at any sound level for frequencies near the middle of the audible range. For example, a sound wave of intensity $10^{-2}\,\text{W/m}^2$ sounds to an average human like it is about twice as loud as one whose intensity is $10^{-3}\,\text{W/m}^2$, and four times as loud as $10^{-4}\,\text{W/m}^2$.

Sound Level

Because of this relationship between the subjective sensation of loudness and the physically measurable quantity "intensity," it is usual to specify sound intensity levels using a logarithmic scale. The unit on this scale is a **bel**, after the inventor, Alexander Graham Bell, or much more commonly, the **decibel** (dB) which is $\frac{1}{10}$ bel (10 dB = 1 bel). The **sound level**, β, of any sound is defined in terms of its intensity, I, as

$$\beta \text{ (in dB)} = 10 \log \frac{I}{I_0}, \qquad (16\text{–}6)$$

Sound level (decibels)

where I_0 is the intensity of some reference level and the logarithm is to the base 10. I_0 is usually taken as the minimum intensity audible to an average person, the "threshold of hearing," which is $I_0 = 1.0 \times 10^{-12} \text{ W/m}^2$. Thus, for example, the sound level of a sound whose intensity $I = 1.0 \times 10^{-10} \text{ W/m}^2$ will be

$$\beta = 10 \log \left(\frac{1.0 \times 10^{-10} \text{ W/m}^2}{1.0 \times 10^{-12} \text{ W/m}^2} \right) = 10 \log 100 = 20 \text{ dB},$$

since log 100 is equal to 2.0. Notice that the sound level at the threshold of hearing is 0 dB. That is, $\beta = 10 \log 10^{-12}/10^{-12} = 10 \log 1 = 0$ since log 1 = 0. Notice too that an increase in intensity by a factor of 10 corresponds to a level increase of 10 dB. An increase in intensity by a factor of 100 corresponds to a level increase of 20 dB. Thus a 50-dB sound is 100 times more intense than a 30-dB sound, and so on.

The intensities and sound levels for a number of common sounds are listed in Table 16–2.

TABLE 16–2 Intensity of Various Sounds

Source of the Sound	Sound Level (dB)	Intensity (W/m^2)
Jet plane at 30 m	140	100
Threshold of pain	120	1
Loud rock concert	120	1
Siren at 30 m	100	1×10^{-2}
Auto interior, at 90 km/h	75	3×10^{-5}
Busy street traffic	70	1×10^{-5}
Conversation, at 50 cm	65	3×10^{-6}
Quiet radio	40	1×10^{-8}
Whisper	20	1×10^{-10}
Rustle of leaves	10	1×10^{-11}
Threshold of hearing	0	1×10^{-12}

EXAMPLE 16–3 Loudspeaker response. A high-quality loudspeaker is advertised to reproduce, at full volume, frequencies from 30 Hz to 18,000 Hz with uniform intensity ± 3 dB. That is, over this frequency range, the sound level does not vary by more than 3 dB from the average. By what factor does the intensity change for the maximum sound level change of 3 dB?

SOLUTION Let us call the average intensity I_1 and the average level β_1. Then the maximum intensity, I_2, corresponds to a level $\beta_2 = \beta_1 + 3$ dB. Thus

$$\beta_2 - \beta_1 = 10 \log \frac{I_2}{I_0} - 10 \log \frac{I_1}{I_0}$$

$$3 \text{ dB} = 10 \left(\log \frac{I_2}{I_0} - \log \frac{I_1}{I_0} \right)$$

$$= 10 \log \frac{I_2}{I_1}$$

because $(\log a - \log b) = \log a/b$. Then

$$\log \frac{I_2}{I_1} = 0.30, \quad \text{or} \quad \frac{I_2}{I_1} = 10^{0.30} = 2.0,$$

so ± 3 dB corresponds to a doubling (or halving) of the intensity.

➡ **PHYSICS APPLIED**

Loudspeaker response (± 3 dB)

It is worth noting that a sound level difference of 3 dB (which corresponds to a doubled intensity as we just saw) corresponds to only a very small change in the subjective sensation of apparent loudness. Indeed, the average human can distinguish a difference in level of only about 1 or 2 dB.

The Ear's Response

Sensitivity of the ear

The ear is not equally sensitive to all frequencies. To hear the same loudness for sounds of different frequencies requires different intensities. Studies averaged over large numbers of people have produced the curves shown in Fig. 16–5. On this graph, each curve represents sounds that seemed to be equally loud. The

Loudness (in "phons")

number labeling each curve represents the **loudness level** (the units are called *phons*), which is numerically equal to the sound level in dB at 1000 Hz. For example, the curve labeled 40 represents sounds that are heard by an average person to have the same loudness as a 1000-Hz sound with a sound level of 40 dB. From this 40-phon curve, we see that a 100-Hz tone must be at a level of about 62 dB to sound as loud as a 1000-Hz tone of only 40 dB. The lowest curve in Fig. 16–5 (labeled 0) represents the sound level, as a function of frequency, for the softest sound that is just audible by a very good ear (the "threshold of hearing"). Note that the ear is most sensitive to sounds of frequency between 2000 and 4000 Hz, which are common in speech and music. Note too that whereas a 1000-Hz sound is audible at a level of 0 dB, a 100-Hz sound must be at least 40 dB to be heard.

The top curve in Fig. 16–5, labeled 120, represents the "threshold of feeling or pain." Sounds above this level can actually be felt and cause pain.

Figure 16–5 shows that at lower intensity levels, our ears are less sensitive to the high and low frequencies relative to middle frequencies. The "loudness" control on stereo systems is intended to compensate for this. As the volume is turned down, the loudness control boosts the high and low frequencies relative to the middle frequencies so that the sound will have a more "normal-sounding" frequency balance. Many listeners, however, find the sound more pleasing or natural without the loudness control.

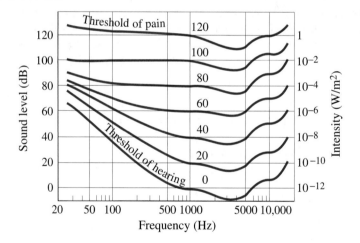

FIGURE 16–5 Sensitivity of the human ear as a function of frequency (see text). Note that the frequency scale is "logarithmic" in order to cover a wide range of frequencies.

Intensity Related to Amplitude

The intensity I is proportional to the square of the amplitude as we saw in Chapter 15. Indeed, using Eq. 15–7, $I = 2\pi^2 v\rho f^2 D_M^2$, we can relate the amplitude quantitatively to the intensity I or level β as the following Example shows.

EXAMPLE 16–4 **How tiny the displacement is.** (*a*) Calculate the maximum displacement of air molecules for a sound at the threshold of hearing, having a frequency of 1000 Hz. (*b*) Determine the maximum pressure variation in such a sound wave.

SOLUTION (*a*) The "threshold of hearing" at 1000 Hz is (Fig. 16–5) about 0 dB or 1.0×10^{-12} W/m². We use Eq. 15–7 of Chapter 15 and solve for D_M:

$$D_M = \frac{1}{\pi f} \sqrt{\frac{I}{2\rho v}}$$

$$= \frac{1}{(3.14)(1.0 \times 10^3 \, \text{s}^{-1})} \sqrt{\frac{1.0 \times 10^{-12} \, \text{W/m}^2}{(2)(1.29 \, \text{kg/m}^3)(343 \, \text{m/s})}} = 1.1 \times 10^{-11} \, \text{m},$$

where we have taken the density of air to be $1.29 \, \text{kg/m}^3$ and the speed of sound in air (assumed 20°C) as 343 m/s. We see how incredibly sensitive the human ear is: It can detect displacements of air molecules which are actually less than the diameter of atoms (about 10^{-10} m).

Ear detects displacements smaller than size of atoms

(*b*) Now we are dealing with sound as a pressure wave (Section 16–2). From Eq. 16–5,

$$\Delta P_M = 2\pi \rho v D_M f = 3.1 \times 10^{-5} \, \text{Pa}$$

or 3.1×10^{-10} atm. Again we see that the human ear is incredibly sensitive.

By combining Eqs. 15–7 and 16–5, we can write the intensity in terms of the pressure amplitude, ΔP_M:

$$I = 2\pi^2 v\rho f^2 D_M^2 = 2\pi^2 v\rho f^2 (\Delta P_M / 2\pi \rho v f)^2$$

$$I = \frac{(\Delta P_M)^2}{2v\rho}. \qquad\qquad \textbf{(16–7)}$$

Intensity related to pressure amplitude

The intensity, when given in terms of pressure amplitude, thus does not depend on frequency.

Normally, the loudness or intensity of a sound decreases as you get farther from the source of the sound. In enclosed rooms, this effect is altered because of absorption or reflection from the walls. However, if a source is in the open so that sound can radiate freely in all directions, the intensity decreases as the inverse square of the distance,

$$I \propto \frac{1}{r^2}$$

as we saw in Section 15–3, Eq. 15–8. Of course, if there is significant reflection from surrounding structures or the ground, the situation will be more complicated.

EXAMPLE 16–5 **Airplane roar.** The sound level of a jet plane at a distance of 30 m is 140 dB. What is the sound level 300 m away? (Ignore reflections from the ground.)

SOLUTION The intensity I at 30 m is found from Eq. 16–6:

$$140 \, \text{dB} = 10 \log \left(\frac{I}{10^{-12} \, \text{W/m}^2} \right).$$

Reversing the log equation to solve for I we have:

$$10^{14} = \frac{I}{10^{-12} \, \text{W/m}^2},$$

so $I = 10^2$ W/m². At 300 m, 10 times as far away, the intensity will be $\left(\frac{1}{10}\right)^2 = 1/100$ as much, or 1 W/m². Hence, the sound level is

$$\beta = 10 \log \left(\frac{1 \, \text{W/m}^2}{10^{-12} \, \text{W/m}^2} \right) = 120 \, \text{dB}.$$

Even at 300 m, the sound is at the threshold of pain. This is why workers at airports wear ear covers to protect their ears from damage (Fig. 16–6).

➡ **PHYSICS APPLIED**

Jet plane noise

FIGURE 16–6 Airport worker with sound-intensity reducing ear covers (headphones).

TABLE 16–3 Equally Tempered Chromatic Scale†

Note	Frequency (Hz)
C	262
C# or Db	277
D	294
D# or Eb	311
E	330
F	349
F# or Gb	370
G	392
G# or Ab	415
A	440
A# or Bb	466
B	494
C'	524

† Only one octave is included.

➡ P H Y S I C S   A P P L I E D

Musical instrument

FIGURE 16–7 Standing waves on a string—only the lowest three frequencies are shown.

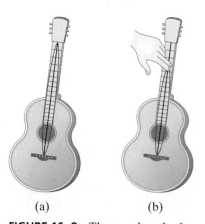

FIGURE 16–8 The wavelength of a fingered string (b) is shorter than that of an unfingered string (a). Hence, the frequency of the fingered string is higher. Only one string is shown on this guitar, and only the simplest standing wave, the fundamental, is shown.

(a)   (b)

16–4 | Sources of Sound: Vibrating Strings and Air Columns

The source of any sound is a vibrating object. Almost any object can vibrate and hence be a source of sound. We now discuss some simple sources of sound, particularly musical instruments. In musical instruments, the source is set into vibration by striking, plucking, bowing, or blowing. Standing waves are produced and the source vibrates at its natural resonant frequencies. The vibrating source is in contact with the air (or other medium) and pushes on it to produce sound waves that travel outward. The frequencies of the waves are the same as the source, but the speed and wavelengths can be different. A drum has a stretched membrane that vibrates. Xylophones and marimbas have metal or wood bars that can be set into vibration. Bells, cymbals, and gongs also make use of a vibrating metal. The most widely used instruments make use of vibrating strings, such as the violin, guitar, and piano, or make use of vibrating columns of air, such as the flute, trumpet, and pipe organ. We have already seen that the pitch of a pure sound is determined by the frequency. Typical frequencies for musical notes on the so-called "equally tempered chromatic scale" are given in Table 16–3 for the octave beginning with middle C. Note that one octave corresponds to a doubling of frequency. For example, middle C has frequency of 262 Hz whereas C' (C above middle C) has twice the frequency, 524 Hz.

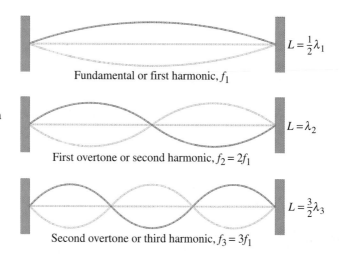

Fundamental or first harmonic, f_1

$L = \frac{1}{2}\lambda_1$

First overtone or second harmonic, $f_2 = 2f_1$

$L = \lambda_2$

Second overtone or third harmonic, $f_3 = 3f_1$

$L = \frac{3}{2}\lambda_3$

Stringed Instruments

We saw in Chapter 15, Fig. 15–27, how standing waves are established on a string, and we show this again here in Fig. 16–7. This is the basis for all stringed instruments. The pitch is normally determined by the lowest resonant frequency, the **fundamental**, which corresponds to nodes occurring only at the ends. The wavelength of the fundamental on the string is equal to twice the length of the string. Therefore, the fundamental frequency is $f = v/\lambda = v/2L$, where v is the velocity of the wave on the string. When a finger is placed on the string of, say, a guitar or violin, the effective length of the string is shortened. So its fundamental frequency, and pitch, is higher since the wavelength of the fundamental is shorter (Fig. 16–8). The strings on a guitar or violin are all the same length. They sound at a different pitch because the strings have different mass per unit length, μ, which affects the velocity as seen in Eq. 15–2, $v = \sqrt{F_T/\mu}$. Thus the velocity on a heavier string is less and the frequency will be less for the same wavelength. The tension may also be different; adjusting the tension is the means for tuning the instrument. In pianos and harps, the strings are each of different length. For the lower notes the strings are not only longer, but heavier as well, and the reason is illustrated in the following Example.

EXAMPLE 16-6 Piano strings. The highest key on a piano corresponds to a frequency about 150 times that of the lowest key. If the string for the highest note is 5.0 cm long, how long would the string for the lowest note have to be if it had the same mass per unit length and was under the same tension?

SOLUTION The velocity would be the same on each string, so the frequency is inversely proportional to the length L of the string ($f = v/\lambda = v/2L$). Thus

$$\frac{L_L}{L_H} = \frac{f_H}{f_L},$$

where the subscripts L and H refer to the lowest and highest notes, respectively. Thus $L_L = L_H(f_H/f_L) = (5.0\text{ cm})(150) = 750\text{ cm}$, or 7.5 m. This would be ridiculously long (≈ 25 ft) for a piano. The longer lower strings are made heavier, so that even on grand pianos the strings are no longer than about 3 m.

EXAMPLE 16-7 Frequencies and wavelengths in the violin. A 0.32-m-long violin string is tuned to play A above middle C at 440 Hz. (*a*) What is the wavelength of the fundamental string vibration, and (*b*) what are the frequency and wavelength of the sound wave produced? (*c*) Why is there a difference?

SOLUTION (*a*) From Fig. 16–7 we see that the wavelength of the fundamental is

$$\lambda = 2L = 0.64\text{ m} = 64\text{ cm}.$$

This is the wavelength of the standing wave on the string.
(*b*) The sound wave that travels outward in the air (to reach our ears) has the same frequency, 440 Hz (why?). Its wavelength is

$$\lambda = \frac{v}{f} = \frac{343\text{ m/s}}{440\text{ Hz}} = 0.78\text{ m} = 78\text{ cm},$$

where v is the speed of sound in air (assumed at 20°C), Section 16–1.
(*c*) The wavelength of the sound wave is different from that of the standing wave on the string because the speed of sound in air (343 m/s at 20°C) is different from the speed of the wave on the string ($= f\lambda = 440\text{ Hz} \times 0.64\text{ m} = 280\text{ m/s}$), which of course depends on the tension in the string and its mass per unit length.

(a)

(b)

FIGURE 16–9 (a) Sounding box (guitar); (b) sounding board (to which the strings are attached, inside of a piano).

FIGURE 16–10 Wind instruments: clarinet (left) and flute.

Stringed instruments would not be very loud if they relied on their vibrating strings to produce the sound waves since the strings are simply too thin to compress and expand much air. Stringed instruments therefore make use of a kind of mechanical amplifier known as a *sounding board* (piano) or *sounding box* (guitar, violin), which acts to amplify the sound by putting a greater surface area in contact with the air (Fig. 16–9). When the strings are set into vibration, the sounding board or box is set into vibration as well. Since it has much greater area in contact with the air, it can produce a more intense sound wave. On an electric guitar, the sounding box is not so important since the vibrations of the strings are amplified electronically.

Wind Instruments

Instruments such as woodwinds, the brasses, and the pipe organ produce sound from the vibrations of standing waves in a column of air within a tube or pipe (Fig. 16–10). Standing waves can occur in the air of any cavity, but the frequencies present are complicated for any but very simple shapes such as a long, narrow tube. In some instruments, a vibrating reed or the vibrating lip of the player helps to set up vibrations of the air column. In others, a stream of air is directed against one edge of the opening or mouthpiece, leading to turbulence which sets up the vibrations. Because of the disturbance, whatever its source, the air within the tube vibrates with a variety of frequencies; but only frequencies that correspond to standing waves will persist.

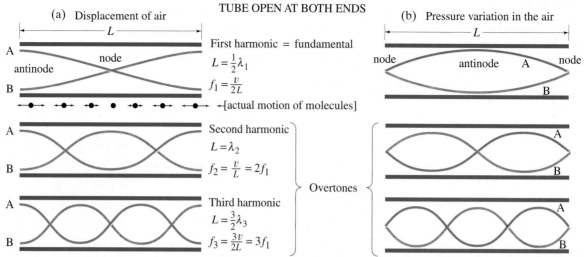

(a) Displacement of air

(b) Pressure variation in the air

First harmonic = fundamental

$L = \frac{1}{2}\lambda_1$

$f_1 = \frac{v}{2L}$

[actual motion of molecules]

Second harmonic

$L = \lambda_2$

$f_2 = \frac{v}{L} = 2f_1$

Overtones

Third harmonic

$L = \frac{3}{2}\lambda_3$

$f_3 = \frac{3v}{2L} = 3f_1$

FIGURE 16–11 Modes of vibration (standing waves) for a tube open at both ends ("open tube"). The simplest modes of vibration are shown in (a), on the left, in terms of the motion of the air (displacement), and in (b), on the right, in terms of air pressure. These are graphs, shown placed within the tube, and are labeled A and B where B represents the wave form a half period after the moment when it has the shape labeled A. The actual motion of molecules for one case is shown just below the tube at upper left.

For a string fixed at both ends, Fig. 16–7, the standing waves have nodes (no movement) at the two ends, and one or more antinodes (large amplitude of vibration) in between; a node separates successive antinodes. The lowest-frequency standing wave, the *fundamental*, corresponds to a single antinode. The higher-frequency standing waves are called **overtones** or **harmonics**, as was discussed in Section 15–9. Specifically, the first harmonic is the fundamental, the second harmonic has twice the frequency of the fundamental,[†] and so on.

The situation is similar for a column of air, but we must remember that it is now air itself that is vibrating. We can describe the waves either in terms of the flow of the air—that is, in terms of the *displacement* of air—or in terms of the *pressure* in the air (see Figs. 16–2 and 16–3). In terms of displacement, the air at the closed end of a tube is a displacement node since the air is not free to move there, whereas near the open end of a tube there will be an antinode since the air can move freely. The air within the tube vibrates in the form of longitudinal standing waves. The possible modes of vibration for a tube open at both ends (called an **open tube**), are shown graphically in Fig. 16–11. They are shown for a tube that is open at one end but closed at the other (called a **closed tube**) in Fig. 16–12. [A tube closed at *both* ends, having no connection to the outside air, would be useless as an instrument.] The graphs in part (a) of each Figure (left sides) represent the displacement amplitude of the vibrating air in the tube. Note that these are graphs, and that the air molecules themselves oscillate *horizontally*, parallel to the tube length, as shown by the small arrows in the top diagram of Fig. 16–11a (on the left). The exact position of the antinode near the open end of a tube depends on the diameter of the tube, but if the diameter is small compared to the length, which is the usual case, the antinode occurs very close to the end as shown. We assume this is the case in what follows. (The position of the antinode may also depend slightly on the wavelength and other factors.)

Let us look in detail at the open tube, in Fig. 16–11a, which might be a flute. An open tube has displacement antinodes at both ends since the air is free to move at open ends. Notice that there must be at least one node within an open tube if there is to be a standing wave at all. A single node corresponds to the *fundamental frequency* of the tube. Since the distance between two successive nodes, or between two successive antinodes, is $\frac{1}{2}\lambda$, there is one-half a wavelength within the length of the tube for the simplest case of the fundamental (top diagram in Fig. 16–11a): $L = \frac{1}{2}\lambda$ or $\lambda = 2L$. So the fundamental frequency is $f_1 = v/\lambda = v/2L$, where v is the velocity of sound in air. The standing

[†]When the resonant frequencies above the fundamental (that is, the overtones) are integral multiples of the fundamental, they are called harmonics. But if the overtones are not integral multiples of the fundamental, as is the case for a vibrating drumhead, for example, they are not harmonics.

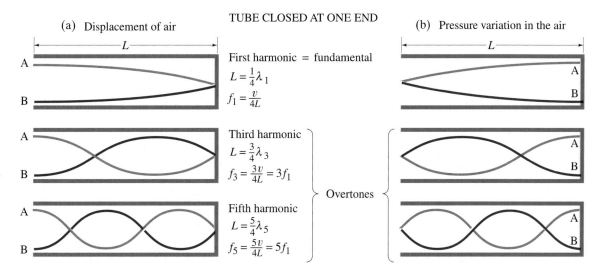

(a) Displacement of air   TUBE CLOSED AT ONE END   (b) Pressure variation in the air

A
B
First harmonic = fundamental
$L = \frac{1}{4}\lambda_1$
$f_1 = \frac{v}{4L}$

A
B
Third harmonic
$L = \frac{3}{4}\lambda_3$
$f_3 = \frac{3v}{4L} = 3f_1$
} Overtones {

A
B
Fifth harmonic
$L = \frac{5}{4}\lambda_5$
$f_5 = \frac{5v}{4L} = 5f_1$

FIGURE 16–12 Modes of vibration (standing waves) for a tube closed at one end ("closed tube"). See caption for Fig. 16–11.

Open tubes produce all harmonics

$\lambda = \frac{v}{f}$

Closed tubes produce only odd harmonics

wave with two nodes is the *first overtone* or *second harmonic* and has half the wavelength ($L = \lambda$) and twice the frequency. Indeed, the frequency of each overtone is an integral multiple of the fundamental frequency, as shown in Fig. 16–11a. This is just what is found for a string.

For a closed tube, shown in Fig. 16–12a, which could be a clarinet, there is always a displacement node at the closed end (because the air is not free to move) and an antinode at the open end (where the air can move freely). Since the distance between a node and the nearest antinode is $\frac{1}{4}\lambda$, we see that the fundamental in a closed tube corresponds to only one-fourth of a wavelength within the length of the tube: $L = \lambda/4$, and $\lambda = 4L$. The fundamental frequency is thus $f_1 = v/4L$, or half what it is for an open pipe of the same length. There is another difference, for as we can see from Fig. 16–12a, only the odd harmonics are present in a closed pipe: the overtones have frequencies equal to $3, 5, 7, \cdots$ times the fundamental frequency. There is no way for waves with $2, 4, 6, \cdots$ times the fundamental frequency to have a node at one end and an antinode at the other, and thus they cannot exist as standing waves in a closed tube.

If this description in terms of displacement seems hard to understand, or you want to understand it from another point of view, then consider a description in terms of the *pressure* in the air, shown in part (b) of Figs. 16–11 and 16–12 (right sides). Where the air in a wave is compressed, the pressure is higher, whereas in a wave expansion (or rarefaction), the pressure is less than normal. The open end of a tube is open to the atmosphere. Hence the pressure variation at an open end must be a *node*: the pressure doesn't alternate, but remains at the outside atmospheric pressure. If a tube has a closed end, the pressure at that closed end can readily alternate to be above or below atmospheric pressure. Hence there is a pressure *antinode* at a closed end of a tube. Of course there can be pressure nodes and antinodes within the tube, and some of the possible vibrational modes in terms of pressure for an open tube are shown in Fig. 16–11b, and for a closed tube are shown in Fig. 16–12b.

Pipe organs (Fig. 16–13) make use of both open and closed pipes. Notes of different pitch are sounded using different pipes with different lengths from a few centimeters to 5 m or more. Other musical instruments act either like a closed tube or an open tube. A flute, for example, is an open tube, for it is open not only where you blow into it, but also at the opposite end as well. The different notes on a flute and many other instruments are obtained by shortening the length of the tube— that is, by uncovering holes along its length. In a trumpet, on the other hand, pushing down on the valves opens additional lengths of tube. In all these instruments, the longer the length of the vibrating air column, the lower the frequency.

FIGURE 16–13 The pipe organ used by J. S. Bach in Leipzig, Germany.

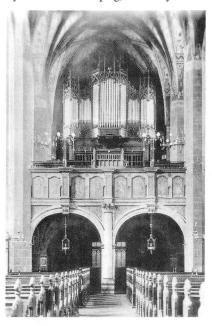

EXAMPLE 16–8 **Open and closed organ pipes.** What will be the fundamental frequency and first three overtones for a 26-cm-long organ pipe at 20°C if it is (a) open and (b) closed?

SOLUTION At 20°C, the speed of sound in air is 343 m/s (Section 16–1). (a) For the open pipe, Fig. 16–11, the fundamental frequency is

$$f_1 = \frac{v}{2L} = \frac{343 \text{ m/s}}{2(0.26 \text{ m})} = 660 \text{ Hz}.$$

The overtones, which include all harmonics, are 1320 Hz, 1980 Hz, 2640 Hz, and so on. (b) For a closed pipe, Fig. 16–12, we have

$$f_1 = \frac{v}{4L} = \frac{343 \text{ m/s}}{4(0.26 \text{ m})} = 330 \text{ Hz}.$$

But only the odd harmonics will be present, so the first three overtones will be 990 Hz, 1650 Hz, and 2310 Hz. (The closed pipe plays 330 Hz, which, from Table 16–3, is E above middle C, whereas the open pipe of the same length plays 660 Hz, an octave higher.)

EXAMPLE 16–9 **Flute.** A flute is designed to play middle C (262 Hz) as the fundamental frequency when all the holes are covered. Approximately how long should the distance be from the mouthpiece to the far end of the flute? (Note: This is only approximate since the antinode does not occur precisely at the mouthpiece.) Assume the temperature is 20°C.

SOLUTION The speed of sound in air at 20°C is 343 m/s. Because a flute is open at both ends, we use Fig. 16–11: the fundamental frequency f_1 is related to the length of the vibrating air column by $f = v/2L$. Solving for L, we find

$$L = \frac{v}{2f} = \frac{343 \text{ m/s}}{2(262 \text{ s}^{-1})} = 0.655 \text{ m}.$$

➡ **PHYSICS APPLIED**

Temperature effect on staying in tune

EXAMPLE 16–10 **A cold flute.** If the temperature is only 10°C, what will be the frequency of the note played when all the openings are covered in the flute of Example 16–9?

SOLUTION The length L is still 65.5 cm. But now the velocity of sound is less since it changes by 0.60 m/s per each C°. For a drop of 10 C°, the velocity decreases by 6 m/s to 337 m/s. The frequency will be

$$f = \frac{v}{2L} = \frac{337 \text{ m/s}}{2(0.655 \text{ m})} = 257 \text{ Hz}.$$

We see why players of wind instruments take time to "warm up" their instruments so they will be in tune. The effect of temperature on stringed instruments is much smaller.

FIGURE 16–14 Example 16–11.

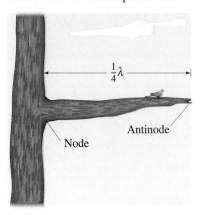

CONCEPTUAL EXAMPLE 16–11 **Wind noise frequencies.** Wind can be noisy—it can "howl" in trees; it can "moan" in chimneys. Why is this so? What is really causing the noise, and about what range of frequencies would you expect to hear?

SOLUTION In each case, jets of air in the wind cause vibrations or oscillations, which produce the sound. The end of a tree limb fixed to the tree trunk is a node, whereas the other end is free to move and therefore is an antinode; the tree limb is thus about $\frac{1}{4}\lambda$ (Fig. 16–14). We estimate $v \approx 4000$ m/s for the speed of sound in wood (Table 16–1). Suppose that a tree limb has length $L \approx 2$ m; then $\lambda = 4L = 8$ m and $f = v/\lambda = (4000 \text{ m/s})/(8 \text{ m}) \approx 500$ Hz.

Wind can excite air oscillations in a chimney, much like in an organ pipe or flute. A chimney is a fairly long tube, perhaps 3 m in length, acting like a tube open at either one end or even both ends. If open at both ends ($\lambda = 2L$), with $v = 340$ m/s, we find $f_1 \approx v/2L \approx 56$ Hz, which is a fairly low note—no wonder chimneys "moan"!

*16–5 Quality of Sound, and Noise

Whenever we hear a sound, particularly a musical sound, we are aware of its loudness, its pitch, and also of a third aspect called "quality." For example, when a piano and then a flute play a note of the same loudness and pitch (say middle C), there is a clear difference in the overall sound. We would never mistake a piano for a flute. This is what is meant by the **quality** of a sound. For musical instruments, the terms *timbre* or *tone color* are also used.

Just as loudness and pitch can be related to physically measurable quantities, so too can quality. The quality of a sound depends on the presence of overtones—their number and their relative amplitudes. Generally, when a note is played on a musical instrument, the fundamental as well as overtones are present simultaneously. We saw in Fig. 15–17 how the superposition of three wave forms, in that case the fundamental and first two overtones (with particular amplitudes), would combine to give a composite *waveform*. Of course, more than two overtones are usually present.

The relative amplitudes of the various overtones are different for different musical instruments, and this is what gives each instrument its characteristic quality or timbre. A graph showing the relative amplitudes of the harmonics produced by an instrument is called a "sound spectrum." Several typical examples for different instruments are shown in Fig. 16–15. Normally, the fundamental has the greatest amplitude and its frequency is what is heard as the pitch, although players can sometimes make the first overtone (for example) sound the loudest.

The manner in which an instrument is played strongly influences the sound quality. Plucking a violin string, for example, makes a very different sound than pulling a bow across it. The sound spectrum at the very start (or end) of a note, as when a hammer strikes a piano string, can be very different from the subsequent sustained tone. This too affects the subjective tone quality of an instrument.

An ordinary sound, like that made by striking two stones together, is a noise that has a certain quality, but a clear pitch is not discernible. A noise such as this is a mixture of many frequencies which bear little relation to one another. If a sound spectrum were made of this noise, it would not show discrete lines like those of Fig. 16–15. Instead it would show a continuous, or nearly continuous, spectrum of frequencies. Such a sound we call "noise" in comparison with the more harmonious sounds which contain frequencies that are simple multiples of the fundamental.

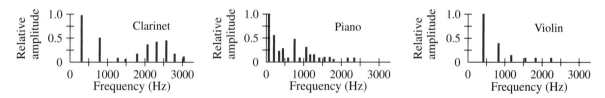

FIGURE 16–15 Sound spectra. The shapes of the spectra change as the instruments play different notes.

16–6 Interference of Sound Waves; Beats

Interference in Space

We saw in Section 15–8 that when two waves simultaneously pass through the same region of space, they interfere with one another. Since this can occur for any kind of wave, we should expect that interference will occur with sound waves, and indeed it does.

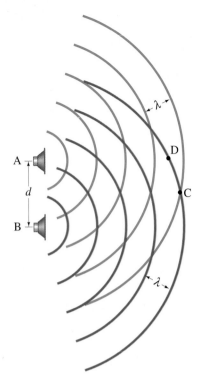

FIGURE 16–16 Sound waves from two loudspeakers interfere.

FIGURE 16–17 Sound waves of a single frequency from loudspeakers A and B (see Fig. 16–16) constructively interfere at C and destructively interfere at D.

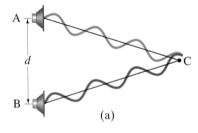

(a)

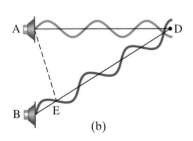

(b)

As a simple example, consider two large loudspeakers, A and B, a distance d apart on the stage of an auditorium as shown in Fig. 16–16. Let us assume the two speakers are emitting sound waves of the same single frequency and that they are in phase: that is, when one speaker is forming a compression, so is the other. (We ignore reflections from walls, floor, etc.) The curved lines in the diagram represent the crests of sound waves from each speaker. Of course, we must remember that for a sound wave, a crest is a compression in the air whereas a trough—which falls between two crests—is a rarefaction. A person or detector at a point such as C, which is the same distance from each speaker, will experience a loud sound because the interference will be constructive. On the other hand, at a point such as D in the diagram, little if any sound will be heard because destructive interference occurs—compressions of one wave meet rarefactions of the other and vice versa (see Fig. 15–24 and related discussion on water waves in Section 15–8).

An analysis of this situation is perhaps clearer if we graphically represent the wave forms as in Fig. 16–17. In Fig. 16–17a it can be seen that at point C, constructive interference occurs since both waves simultaneously have crests or simultaneously have troughs. In Fig. 16–17b we see the situation for point D. The wave from speaker B must travel a greater distance than the wave from A. Thus the wave from B lags behind that from A. In this diagram, point E is chosen so that the distance ED is equal to AD. Thus we see that if the distance BE is equal to precisely one-half the wavelength of the sound, the two waves will be exactly out of phase when they reach D, and destructive interference occurs. This then is the criterion for determining at what points destructive interference occurs: destructive interference occurs at any point whose distance from one speaker is greater than its distance from the other speaker by exactly one-half wavelength. Notice that if this extra distance (BE in Fig. 16–17b) is equal to a whole wavelength (or 2, 3, ⋯ wavelengths), then the two waves will be in phase and *constructive interference* occurs. If the distance BE equals $\frac{1}{2}, 1\frac{1}{2}, 2\frac{1}{2}, \cdots$ wavelengths, *destructive interference* occurs.

It is important to realize that a person sitting at point D hears nothing at all at this particular frequency, yet sound is coming from both speakers. Indeed, if one of the speakers is turned off, the sound from the other speaker would be clearly heard.

If a loudspeaker emits a whole range of frequencies, only specific ones will destructively interfere completely at a given point.

EXAMPLE 16–12 **Loudspeakers' interference.** Two loudspeakers are 1.00 m apart. A person stands 4.00 m from one speaker. How far must she be from the second speaker in order to detect destructive interference when the speakers emit 1150-Hz sound waves in phase with each other? Assume the temperature is 20°C.

SOLUTION The wavelength of this sound is

$$\lambda = \frac{v}{f} = \frac{343 \text{ m/s}}{1150 \text{ Hz}} = 0.30 \text{ m}.$$

For destructive interference to occur, the person must be one-half wavelength farther from one loudspeaker than from the other, or 0.15 m. Thus the person must be 4.15 m (or 3.85 m) from the second speaker. If the speakers are less than 0.15 m apart, there would be no point that was 0.15 m farther from one speaker than the other, and there would be no point where destructive interference would occur.

Beats—Interference in Time

Beats

We have been discussing interference of sound waves that takes place in space. An interesting and important example of interference that occurs in time is the phenomenon known as **beats**: two sources of sound—say, two tuning forks—are close in frequency but not exactly the same. Sound waves from the two sources interfere with each other and the sound level at a given position alternately rises and falls; the regularly spaced intensity changes are called beats.

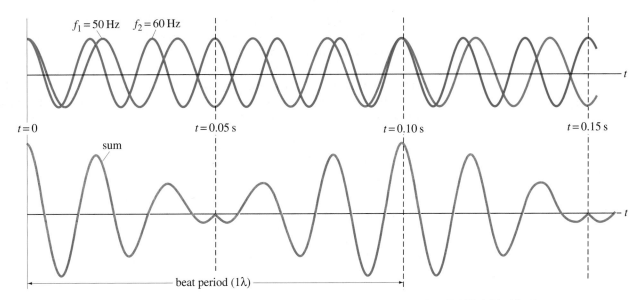

To see how beats arise, consider two equal-amplitude sound waves of frequency $f_1 = 50$ Hz and $f_2 = 60$ Hz, respectively. In 1.00 s, the first source makes 50 vibrations whereas the second makes 60. We now examine the waves at one point in space equidistant from the two sources. The waveforms for each wave as a function of time are shown on the top graph of Fig. 16–18; the magenta line represents the 50 Hz wave, the blue line represents the 60 Hz wave. The lower graph in Fig. 16–18 shows the sum of the two waves. At time $t = 0$ the two waves are shown to be in phase and interfere constructively. Because the two waves vibrate at different rates, at time $t = 0.05$ s they are completely out of phase and destructive interference occurs as shown in the figure. At $t = 0.10$ s, they are again in phase and the resultant amplitude again is large. Thus the resultant amplitude is large every 0.10 s and in between it drops drastically. This rising and falling of the intensity is what is heard as beats.[†] In this case the beats are 0.10 s apart. That is, the **beat frequency** is ten per second or 10 Hz. This result, that the beat frequency equals the difference in frequency of the two waves can be shown in general as follows.

Beat frequency = difference in the two wave frequencies

Let the two waves, of frequencies f_1 and f_2, be represented at a fixed point in space by

$$D_1 = D_M \sin 2\pi f_1 t$$

and

$$D_2 = D_M \sin 2\pi f_2 t.$$

The resultant displacement, by the principle of superposition, is

$$D = D_1 + D_2 = D_M(\sin 2\pi f_1 t + \sin 2\pi f_2 t).$$

Using the trigonometric identity $\sin A + \sin B = 2 \sin \frac{1}{2}(A + B) \cos \frac{1}{2}(A - B)$, we have

$$D = \left[2D_M \cos 2\pi \left(\frac{f_1 - f_2}{2} \right) t \right] \sin 2\pi \left(\frac{f_1 + f_2}{2} \right) t. \tag{16–8}$$

We can interpret Eq. 16–8 as follows. The superposition of the two waves results in a wave that vibrates at the average frequency of the two components, $(f_1 + f_2)/2$. This vibration has an amplitude given by the expression in brackets, and this amplitude varies in time, from zero to a maximum of $2D_M$ (the sum of the separate amplitudes), with a frequency of $(f_1 - f_2)/2$. A beat occurs whenever $\cos 2\pi[(f_1 - f_2)/2]t$ equals $+1$ or -1 (see Fig. 16–18); that is, two beats occur per cycle, so the beat frequency is twice $(f_1 - f_2)/2$ which is just $f_1 - f_2$, the difference in frequency of the component waves.

[†]Beats will be heard even if the amplitudes are not equal, as long as the difference is not great.

The phenomenon of beats can occur with any kind of wave and is a very sensitive method for comparing frequencies. For example, to tune a piano, a piano tuner listens for beats produced between his standard tuning fork and that of a particular string on the piano, and knows it is in tune when the beats disappear. The members of an orchestra can tune up by listening for beats between their instruments and that of a standard tone (usually A above middle C at 440 Hz) produced by a piano or an oboe.

EXAMPLE 16–13 Beats. A tuning fork produces a steady 400-Hz tone. When this tuning fork is struck and held near a vibrating guitar string, twenty beats are counted in five seconds. What are the possible frequencies produced by the guitar string?

SOLUTION The beat frequency is

$$f_{\text{beat}} = 20 \text{ vibrations}/5 \text{ s} = 4 \text{ Hz}.$$

This is the difference of the frequencies of the two waves, and because one wave is known to be 400 Hz, the other must be either 404 Hz or 396 Hz.

16–7 Doppler Effect

You may have noticed that the pitch of the siren on a speeding firetruck drops abruptly as it passes you. Or you may have noticed the change in pitch of a blaring horn on a fast-moving car as it passes by. The pitch of the sound from the engine of a race car changes as it passes an observer. When a source of sound is moving toward an observer, the pitch is higher than when the source is at rest; and when the source is traveling away from the observer, the pitch is lower. This phenomenon is known as the **Doppler effect**[†] and occurs for all types of waves. Let us now see why it occurs, and calculate the change in frequency for sound waves.

To be concrete, consider the siren of a firetruck at rest, which is emitting sound of a particular frequency in all directions as shown in Fig. 16–19a. The wave velocity depends only on the medium in which it is traveling, and is independent of the velocity of the source or observer. If our source, the firetruck, is moving, the siren emits sound at the same frequency as it does at rest. But the sound wavefronts it emits forward are closer together than when the firetruck is at rest, as shown in Fig. 16–19b. This is because the firetruck, as it moves, is "chasing after" the previously emitted wavefronts. Thus an observer on the sidewalk will detect more wave crests passing per second, so the frequency heard is higher. The wavefronts emitted behind the truck, on the other hand, are farther apart than when the truck is at rest because the truck is speeding away from them. Hence, fewer wave crests per second pass by an observer behind the truck and the pitch is lower.

[†] After J. C. Doppler (1803–1853).

FIGURE 16–19 (a) Both observers on the sidewalk hear the same frequency from the firetruck at rest. (b) Doppler effect: observer toward whom the firetruck moves hears a higher-frequency sound, and observer behind the firetruck hears a lower frequency.

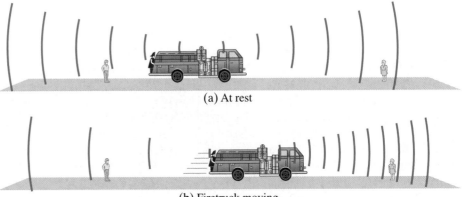

(a) At rest

(b) Firetruck moving

To calculate the change in frequency, we make use of Fig. 16–20, and we assume the air (or other medium) is at rest in our reference frame. In Fig. 16–20a, the source of the sound, shown as a dot, is at rest; two successive wave crests are shown, the second of which is just in the process of being emitted. The distance between these crests is λ, the wavelength. If the frequency of the source is f, then the time between emissions of wave crests is

$$T = \frac{1}{f}.$$

Frequency change, moving source

In Fig. 16–20b, the source is moving with a velocity v_S. In a time T (as just defined), the first wave crest has moved a distance $d = vT$, where v is the velocity of the sound wave in air (which is, of course, the same whether the source is moving or not). In this same time, the source has moved a distance $d_S = v_S T$. Then the distance between successive wave crests, which is the new wavelength λ', is (since $d = \lambda$)

$$\lambda' = d - d_S$$
$$= \lambda - v_S T$$
$$= \lambda - v_S \frac{\lambda}{v} = \lambda\left(1 - \frac{v_S}{v}\right).$$

The change in wavelength, $\Delta\lambda$, is

$$\Delta\lambda = \lambda' - \lambda = -v_S \frac{\lambda}{v}.$$

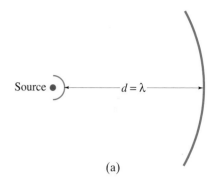

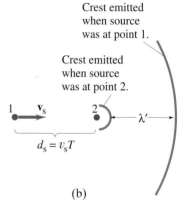

FIGURE 16–20 Determination of frequency change in the Doppler effect (see text). Red dot is the source.

So the shift in wavelength is directly proportional to the speed v_S of the source. The new frequency, on the other hand, is given by

$$f' = \frac{v}{\lambda'} = \frac{v}{\lambda\left(1 - \dfrac{v_S}{v}\right)}$$

or, since $v/\lambda = f$,

$$f' = \frac{f}{\left(1 - \dfrac{v_S}{v}\right)}. \qquad \begin{bmatrix} \text{source moving toward} \\ \text{stationary observer} \end{bmatrix} \quad \textbf{(16–9a)}$$

Because the denominator is less than 1, $f' > f$. For example, if a source emits a sound of frequency 400 Hz when at rest, then when the source moves toward a fixed observer with a speed of 30 m/s, the observer hears a frequency (at 20°C) of

$$f' = \frac{400\,\text{Hz}}{1 - \dfrac{30\,\text{m/s}}{343\,\text{m/s}}} = 438\,\text{Hz}.$$

For a source moving *away* from the observer at a speed v_S, the new wavelength will be

$$\lambda' = d + d_S,$$

and the change in wavelength will be

$$\Delta\lambda = \lambda' - \lambda = +v_S \frac{\lambda}{v}.$$

The frequency of the wave will be

$$f' = \frac{f}{\left(1 + \dfrac{v_S}{v}\right)}. \qquad \begin{bmatrix} \text{source moving away from} \\ \text{stationary observer} \end{bmatrix} \quad \textbf{(16–9b)}$$

In this case, if a source vibrating at 400 Hz is moving away from a fixed observer at 30 m/s, the observer hears a frequency of about 368 Hz.

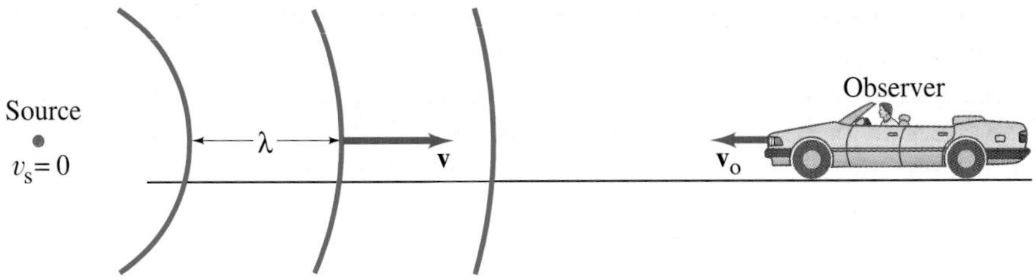

FIGURE 16–21 Observer moving
with speed v_O toward a stationary
source "sees" wave crests pass at
speed $v' = v + v_O$ where v is the
speed of the sound waves in air.

*Frequency change,
moving observer*

The Doppler effect also occurs when the source is at rest and the observer
is in motion. If the observer is traveling toward the source, the pitch is higher;
and if the observer is traveling away from the source, the pitch is lower. Quantitatively the change in frequency is slightly different than for the case of a moving
source. With a fixed source and a moving observer, the distance between wave
crests, the wavelength λ, is not changed. But the velocity of the crests with
respect to the observer is changed. If the observer is moving toward the source,
Fig. 16–21, the speed of the waves relative to the observer is $v' = v + v_O$, where
v is the velocity of sound in air (we assume the air is still) and v_O is the velocity
of the observer. Hence, the new frequency is

$$f' = \frac{v'}{\lambda} = \frac{v + v_O}{\lambda}$$

or, since $\lambda = v/f$,

$$f' = \left(1 + \frac{v_O}{v}\right)f. \qquad \begin{bmatrix} \text{observer moving toward} \\ \text{stationary source} \end{bmatrix} \quad \textbf{(16–10a)}$$

If the observer is moving away from the source, the relative velocity is $v' = v - v_O$,
so

$$f' = \left(1 - \frac{v_O}{v}\right)f. \qquad \begin{bmatrix} \text{observer moving away} \\ \text{from stationary source} \end{bmatrix} \quad \textbf{(16–10b)}$$

EXAMPLE 16–14 **A moving siren.** The siren of a police car at rest emits at
a predominant frequency of 1600 Hz. What frequency will you hear if you are
at rest and the police car moves at 25.0 m/s (*a*) toward you, (*b*) away from you?

SOLUTION (*a*) We use Eq. 16–9a:

$$f' = \frac{f}{\left(1 - \dfrac{v_S}{v}\right)} = \frac{1600 \text{ Hz}}{\left(1 - \dfrac{25.0 \text{ m/s}}{343 \text{ m/s}}\right)} = 1726 \text{ Hz}.$$

(*b*) We use Eq. 16–9b:

$$f' = \frac{f}{\left(1 + \dfrac{v_S}{v}\right)} = \frac{1600 \text{ Hz}}{\left(1 + \dfrac{25.0 \text{ m/s}}{343 \text{ m/s}}\right)} = 1491 \text{ Hz}.$$

When a sound wave is reflected from a moving obstacle, the frequency of
the reflected wave will, because of the Doppler effect, be different from that of the
incident wave. This is illustrated in the following Example.

EXAMPLE 16–15 **Two Doppler shifts.** A 5000-Hz sound wave is emitted by
a stationary source toward an object moving 3.50 m/s toward the source
(Fig. 16–22). What is the frequency of the wave reflected by the moving object as
detected by a detector at rest near the source?

SOLUTION There are actually two Doppler shifts in this situation. First, the emitted wave strikes the moving object which is in effect a moving observer (Fig. 16–22a) that "detects" a sound wave of frequency (Eq. 16–10a):

$$f' = \left(1 + \frac{v_O}{v}\right)f = \left(1 + \frac{3.50\,\text{m/s}}{343\,\text{m/s}}\right)(5000\,\text{Hz}) = 5051\,\text{Hz}.$$

Second, the moving object takes this wave of frequency f' and reemits (or reflects) it, acting effectively as a moving source, so the frequency detected, f'', will be given by Eq. 16–9a:

$$f'' = \frac{f'}{\left(1 - \dfrac{v_S}{v}\right)} = \frac{5051\,\text{Hz}}{\left(1 - \dfrac{3.50\,\text{m/s}}{343\,\text{m/s}}\right)} = 5103\,\text{Hz}.$$

Thus the frequency shifts by 103 Hz.

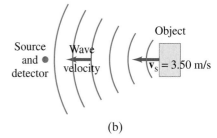

FIGURE 16–22 Example 16–15.

The incident wave and the reflected wave in Example 16–15, when mixed together (say, electronically), interfere with one another and beats are produced. The beat frequency is equal to the difference in the two frequencies, 103 Hz. This Doppler technique is used in a variety of medical applications, usually with ultrasonic waves in the megahertz frequency range. For example, ultrasonic waves reflected from red blood cells can be used to determine the velocity of blood flow. Similarly, the technique can be used to detect the movement of the chest of a young fetus and to monitor its heartbeat.

For convenience, we can write Eqs. 16–9 and 16–10 as a single equation that covers all cases of both source and observer in motion:

$$f' = f\left(\frac{v \pm v_O}{v \mp v_S}\right). \tag{16–11}$$

The upper signs apply if source and/or observer move toward each other; the lower signs apply if they are moving apart.

Doppler Effect for Light

The Doppler effect occurs for other types of waves as well. Light and other types of electromagnetic waves (such as radar) exhibit the Doppler effect: although the formulas for the frequency shift are not identical to Eqs. 16–9 and 16–10, the effect is similar. One important application is for weather forecasting using radar. The time delay between the emission of radar pulses and their reception after being reflected off raindrops gives the position of precipitation. Measuring the Doppler shift in frequency (as in Example 16–15) tells how fast the storm is moving and in which direction.

Another important application is to astronomy where the velocities of distant galaxies can be determined from the Doppler shift. Light from such galaxies is shifted toward lower frequencies, indicating that the galaxies are moving away from us. This is called the **red shift** since red has the lowest frequency of visible light. The greater the frequency shift, the greater the velocity of recession. It is found that the farther the galaxies are from us, the faster they move away. This observation is the basis for the idea that the universe is expanding, and is one basis for the idea that the universe began with a great explosion, affectionately called the "Big Bang" (see Chapter 45).

*16–8 Shock Waves and the Sonic Boom

An object such as an airplane traveling faster than the speed of sound is said to have a **supersonic speed**. Such a speed is often given as a **Mach**[†] **number**, which is defined as the ratio of the object's speed to that of sound in the medium at that location. For example, a plane traveling 600 m/s high in the atmosphere, where the speed of sound is only 300 m/s, has a speed of Mach 2.

[†] After the Austrian physicist Ernst Mach (1838–1916).

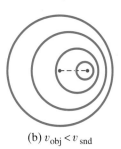 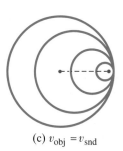 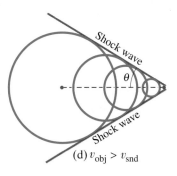

(a) $v_{\text{obj}} = 0$ (b) $v_{\text{obj}} < v_{\text{snd}}$ (c) $v_{\text{obj}} = v_{\text{snd}}$ (d) $v_{\text{obj}} > v_{\text{snd}}$

FIGURE 16–23 Sound waves emitted by an object at rest (a) or moving (b, c, and d). If the object's velocity is less than the velocity of sound, the Doppler effect occurs (b); if its velocity is greater than the velocity of sound, a shock wave is produced (d).

Shock wave

Sonic boom

FIGURE 16–24 Bow waves produced by a boat.

FIGURE 16–25 (a) The (double) sonic boom has already been heard by person A on the left. It is just being heard by person B in the center. And it will shortly be heard by person C on the right. (b) Special photo of supersonic aircraft showing shock waves produced in the air. (Several closely spaced shock waves are produced by different parts of the aircraft.)

When a source of sound moves at subsonic speeds, the pitch of the sound is altered, as we have seen (the Doppler effect); see also Fig. 16–23a and b. But if a source of sound moves faster than the speed of sound, a more dramatic effect known as a **shock wave** occurs. In this case the source is actually "outrunning" the waves it produces. As shown in Fig. 16–23c, when the source is traveling at the speed of sound, the wave fronts it emits in the forward direction "pile up" directly in front of it. When the object moves at a supersonic speed, the wave fronts pile up on one another along the sides, as shown in Fig. 16–23d. The different wave crests overlap one another and form a single very large crest which is the shock wave. Behind this very large crest there is usually a very large trough. A shock wave is essentially the result of constructive interference of a large number of wave fronts. A shock wave in air is analogous to the bow wave of a boat traveling faster than the speed of the water waves it produces, Fig. 16–24.

When an airplane travels at supersonic speeds, the noise it makes and its disturbance of the air form into a shock wave containing a tremendous amount of sound energy. When the shock wave passes a listener, it is heard as a loud "sonic boom." A sonic boom lasts only a fraction of a second, but the energy it contains is often sufficient to break windows and cause other damage. It can be psychologically unnerving as well. Actually, a sonic boom is made up of two or more booms since major shock waves can form at the front and the rear of the aircraft, as well as at the wings, etc. (Fig. 16–25). Bow waves of a boat are also multiple, as can be seen in Fig. 16–24.

When an aircraft approaches the speed of sound, it encounters a barrier of sound waves in front of it (see Fig. 16–23c). In order to exceed the speed of sound, extra thrust is needed to pass through this "sound barrier." This is called "breaking the sound barrier." Once a supersonic speed is attained, this barrier no longer impedes the motion. It is sometimes erroneously thought that a sonic boom is produced only at the moment an aircraft is breaking through the sound barrier. Actually, a shock wave follows the aircraft at all times it is traveling at supersonic speeds. A series of observers on the ground will each hear a loud "boom" as the shock wave passes, Fig. 16–25. The shock wave consists of a cone whose apex is at the aircraft. The angle of this cone, θ, (see Fig. 16–23 d) is given by

$$\sin \theta = \frac{v_{\text{snd}}}{v_{\text{obj}}}, \qquad\qquad \textbf{(16–12)}$$

where v_{obj} is the velocity of the object (the aircraft) and v_{snd} is the velocity of sound in the medium (the proof is left as a Problem).

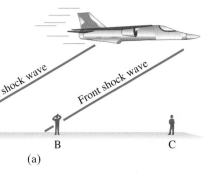

A   B   C

(a)

(b)

Applications; Sonar, Ultrasound, and Ultrasound Imaging

The reflection of sound is used in many applications to determine distance. The **sonar**[†] or pulse-echo technique is used to locate underwater objects. A transmitter sends out a sound pulse through the water, and a detector receives its reflection, or echo, a short time later. This time interval is carefully measured, and from it the distance to the reflecting object can be determined since the speed of sound in water is known. The depth of the sea and the location of reefs, sunken ships, submarines, or schools of fish can be determined in this way. The interior structure of the Earth is studied in a similar way by detecting reflections of waves traveling through the Earth whose source was a deliberate explosion (called "soundings"). An analysis of waves reflected from various structures and boundaries within the Earth reveals characteristic patterns that are also useful in the exploration for oil and minerals.

Sonar generally makes use of **ultrasonic** frequencies: that is, waves whose frequencies are above 20 kHz, beyond the range of human detection. For sonar, the frequencies are typically in the range 20 kHz to 100 kHz. One reason for using ultrasound waves, other than the fact that they are inaudible, is that for shorter wavelengths there is less diffraction, so the beam spreads less and smaller objects can be detected.

The diagnostic use of ultrasound in medicine, in the form of images (sometimes called "sonograms") is an important and interesting application of physical principles. A **pulse-echo technique** is used, much like sonar. A high-frequency sound pulse is directed into the body, and its reflections from boundaries or interfaces between organs and other structures and lesions in the body are then detected. It is even possible to produce "real-time" ultrasound images, as if one were watching a movie of a section of the interior of the body.

The pulse-echo technique for medical imaging works as follows. A brief pulse of ultrasound is emitted by a transducer that transforms an electrical pulse into a sound-wave pulse. Part of the pulse is reflected at each interface surface in the body, and most (usually) continues on. The detection of the reflected pulses by the same transducer can then be displayed on the screen of a display terminal or monitor, as shown in Fig. 16–26a. The time elapsed from when the pulse is emitted to when each reflection (echo) is received is proportional to the distance to the reflecting surface. For example, if the distance from transducer to the vertebra is 25 cm, the pulse travels a round-trip distance of 2×25 cm $= 0.50$ m; the speed of sound in human tissue is about 1540 m/s (close to water), so the time taken is $t = d/v = (0.50 \text{ m})/(1540 \text{ m/s}) = 320 \ \mu\text{s}$.

The *strength* of a reflected pulse depends mainly on the difference in density of the two materials on either side of the interface and can be displayed as a pulse or as a dot (Figs. 16–26b and c). Each echo dot (Fig. 16–26c) can be represented as a point, whose position is given by the time delay and whose brightness depends on the strength of the echo. A two-dimensional image can then be formed out of these dots from a series of scans. The transducer is moved, and at each position it sends out a pulse and receives echoes as shown in Fig. 16–27. Each trace can be plotted, spaced appropriately one below the other, to form an image on a display terminal as shown in Fig. 16–27b. Only 10 lines are shown in Fig. 16–27, so the image is crude. More lines give a more precise image. Photographs of ultrasound images are shown in Fig. 16–28.

[†] Sonar stands for "*so*und *na*vigation *r*anging."

Sonar and imaging

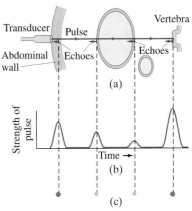

FIGURE 16–26 (a) Ultrasound pulse passes through abdomen, reflecting from surfaces in its path. (b) Reflected pulses plotted as a function of time when received by transducer. The vertical dashed lines point out which reflected pulse goes with which surface. (c) Dot display for the same echoes: brightness of each dot is related to signal strength.

FIGURE 16–27 (a) Ten traces are made across the abdomen by moving the transducer, or by using an array of transducers. (b) The echoes are plotted to produce the image. More closely spaced traces would give a more detailed image.

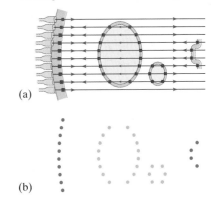

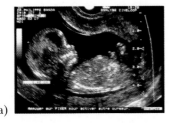

(a)

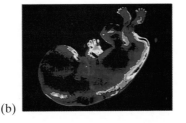

(b)

FIGURE 16–28 (a) Ultrasound image of a human fetus (with head at the left) within the uterus. (b) False-color high-resolution ultrasound image of fetus (different colors represent different intensities of reflected pulses).

Summary

Sound travels as a longitudinal wave in air and other materials. In air, the speed of sound increases with temperature; at 20°C, it is about 343 m/s.

The **pitch** of a sound is determined by the frequency; the higher the frequency, the higher the pitch.

The **audible range** of frequencies for humans is roughly 20 Hz to 20,000 Hz (1 Hz = 1 cycle per second).

The **loudness** or **intensity** of a sound is related to the amplitude of the wave. Because the human ear can detect sound intensities from 10^{-12} W/m² to over 1 W/m², intensity levels are specified on a logarithmic scale. The **sound level** β, specified in decibels, is defined in terms of intensity I as

$$\beta = 10 \log \left(I/I_0 \right),$$

where the reference intensity I_0 is usually taken to be 10^{-12} W/m².

Musical instruments are simple sources of sound in which *standing waves* are produced.

The strings of a stringed instrument may vibrate as a whole with nodes only at the ends; the frequency at which this occurs is called the **fundamental**. The string can also vibrate at higher frequencies, called **overtones** or **harmonics**, in which there are one or more additional nodes. The frequency of each harmonic is a whole-number multiple of the fundamental.

In wind instruments, standing waves are set up in the column of air within the tube.

The vibrating air in an **open tube** (open at both ends) has displacement antinodes at both ends. The fundamental frequency corresponds to a wavelength equal to twice the tube length: $\lambda_1 = 2L$. The harmonics have frequencies that are $2, 3, 4, \cdots$ times the fundamental frequency, just as for strings.

For a **closed tube** (closed at one end), the fundamental corresponds to a wavelength four times the length of the tube: $\lambda_1 = 4L$. Only the odd harmonics are present, equal to $1, 3, 5, 7, \cdots$ times the fundamental frequency.

Sound waves from different sources can interfere with each other. If two sounds are at slightly different frequencies, **beats** can be heard at a frequency equal to the difference in frequency of the two sources.

The **Doppler effect** refers to the change in pitch of a sound due to the motion either of the source or of the listener. If they are approaching each other, the pitch is higher; if they are moving apart, the pitch is lower.

Questions

1. What is the evidence that sound travels as a wave?

2. What is the evidence that sound is a form of energy?

3. Children sometimes play with a homemade "telephone" by attaching a string to the bottoms of two paper cups. When the string is stretched and a child speaks into one of the cups, the sound can be heard at the other cup. Explain clearly how the sound wave travels from one cup to the other. (See Fig. 16–29.)

FIGURE 16–29 Question 3.

4. When a sound wave passes from air into water, do you expect the frequency or wavelength to change?

5. What evidence can you give that the speed of sound in air does not depend significantly on frequency?

6. The voice of a person who has inhaled helium sounds very high-pitched. Why?

7. What is the main reason the speed of sound in hydrogen is greater than the speed of sound in air?

8. The molecules of a gas, such as air, move around randomly at fairly high speeds (Chapter 18). The distance between molecules, on the average, is many times their diameter. When a wave passes through a gas the impulse given to one molecule is given to another only when this distance is traveled and the two collide. Would you therefore expect the speed of sound in a gas to be limited by the average molecular speed?

9. Two tuning forks oscillate with the same amplitude, but one has twice the frequency. Which (if either) produces the more intense sound?

10. How does a rise in air temperature affect the loudness of sound coming from a source of fixed frequency and amplitude? (Assume that atmospheric pressure doesn't change.)

11. What is the reason that catgut strings on some musical instruments are wrapped with fine wire?

12. Explain how a tube might be used as a filter to reduce the amplitude of sounds in various frequency ranges. (An example is a car muffler.)

13. How will the air temperature in a room affect the pitch of organ pipes?

14. Why are the frets on a guitar spaced closer together as you move up the fingerboard toward the bridge?

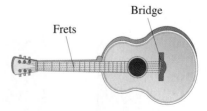

FIGURE 16–30 Question 14.

15. Standing waves can be said to be due to "interference in space," whereas beats can be said to be due to "interference in time." Explain.

16. In Fig. 16–16, if the frequency of the speakers were lowered, would the points D and C (where destructive and constructive interference occur) move farther apart or closer together?

17. Traditional methods of protecting the hearing of people who work in areas with very high noise levels have consisted mainly of efforts to block or reduce noise levels. With a relatively new technology, headphones are worn that do not block the ambient noise. Instead, a device is used which detects the noise, inverts it electronically, then feeds it to the headphones *in addition to* the ambient noise. How could adding *more* noise actually reduce the sound levels reaching the ears?

18. Suppose a source of sound moves at right angles to the line of sight of a listener at rest in still air. Will there be a Doppler effect? Explain.

19. If a wind is blowing, will this alter the frequency of the sound heard by a person at rest with respect to the source? Is the wavelength or velocity changed?

20. Figure 16–31 shows various positions of a child in motion on a swing. A monitor is blowing a whistle in front of the child on the ground. At which position will the child hear the highest frequency for the sound of the whistle? Explain your reasoning.

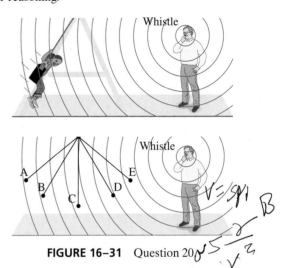

FIGURE 16–31 Question 20.

Problems

[Unless stated otherwise, assume $T = 20°C$ and $v_{sound} = 343$ m/s in air.]

Section 16–1

1. (I) A hiker determines the length of a lake by listening for the echo of her shout reflected by a cliff at the far end of the lake. She hears the echo 1.5 s after shouting. Estimate the length of the lake.

2. (I) (a) Calculate the wavelengths in air at 20°C for sounds in the maximum range of human hearing, 20 Hz to 20,000 Hz. (b) What is the wavelength of a 10-MHz ultrasonic wave?

3. (II) A person sees a heavy stone strike the concrete pavement. A moment later two sounds are heard from the impact: one travels in the air and the other in the concrete, and they are 1.4 s apart. How far away did the impact occur?

4. (II) A fishing boat is drifting just above a school of tuna on a foggy day. Without warning, an engine backfire occurs on another boat 1.0 km away (Fig. 16–32). How much time elapses before the backfire is heard (a) by the fish and (b) by the fishermen?

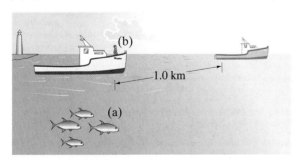

FIGURE 16–32 Problem 4.

5. (II) The sound from a very high burst of fireworks takes 4.5 s to arrive at your eardrums. The burst occurred 1500 m above you and traveled vertically through two stratified layers of air, the top one at 0°C and the bottom one at 20°C. How thick is each layer of air?

6. (II) If the camera of Example 16–2 focuses precisely at 20°C, what percent error will there be in distances when used at 0°C?

Section 16–2

7. (I) The pressure amplitude of a sound wave in air $(\rho = 1.29 \text{ kg/m}^3)$ at 0°C is 3.0×10^{-3} Pa. What is the displacement amplitude if the frequency is (a) 100 Hz and (b) 10 kHz?

8. (I) What must be the pressure amplitude in a sound wave in air (0°C) if the air molecules undergo a maximum displacement equal to the diameter of an oxygen molecule, about 3×10^{-10} m? Assume a sound wave frequency of (a) 50 Hz and (b) 5.0 kHz.

9. (II) Write an expression that describes the pressure variation as a function of x and t for the waves described in Problem 8.

10. (II) The pressure variation in a sound wave is given by

$$\Delta P = 0.0025 \sin\left(\frac{\pi}{3}x - 1700\pi t\right),$$

where ΔP is in pascals, x in meters, and t in seconds. Determine (a) the wavelength, (b) the frequency, (c) the speed, and (d) the displacement amplitude of the wave. Assume the density of the medium to be $\rho = 2.7 \times 10^3$ kg/m^3.

Section 16–3

11. (I) (a) What is the sound level of a sound whose intensity is 8.5×10^{-8} W/m²? (b) What is the intensity of a sound whose sound level is 25 dB?

12. (I) A 6000-Hz tone must have what sound level to seem as loud as a 100-Hz tone that has a 50-dB sound level? (See Fig. 16–5.)

13. (I) What are the lowest and highest frequencies that an average ear can detect when the sound level is 30 dB? (See Fig. 16–5)

14. (I) A stereo cassette player is said to have a signal-to-noise ratio of 63 dB. What is the ratio of intensities of the signal and the background noise?

15. (II) If the amplitude of a sound wave is tripled, (a) by what factor will the intensity increase? (b) By how many dB will the sound level increase?

16. (II) Human beings can typically detect a difference in sound intensity level of 2.0 dB. What is the ratio of the amplitudes of two sounds whose levels differ by this amount?

17. (II) Two sound waves have equal displacement amplitudes, but one has twice the frequency of the other. (a) Which has the greater pressure amplitude and by what factor is it greater? (b) What is the ratio of their intensities?

18. (II) A person standing a certain distance from an airplane with four equally noisy jet engines is experiencing a sound level bordering on pain, 120 dB. What sound level would this person experience if the captain shut down all but one engine?

19. (II) A 40-dB sound wave strikes an eardrum whose area is 5.0×10^{-5} m². (a) How much energy is absorbed by the eardrum per second? (b) At this rate, how long would it take your eardrum to receive a total energy of 1.0 J?

20. (II) (a) Estimate the power output of sound from a person speaking in normal conversation. Use Table 16–2. Assume the sound spreads roughly uniformly over a hemisphere in front of the mouth. (b) How many people would produce a total sound output of 100 W of ordinary conversation?

21. (II) If two firecrackers produce a sound level of 90 dB when fired simultaneously at a certain place, what will be the sound level if only one is exploded?

22. (II) What would be the sound level (in dB) of a sound wave in air that corresponds to a displacement amplitude of vibrating air molecules of 1.3 mm at 330 Hz?

23. (II) (a) Calculate the maximum displacement of air molecules when a 210-Hz sound wave passes whose intensity is at the threshold of pain (120 dB). (b) What is the pressure amplitude in this wave?

24. (II) Expensive stereo amplifier A is rated at 250 W per channel, while the more modest amplifier B is rated at 40 W per channel. (a) Estimate the sound level in decibels you would expect at a point 2.5 m from a loudspeaker connected in turn to each amp. (b) Will the expensive amp sound twice as loud as the cheaper one?

25. (II) At a rock concert, a dB meter registered 130 dB when placed 3.4 m in front of a loudspeaker on the stage. (a) What was the power output of the speaker, assuming uniform spherical spreading of the sound and neglecting absorption in the air? (b) How far away would the sound level be a somewhat reasonable 90 dB?

26. (II) A jet plane emits 5.0×10^{5} J of sound energy per second. (a) What is the sound level 30 m away? Air absorbs sound at a rate of about 7.0 dB/km; calculate what the sound level will be (b) 1.0 km and (c) 5.0 km away from this jet plane, taking into account air absorption.

27. (II) (a) Show that the sound level, β, can be written in terms of the pressure amplitude, ΔP_M, as

$$\beta(\text{dB}) = 20 \log \frac{\Delta P_M}{\Delta P_{M0}},$$

where ΔP_{M0} is the pressure amplitude at some reference level. (b) The reference pressure amplitude ΔP_{M0} is often taken to be 3.0×10^{-5} N/m² corresponding to an intensity of 1.0×10^{-12} W/m². What would the sound level be if ΔP_M were 1 atm?

Section 16–4

28. (I) The G string on a violin has a fundamental frequency of 196 Hz. The length of the vibrating portion is 32 cm and has a mass of 0.68 g. Under what tension must the string be placed?

29. (I) (a) What resonant frequency would you expect from blowing across the top of an empty soda bottle that is 15 cm deep? (b) How would that change if it was one-third full of soda?

30. (I) How far from the end of the flute in Example 16–9 should the hole be that must be uncovered to play D above middle C at 294 Hz?

31. (I) If you were to build a pipe organ with open-tube pipes spanning the range of human hearing (20 Hz to 20 kHz), what would be the range of the lengths of pipes required?

32. (I) At a science museum there is a display called a sewer pipe symphony. It consists of many plastic pipes of various lengths, which are open on both ends. (a) If the pipes have lengths of 3.0 m, 2.5 m, 2.0 m, 1.5 m and 1.0 meters, what frequencies will be heard by a visitor's ear placed near the ends of the pipes? (b) Why does this display work better on a noisy day rather than a quiet day?

33. (I) An organ pipe is 78.0 cm long. What are the fundamental and first three audible overtones if the pipe is (a) closed at one end, and (b) open at both ends?

34. (II) An unfingered guitar string is 0.73 m long and is tuned to play E above middle C (330 Hz). (a) How far from the end of this string must the finger be placed to play A above middle C (440 Hz)? (b) What are the frequency and wavelength of the sound wave produced in air at 20°C?

35. (II) (a) Determine the length of an open organ pipe that emits middle C (262 Hz) when the temperature is 21°C. (b) What are the wavelength and frequency of the fundamental standing wave in the tube? (c) What are λ and f in the traveling sound wave produced in the outside air?

36. (II) The human ear canal is approximately 2.5 cm long. It is open to the outside and is closed at the other end by the tympanic membrane. Estimate the frequencies of the standing wave vibrations in the ear canal. What is the relationship of your answer to the information in the graph of Fig. 16–5?

37. (II) An organ is in tune at 20°C. By what percent will the frequency be off at 5.0°C?

38. (II) A particular organ pipe can resonate at 264 Hz, 440 Hz, and 616 Hz, but not at any other intermediate frequencies. (a) Is this an open or closed pipe? (b) What is the fundamental frequency of this pipe?

39. (II) (a) At $T = 15°C$, how long must an open organ pipe be if it is to have a fundamental frequency of 294 Hz? (b) If this pipe were filled with helium, what would its fundamental frequency be?

40. (II) A pipe in air at 20°C is to be designed to produce two successive harmonics at 240 Hz and 280 Hz. How long must the pipe be, and is it open or closed?

41. (II) A uniform narrow tube 1.95 m long is open at both ends. It resonates at two successive harmonics of frequency 275 Hz and 330 Hz. What is the speed of sound in the gas in the tube?

42. (II) How many overtones are present within the audible range for a 2.16 m-long organ pipe at 20°C (a) if it is open, and (b) if it is closed?

* Section 16–5

*43. (II) Approximately what are the intensities of the first two overtones of a violin compared to the fundamental? How many decibels softer than the fundamental are the first and second overtones? (See Fig. 16–15.)

Section 16–6

44. (I) A piano tuner hears one beat every 2.0 s when trying to adjust two strings, one of which is sounding 440 Hz. How far off in frequency is the other string?

45. (I) A certain dog whistle operates at 23.5 kHz, while another (brand X) operates at an unknown frequency. If neither whistle can be heard by humans when played separately, but a shrill whine of frequency 5000 Hz occurs when they are played simultaneously, estimate the operating frequency of brand X.

46. (II) Two violin strings are tuned to the same frequency, 294 Hz. The tension in one string is then decreased by 2.0 percent. What will be the beat frequency heard when the two strings are played together?

47. (II) Two piano strings are supposed to be vibrating at 132 Hz, but a piano tuner hears three beats every 2.0 s when they are played together. (a) If one is vibrating at 132 Hz, what must be the frequency of the other (is there only one answer)? (b) By how much (in percent) must the tension be increased or decreased to bring them in tune?

48. (II) How many beats will be heard if two identical flutes each try to play middle C (262 Hz), but one is at 5.0°C and the other at 25.0°C?

49. (II) Two loudspeakers are 1.8 m apart. A person stands 3.0 m from one speaker and 3.5 m from the other. (a) What is the lowest frequency at which destructive interference will occur at this point? (b) Calculate two other frequencies that also result in destructive interference at this point (give the next two highest). Let $T = 20°C$.

50. (II) The two sources of sound in Fig. 16–16 face each other and emit sounds of equal amplitude and equal frequency (250 Hz) but 180° out of phase. For what minimum separation of the two speakers will there be some point at which (a) complete constructive interference occurs and (b) complete destructive interference occurs. (Assume $T = 20°C$)

51. (II) The two sources shown in Fig. 16–16 emit sound waves, in phase, each of wavelength λ and amplitude D_M. Consider a point such as C or D in the diagram, and let r_A and r_B be the distances of this point from source A and source B, respectively. Show that if r_A and r_B are nearly equal $(r_A - r_B \ll r_A)$ then the amplitude varies approximately with position as

$$\left(\frac{2D_M}{r_A}\right) \cos \frac{\pi}{\lambda}(r_A - r_B).$$

52. (II) Two loudspeakers are placed 3.00 m apart, as shown in Fig. 16–33. They emit 440-Hz sounds, in phase. A microphone is placed 3.20 m distant from a point midway between the two speakers, where an intensity maximum is recorded. (a) How far must the microphone be moved to the right to find the first intensity minimum? (b) Suppose the speakers are reconnected so that the 440-Hz sounds they emit are exactly out of phase. At what positions are the intensity maximum and minimum now?

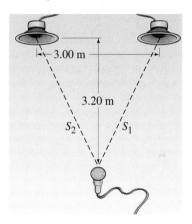

FIGURE 16–33
Problem 52.

53. (II) A guitar string produces 4 beats/s when sounded with a 350-Hz tuning fork and 9 beats/s when sounded with a 355-Hz tuning fork. What is the vibrational frequency of the string? Explain your reasoning.

54. (II) You have three tuning forks, A, B, and C. Fork B has a frequency of 440 Hz; when A and B are sounded together, a beat frequency of 3 Hz is heard. When B and C are sounded together, the beat frequency is 4 Hz. What are the possible frequencies of A and C? What beat frequencies are possible when A and C are sounded together?

55. (II) Show that the two speakers in Fig. 16–16 must be separated by at least a distance d equal to one-half the wavelength λ of sound if there is to be any place where complete destructive interference occurs. The speakers are in phase.

56. (II) A source emits sound of wavelengths 2.64 m and 2.76 m in air. (a) How many beats per second will be heard (assume $T = 20°C$)? (b) How far apart in space are the regions of maximum intensity?

57. (I) The predominant frequency of a certain police car's siren is 1550 Hz when at rest. What frequency do you detect if you move with a speed of 30.0 m/s (a) toward the car, and (b) away from the car?

58. (I) A bat at rest sends out ultrasonic sound waves at 50,000 Hz and receives them returned from an object moving radially away from it at 25.0 m/s. What is the received sound frequency?

59. (II) A bat flies toward a wall at a speed of 5.0 m/s. As it flies, the bat emits an ultrasonic sound wave with frequency 30,000 Hz. What frequency does the bat hear in the reflected wave?

60. (II) In one of the original Doppler experiments, one tuba was played on a moving train car at a frequency of 75 Hz, and a second identical tuba played the same tone while at rest in the railway station. What beat frequency was heard if the train car approached the station at a speed of 10.0 m/s?

61. (II) Two automobiles are equipped with the same single-frequency horn. When one is at rest and the other is moving toward an observer at 15 m/s, a beat frequency of 5.5 Hz is heard. What is the frequency the horns emit? Assume $T = 20°C$.

62. (II) Compare the shift in frequency if a 2000-Hz source is moving toward you at 15 m/s versus if you are moving toward it at 15 m/s. Are the two frequencies exactly the same? Are they close? Repeat the calculation for 150 m/s and then again for 300 m/s. What can you conclude about the asymmetry of the Doppler formulas? Show that at low speeds (relative to the speed of sound), the two formulas—source approaching and detector approaching—yield the same result.

63. (II) A Doppler flow meter uses ultrasound waves to measure blood-flow speeds. Suppose the device emits sound at 3.5 MHz, and the speed of sound in human tissue is taken to be 1540 m/s. What is the expected beat frequency if blood is flowing normally in large leg arteries at 2.0 cm/s directly away from the sound source?

64. (II) The Doppler effect using ultrasonic waves of frequency 2.25×10^6 Hz is used to monitor the heartbeat of a fetus. A (maximum) beat frequency of 500 Hz is observed. Assuming that the speed of sound in tissue is 1.54×10^3 m/s, calculate the maximum velocity of the surface of the beating heart.

65. (II) In Problem 64, the beat frequency is found to appear and then disappear 180 times per minute, which reflects the fact that the heart is beating and its surface changes speed. What is the heartbeat rate?

66. (II) (a) Use the binomial expansion to show that Eqs. 16–9a and 16–10a become essentially the same for small relative velocity between source and observer. (b) What percent error would result if Eq. 16–10a were used instead of Eq. 16–9a for a relative velocity of 22 m/s?

67. (III) A factory whistle emits sound of frequency 570 Hz. On a day when the wind velocity is 12.0 m/s from the north, what frequency will observers hear who are located, at rest, (a) due north, (b) due south, (c) due east, and (d) due west, of the whistle? What frequency is heard by a cyclist heading (e) north or (f) west, toward the whistle at 15.0 m/s? Assume $T = 20°C$.

* 68. (I) (a) How fast is an object moving on land if it is moving at Mach 0.33? (b) A high-flying Concorde passenger jet displays its Mach number on a screen while cruising at 3000 km/h to be 3.2. What is the speed of sound at that altitude?

* 69. (II) Show that the angle θ a sonic boom makes with the path of a supersonic object is given by Eq. 16–12.

* 70. (II) An airplane travels at Mach 2.3 where the speed of sound is 310 m/s. (a) What is the angle the shock wave makes with the direction of the airplane's motion? (b) If the plane is flying at a height of 7100 m, how long after it is directly overhead will a person on the ground hear the shock wave?

* 71. (II) A space probe enters the thin atmosphere of another planet where the speed of sound is only about 35 m/s. (a) What is the probe's Mach number if its initial speed is 15,000 km/h? (b) What is the apex angle of the shock wave it produces?

* 72. (II) A meteorite traveling 8000 m/s strikes the ocean. Determine the shock wave angle it produces (a) in the air just before entering the ocean, and (b) in the water just after entering. Assume $T = 20°C$.

* 73. (II) You look directly overhead and see a plane exactly 1.5 km above the ground flying faster than the speed of sound. By the time you hear the sonic boom, the plane has traveled a horizontal distance of 2.0 km. See Fig. 16–34. Determine (a) the angle of the shock cone, θ, and (b) the speed of the plane (the Mach number). Assume the speed of sound is 330 m/s.

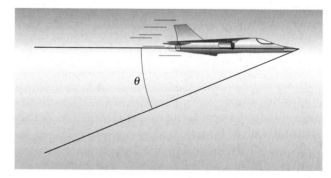

FIGURE 16–34 Problem 73.

* 74. (II) A supersonic jet traveling at Mach 1.8 at an altitude of 10,000 m passes directly over an observer on the ground. Where will the plane be relative to the observer when the latter hears the sonic boom? (See Fig. 16–35.)

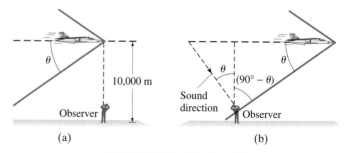

FIGURE 16–35 Problem 74.

General Problems

75. (II) A fish finder uses a sonar device that sends 20,000 Hz sound pulses downwards from the bottom of the boat, and then detects echoes. If the maximum depth for which it is designed to work is 200 meters, what is the minimum time between pulses?

76. Approximately how many octaves are there in the human audible range?

77. A stone is dropped from the top of a cliff. The splash it makes when striking the water below is heard 3.5 s later. How high is the cliff?

78. A single mosquito 5.0 m from a person makes a sound close to the threshold of human hearing (0 dB). What will be the sound level of 1000 such mosquitoes?

79. At the Indianapolis 500, you can estimate the speed of cars just by listening to the difference in pitch of the engine noise between approaching and receding cars. Suppose the sound of a certain car drops by a full octave as it goes by on the straightaway. How fast is it going?

80. A tight guitar string has a frequency of 540 Hz as its third harmonic. What will be its fundamental frequency if it is fingered at a length of only 60 percent of its original length?

81. Each string on a violin is tuned to a frequency $1\frac{1}{2}$ times that of its neighbor. If all the strings are to be placed under the same tension, what must be the mass per unit length of each string relative to that of the lowest string?

82. What is the resultant sound level when an 80-dB sound and an 85-dB sound are heard simultaneously?

83. The sound level 12.0 m from a loudspeaker, placed in the open, is 100 dB. What is the acoustic power output (W) of the speaker, assuming it radiates equally in all directions?

84. The A string of a violin is 32 cm long between fixed points with a fundamental frequency of 440 Hz and a linear density of 6.1×10^{-4} kg/m. (a) What are the wave speed and tension in the string? (b) What is the length of the tube of a simple wind instrument (say, an organ pipe) closed at one end whose fundamental is also 440 Hz if the speed of sound is 343 m/s in air? (c) What is the frequency of the first overtone of each instrument?

85. A stereo amplifier is rated at 150 W output at 1000 Hz. The power output drops by 10 dB at 15 kHz. What is the power output in watts at 15 kHz?

86. A tuning fork is set into vibration above a vertical open tube filled with water (Fig. 16–36). The water level is allowed to drop slowly. As it does so, the air in the tube above the water level is heard to resonate with the tuning fork when the distance from the tube opening to the water level is 0.125 m and again at 0.395 m. What is the frequency of the tuning fork?

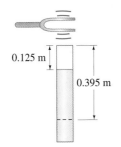

0.125 m

0.395 m

FIGURE 16–36
Problem 86.

87. Two loudspeakers face each other at opposite ends of a long corridor. They are connected to the same source which produces a pure tone of 280 Hz. A person walks from one speaker toward the other at a speed of 1.4 m/s. What "beat" frequency does the person hear?

88. Workers around jet aircraft typically wear protective devices over their ears. Assume that the intensity level of a jet airplane engine, at a distance of 30 m, is 140 dB, and that the average human ear has an effective radius of 2.0 cm. What would be the power intercepted by an unprotected ear at a distance of 30 m from a jet airplane engine?

89. The intensity at the "threshold of hearing" for the human ear at a frequency of about 1000 Hz is $I_0 = 1.0 \times 10^{-12}$ W/m², for which β, the sound level, is 0 dB. The "threshold of pain" at the same frequency is about 120 dB, or $I = 1.0$ W/m², corresponding to an increase of intensity by a factor of 10^{12}. By what factors do the displacement amplitude, D_M, and the pressure amplitude ΔP_M, vary?

90. As Fig. 16–5 shows, the human ear is not equally sensitive to all frequencies; the threshold of hearing is about 60 dB at 35 Hz, 0 dB at 1000 Hz and at 5000 Hz, and 20 dB at 15,000 Hz. Thus, near threshold, the ear is especially insensitive to low frequencies. Estimate the displacement amplitude for each of these four points. At which frequency is the ear most sensitive to displacement?

91. In audio and communications systems, the *gain*, β, in decibels is defined as $\beta = 10 \log(P_{out}/P_{in})$ where P_{in} is the power input to the system and P_{out} is the power output. A particular stereo amplifier puts out 100 W of power for an input of 1 mW. What is its gain in dB?

92. Two loudspeakers are at opposite ends of a railroad car as it moves past a stationary observer at 10.0 m/s, as shown in Fig. 16–37. If they have identical sound frequencies of 200 Hz, what is the beat frequency heard by the observer when (a) he listens from the position A, in front of the car, (b) he is between the speakers, at B, and (c) he hears the speakers after they have passed him, at C?

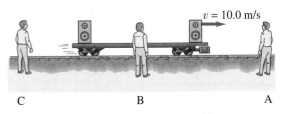

$v = 10.0$ m/s

C   B   A

FIGURE 16–37   Problem 92.

93. The frequency of a steam train whistle as it approaches you is 538 Hz. After it passes you, its frequency is measured as 486 Hz. How fast was the train moving (assume constant velocity)?

94. A 75-cm-long guitar string of mass 2.10 g is placed near a tube open at one end, and also 75 cm long. How much tension should be in the string if it is to produce resonance (in its fundamental mode) with the third harmonic in the tube?

95. If the velocity of blood flow in the aorta is normally about 0.32 m/s, what beat frequency would you expect if 5.50-MHz ultrasound waves were directed along the flow and reflected from the red blood cells? Assume that the waves travel with a speed of 1.54×10^3 m/s.

96. A source of sound waves (wavelength λ) is a distance l from a detector. Sound reaches the detector directly, and also by reflecting off an obstacle, as shown in Fig. 16–38. The obstacle is equidistant from source and detector. When the obstacle is a distance d to the right of the line of sight between source and detector, as shown, the two waves arrive in phase. How much farther to the right must the obstacle be moved if the two waves are to be out of phase by $\frac{1}{2}$ wavelength, so destructive interference occurs? (Assume $\lambda \ll l, d$.)

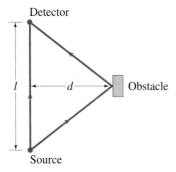

FIGURE 16–38 Problem 96.

97. A person hears a pure tone in the 500–1000 Hz range coming from two sources. The sound is loudest at points equidistant from the two sources. In order to determine exactly what the frequency is, the person moves about and finds that the sound level is minimal at a point 0.31 m farther from one source than the other. What is the frequency of the sound?

98. A bat flies toward a moth at speed 6.5 m/s while the moth is flying toward the bat at speed 5.0 m/s. The bat emits a sound wave of 51.35 kHz. What is the frequency of the wave detected by the bat after it reflects off the moth?

99. A dramatic demonstration, called "singing rods," involves a long, slender aluminum rod held in the hand near the rod's midpoint. The rod is stroked with the other hand. With a little practice, the rod can be made to "sing," or emit a clear, loud, ringing sound. For a 90-cm-long rod: (a) what is the fundamental frequency of the sound? (b) What is its wavelength in the rod, and (c) what is the traveling wavelength in air at 20°C?

100. Room acoustics for stereo listening can be compromised by the presence of standing waves, which can cause acoustic "dead spots" at the locations of the pressure nodes. Consider a living room with dimensions 5.0 m long, 4.0 m wide, and 2.8 m high. Calculate the fundamental frequencies for the standing waves in this room.

101. Assuming that the maximum displacement of the air molecules in a sound wave is about the same as that of the speaker cone that produces the sound (Fig. 16–39), estimate by how much a loudspeaker cone moves for a fairly loud (100 dB) sound of (a) 10 kHz, and (b) 40 Hz.

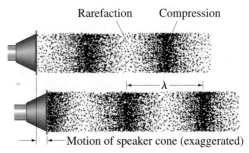

FIGURE 16–39 Problem 101.

102. A Doppler flow meter is used to measure the speed of blood flow. Transmitting and receiving elements are placed on the skin, as shown in Fig. 16–40. Typical sound-wave frequencies of about 5.0 MHz are used, which have a reasonable chance of being reflected from red blood cells. By measuring the frequency of the reflected waves, which are Doppler-shifted because the red blood cells are moving, the speed of the blood flow can be deduced. "Normal" blood flow speed is about 0.1 m/s. Suppose that an artery is partly constricted, so that the speed of the blood flow is increased, and the flow meter measures a Doppler shift of 900 Hz. What is the speed of blood flow in the constricted region? The effective angle between the sound waves (both transmitted and reflected) and the direction of blood flow is 45°. Assume the velocity of sound in tissue is 1540 m/s.

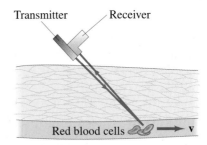

FIGURE 16–40 Problem 102.

The balloon in the foreground is about to leave on the first successful nonstop trip around the world in a balloon. To the rear, we can see the flame heating the air in another balloon. Heating the air inside a "hot-air" balloon raises the air's temperature and forces some of the air to escape from the opening at the bottom. This is an example of Charles's law: the volume of an enclosed gas at constant pressure increases directly with the Kelvin temperature. The balloon is open at the bottom, so the pressure is atmospheric. The volume increase caused by the heating forces some gas to escape through the opening. The reduced amount of gas inside means its density is lower, so there is a net buoyant force upward on the balloon. In this chapter we study temperature and its effects on matter: thermal expansion, thermal stresses, and the gas laws.

CHAPTER 17

Temperature, Thermal Expansion, and the Ideal Gas Law

The topics of temperature, heat and thermodynamics are the subject of the next four chapters, Chapters 17 through 20. Also included, and closely related, as a means of understanding, is the kinetic theory of gases.

We will often consider a particular **system**, by which we mean a particular object or set of objects; everything else in the universe is called the "environment." We can describe the **state** (or condition) of a particular system—such as a gas in a container—from either a microscopic or macroscopic point of view. A **microscopic** description would involve details of the motion of all the atoms or molecules making up the system, which could be very complicated. A **macroscopic** description is given in terms of quantities that are detectable directly by our senses, such as volume, mass, pressure, and temperature.

Microscopic vs. macroscopic properties

The description of processes in terms of macroscopic quantities is the field of **thermodynamics**. The number of macroscopic variables required to describe the state of a system at any time depends on the type of system. To describe the state of a pure gas in a container, for example, we need only three variables, which could be the volume, the pressure, and the temperature. Quantities such as these that can be used to describe the state of a system are called **state variables**.

The emphasis in this chapter is on the concept of temperature. We will begin, however, with a brief discussion of the theory that matter is made up of atoms and that these atoms are in continuous random motion. This theory is called the *kinetic theory* ("kinetic," you may remember, comes from the Greek word for "moving"), and we will discuss it in more detail in the next chapter.

17–1 | Atomic Theory of Matter

The idea that matter is made of atoms dates back to the ancient Greeks. According to the Greek philosopher Democritus, if a given substance—say, a piece of iron—were cut into smaller and smaller bits, eventually a smallest piece of that substance would be obtained which could not be divided further. This smallest piece was called an **atom**, which comes from the Greek *atomos*, "indivisible."[†] The only real alternative to the atomic theory of matter was the idea that matter is continuous and can be subdivided indefinitely.

Atomic and molecular masses

Today, we often speak of the relative masses of atoms and molecules—what we call the **atomic mass** or **molecular mass**, respectively.[‡] These are based on assigning the abundant carbon atom, ^{12}C, the value of exactly 12.0000 **unified atomic mass units** (u). In terms of kilograms,

$$1\,u = 1.6605 \times 10^{-27}\,kg.$$

The atomic mass of hydrogen is then 1.0078 u, and the values for other atoms are as listed in the periodic table inside the back cover of this book, and also in Appendix D. The molecular mass of a compound is the sum of the atomic masses of the atoms making up the molecule.

An important piece of evidence for the atomic theory is the so-called **Brownian movement**, named after the biologist Robert Brown, who is credited with its discovery in 1827. While he was observing tiny pollen grains suspended in water under his microscope, Brown noticed that the tiny grains moved about in tortuous paths (Fig. 17–1), even though the water appeared to be perfectly still. The atomic theory easily explains Brownian movement if the further reasonable assumption is made that the atoms of any substance are continually in motion. Then Brown's tiny pollen grains are jostled about by the vigorous barrage of rapidly moving molecules of water.

In 1905, Albert Einstein examined Brownian movement from a theoretical point of view and was able to calculate from the experimental data the approximate size and mass of atoms and molecules. His calculations showed that the diameter of a typical atom is about 10^{-10} m.

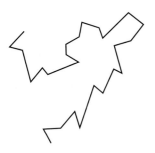

FIGURE 17–1 Path of a tiny particle (pollen grain, for example) suspended in water. The straight lines connect observed positions of the particle at equal time intervals. [The plot has the same general shape independent of the time interval—whether it be every 60 s or 0.1 s, a phenomenon referred to as *fractal* behavior.]

Phases of matter

At the start of Chapter 13, we distinguished the three common states of matter—solid, liquid, gas—based on macroscopic, or "large-scale," properties. Now let us see how these three phases of matter differ from the atomic, or microscopic, point of view. Clearly, atoms and molecules must exert attractive forces on each other. For how else could a brick or a piece of aluminum stay together in one piece? The attractive forces between molecules are of an electrical nature (more on this in later chapters). If the molecules come too close together, the force between them becomes repulsive (electric repulsion between their outer electrons). Thus molecules maintain a minimum distance from each other. In a solid material, the attractive forces are strong enough that the atoms or molecules are held in more or less fixed positions, often in an array known as a crystal lattice, as shown in Fig. 17–2a. The atoms or molecules in a solid are in motion—they vibrate about their nearly fixed positions. In a liquid, the atoms or molecules are moving more rapidly, or the forces between them are weaker, so that they are sufficiently free to pass over one another, as in Fig. 17–2b. In a gas, the forces are so weak, or the speeds so high, that the molecules do not even stay close together. They move rapidly every which way, Fig. 17–2c, filling any container and occasionally colliding with one another. On the average, the speeds are sufficiently high in a gas that

[†]Today, of course, we don't consider the atom as indivisible, but rather as consisting of a nucleus (containing protons and neutrons) and electrons.

[‡]The terms *atomic weight* and *molecular weight* are popularly used for these quantities, but properly speaking we are comparing masses.

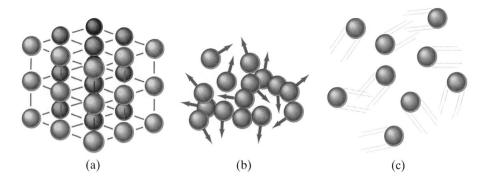

(a)   (b)   (c)

FIGURE 17–2 Atomic arrangements in (a) a crystalline solid, (b) a liquid, and (c) a gas.

when two molecules collide, the force of attraction is not strong enough to keep them close together and they fly off in new directions.

EXAMPLE 17–1 ESTIMATE **Distance between atoms.** The density of copper is $8.9 \times 10^3 \, \text{kg/m}^3$ and each copper atom has a mass of 63 u, where $1 \, \text{u} = 1.66 \times 10^{-27} \, \text{kg}$. Estimate the average distance between neighboring atoms.

SOLUTION The mass of 1 copper atom is $63 \times 1.66 \times 10^{-27} \, \text{kg} = 1.04 \times 10^{-25} \, \text{kg}$. This means that in a cube of copper 1 m on a side $(\text{volume} = 1 \, \text{m}^3)$, there are

$$\frac{8.9 \times 10^3 \, \text{kg/m}^3}{1.04 \times 10^{-25} \, \text{kg/atom}} = 8.5 \times 10^{28} \, \text{atoms/m}^3.$$

The volume of a cube of side l is $V = l^3$, so on one edge of the 1-m-long cube there are $(8.5 \times 10^{28})^{\frac{1}{3}} \, \text{atoms} = 4.4 \times 10^9 \, \text{atoms}$. Hence the distance between neighboring atoms is

$$\frac{1 \, \text{m}}{4.4 \times 10^9 \, \text{atoms}} = 2.3 \times 10^{-10} \, \text{m between atoms}.$$

17–2 Temperature and Thermometers

In everyday life, **temperature** is a measure of how hot or cold an object is. A hot oven is said to have a high temperature, whereas the ice of a frozen lake is said to have a low temperature.

Many properties of matter change with temperature. For example, most materials expand when heated.[†] An iron beam is longer when hot than when cold. Concrete roads and sidewalks expand and contract slightly according to temperature, which is why compressible spacers or expansion joints (Fig. 17–3) are placed at regular intervals. The electrical resistance of matter changes with temperature (see Chapter 25). So too does the color radiated by objects, at least at high temperatures: you may have noticed that the heating element of an electric stove glows with a red color when hot. At higher temperatures, solids such as iron glow orange or even white. The white light from an ordinary incandescent lightbulb comes from an extremely hot tungsten wire. The surface temperatures of the Sun and other stars can be measured by the predominant color (more precisely, wavelengths) of light they emit.

Instruments designed to measure temperature are called **thermometers**. There are many kinds of thermometers, but their operation always depends on some property of matter that changes with temperature. Most common thermometers

FIGURE 17–3 Expansion joint on a bridge.

Thermometers: measuring temperature

[†] Most materials expand when their temperature is raised, but not all. Water, for example, in the range 0°C to 4°C contracts with an increase in temperature (see Section 17–4).

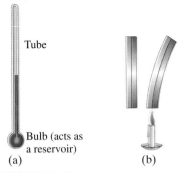

FIGURE 17–4
(a) Mercury- or alcohol-in-glass thermometer; (b) bimetallic strip.

Temperature scales

FIGURE 17–5 Photograph of a thermometer using a coiled bimetallic strip.

rely on the expansion of a material with an increase in temperature. The first idea for a thermometer, by Galileo, made use of the expansion of a gas.

Common thermometers today consist of a hollow glass tube filled with mercury or with alcohol colored with a red dye. In such liquid-in-glass thermometers, the liquid expands more than the glass when the temperature is increased, so the liquid level rises in the tube (Fig. 17–4a). Although metals also expand with temperature, the change in length of a metal rod, say, is generally too small to measure accurately for ordinary changes in temperature. However, a useful thermometer can be made by bonding together two dissimilar metals whose rates of expansion are different (Fig. 17–4b). When the temperature is increased, the slightly different amounts of expansion cause the bimetallic strip to bend. Often the bimetallic strip is in the form of a coil, one end of which is fixed while the other is attached to a pointer, Fig. 17–5. This kind of thermometer is used as ordinary air thermometers, oven thermometers, automatic-off switch in electric coffeepots, and in room thermostats for determining when the heater or air conditioner should go on or off. Very precise thermometers make use of electrical properties (Chapter 25), such as resistance thermometers, thermocouples, and thermistors, which may have a digital readout.

In order to measure temperature quantitatively, some sort of numerical scale must be defined. The most common scale today is the **Celsius** scale, sometimes called the **centigrade** scale. In the United States, the **Fahrenheit** scale is also common. The most important scale in scientific work is the absolute, or Kelvin, scale, and it will be discussed later in this chapter.

One way to define a temperature scale is to assign arbitrary values to two readily reproducible temperatures. For both the Celsius and Fahrenheit scales these two fixed points are chosen to be the freezing point and the boiling point[†] of water, both taken at atmospheric pressure. On the Celsius scale, the freezing point of water is chosen to be 0°C ("zero degrees Celsius") and the boiling point 100°C. On the Fahrenheit scale, the freezing point is defined as 32°F and the boiling point 212°F. A practical thermometer is calibrated by placing it in carefully prepared environments at each of the two temperatures and marking the position of the mercury or pointer. For a Celsius scale, the distance between the two marks is then divided into one hundred equal intervals separated by small marks representing each degree between 0°C and 100°C (hence the name "centigrade scale" meaning "hundred steps"). For a Fahrenheit scale, the two points are labeled 32°F and 212°F and the distance between them is divided into 180 equal intervals. For temperatures below the freezing point of water and above the boiling point of water the scales can be extended using the same equally spaced intervals. However, ordinary thermometers can be used only over a limited temperature range because of their own limitations—for example, the mercury in a mercury-in-glass thermometer solidifies at some point, below which the thermometer will be useless. It is also rendered useless above temperatures where the fluid vaporizes. For very low or very high temperatures, specialized thermometers are required, some of which we will mention later.

Every temperature on the Celsius scale corresponds to a particular temperature on the Fahrenheit scale, Fig. 17–6. It is easy to convert from one to the other if you remember that 0°C corresponds to 32°F and that a range of 100° on the Celsius scale corresponds to a range of 180° on the Fahrenheit scale. Thus, one Fahrenheit degree (1 F°) corresponds to $100/180 = \frac{5}{9}$ of a Celsius degree (1 C°).

[†] The freezing point of a substance is defined as that temperature at which the solid and liquid phases coexist in equilibrium—that is, without any net liquid changing into the solid or vice versa. Experimentally, this is found to occur at only one definite temperature, for a given pressure. Similarly, the boiling point is defined as that temperature at which the liquid and gas coexist in equilibrium. Since these points vary with pressure, the pressure must be specified (usually it is 1 atm).

That is, $1 \text{ F}° = \frac{5}{9} \text{ C}°$. (Notice that when we refer to a specific temperature, we say "degrees Celsius," as in 20°C; but when we refer to a *change* in temperature or a temperature interval, to avoid misunderstanding we will say "Celsius degrees," as in "1 C°.") The conversion between the two temperature scales can be written

$$T(°\text{C}) = \tfrac{5}{9}[T(°\text{F}) - 32] \qquad \text{or} \qquad T(°\text{F}) = \tfrac{9}{5}T(°\text{C}) + 32.$$

Rather than memorizing these relations (it would be easy to confuse them), it is better to simply remember that 0°C = 32°F and 5 C° = 9 F° (or 100°C = 212°F).

EXAMPLE 17–2 **Taking your temperature.** Normal body temperature is 98.6°F. What is this on the Celsius scale?

SOLUTION First we relate the given temperature to the freezing point of water (0°C). That is, 98.6°F is 98.6 − 32.0 = 66.6 F° above the freezing point of water. Since each F° is equal to $\frac{5}{9}$ C°, this corresponds to 66.6 × $\frac{5}{9}$ = 37.0 Celsius degrees above the freezing point. Since the freezing point is 0°C, the temperature is 37.0°C.

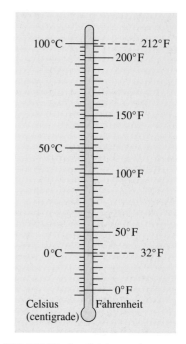

FIGURE 17–6 Celsius and Fahrenheit scales compared.

Different materials do not expand in quite the same way over a wide temperature range. Consequently, if we calibrate different kinds of thermometers exactly as described above, they will not usually agree precisely. Because of how we calibrated them, they will agree at 0°C and at 100°C. But because of different expansion properties, they may not agree precisely at intermediate temperatures (remember we arbitrarily divided the thermometer scale into 100 equal divisions between 0°C and 100°C). Thus a carefully calibrated mercury-in-glass thermometer might register 52.0°C, whereas a carefully calibrated thermometer of another type might read 52.6°C.

Because of this discrepancy, some standard kind of thermometer must be chosen so that these intermediate temperatures can be precisely defined. The chosen standard for this purpose is the so-called **constant-volume gas thermometer**. As shown in the simplified diagram of Fig. 17–7, this thermometer consists of a bulb filled with a dilute gas connected by a thin tube to a mercury manometer. The volume of the gas is kept constant by raising or lowering the right-hand tube of the manometer so that the mercury in the left tube coincides with the reference mark. An increase in temperature causes a proportional increase in pressure in the bulb or in volume of the gas. The tube must thus be lifted higher to keep the gas volume constant. The height of the mercury in the right-hand column is then a measure of the temperature. This thermometer can be calibrated and gives the same results for all gases in the limit of reducing the gas pressure in the bulb toward zero. The resulting scale is defined as the standard temperature scale (Section 17–10).

Constant-volume gas thermometer

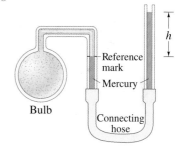

FIGURE 17–7 Constant-volume gas thermometer.

17–3 Thermal Equilibrium and the Zeroth Law of Thermodynamics

We are all familiar with the fact that if two objects at different temperatures are placed in thermal contact (meaning thermal energy can pass from one to the other), the two objects will eventually reach the same temperature. They will then be said to be in **thermal equilibrium**. For example, an ice cube placed in a large glass of hot water melts to water, all of which eventually comes to the same temperature. If you put your hand in the water of an icy lake, you can *feel* the temperature of your hand drop. (It's best to pull your hand out before thermal equilibrium is reached!) Two objects are defined to be in *thermal equilibrium* if, when placed in thermal contact, their temperatures don't change.

Thermal equilibrium

Suppose you wanted to determine if two systems, A and B, are in thermal equilibrium, but without putting them in contact. You could do so by making use of a third system, C (which could be considered a thermometer). Suppose that C and A are in thermal equilibrium and that C and B are in thermal equilibrium. Does this imply that A and B are necessarily in thermal equilibrium with each other? Actually, it isn't completely obvious unless you do some experiments, and all experiments indicate that

Zeroth law of thermodynamics

if two systems are in thermal equilibrium with a third system, then they are in thermal equilibrum with each other.

This postulate is called the **zeroth law of thermodynamics**. It has this rather odd name because it was not until after the great first and second laws of thermodynamics (Chapters 19 and 20) were worked out that scientists realized that this apparently obvious postulate needed to be stated first.

Temperature is a property of a system that determines whether the system will be in thermal equilibrium with other systems. When two systems are in thermal equilibrium, their temperatures are, by definition, equal. This is consistent with our everyday notion of temperature, since when a hot body and a cold one are put into contact, they eventually come to the same temperature. Thus the importance of the zeroth law is that it allows a useful definition of temperature.

17–4 | Thermal Expansion

Most substances expand when heated and contract when cooled. However, the amount of expansion or contraction varies, depending on the material.

Linear Expansion

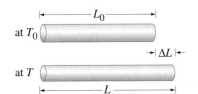

FIGURE 17–8 A thin rod of length L_0 at temperature T_0 is heated to a new uniform temperature T and acquires length L, where $L = L_0 + \Delta L$.

Linear expansion

Experiments indicate that the change in length ΔL of almost all solids is, to a very good approximation, directly proportional to the change in temperature ΔT. As might be expected, the change in length is also proportional to the original length of the object, L_0, Fig. 17–8. That is, for the same temperature change, a 4-m-long iron rod will increase in length twice as much as a 2-m-long iron rod. We can write this proportionality as an equation:

$$\Delta L = \alpha L_0 \, \Delta T, \tag{17–1a}$$

where α, the proportionality constant, is called the *coefficient of linear expansion* for the particular material and has units of $(C°)^{-1}$. This equation can also be written as

$$L = L_0(1 + \alpha \Delta T), \tag{17–1b}$$

where L_0 is the length initially, at temperature T_0, and L is the length after heating or cooling to a temperature T. If the temperature change $\Delta T = T - T_0$ is negative, then $\Delta L = L - L_0$ is also negative, so the length decreases.

The values of α for various materials at 20°C are listed in Table 17–1. It should be noted that α does vary slightly with temperature (which is why thermometers made of different materials do not agree precisely). However, if the temperature range is not too great, the variation can usually be ignored.

➡ PHYSICS APPLIED

Expansion in structures

EXAMPLE 17–3 **Bridge expansion.** The steel bed of a suspension bridge is 200 m long at 20°C. If the extremes of temperature to which it might be exposed are −30°C to +40°C, how much will it contract and expand?

SOLUTION From Table 17–1, we find that $\alpha = 12 \times 10^{-6}(C°)^{-1}$. The increase in length when it is at 40°C will be

$$\Delta L = (12 \times 10^{-6}/C°)(200 \text{ m})(40°C - 20°C) = 4.8 \times 10^{-2} \text{ m},$$

or 4.8 cm. When the temperature decreases to −30°C, $\Delta T = -50 \, C°$. Then

$$\Delta L = (12 \times 10^{-6}/C°)(200 \text{ m})(-50 \, C°) = -12.0 \times 10^{-2} \text{ m},$$

or a decrease in length of 12 cm.

TABLE 17-1 Coefficients of Expansion, at 20°C

Material	Coefficient of Linear Expansion, α $(C°)^{-1}$	Coefficient of Volume Expansion, β $(C°)^{-1}$
Solids		
Aluminum	25×10^{-6}	75×10^{-6}
Brass	19×10^{-6}	56×10^{-6}
Copper	17×10^{-6}	50×10^{-6}
Iron or steel	12×10^{-6}	35×10^{-6}
Lead	29×10^{-6}	87×10^{-6}
Glass (Pyrex)	3×10^{-6}	9×10^{-6}
Glass (ordinary)	9×10^{-6}	27×10^{-6}
Quartz	0.4×10^{-6}	1×10^{-6}
Concrete and brick	$\approx 12 \times 10^{-6}$	$\approx 36 \times 10^{-6}$
Marble	$1.4–3.5 \times 10^{-6}$	$4–10 \times 10^{-6}$
Liquids		
Gasoline		950×10^{-6}
Mercury		180×10^{-6}
Ethyl alcohol		1100×10^{-6}
Glycerin		500×10^{-6}
Water		210×10^{-6}
Gases		
Air (and most other gases at atmospheric pressure)		3400×10^{-6}

CONCEPTUAL EXAMPLE 17–4 **Do holes expand or contract?** A circular hole is cut from sheet metal, as shown in Fig. 17–9. When the metal is heated in the oven, does the hole get larger or smaller?

RESPONSE You might guess that the metal expands into the hole, making the hole smaller. But instead of cutting a hole, imagine drawing a circle on the sheet metal with a pencil. When the metal expands, the material inside the circle will expand along with the rest of the metal; so the circle expands. Cutting the metal where the circle is makes clear to us that the hole grows as well. The material does *not* expand inward to fill the hole. In a solid object, *all* sections expand outward with increased temperature.

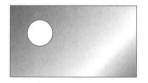

FIGURE 17–9 Conceptual Example 17–4.

EXAMPLE 17–5 **Ring on a rod.** An iron ring is to fit snugly on a cylindrical iron rod. At 20°C, the diameter of the rod is 6.445 cm and the inside diameter of the ring is 6.420 cm. To slip over the rod, the ring must be slightly larger than the rod diameter by about 0.008 cm. To what temperature must the ring be brought if its hole is to be large enough so it will slip over the rod?

SOLUTION The hole in the ring must be increased from a diameter of 6.420 cm to 6.445 cm + 0.008 cm = 6.453 cm. The ring must be heated since the hole diameter will increase linearly with temperature (as in Example 17–4). We solve for ΔT in Eq. 17–1a and find

$$\Delta T = \frac{\Delta L}{\alpha L_0} = \frac{6.453 \text{ cm} - 6.420 \text{ cm}}{(12 \times 10^{-6} \text{ C}^{°-1})(6.420 \text{ cm})} = 430 \text{ C}°.$$

So the ring's temperature must be raised at least to $T = (20°C + 430C°) = 450°C$.

Volume Expansion

The change in *volume* of a material which undergoes a temperature change is given by a relation similar to Eq. 17–1, namely,

Volume expansion

$$\Delta V = \beta V_0 \, \Delta T, \tag{17–2}$$

where ΔT is the change in temperature, V_0 is the original volume, ΔV is the change in volume, and β is the *coefficient of volume expansion*. The units of β are $(\text{C}°)^{-1}$.

$\beta \approx 3\alpha$

Values of β for various materials are given in Table 17–1. Notice that for solids, β is normally equal to approximately 3α. To see why, consider a rectangular solid of length L_0, width W_0, and height H_0. When its temperature is changed by ΔT, its volume changes from $V_0 = L_0 W_0 H_0$ to

$$V = L_0(1 + \alpha \Delta T) W_0(1 + \alpha \Delta T) H_0(1 + \alpha \Delta T),$$

using Eq. 17–1b and assuming α is the same in all directions. Thus,

$$\Delta V = V - V_0 = V_0(1 + \alpha \Delta T)^3 - V_0 = V_0 \big[3\alpha \Delta T + 3(\alpha \Delta T)^2 + (\alpha \Delta T)^3 \big].$$

If the amount of expansion is much smaller than the original size of the object, then $\alpha \Delta T \ll 1$ and we can ignore all but the first term and obtain

$$\Delta V \approx (3\alpha) V_0 \, \Delta T.$$

This is Eq. 17–2 with $\beta \approx 3\alpha$. For solids that are not isotropic (having the same properties in all directions), however, the relation $\beta \approx 3\alpha$ is not valid. Note also that linear expansion has no meaning for liquids and gases since they do not have fixed shapes.

Equations 17–1 and 17–2 are accurate only if ΔL (or ΔV) is small compared to L_0 (or V_0). This is of particular concern for liquids and even more so for gases because of the large values of β. Furthermore, β itself varies substantially with temperature for gases. Therefore, a more convenient way of dealing with gases is needed, and this will be discussed starting in Section 17–6.

→ PHYSICS APPLIED

Gas tank overflow

EXAMPLE 17–6   Gas tank in the sun.   The 70-L steel gas tank of a car is filled to the top with gasoline at 20°C. The car is then left to sit in the sun, and the tank reaches a temperature of 40°C (104°F). How much gasoline do you expect to overflow from the tank?

SOLUTION   The gasoline expands (Eq. 17–2) by

$$\Delta V = \beta V_0 \, \Delta T = \left(950 \times 10^{-6}\, \text{C}^{°-1} \right)(70\,\text{L})(20\,\text{C}°) = 1.3\,\text{L}.$$

The tank also expands. We can think of it as a steel shell that undergoes volume expansion $\left(\beta \approx 3\alpha = 36 \times 10^{-6}\, \text{C}^{°-1} \right)$. If the tank were solid, the exterior surface layer (the shell) would expand just the same. Thus the tank increases in volume by

$$\Delta V = \left(36 \times 10^{-6}\, \text{C}^{°-1} \right)(70\,\text{L})(20\,\text{C}°) = 0.050\,\text{L},$$

so the tank expansion has little effect. If this full tank were left in the sun, over a liter of gas could spill out into the road.

Want to save a few pennies? Fill your gas tank early in the morning when it is cool and the gas is denser—but don't fill the tank quite all the way.

*Atomic Theory of Expansion

How can we understand thermal expansion from a microscopic point of view? Let us assume that the atoms in a solid are always in motion, vibrating about their equilibrium positions. We also assume that their average kinetic energy increases with temperature, as we shall discuss in the next chapter. But as temperature increases, does this mean that the average distance between atoms increases?

Experimentally, a solid rod gets longer when you raise its temperature, so we conclude that the average distance between atoms does increase. To understand this, let us look at a simplified potential-energy diagram as shown in Fig. 17–10, which represents the potential energy of two atoms versus their separation r. At large r, we assume the potential energy ≈ 0, and as r decreases, the potential energy decreases, indicating an attractive force as discussed in Section 8–9. For r less than r_0 (the equilibrium position) the potential-energy curve rises, indicating a repulsive force between atoms as they approach each other. The horizontal blue lines in Fig. 17–10 labeled E_2 and E_1 represent the total energy for two different temperatures, T_2 and T_1, where $T_2 > T_1$. The short vertical lines for E_1 and E_2 on the diagram represent the midpoints of the motion at these two temperatures. Because the potential-energy curve is not symmetrical, the average separation of atoms is greater for the higher temperature, as shown. Thus thermal expansion is due to the nonsymmetry of the potential-energy function. If the potential-energy curve were symmetrical, there would be no thermal expansion at all. Indeed, the fact that most substances expand when heated *implies* that the potential-energy curve must be asymmetrical, as in Fig. 17–10.

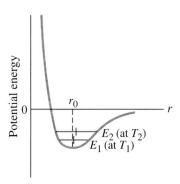

FIGURE 17–10 Typical curve of potential energy versus separation of atoms, r, for atoms in a crystal solid (simplified). Note that the midpoint (short vertical line) of the oscillatory motion of atoms is greater at the higher temperature T_2.

Anomalous Behavior of Water Below 4°C

Most substances expand more or less uniformly with an increase in temperature, as long as no phase change occurs. Water, however, does not follow the usual pattern. If water at 0°C is heated, it actually *decreases* in volume until it reaches 4°C. Above 4°C water behaves normally and expands in volume as the temperature is increased, Fig. 17–11. Water thus has its greatest density at 4°C. This anomalous behavior of water is of great importance for the survival of aquatic life during cold winters. When the water in a lake or river is above 4°C and begins to cool by contact with cold air, the water at the surface sinks because of its greater density and it is replaced by warmer water from below. This mixing continues until the temperature reaches 4°C. As the surface water cools further, it remains on the surface because it is less dense than the 4°C water below. Water then freezes first at the surface, and the ice remains on the surface since ice (specific gravity = 0.917) is less dense than water. The water at the bottom remains at 4°C until almost the whole body of water is frozen. If water were like most substances, becoming more dense as it cools, the water at the bottom of a lake would be frozen first. Lakes would freeze solid more easily since circulation would bring the warmer water to the surface to be efficiently cooled. The complete freezing of a lake would cause severe damage to its plant and animal life. Because of the unusual behavior of water below 4°C, it is rare for any large body of water to freeze completely, and this is helped by the layer of ice on the surface which acts as an insulator to reduce the flow of heat out of the water into the cold air above. Without this peculiar but wonderful property of water, life on this planet as we know it might not have been possible.

Not only does water expand as it cools from 4°C to 0°C, it expands even more as it freezes to ice. This is why ice cubes float in water and pipes break when water inside them freezes.

Water is unusual: it expands *when cooled from 4°C to 0°C*

➡ **P H Y S I C S   A P P L I E D**

Life under ice

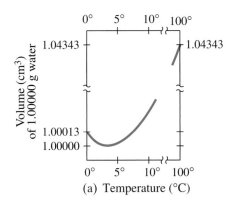

(a) Temperature (°C)

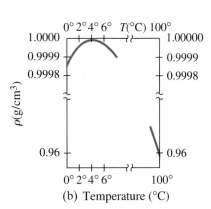

(b) Temperature (°C)

FIGURE 17–11 Behavior of water as a function of temperature near 4°C. (a) Volume, of 1.00000 gram of water, as a function of temperature. (b) Density vs. temperature. [Note the break in each axis.]

*17–5 | Thermal Stresses

In many situations, such as in buildings and roads, the ends of a beam or slab of material are rigidly fixed, which greatly limits expansion or contraction. If the temperature should change, large compressive or tensile stresses, called *thermal stresses*, will occur. The magnitude of such stresses can be calculated using the concept of elastic modulus developed in Chapter 12. To calculate the internal stress, we can think of this process as occurring in two steps. The rod expands (or contracts) by an amount ΔL given by Eq. 17–1; then a force is applied to compress (or expand) the material back to its original length. The force F required is given by Eq. 12–4:

$$\Delta L = \frac{1}{E} \frac{F}{A} L_0,$$

where E is Young's modulus for the material. To calculate the internal stress, F/A, we then set ΔL in Eq. 17–1a equal to ΔL in the equation above and find

$$\alpha L_0 \, \Delta T = \frac{1}{E} \frac{F}{A} L_0.$$

Hence, $F/A = \alpha E \, \Delta T$.

⇒ PHYSICS APPLIED

Highway buckling

EXAMPLE 17–7 **Stress in concrete on a hot day.** A highway is to be made of blocks of concrete 10 m long placed end to end with no space in between them to allow for expansion. If the blocks were placed at a temperature of 10°C, what force of compression would occur if the temperature reached 40°C? The contact area between each block is $0.20 \, \text{m}^2$. Will fracture occur?

SOLUTION We solve for F in the equation above and use the value of E from Table 12–1:

$$F = \alpha \, \Delta T \, E \, A$$
$$= (12 \times 10^{-6}/\text{C}°)(30 \, \text{C}°)(20 \times 10^9 \, \text{N/m}^2)(0.20 \, \text{m}^2) = 1.4 \times 10^6 \, \text{N}.$$

The stress, F/A, is $(1.4 \times 10^6 \, \text{N})/(0.20 \, \text{m}^2) = 7.0 \times 10^6 \, \text{N/m}^2$. This is not far from the ultimate strength of concrete (Table 12–2) under compression and exceeds it for tension and shear. Hence, assuming the concrete is not perfectly aligned, part of the force will act in shear, and fracture is likely. How much space would you allow between blocks if you expected a temperature range of 0°F to 110°F?

17–6 | The Gas Laws and Absolute Temperature

Equation 17–2 is not very useful for describing the expansion of a gas, partly because the expansion can be so great, and partly because gases generally expand to fill whatever container they are in. Indeed, Eq. 17–2 is meaningful only if the pressure is kept constant. The volume of a gas depends very much on the pressure as well as on the temperature. It is therefore valuable to determine a relation between the volume, the pressure, the temperature, and the mass of a gas. Such a relation is called an **equation of state**. (By the word *state*, we mean the physical condition of the system.)

If the state of a system is changed, we will always wait until the pressure and temperature have reached the same values throughout. We thus consider only **equilibrium states** of a system—when the variables that describe it (such as temperature and pressure) are the same throughout the system and are not changing in time. We also note that the results of this Section are accurate only for gases that are not too dense (the pressure is not too high, on the order of an atmosphere or so) and not close to the liquefaction (boiling) point.

For a given quantity of gas it is found experimentally that, to a good approximation, *the volume of a gas is inversely proportional to the pressure applied to it when the temperature is kept constant.* That is,

$$V \propto \frac{1}{P}. \qquad\qquad \text{[constant } T\text{]}$$

Boyle's law

where P is the absolute pressure (not "gauge pressure"—see Chapter 13). For example, if the pressure on a gas is doubled, the volume is reduced to half its original volume. This relation is known as **Boyle's law**, after Robert Boyle (1627–1691), who first stated it on the basis of his own experiments. A graph of P vs. V for a fixed temperature is shown in Fig. 17–12. Boyle's law can also be written

$$PV = \text{constant.} \qquad\qquad \text{[constant } T\text{]}$$

That is, at constant temperature, if either the pressure or volume of the gas is allowed to vary, the other variable also changes so that the product PV remains constant.

Temperature also affects the volume of a gas, but a quantitative relationship between V and T was not found until more than a century after Boyle's work. The Frenchman Jacques Charles (1746–1823) found that when the pressure is not too high and is kept constant, the volume of a gas increases with temperature at a nearly constant rate, as in Fig. 17–13a. However, all gases liquefy at low temperatures (for example, oxygen liquefies at $-183°C$) and so the graph cannot be extended below the liquefaction point. Nonetheless, the graph is essentially a straight line and if projected to lower temperatures, as shown by the dashed line, it crosses the axis at about $-273°C$.

Such a graph can be drawn for any gas, and the straight line always projects back to $-273°C$ at zero volume. This seems to imply that if a gas could be cooled to $-273°C$ it would have zero volume, and at lower temperatures a negative volume, which makes no sense, of course. It could be argued that $-273°C$ is the lowest temperature possible, and many other more recent experiments indicate that it is so. This temperature is called the **absolute zero** of temperature. Its value has been determined to be $-273.15°C$.

Absolute zero

Absolute zero forms the basis of a temperature scale known as the **absolute** or **Kelvin scale**, and it is used extensively in scientific work. On this scale the temperature is specified as degrees Kelvin or, preferably, simply as kelvins (K) without the degree sign. The intervals are the same as for the Celsius scale, but the zero on this scale (0 K) is chosen as absolute zero itself. Thus the freezing point of water (0°C) is 273.15 K and the boiling point of water is 373.15 K. Indeed, any temperature on the Celsius scale can be changed to kelvins by adding 273.15 to it:

Kelvin scale

$$T(\text{K}) = T(°\text{C}) + 273.15.$$

Now let us look at Fig. 17–13b, where we see that the graph of the volume of a gas versus absolute temperature is a straight line that passes through the origin. Thus,

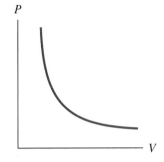

FIGURE 17–12 Pressure vs. volume of a gas at a constant temperature, showing the inverse relationship as given by Boyle's law: as the pressure increases, the volume decreases.

FIGURE 17–13 Volume of a gas as a function of (a) Celsius temperature, and (b) Kelvin temperature, when the pressure is kept constant.

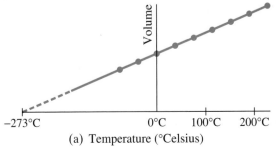

(a) Temperature (°Celsius)

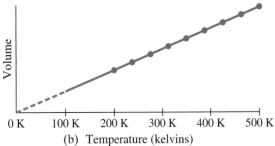

(b) Temperature (kelvins)

to a good approximation, *the volume of a given amount of gas is directly proportional to the absolute temperature when the pressure is kept constant*. This is known as **Charles's law**, and is written

Charles's law
$$V \propto T.$$
[constant P]

A third gas law, known as **Gay-Lussac's law**, after Joseph Gay-Lussac (1778–1850), states that *at constant volume, the pressure of a gas is directly proportional to the absolute temperature*:

Gay-Lussac's law
$$P \propto T.$$
[constant V]

A familiar example is that a closed jar, or aerosol can, thrown into a fire will explode due to the increase in gas pressure inside.

The laws of Boyle, Charles, and Gay-Lussac are not really laws in the sense that we use this term today (precise, deep, wide-ranging validity). They are really only approximations that are accurate for real gases only as long as the pressure and density of the gas are not too high, and the gas is not too close to condensation. The term *law* applied to these three relationships has become traditional, however, so we have stuck with that usage.

17–7 The Ideal Gas Law

The gas laws of Boyle, Charles, and Gay-Lussac were obtained by means of a technique that is very useful in science: namely, to hold one or more variables constant in order to see clearly the effects of changing only one of the variables. These laws can now be combined into a single more general relation between the pressure, volume, and temperature of a fixed quantity of gas:

$$PV \propto T.$$

This relation indicates how any of the quantities P, V, or T will vary when the other two quantities change. This relation reduces to Boyle's, Charles's, or Gay-Lussac's law when either the temperature, the pressure, or the volume, respectively, is held constant.

Finally, we must incorporate the effect of the amount of gas present. Anyone who has blown up a balloon knows that the more air forced into the balloon, the bigger it gets (Figure 17–14). Indeed, careful experiments show that at constant temperature and pressure, the volume V of an enclosed gas increases in direct proportion to the mass m of gas present. Hence we write

$$PV \propto mT.$$

This proportion can be made into an equation by inserting a constant of proportionality. Experiment shows that this constant has a different value for different gases. However, the constant of proportionality turns out to be the same for all gases if, instead of the mass m, we use the number of moles.

Mole (unit)

One **mole** (abbreviated mol) is defined as the amount of substance that contains as many atoms or molecules as there are in 12.00 grams of carbon 12 (whose atomic mass is exactly 12 u). A simpler but equivalent definition is: 1 mol is that number of grams of a substance numerically equal to the molecular mass (Section 17–1) of the substance. For example, the molecular mass of hydrogen gas (H_2) is 2.0 u (since each molecule contains two atoms of hydrogen and each atom has an atomic mass of 1.0 u). Thus 1 mol of H_2 has a mass of 2.0 g. Similarly, 1 mol of neon gas has a mass of 20 g, and 1 mol of CO_2 has a mass of $\left[12 + (2 \times 16)\right] = 44$ g.

FIGURE 17–14 Blowing up a balloon means putting more air (more air molecules) into the balloon, which increases its volume. The pressure is nearly constant (atmospheric) except for the small effect of the balloon's elasticity.

The mole is the official unit in the SI system. In general, the number of moles, n, in a given sample of a pure substance is equal to its mass in grams divided by its molecular mass specified as grams per mole:

$$n \, (\text{mol}) = \frac{\text{mass (grams)}}{\text{molecular mass (g/mol)}}.$$

For example, the number of moles in 132 g of CO_2 is

$$n = \frac{132 \, \text{g}}{44 \, \text{g/mol}} = 3.0 \, \text{mol}.$$

We can now write the proportion discussed above as an equation:

$$PV = nRT, \qquad \text{(17–3)} \qquad \boxed{\textit{IDEAL GAS LAW}}$$

where n represents the number of moles and R is the constant of proportionality. R is called the **universal gas constant** because its value is found experimentally to be the same for all gases. The value of R, in several sets of units (only the first is the proper SI unit), is:

$$R = 8.315 \, \text{J/(mol} \cdot \text{K)} \qquad \text{[SI units]}$$
$$= 0.0821 \, \text{(L} \cdot \text{atm)/(mol} \cdot \text{K)}$$
$$= 1.99 \, \text{calories/(mol} \cdot \text{K)}.^{\dagger}$$

Equation 17–3 is called the **ideal gas law**, or the **equation of state for an ideal gas**. We use the term "ideal" because real gases do not follow Eq. 17–3 precisely, particularly at high pressure (and density) or when the gas is near the liquefaction point (= boiling point). However, at pressures less than an atmosphere or so, and when T is not close to the liquefaction point of the gas, Eq. 17–3 is quite accurate and useful for real gases.

17–8 Problem Solving with the Ideal Gas Law

The ideal gas law is an extremely useful tool, and we now consider some Examples. We will often refer to "standard conditions" or "standard temperature and pressure" (STP), which means $T = 273 \, \text{K}$ ($0°\text{C}$) and $P = 1.00 \, \text{atm} = 1.013 \times 10^5 \, \text{N/m}^2 = 101.3 \, \text{kPa}$.

STP = 273 K, 1 atm

EXAMPLE 17–8 **Volume of one mol at STP.** Determine the volume of 1.00 mol of any gas, assuming it behaves like an ideal gas, (a) at STP, (b) 20°C.

SOLUTION (a) We solve for V in Eq. 17–3:

$$V = \frac{nRT}{P} = \frac{(1.00 \, \text{mol})(8.315 \, \text{J/mol} \cdot \text{K})(273 \, \text{K})}{(1.013 \times 10^5 \, \text{N/m}^2)}$$
$$= 22.4 \times 10^{-3} \, \text{m}^3.$$

Since 1 liter is $1000 \, \text{cm}^3 = 1 \times 10^{-3} \, \text{m}^3$, 1 mol of any gas has a volume of 22.4 L at STP. [How big is 22.4 L? About the size of a cube one foot on a side; more precisely, $\sqrt[3]{22.4 \times 10^{-3} \, \text{m}^3} = 0.28 \, \text{m} = 28 \, \text{cm}$ on a side.]
(b) At 20°C, $T = 293 \, \text{K}$ and $V = 24.0 \, \text{L}$.

1 mol of gas at STP has volume = 22.4 L

The value for the volume of 1 mol of an ideal gas at STP (22.4 L) is worth remembering, for it sometimes makes calculation simpler.

---

† Calories are defined in Chapter 19, and sometimes it is useful to use R as given in terms of calories.

➥ **PROBLEM SOLVING**

*Always give T in K, and
P as absolute, not gauge, pressure*

➥ **PHYSICS APPLIED**

*What is the mass (and weight)
of the air in a small room?*

Always remember, when using the ideal gas law, that temperatures must be given in kelvins (K) and that the pressure P must always be *absolute* pressure, not gauge pressure.

EXAMPLE 17–9 **ESTIMATE** **Mass of air in a room.** Estimate the mass of air in a room whose dimensions are 5 m × 3 m × 2.5 m high, at 20°C.

SOLUTION First we determine the number of moles n, and then we can multiply by the mass of one mole to get the total mass. Example 17–8 told us that 1 mol at 20°C has a volume of 24.0 L. The room's volume is 3 m × 5 m × 2.5 m, so

$$n = \frac{(3\,\text{m})(5\,\text{m})(2.5\,\text{m})}{24.0 \times 10^{-3}\,\text{m}^3} \approx 1600\,\text{mol}.$$

Air is a mixture of about 20% oxygen (O_2) and 80% nitrogen (N_2) whose molecular masses are $2 \times 16\,\text{u} = 32\,\text{u}$ and $2 \times 14\,\text{u} = 28\,\text{u}$ respectively, for an average of about 29 u. Thus, 1 mol of air has a mass of about 29 g = 0.029 kg, so our room has a mass of air

$$m \approx (1600\,\text{mol})(0.029\,\text{kg/mol}) \approx 50\,\text{kg}.$$

That is roughly 100 lbs of air!

Frequently, volume is specified in liters and pressure in atmospheres. Rather than convert these to SI units, we can instead use the value of R given above as 0.0821 L·atm/mol·K.

In many situations it is not necessary to use the value of R at all. For example, many problems involve a change in the pressure, temperature, and volume of a fixed amount of gas. In this case, $PV/T = nR = $ constant, since n and R remain constant. If we now let P_1, V_1, and T_1 represent the appropriate variables initially, and P_2, V_2, T_2 represent the variables after the change is made, then we can write

$$\frac{P_1 V_1}{T_1} = \frac{P_2 V_2}{T_2}.$$

If we know any five of the quantities in this equation, we can solve for the sixth. Or, if one of the three variables is constant $(V_1 = V_2, \text{ or } P_1 = P_2, \text{ or } T_1 = T_2)$ then we can use this equation to solve for one unknown when given the other three quantities.

EXAMPLE 17–10 **Check tires cold.** An automobile tire is filled to a gauge pressure of 200 kPa at 10°C. After driving 100 km, the temperature within the tire rises to 40°C. What is the pressure within the tire now?

SOLUTION Since the volume remains essentially constant, $V_1 = V_2$, and therefore

$$\frac{P_1}{T_1} = \frac{P_2}{T_2}.$$

This is, incidentally, a statement of Gay-Lussac's law. Since the pressure given is the gauge pressure (Section 13–3), we must add atmospheric pressure (= 101 kPa) to get the absolute pressure $P_1 = (200\,\text{kPa} + 101\,\text{kPa}) = 301\,\text{kPa}$. And we convert temperatures to kelvin by adding 273:

$$P_2 = \frac{P_1}{T_1} T_2 = \frac{(3.01 \times 10^5\,\text{Pa})(313\,\text{K})}{(283\,\text{K})} = 333\,\text{kPa}.$$

Subtracting atmospheric pressure, we find the resulting gauge pressure to be 232 kPa, which is a 15 percent increase. This Example shows why car manuals suggest checking the pressure when the tires are cold.

17–9 Ideal Gas Law in Terms of Molecules: Avogadro's Number

The fact that the gas constant, R, has the same value for all gases is a remarkable reflection of simplicity in nature. It was first recognized, although in a slightly different form, by the Italian scientist Amedeo Avogadro (1776–1856). Avogadro stated that *equal volumes of gas at the same pressure and temperature contain equal numbers of molecules*. This is sometimes called **Avogadro's hypothesis**. That this is consistent with R being the same for all gases can be seen as follows. First of all, from Eq. 17–3 we see that for the same number of moles, n, and the same pressure and temperature, the volume will be the same for all gases as long as R is the same. Second, the number of molecules in 1 mole is the same for all gases.[†] Thus Avogadro's hypothesis is equivalent to R being the same for all gases.

Avogadro's hypothesis

The number of molecules in a mole is known as **Avogadro's number**, N_A. Although Avogadro conceived the notion, he was not able to actually determine the value of N_A. Indeed, precise measurements were not done until the twentieth century.

A number of methods have been devised to measure N_A and the accepted value today is

$$N_A = 6.02 \times 10^{23}. \qquad \text{[molecules/mole]}$$

Avogadro's number

Since the total number of molecules, N, in a gas is equal to the number per mole times the number of moles $(N = nN_A)$, the ideal gas law, Eq. 17–3, can be written in terms of the number of molecules present:

$$PV = nRT = \frac{N}{N_A} RT,$$

or

$$PV = NkT, \qquad\qquad (17\text{–}4)$$

IDEAL GAS LAW (IN TERMS OF MOLECULES)

where $k = R/N_A$ is called **Boltzmann's constant** and has the value

$$k = \frac{R}{N_A} = 1.38 \times 10^{-23} \, \text{J/K}.$$

EXAMPLE 17–11 **Hydrogen atom mass.** Use Avogadro's number to determine the mass of a hydrogen atom.

SOLUTION One mole of hydrogen (atomic mass = 1.008 u, Section 17–1) has a mass of 1.008×10^{-3} kg and contains 6.02×10^{23} atoms. Thus one atom has a mass

$$m = \frac{1.008 \times 10^{-3} \, \text{kg}}{6.02 \times 10^{23}} = 1.67 \times 10^{-27} \, \text{kg}.$$

Historically, the reverse process was one method used to obtain N_A: that is, from the measured mass of the hydrogen atom.

[†] For example, the molecular mass of H_2 gas is 2.0 atomic mass units (u), whereas that of O_2 gas is 32.0 u. Thus 1 mol of H_2 has a mass of 0.0020 kg and 1 mol of O_2 gas, 0.032 kg. The number of molecules in a mole is equal to the total mass M of a mole divided by the mass m of one molecule; since this ratio (M/m) is the same for all gases by definition of the mole, a mole of any gas must contain the same number of molecules.

Ideal Gas Temperature Scale—a Standard

It is important to have a very precisely defined temperature scale so that measurements of temperature made at different laboratories around the world can be accurately compared. We now discuss such a scale that has been accepted by the general scientific community.

The standard thermometer for this scale is the constant-volume gas thermometer already discussed in Section 17–2. The scale itself is called the **ideal gas temperature scale**, since it is based on the property of an ideal gas that the pressure is directly proportional to the absolute temperature (Gay-Lussac's law). A real gas, which would have to be used in any real constant-volume gas thermometer, approaches this ideal at low density. In other words, the temperature at any point in space is *defined* as being proportional to the pressure in the (nearly) ideal gas used in the thermometer. To set up a scale we need two fixed points. One fixed point will be $P = 0$ at $T = 0\,$K. The second fixed point is chosen to be the **triple point** of water, which is that point where water in the solid, liquid, and gas states can coexist in equilibrium. This occurs only at a unique temperature and pressure,[†] and can be reproduced at different laboratories with great precision. The pressure at the triple point of water is 4.58 torr and the temperature is 0.01°C. This temperature corresponds to 273.16 K, since absolute zero is about −273.15°C. In fact, the triple point is now *defined* to be exactly 273.16 K.

The absolute or Kelvin temperature T at any point is then defined, using a constant-volume gas thermometer for an ideal gas, as

$$T = (273.16\,\text{K})\left(\frac{P}{P_{tp}}\right). \qquad \text{[ideal gas; constant volume]} \quad \textbf{(17–5a)}$$

In this relation, P_{tp} is the pressure in the thermometer at the triple point temperature of water, and P is the pressure in the thermometer when it is at the point where T is being determined. Note that if we let $P = P_{tp}$ in this relation, then $T = 273.16\,$K, as it must.

The definition of temperature, Eq. 17–5a, with a constant-volume gas thermometer filled with a real gas is only approximate because we find that we get different results for the temperature depending on the type of gas that is used in the thermometer. Temperatures determined in this way also vary depending on the amount of gas in the bulb of the thermometer: for example, the boiling point of water at 1.00 atm is found from Eq. 17–5a to be 373.87 K when the gas is O_2 and $P_{tp} = 1000$ torr. If the amount of O_2 in the bulb is reduced so that at the triple point $P_{tp} = 500$ torr, the boiling point of water from Eq. 17–5a is then found to be 373.51 K. If H_2 gas is used instead, the corresponding values are 373.07 K and 373.11 K (see Fig. 17–15). But now suppose we use a particular real gas and make a series of measurements in which the amount of gas in the thermometer bulb is reduced to smaller and smaller amounts, so that P_{tp} becomes smaller and smaller. It is found experimentally that an extrapolation of such data to $P_{tp} = 0$ always gives the *same value* for the temperature of a given system (such as $T = 373.15$ K for the boiling point of water at 1.00 atm) as shown in Fig. 17–15. Thus the temperature T at any point in space, determined using a constant-volume gas thermometer containing a real gas, is defined using this limiting process:

$$T = (273.16\,\text{K})\lim_{P_{tp}\to 0}\left(\frac{P}{P_{tp}}\right). \qquad \text{[constant volume]} \quad \textbf{(17–5b)}$$

This defines the **ideal gas temperature scale**. One of the great advantages of this

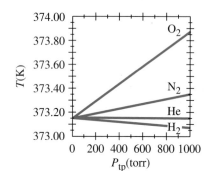

FIGURE 17–15 Temperature readings of a constant-volume gas thermometer for the boiling point of water at 1.00 atm are plotted, for different gases, as a function of the gas pressure in the thermometer at the triple point (P_{tp}). Note that as the amount of gas in the thermometer is reduced, so that $P_{tp} \to 0$, all gases give the same reading, 373.15 K. For pressure less than 0.10 atm (76 torr), the variation shown is less than 0.07 K.

Ideal gas temperature scale

[†]Liquid water and steam can coexist (the boiling point) at a range of temperatures depending on the pressure. Water boils at a lower temperature when the pressure is less, such as high in the mountains. The triple point represents a more precisely reproducible fixed point than does either the freezing point or boiling point of water at, say, 1 atm. See Section 18–3 for further discussion.

scale is that the value for T does not depend on the kind of gas used. But the scale does depend on the properties of gases in general. Helium has the lowest condensation point of all gases; at very low pressures it liquefies at about 1 K, so temperatures below this cannot be defined on this scale.

Summary

The atomic theory of matter postulates that all matter is made up of tiny entities called **atoms**, which are typically 10^{-10} m in diameter.

Atomic and **molecular masses** are specified on a scale where ordinary carbon (^{12}C) is arbitrarily given the value 12.0000 u (atomic mass units).

The distinction between solids, liquids, and gases can be attributed to the strength of the attractive forces between the atoms or molecules and to their average speed.

Temperature is a measure of how hot or cold a body is. **Thermometers** are used to measure temperature on the **Celsius** (°C), **Fahrenheit** (°F), and **Kelvin** (K) scales. Two standard points on each scale are the freezing point of water (0°C, 32°F, 273.15 K) and the boiling point of water (100°C, 212°F, 373.15 K). A one kelvin change in temperature equals a change of one Celsius degree or $\frac{9}{5}$ Fahrenheit degrees.

The change in length, ΔL, of a solid, when its temperature changes by an amount ΔT, is directly proportional to the temperature change and to its original length L_0. That is,

$$\Delta L = \alpha L_0 \, \Delta T,$$

where α is the *coefficient of linear expansion*.

The change in volume of most solids, liquids, and gases is proportional to the temperature change and to the original volume V_0: $\Delta V = \beta V_0 \, \Delta T$. The *coefficient of volume expansion*, β, is approximately equal to 3α for most solids.

Water is unusual because, unlike most materials whose volume increases with temperature, its volume actually decreases as the temperature increases in the range from 0°C to 4°C.

The **ideal gas law**, or **equation of state for an ideal gas**, relates the pressure P, volume V, and temperature T (in kelvins) of n moles of gas by the equation

$$PV = nRT,$$

where $R = 8.315$ J/mol·K for all gases. Real gases obey the ideal gas law quite accurately if they are not at too high a pressure or near their liquefaction point.

One **mole** of a substance is defined as the number of grams which is numerically equal to the atomic or molecular mass.

Avogadro's number, $N_A = 6.02 \times 10^{23}$, is the number of atoms or molecules in 1 mol of any pure substance.

The ideal gas law can be written in terms of the number of molecules N in the gas as

$$PV = NkT,$$

where $k = R/N_A = 1.38 \times 10^{-23}$ J/K is Boltzmann's constant.

Questions

1. Which has more atoms: 1 kg of iron or 1 kg of aluminum? See the Periodic Table or Appendix D.
2. Name several properties of materials that could be exploited to make a thermometer.
3. Suppose system C is not in equilibrium with system A nor in equilibrium with system B. Does this imply that A and B are not in equilibrium? What can you infer regarding the temperatures of A, B, and C?
4. If system A is in equilibrium with system B, but B is not in equilibrium with system C, what can you say about the temperatures of A, B, and C?
5. A flat bimetallic strip consists of aluminum riveted to a strip of iron. When heated, which metal will be on the outside of the curve?
6. In the relation $\Delta L = \alpha L_0 \Delta T$, should L_0 be the initial length, the final length, or does it matter?
7. Why is it sometimes easier to remove the lid from a tightly closed jar after warming it under hot running water?
8. Long steam pipes often have a section in the shape of a U. Why?
9. Figure 17–16 shows a diagram of a simple *thermostat* used to control a furnace (or other heating or cooling system). The bimetallic strip consists of two strips of different metals bonded together. The electric switch is a glass vessel containing liquid mercury that conducts electricity when it can flow to touch both contact wires. Explain how this device controls the furnace and how it can be set at different temperatures.

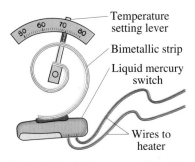

FIGURE 17–16 A thermostat (Question 9).

10. The units for the coefficients of expansion α are $(C°)^{-1}$, and there is no mention of a length unit such as meters. Would the expansion coefficient change if we used feet or millimeters instead of meters?

11. Explain why it is advisable to add water to an overheated automobile engine only slowly, and only with the engine running.

12. A glass container may break if one part of it is heated or cooled more rapidly than adjacent parts. Explain.

13. When a cold mercury-in-glass thermometer is first placed in a hot tub of water, the mercury initially descends a bit and then rises. Explain.

14. The principal virtue of Pyrex glass is that its coefficient of linear expansion is much smaller than that for ordinary glass (Table 17–1). Explain why this gives rise to the high "heat resistance" of Pyrex.

15. Will a grandfather clock, accurate at 20°C, run fast or slow on a hot day (30°C)? The clock uses a pendulum supported on a long thin brass rod.

16. Freezing a can of soda will cause its bottom and top to bulge so badly the can will not stand up. What has happened?

17. Why might you expect an alcohol-in-glass thermometer to be more precise than a mercury-in-glass thermometer?

18. Will the buoyant force on an aluminum sphere submerged in water increase or decrease if the temperature is increased from 20°C to 40°C?

19. A flat, uniform cylinder of lead floats in mercury at 0°C. Will the lead float higher or lower when the temperature is raised?

20. Which scale, Fahrenheit, Celsius, or Kelvin, might be considered most "natural" from a scientific point of view? Discuss.

21. If an atom is measured to have a mass of 6.7×10^{-27} kg, what atom do you think it is?

__*__ **22.** From a practical point of view, does it really matter what gas is used in a constant-volume gas thermometer? If so, explain. [*Hint*: See Fig. 17–15.]

☐ Problems

Section 17–1

1. (I) How does the number of atoms in a 26.5-gram gold ring compare to the number in a silver ring of the same mass?

2. (I) How many atoms are there in a 3.4-gram copper penny?

Section 17–2

3. (I) (*a*) "Room temperature" is often taken to be 68°F; what is this on the Celsius scale? (*b*) The temperature of the filament in a lightbulb is about 1800°C; what is this on the Fahrenheit scale?

4. (I) (*a*) 15° below zero on the Celsius scale is what Fahrenheit temperature? (*b*) 15° below zero on the Fahrenheit scale is what Celsius temperature?

5. (I) In a foreign country, a thermometer tells you that you have a fever of 40.0°C. What is this in Fahrenheit?

6. (I) In an alcohol-in-glass thermometer, the alcohol column has length 11.82 cm at 0.0°C and length 22.85 cm at 100.0°C. What is the temperature if the column has length (*a*) 16.70 cm, and (*b*) 20.50 cm?

7. (II) At what temperature will the Fahrenheit and Centigrade scales yield the same numerical value?

Section 17–4

8. (I) A concrete highway is built of slabs 12 m long (20°C). How wide should the expansion cracks between the slabs be (at 20°C) to prevent buckling if the range of temperature is −30°C to +50°C?

9. (I) Super Invar, an alloy of iron and nickel, is a strong material with a very low coefficient of thermal expansion $[0.2 \times 10^{-6}(C°)^{-1}]$. A 2.0-m-long-tabletop of this alloy is used for sensitive laser measurements where extremely high tolerances are required. How much will this alloy table expand along its length if the temperature increases 5.0 C°? Compare to tabletops made of steel and marble.

10. (I) The Eiffel Tower (Fig. 17–17) is built of wrought iron approximately 300 m tall. Estimate how much its height changes between July (average temperature of 25°C) and January (average temperature of 2°C). Ignore the angles of the iron beams and treat the tower as a vertical beam.

FIGURE 17–17 Problem 10: The Eiffel Tower in Paris.

11. (II) To make a secure fit, rivets that are larger than the rivet hole are often used and the rivet is cooled (usually in dry ice) before it is placed in the hole. A steel rivet 1.871 cm in diameter is to be placed in a hole 1.869 cm in diameter. To what temperature must the rivet be cooled if it is to fit in the hole at 20°C?

12. (II) A uniform rectangular plate of length l and width w has coefficient of linear expansion α. Show that, if we neglect very small quantities, the change in area of the plate due to a temperature change ΔT is $\Delta A = 2\alpha l w \, \Delta T$. See Fig. 17–18.

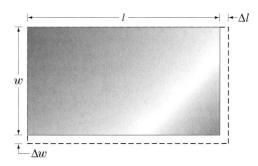

FIGURE 17–18 Rectangular plate is heated. Problem 12.

13. (II) An ordinary glass is filled to the brim with 350.0 mL of water at 100.0°C. If the temperature decreased to 20.0°C, how much water could be added to the glass?

14. (II) It is observed that 55.50 mL of water at 20°C completely fills a container to the brim. When the container and the water are heated to 60°C, 0.35 g of water is lost. (a) What is the coefficient of volume expansion of the container? (b) What is the most likely material of the container? Density of water at 60°C is 0.98324 g/mL.

15. (II) A quartz sphere is 8.75 cm in diameter. What will be its change in volume if it is heated from 30°C to 200°C?

16. (II) A brass plug is to be placed in a ring made of iron. At room temperature, the diameter of the plug is 8.753 cm and that of the inside of the ring is 8.743 cm. They must both be brought to what common temperature in order to fit?

17. (II) If a fluid is contained in a long narrow vessel so it can expand in essentially one direction only, show that the effective coefficient of linear expansion α is approximately equal to the coefficient of volume expansion β.

18. (II) (a) Show that the change in the density ρ of a substance, when the temperature changes by ΔT, is given by $\Delta\rho = -\beta\rho \, \Delta T$. (b) What is the fractional change in density of a lead sphere whose temperature decreases from 25°C to −40°C?

19. (III) The pendulum in a grandfather clock is made of brass and keeps perfect time at 17°C. How much time is gained or lost in a year if the clock is kept at 25°C? (Assume the frequency dependence on length for a simple pendulum applies.)

20. (III) (a) Determine a formula for the change in surface area of a uniform solid sphere of radius r if its coefficient of linear expansion is α (assumed constant) and its temperature is changed by ΔT. (b) What is the increase in area of a solid iron sphere of radius 60.0 cm if its temperature is raised from 20°C to 310°C?

21. (III) A 23.4-kg solid aluminum cylindrical wheel of radius 0.41 m is rotating about its axle in frictionless bearings with angular velocity $\omega = 32.8$ rad/s. If its temperature is now raised from 20.0°C to 75.0°C, what is the fractional change in ω?

* Section 17–5

*22. (I) At what temperature will the ultimate compressive strength of concrete be exceeded for the blocks discussed in Example 17–7?

*23. (I) An aluminum bar has the precisely desired length when at 15°C. How much stress is required to keep it at this length if the temperature increases to 35°C?

*24. (II) (a) A horizontal steel I-beam of cross-sectional area 0.041 m² is rigidly connected to two vertical steel girders. If the beam was installed when the temperature was 30°C, what stress is developed in the beam when the temperature drops to −30°C? (b) Is the ultimate strength of the steel exceeded? (c) What stress is developed if the beam is concrete and has a cross-sectional area of 0.13 m²? Will it fracture?

*25. (III) A barrel of diameter 134.122 cm at 20°C is to be enclosed by an iron band. The circular band has an inside diameter of 134.110 cm at 20°C. It is 7.4 cm wide and 0.65 cm thick. (a) To what temperature must the band be heated so that it will fit over the barrel? (b) What will be the tension in the band when it cools to 20°C?

Section 17–6

26. (I) What are the following temperatures on the Kelvin scale: (a) 86°C, (b) 78°F, (c) −100°C, (d) 5500°C?

27. (I) Absolute zero is what temperature on the Fahrenheit scale?

28. (II) Typical temperatures in the interior of the Earth and Sun are about 4000°C and 15×10^6 °C, respectively. (a) What are these temperatures in kelvins? (b) What percent error is made in each case if a person forgets to change °C to K?

Sections 17–7 and 17–8

29. (I) If 3.00 m³ of a gas initially at STP is placed under a pressure of 3.20 atm, the temperature of the gas rises to 38.0°C. What is the volume?

30. (I) In an internal combustion engine, air at atmospheric pressure and a temperature of about 20°C is compressed in the cylinder by a piston to $\frac{1}{9}$ of its original volume (compression ratio = 9.0). Estimate the temperature of the compressed air, assuming the pressure reaches 40 atm.

31. (II) Calculate the density of oxygen at STP using the ideal gas law.

32. (II) A storage tank contains 21.6 kg of nitrogen (N_2) at an absolute pressure of 3.65 atm. What will the pressure be if the nitrogen is replaced by an equal mass of CO_2?

33. (II) A storage tank at STP contains 18.5 kg of nitrogen (N_2). (a) What is the volume of the tank? (b) What is the pressure if an additional 15.0 kg of nitrogen is added without changing the temperature?

34. (II) If 18.75 mol of helium gas is at 10.0°C and a gauge pressure of 0.350 atm, calculate (a) the volume of the helium gas under these conditions, and (b) the temperature if the gas is compressed to precisely half the volume at a gauge pressure of 1.00 atm.

35. (II) What is the pressure inside a 35.0-L container holding 105.0 kg of argon gas at 20.0°C?

36. (II) A tank contains 30.0 kg of O_2 gas at a gauge pressure of 8.70 atm. If the oxygen is replaced by helium, how many kilograms of the latter will be needed to produce a gauge pressure of 7.00 atm?

37. (II) A hot-air balloon achieves its buoyant lift by heating the air inside the balloon which makes it less dense than the air outside. Suppose the volume of a balloon is 1800 m³ and the required lift is 2700 N (rough estimate of the weight of the equipment and passenger). Calculate the temperature of the air inside the balloon which will produce the required lift. Assume that the outside air temperature is 0° and that air is an ideal gas under these conditions. What factors limit the maximum altitude attainable by this method for a given load? (Neglect variables like wind.)

38. (II) A tire is filled with air at 15°C to a gauge pressure of 220 kPa. If the tire reaches a temperature of 38°C, what fraction of the original air must be removed if the original pressure of 220 kPa is to be maintained?

39. (II) If 61.5 L of oxygen at 18.0°C and an absolute pressure of 2.45 atm are compressed to 48.8 L and at the same time the temperature is raised to 50.0°C, what will the new pressure be?

40. (II) A child's helium-filled balloon escapes at sea level and 20.0°C. When it reaches an altitude of 3000 m, where the temperature is 5.0°C and the pressure only 0.70 atm, how will its volume compare to that at sea level?

41. (III) Compare the value for the density of water vapor at 100°C and 1 atm (Table 13–1) with the value predicted from the ideal gas law. Why would you expect a difference?

42. (III) An air bubble at the bottom of a lake 37.0 m deep has a volume of 1.00 cm³. If the temperature at the bottom is 5.5°C and at the top 21.0°C, what is the volume of the bubble just before it reaches the surface?

Section 17–9

43. (I) Calculate the number of molecules/m³ in an ideal gas at STP.

44. (I) How many moles of water are there in 1.000 L? How many molecules?

45. (I) Estimate the number of molecules you inhale in a 2.0-L breath.

46. (II) Estimate the number of (a) moles, and (b) molecules of water in all the Earth's oceans. Assume water covers 75 percent of the Earth to an average depth of 3 km.

47. (II) A cubic box of volume 5.1×10^{-2} m³ is filled with air at atmospheric pressure at 20°C. The box is closed and heated to 180°C. What is the net force on each side of the box?

48. (III) Estimate how many molecules of air are in each 2.0-L breath you inhale that were also in the last breath Galileo took. [Hint: Assume the atmosphere is about 10 km high and of constant density.]

* Section 17–10

* 49. (I) At the boiling point of sulfur (444.6°C) the pressure in a constant-volume gas thermometer is 187 torr. Estimate (a) the pressure at the triple point of water, (b) the temperature when the pressure in the thermometer is 112 torr.

* 50. (I) In a constant-volume gas thermometer, what is the limiting ratio of the pressure at the boiling point of water at 1 atm to that at the triple point? (Keep five significant figures.)

* 51. (II) Use Fig. 17–15 to determine the inaccuracy of a constant-volume gas thermometer using oxygen if it reads a pressure $P = 268$ torr at the boiling point of water at 1 atm. Express answer (a) in kelvins and (b) as a percentage.

* 52. (II) A constant-volume gas thermometer is being used to determine the temperature of the melting point of a substance. The pressure in the thermometer at this temperature is 218 torr; at the triple point of water, the pressure is 286 torr. Some gas is now released from the thermometer bulb so that the pressure at the triple point of water becomes 163 torr. At the temperature of the melting substance, the pressure is 128 torr. Estimate, as accurately as possible, the melting-point temperature of the substance.

General Problems

53. A precise steel tape measure has been calibrated at 20°C. At 34°C, (a) will it read high or low, and (b) what will be the percentage error?

54. A Pyrex measuring cup was calibrated at normal room temperature. How much error will be made in a recipe calling for 300 mL of cool water, if the water and the cup are hot, at 80°C, instead of at room temperature? Neglect the glass expansion.

55. The pressure in a helium gas cylinder is initially 35 atmospheres. After many balloons have been blown up, the pressure has decreased to 5 atm. What fraction of the original gas remains in the cylinder? Pressures are gauge pressures.

56. Write the ideal gas law in terms of the density of the gas.

57. Estimate the number of air molecules in a room of length 8.0 m, width 6.0 m, and height 4.2 m. Assume the temperature is 20°C. How many moles does that correspond to?

58. The lowest pressure attainable using the best available vacuum techniques is about 10^{-12} N/m². At such a pressure, how many molecules are there per cm³ at 0°C?

59. If a scuba diver fills his lungs to full capacity of 5.5 L when 10 m below the surface, to what volume would his lungs expand if he quickly rose to the surface? Is this advisable?

60. If a rod of original length L_1 has its temperature changed from T_1 to T_2, determine a formula for its new length L_2 in terms of T_1, T_2, and α. Assume (a) $\alpha = $ constant, (b) $\alpha = \alpha(T)$ is some function of temperature, and (c) $\alpha = \alpha_0 + bT$ where α_0 and b are constants.

61. A house has a volume of 770 m³. (a) What is the total mass of air inside the house at 20°C? (b) If the temperature drops to −10°C, what mass of air enters or leaves the house?

62. (a) Use the ideal gas law to show that, for an ideal gas at constant pressure, the coefficient of volume expansion is equal to $\beta = 1/T$ where T is the kelvin temperature. Compare to Table 17–1 for gases at $T = 293$ K. (b) Show that the bulk modulus (Section 12–5) for an ideal gas held at constant temperature is $B = P$ where P is the pressure.

63. From the known value of atmospheric pressure at the surface of the Earth, estimate the total number of air molecules in the Earth's atmosphere.

64. (a) The tube of a mercury thermometer has an inside diameter of 0.140 mm. The bulb has a volume of 0.315 cm³. How far will the thread of mercury move when the temperature changes from 11.5°C to 33.0°C? Take into account expansion of the Pyrex glass. (b) Determine a formula for the length of the mercury column in terms of relevant variables.

65. What is the average distance between oxygen molecules at STP?

66. An iron cube floats in a bowl of liquid mercury at 0°C. (a) If the temperature is raised to 25°C, will the cube float higher or lower in the mercury? (b) By what percent will the fraction of volume submerged change?

67. If a steel band were to fit snugly around the Earth's equator at 20°C, but then was heated to 35°C, how high above the Earth would the band be (assume equal everywhere)?

68. Estimate the percent change in density of iron when it is still a solid, but deep in the Earth where the temperature is 2000°C and it is under 5000 atm of pressure. Take into account both thermal expansion and changes due to increased outside pressure. Assume both the bulk modulus and the volume coefficient of expansion do not vary with temperature and are the same as at normal room temperature. The bulk modulus for iron is about 90×10^9 N/m².

69. A standard cylinder of oxygen used in a hospital has the following characteristics at room temperature (300 K): gauge pressure = 2000 psi (13,800 kPa), volume = 16 liters (0.016 m^3). How long will the cylinder last if the flow rate, measured at atmospheric pressure, is constant at 2.4 liters/min?

70. A helium party balloon, assumed to be a perfect sphere, has a radius of 18.0 cm. At room temperature (20°C), its internal pressure is 1.05 atm. Find the number of moles of helium in the balloon and the mass of helium needed to inflate the balloon to these values.

71. The density of gasoline at 0°C is 0.68×10^3 kg/m³. What is the density on a hot day, when the temperature is 32°C?

72. A brass lid screws tightly onto a glass jar at 20°C. To help open the jar, it can be placed into a bath of hot water. After this treatment, the temperatures of the lid and the jar are both 60°C. The inside diameter of the lid is 8.0 cm. Find the size of the gap (difference in radius) that develops by this procedure.

73. The first length standard, adopted in the 18th century, was a platinum bar with two very fine marks separated by what was defined to be exactly one meter. If this standard bar was to be accurate to within $\pm 1.0 \ \mu$m, how carefully would the trustees have needed to control the temperature? The coefficient of linear expansion is $9 \times 10^{-6} \ \text{C}^{\circ-1}$.

74. Reinforcing rods in concrete are made of steel, which has nearly the same coefficient of thermal expansion as concrete. What would happen if brass were used instead? Imagine a 2.5-cm diameter brass rod, embedded in a concrete matrix. If the temperature rises by 20 C°, the brass will be under compression and the concrete under tension. Will the concrete stay in one piece? (You will need data from Tables 12–1 and 12–2. Assume the same magnitude of stress in both materials.)

75. A helium balloon has volume V_0 and temperature T_0 at sea level where the pressure is P_0 and the air density is ρ_0. The balloon is allowed to float up in the air to altitude y where the temperature is T_1. (a) Show that the volume occupied by the balloon is then $V = V_0 (T_1/T_0)e^{+cy}$ where $c = \rho_0 g/P_0 = 1.25 \times 10^{-4} \ \text{m}^{-1}$. (b) Show that the buoyant force does not depend on altitude y. Assume that the skin of the balloon maintains the helium pressure at a constant factor of 1.05 times greater than the outside pressure. [Hint: Assume that the pressure change with altitude is $P = P_0 e^{-cy}$, as in Example 13–4, Chapter 13.]

76. A scuba tank when fully charged has a pressure of 200 atmospheres at 20°C. The volume of the tank is 11.3 liters. (a) What would the volume of the air be at 1.00 atmosphere and at the same temperature? (b) Before entering the water, a person consumes 2.0 liters of air in each breath, and breathes 12 times a minute. At this rate, how long would the tank last? (c) At a depth of 20.0 m of sea water and temperature of 10°C, how long would the same tank last assuming the breathing rate does not change?

77. A temperature controller, designed to work in a steam environment, involves a bimetallic strip constructed of brass and steel, connected at their ends by rivets. Each of the metals is 2.0 mm thick. At 20°C, the strip is 10.0 cm long and straight. Find the radius of curvature of the assembly at 100°C. See Fig. 17–19.

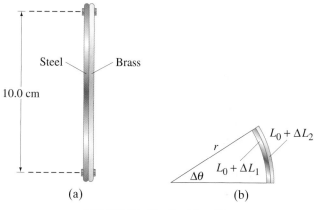

FIGURE 17–19 Problem 77.

Idaho's Salmon River, famous for rafting and fishing, in December. We see the three states of matter in the form of water as a liquid, as a solid (snow and ice), and as a gas (steam or fog). In this chapter we will examine the microscopic theory of matter as atoms or molecules that are always in motion, which we call kinetic theory. We will see that the temperature of a gas is directly related to the average kinetic energy of its molecules. We will consider ideal gases, but we'll also look at real gases and how they change phase including evaporation, vapor pressure, and humidity.

Kinetic Theory of Gases

The analysis of matter in terms of atoms in continuous random motion is called the **kinetic theory**. We now investigate the properties of a gas from the point of view of kinetic theory, which is based on the laws of classical mechanics. But to apply Newton's laws to each one of the vast number of molecules in a gas ($>10^{25}/m^3$ at STP) is far beyond the capability of any present computer. Instead we take a statistical approach and determine averages of certain quantities, and these averages correspond to macroscopic variables. We will, of course, demand that our microscopic description correspond to the macroscopic properties of gases; otherwise our theory would be of little value. Most importantly, we will arrive at an important relation between the average kinetic energy of molecules in a gas and the absolute temperature.

18–1 The Ideal Gas Law and the Molecular Interpretation of Temperature

We make the following assumptions about the molecules in a gas. These assumptions reflect a simple view of a gas, but nonetheless the results they predict correspond well to the essential features of real gases that are at low pressures and far from the liquefaction point. Under these conditions real gases follow the ideal gas law quite closely, and indeed the gas we now describe is referred to as an

ideal gas. The assumptions, which represent the basic postulates of the kinetic theory, are:

1. There are a large number of molecules, N, each of mass m, moving in random directions with a variety of speeds. This assumption is in accord with our observation that a gas fills its container and, in the case of air on the Earth, is kept from escaping only by the force of gravity.

2. The molecules are, on the average, far apart from one another. That is, their average separation is much greater than the diameter of each molecule.

3. The molecules are assumed to obey the laws of classical mechanics, and are assumed to interact with one another only when they collide. Although molecules exert weak attractive forces on each other between collisions, the potential energy associated with these forces is small compared to the kinetic energy, and we ignore it for now.

4. Collisions with another molecule or the wall of the vessel are assumed to be perfectly elastic, like the collisions of perfectly elastic billiard balls (Chapter 9). We assume the collisions are of very short duration compared to the time between collisions. Then we can ignore the potential energy associated with collisions in comparison to the kinetic energy between collisions.

Postulates of kinetic theory

We can see immediately how this kinetic view of a gas can explain Boyle's law (Section 17–6). The pressure exerted on a wall of a container of gas is due to the constant bombardment of molecules. If the volume is reduced by (say) half, the molecules are closer together and twice as many will be striking a given area of the wall per second. Hence we expect the pressure to be twice as great, in agreement with Boyle's law.

Boyle's law explained

Now let us calculate quantitatively the pressure in a gas based on kinetic theory. For purposes of argument, we imagine that the molecules are contained in a rectangular vessel whose ends have area A and whose length is l, as shown in Fig. 18–1a. The pressure exerted by the gas on the walls of its container is, according to our model, due to the collisions of the molecules with the walls. Let us focus our attention on the wall, of area A, at the left end of the container and examine what happens when one molecule strikes this wall, as shown in Fig. 18–1b. This molecule exerts a force on the wall, and the wall exerts an equal and opposite force back on the molecule. The magnitude of this force, according to Newton's second law, is equal to the molecule's rate of change of momentum, $F = dp/dt$. Assuming the collision is elastic, only the x component of the molecule's momentum changes, and it changes from $-mv_x$ (it is moving in the negative x direction) to $+mv_x$. Thus the change in momentum, $\Delta(mv)$, which is the final momentum minus the initial momentum, is

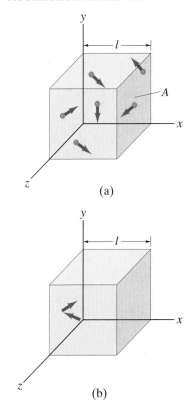

FIGURE 18–1 (a) Molecules of a gas moving about in a cubical container. (b) Arrows indicate the momentum of one molecule as it rebounds from the end wall.

(a)

(b)

$$\Delta(mv) = mv_x - (-mv_x) = 2mv_x$$

for one collision. This molecule will make many collisions with the wall, each separated by a time Δt, which is the time it takes the molecule to travel across the box and back again, a distance (x component) equal to $2l$. Thus $2l = v_x \Delta t$ or

$$\Delta t = 2l/v_x.$$

The time Δt between collisions is very small, so the number of collisions per second is very large. Thus the average force—averaged over many collisions—will be equal to the force exerted during one collision divided by the time between collisions (Newton's second law):

$$F = \frac{\Delta(mv)}{\Delta t} = \frac{2mv_x}{2l/v_x} = \frac{mv_x^2}{l}.$$ [due to one molecule]

During its passage back and forth across the container, the molecule may collide

with the tops and sides of the container, but this does not alter its x component of momentum and thus does not alter our result. It may also collide with other molecules, which may change its v_x. However, any loss (or gain) of momentum is acquired by the other molecule, and because we will eventually sum over all the molecules, this effect will be included. So our result above is not altered.

Of course, the actual force due to one molecule is intermittent, but because a huge number of molecules are striking the wall per second, the force is, on average, nearly constant. To calculate the force due to *all* the molecules in the box, we have to add the contributions of each. Thus the net force on the wall is

$$F = \frac{m}{l}\left(v_{x1}^2 + v_{x2}^2 + \cdots + v_{xN}^2\right),$$

where v_{x1} means v_x for molecule number 1 (we arbitrarily assign each molecule a number) and the sum extends over the total of N molecules. Now the average value of the square of the x component of velocity is

$$\overline{v_x^2} = \frac{v_{x1}^2 + v_{x2}^2 + \cdots + v_{xN}^2}{N}. \tag{18–1}$$

Thus we can write the force as

$$F = \frac{m}{l}N\overline{v_x^2}.$$

We know that the square of any vector is equal to the sum of the squares of its components (theorem of Pythagoras). Thus $v^2 = v_x^2 + v_y^2 + v_z^2$ for any velocity v. Taking averages, we obtain

$$\overline{v^2} = \overline{v_x^2} + \overline{v_y^2} + \overline{v_z^2}.$$

Since the velocities of the molecules in our gas are assumed to be random, there is no preference to one direction or another. Hence

$$\overline{v_x^2} = \overline{v_y^2} = \overline{v_z^2}.$$

Combining this relation with the one just above, we get

$$\overline{v^2} = 3\overline{v_x^2}.$$

We substitute this into the equation for net force F:

$$F = \frac{m}{l}N\frac{\overline{v^2}}{3}.$$

The pressure on the wall is then

$$P = \frac{F}{A} = \frac{1}{3}\frac{Nm\overline{v^2}}{Al}$$

or

Pressure in a gas

$$P = \frac{1}{3}\frac{Nm\overline{v^2}}{V}, \tag{18–2}$$

where $V = lA$ is the volume of the container. This is the result we were seeking, the pressure in a gas expressed in terms of molecular properties.

Equation 18–2 can be rewritten in a clearer form by multiplying both sides by V and slightly rearranging the right side:

$$PV = \tfrac{2}{3}N\left(\tfrac{1}{2}m\overline{v^2}\right). \tag{18–3}$$

The quantity $\frac{1}{2}m\overline{v^2}$ is the average kinetic energy $(\overline{K})$ of the molecules in the gas. If

we compare Eq. 18–3 with Eq. 17–4, the ideal gas law $PV = NkT$, we see that the two agree if

$$\tfrac{2}{3}\left(\tfrac{1}{2}m\overline{v^2}\right) = kT,$$

or

$$\overline{K} = \tfrac{1}{2}m\overline{v^2} = \tfrac{3}{2}kT. \qquad \text{[Ideal gas]} \quad (18\text{–}4)$$

TEMPERATURE RELATED TO AVERAGE KINETIC ENERGY OF MOLECULES

This equation tells us that

> the average translational kinetic energy of molecules in an ideal gas is directly proportional to the absolute temperature.

The higher the temperature, according to kinetic theory, the faster the molecules are moving on the average. This relation is one of the triumphs of the kinetic theory.

EXAMPLE 18–1 **Molecular kinetic energy.** What is the average translational kinetic energy of molecules in an ideal gas at 37°C?

SOLUTION We use Eq. 18–4 and change 37°C to 310 K:

$$\overline{K} = \tfrac{3}{2}kT$$
$$= \tfrac{3}{2}(1.38 \times 10^{-23}\,\text{J/K})(310\,\text{K}) = 6.42 \times 10^{-21}\,\text{J}.$$

Note that a mole of molecules would have a total of translational kinetic energy equal to $(6.42 \times 10^{-21}\,\text{J})(6.02 \times 10^{23}) = 3900\,\text{J}$, which equals the kinetic energy of a 1-kg stone traveling faster that 85 m/s.

Equation 18–4 holds not only for gases, but also applies reasonably accurately to liquids and solids. Thus the result of Example 18–1 would apply to molecules within living cells at body temperature (37°C).

We can use Eq. 18–4 to calculate how fast molecules are moving on the average. Notice that the average in Eqs. 18–1 through 18–4 is over the *square* of the velocity. The square root of $\overline{v^2}$ is called the **root-mean-square** velocity, v_{rms} (since we are taking the square *root* of the *mean* of the *square* of the velocity):

Root-mean-square (rms) velocity

$$v_{\text{rms}} = \sqrt{\overline{v^2}} = \sqrt{\frac{3kT}{m}}. \qquad (18\text{–}5)$$

rms speed of molecules

The **mean speed**, $\overline{v}$, is the average of the magnitudes of the speeds themselves; $\overline{v}$ is generally not equal to v_{rms}. To see the difference between the mean speed and the rms speed, consider the following Example.

Mean speed

EXAMPLE 18–2 **Mean speed and rms speed.** Eight particles have the following speeds, given in m/s: 1.0, 6.0, 4.0, 2.0, 6.0, 3.0, 2.0, 5.0. Calculate (*a*) the mean speed and (*b*) the rms speed.

SOLUTION (*a*) The mean speed is

$$\overline{v} = \frac{1.0 + 6.0 + 4.0 + 2.0 + 6.0 + 3.0 + 2.0 + 5.0}{8} = 3.6\,\text{m/s}.$$

(*b*) The rms speed is (Eq. 18–1):

$$v_{\text{rms}} = \sqrt{\frac{(1.0)^2 + (6.0)^2 + (4.0)^2 + (2.0)^2 + (6.0)^2 + (3.0)^2 + (2.0)^2 + (5.0)^2}{8}}\,\text{m/s}$$
$$= 4.0\,\text{m/s}.$$

We see in this Example that $\overline{v}$ and v_{rms}, are not necessarily equal. In fact, for an ideal gas they differ by about 8 percent. We will see in the next Section how to calculate $\overline{v}$ for an ideal gas. We already have the tool to calculate v_{rms} (Eq. 18–5).

EXAMPLE 18–3 **Speeds of air molecules.** What is the rms speed of air molecules $(O_2$ and $N_2)$ at room temperature (20°C)?

SOLUTION We must apply Eq. 18–5 to oxygen and nitrogen separately since they have different masses. The masses of one molecule of O_2 (molecular mass = 32 u) and N_2 (molecular mass = 28 u) are $(1\,u = 1.66 \times 10^{-27}\,kg)$

$$m(O_2) = (32)(1.66 \times 10^{-27}\,kg) = 5.3 \times 10^{-26}\,kg$$
$$m(N_2) = (28)(1.66 \times 10^{-27}\,kg) = 4.7 \times 10^{-26}\,kg.$$

Thus, for oxygen

$$v_{\text{rms}} = \sqrt{\frac{3kT}{m}} = \sqrt{\frac{(3)(1.38 \times 10^{-23}\,J/K)(293\,K)}{(5.3 \times 10^{-26}\,kg)}}$$
$$= 480\,m/s,$$

and for nitrogen the result is $v_{\text{rms}} = 510\,m/s$. These speeds are more than 1500 km/h or 1000 mi/h.

Equation 18–4, $\overline{K} = \frac{3}{2}kT$, implies that as the temperature approaches absolute zero the kinetic energy of molecules approaches zero. Modern quantum theory, however, tells us this is not quite so. Instead, as absolute zero is approached, the kinetic energy approaches a very small nonzero minimum value. Even though all real gases become liquid or solid near 0 K, molecular motion does not cease, even at absolute zero.

18–2 | Distribution of Molecular Speeds

The Maxwell Distribution

The molecules in a gas are assumed to be in random motion, which means that many molecules have speeds less than the average speed and others have speeds greater than the average. In 1859, James Clerk Maxwell (1831–1879) worked out a formula for the most probable distribution of speeds in a gas containing N molecules. We will not give a derivation here but merely quote his result:

Maxwell distribution

$$f(v) = 4\pi N \left(\frac{m}{2\pi kT}\right)^{\frac{3}{2}} v^2 e^{-\frac{1}{2}\frac{mv^2}{kT}}. \tag{18–6}$$

$f(v)$ is called the **Maxwell distribution of speeds**, and is plotted in Fig. 18–2. The quantity $f(v)\,dv$ represents the number of molecules that have speed between v and $v + dv$. Notice that $f(v)$ does not give the number of molecules with speed v; $f(v)$ must be multiplied by dv to give the number of molecules (clearly the number of molecules must depend on the "width" or "range" of velocities, dv). In the formula for $f(v)$, m is the mass of a single molecule, T is the absolute temperature, and k is Boltzmann's constant. Since N is the total number of molecules in the gas, when we sum over all the molecules in the gas we must get N; thus we must have

$$\int_0^\infty f(v)\,dv = N.$$

(Problem 15 is an exercise to show that this is true.)

Experiments to determine the distribution of speeds in real gases, starting in the 1920s, confirmed with considerable accuracy the Maxwell distribution (for gases at not too high a pressure) and the direct proportion between average kinetic energy and absolute temperature, Eq. 18–4.

FIGURE 18–2 Distribution of speeds of molecules in an ideal gas. Note that $\overline{v}$ and v_{rms} are not at the peak of the curve (that speed is called the "most probable speed," v_{p}). This is because the curve is skewed to the right: it is not symmetrical.

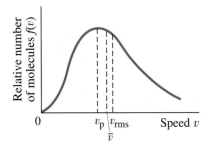

The Maxwell distribution for a given gas depends only on the absolute temperature. Figure 18–3 shows the distributions for two different temperatures. Just as v_{rms} increases with temperature, so the whole distribution curve shifts to the right at higher temperatures.

Figure 18–3 illustrates how kinetic theory can be used to explain why many chemical reactions, including those in biological cells, take place more rapidly as the temperature increases. Most chemical reactions take place in a liquid solution, and the molecules in a liquid have a distribution of speeds close to the Maxwell distribution. Two molecules may chemically react only if their kinetic energy is great enough so that when they collide, they penetrate into each other somewhat. The minimum energy required is called the *activation energy*, E_A, and it has a specific value for each chemical reaction. The molecular speed corresponding to a kinetic energy of E_A for a particular reaction is indicated in Fig. 18–3. The relative number of molecules with energy greater than this value is given by the area under the curve to the right of $v(E_A)$. In Fig. 18–3, the respective areas for two different temperatures are indicated by the two different shadings in the figure. It is clear that the number of molecules that have kinetic energies in excess of E_A increases greatly for only a small increase in temperature. The rate at which a chemical reaction occurs is proportional to the number of molecules with energy greater than E_A, and thus we see why reaction rates increase rapidly with increased temperature.

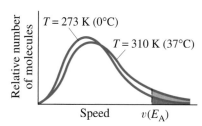

FIGURE 18–3 Distribution of molecular speeds for two different temperatures.

➡ PHYSICS APPLIED

How chemical reactions depend on temperature

* Calculations Using the Maxwell Distribution

Let us see how the Maxwell distribution can be used to obtain some interesting results.

EXAMPLE 18–4 **Determining $\bar{v}$ and v_p.** Determine formulas for (a) the average speed, $\bar{v}$, and (b) the most probable speed, v_p, of molecules in an ideal gas at temperature T.

SOLUTION (a) The average value of any quantity is found by multiplying each possible value of the quantity (say, speed) by the number of molecules that have that value, and then summing all these numbers and dividing by N (the total number). We are given a continuous distribution of speeds (Eq. 18–6), so the sum becomes an integral over the product of v and the number $f(v)\,dv$ that have speed v:

$$\bar{v} = \frac{\int_0^\infty v f(v)\,dv}{N} = 4\pi\left(\frac{m}{2\pi kT}\right)^{\frac{3}{2}}\int_0^\infty v^3 e^{-\frac{1}{2}\frac{mv^2}{kT}}\,dv.$$

We can look up the definite integral in the tables, or integrate by parts, and obtain

$$\bar{v} = 4\pi\left(\frac{m}{2\pi kT}\right)^{\frac{3}{2}}\left(\frac{2k^2T^2}{m^2}\right) = \sqrt{\frac{8}{\pi}\frac{kT}{m}} \approx 1.60\sqrt{\frac{kT}{m}}.$$

Average speed, $\bar{v}$

(b) The *most probable speed* is that speed which occurs more than any others, and thus is that speed where $f(v)$ has its maximum value. Since $df(v)/dv = 0$ at the maximum of the curve, we have

$$\frac{df(v)}{dv} = 4\pi\left(\frac{m}{2\pi kT}\right)^{\frac{3}{2}}\left(2ve^{-\frac{mv^2}{2kT}} - \frac{2mv^3}{2kT}e^{-\frac{mv^2}{2kT}}\right) = 0.$$

Solving for v, we get

$$v_p = \sqrt{\frac{2kT}{m}} \approx 1.41\sqrt{\frac{kT}{m}}.$$

Most probable speed, v_p

(Another solution is $v = 0$, but this corresponds to a minimum, not a maximum.)

In summary,

Most probable speed, v_p

$$v_p = \sqrt{2\frac{kT}{m}} \approx 1.41\sqrt{\frac{kT}{m}} \qquad \text{(18–7a)}$$

Mean speed, $\bar{v}$

$$\bar{v} = \sqrt{\frac{8}{\pi}\frac{kT}{m}} \approx 1.60\sqrt{\frac{kT}{m}} \qquad \text{(18–7b)}$$

and from Eq. 18–5

rms speed, v_{rms}

$$v_{rms} = \sqrt{3\frac{kT}{m}} \approx 1.73\sqrt{\frac{kT}{m}} .$$

These are all indicated in Fig. 18–2. From Eq. 18–6 and Fig. 18–2, it is clear that the speeds of molecules in a gas vary from zero up to many times the average speed, but as can be seen from the graph, most molecules have speeds that are not far from the average. Less than 1 percent of the molecules exceed four times v_{rms}.

EXAMPLE 18–5 **ESTIMATE** **Using $f(v)$.** Suppose a sample of helium gas at temperature $T = 300\,\text{K}$ contains $N = 10^6$ atoms, each of mass $m_{He} = 6.65 \times 10^{-27}\,\text{kg}$. (a) What is the most probable speed, v_p, of a helium atom in the sample? (b) Estimate how many atoms in the sample have speeds in the range between v_p and $v_p + 40\,\text{m/s}$. (c) Now let $v = 10\,v_p$, and calculate the number of molecules with speeds between v and $v + 40\,\text{m/s}$.

SOLUTION (a) The most probable speed v_p is (Eq. 18–7a):

$$v_p = \sqrt{2\frac{kT}{m}} = \sqrt{\frac{(2)(1.38 \times 10^{-23}\,\text{J/K})(300\,\text{K})}{(6.65 \times 10^{-27}\,\text{kg})}} = 1.12 \times 10^3\,\text{m/s}.$$

(b) The number of molecules having a speed between v and $v + dv$ is equal to $f(v)\,dv$, as we discussed just after Eq. 18–6. For our estimate over a small finite range, $\Delta v = 40\,\text{m/s}$, we replace the differential dv by Δv and write

$$\Delta N = f(v)\,\Delta v.$$

As shown in Fig. 18–4, ΔN is equal to the area under the $f(v)$ curve between v and $v + \Delta v$. We approximate this area by a rectangle $f(v)$ high by Δv wide. A better estimate is obtained using the midpoint of the range between $v = 1120\,\text{m/s}$ and $v + \Delta v = 1160\,\text{m/s}$, namely $1140\,\text{m/s}$, although it doesn't make a great deal of difference here. Thus

$$\Delta N = f(v)\Delta v = 4\pi N\left(\frac{m}{2\pi kT}\right)^{\frac{3}{2}} v^2 e^{-\frac{mv^2}{2kT}}\Delta v$$

$$= (4\pi)(10^6)\left(\frac{6.65 \times 10^{-27}\,\text{kg}}{2\pi(1.38 \times 10^{-23}\,\text{J/K})(300\,\text{K})}\right)^{\frac{3}{2}}$$

$$\times (1.14 \times 10^3\,\text{m/s})^2 e^{-\frac{(6.65\times10^{-27}\,\text{kg})(1.14\times10^3\,\text{m/s})^2}{2(1.38\times10^{-23}\,\text{J/K})(300\,\text{K})}}\,(40\,\text{m/s})$$

$$\approx 30{,}000$$

or about 3% of all molecules.

(c) If we now replace $v = v_p$ by $v = 10\,v_p = 1.12 \times 10^4\,\text{m/s}$, and repeat the calculation with all else the same, we obtain $\Delta N = f(v)\Delta v = 4 \times 10^{-39}$, which means essentially that *no* molecules have speeds this high.

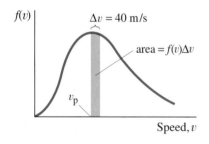

FIGURE 18–4 Calculating the number of molecules ΔN having speeds between v_p and $v_p + 40\,\text{m/s}$. $\Delta N = f(v)\,\Delta v$ is equal to the area shown shaded.

18–3 | Real Gases and Changes of Phase

The ideal gas law

$$PV = NkT$$

is an accurate description of the behavior of a real gas as long as the pressure is not too high and as long as the temperature is far from the liquefaction point. But what happens when these two criteria are not satisfied? First we discuss real gas behavior, and then we will look at how kinetic theory can help us understand this behavior.

Let us look at a graph of pressure plotted against volume for a given amount of gas. On such a "PV diagram," Fig. 18–5, each point represents an equilibrium state of the given substance. The various curves (labeled A, B, C, and D) show how the pressure varies as the volume is changed at constant temperature for several different values of the temperature. The dashed curve A' represents the behavior of a gas as predicted by the ideal gas law; that is, PV = constant. The solid curve A represents the behavior of a real gas at the same temperature. Notice that at high pressure, the volume of a real gas is less than that predicted by the ideal gas law. The curves B and C in Fig. 18–5 represent the gas at successively lower temperatures, and we see that the behavior deviates even more from the curves predicted by the ideal gas law (for example, B'), and the deviation is greater the closer the gas is to liquefying.

To explain this, we note that at higher pressure we expect the molecules to be closer together. And, particularly at lower temperatures, the potential energy associated with the attractive forces between the molecules (which we ignored before) is no longer negligible compared to the now reduced kinetic energy of the molecules. These attractive forces tend to pull the molecules closer together so that at a given pressure, the volume is less than expected from the ideal gas law. At still lower temperatures, these forces cause liquefaction, and the molecules become very close together.

Curve D represents the situation when liquefaction occurs. At low pressure on curve D (on the right in Fig. 18–5), the substance is a gas and occupies a large volume. As the pressure is increased, the volume decreases until point b is reached. Beyond b, the volume decreases with no change in pressure; the substance is gradually changing from the gas to the liquid phase. At point a, all of the substance has changed to liquid. Further increase in pressure reduces the volume only slightly—liquids are nearly incompressible—so the curve is very steep as shown. The area which is colored yellow in Fig. 18–5, within the orange dashed line, represents the region where the gas and liquid phases exist together in equilibrium.

Curve C in Fig. 18–5 represents the behavior of the substance at its **critical temperature**; and the point c (the one point where this curve is horizontal) is called the **critical point**. At temperatures less than the critical temperature (and this is the definition of the term), a gas will change to the liquid phase if sufficient pressure is applied. Above the critical temperature, no amount of pressure can cause a gas to change phase and become a liquid. What happens instead is that the gas becomes denser and denser as the pressure is increased and gradually it acquires properties resembling a liquid, but no liquid surface forms. The critical temperatures for various gases are given in Table 18–1. Scientists tried for many years to liquefy oxygen without success. Only after the discovery of the behavior of substances associated with the critical point was it realized that oxygen can be liquefied only if first cooled below its critical temperature of −118°C.

Often a distinction is made between the terms "gas" and "vapor": a substance below its critical temperature in the gaseous state is called a **vapor**; when above the critical temperature, it is called a **gas**. This is indicated in Fig. 18–5.

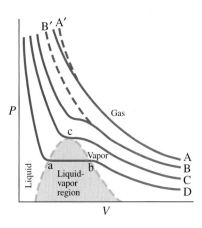

FIGURE 18–5 *PV* diagram for a real substance. Curves A, B, C, and D represent the same gas at different temperatures.

PV diagram

TABLE 18–1
Critical Temperatures and Pressures

Substance	Critical Temperature °C	Critical Temperature K	Critical Pressure (atm)
Water	374	647	218
CO₂	31	304	72.8
Oxygen	−118	155	50
Nitrogen	−147	126	33.5
Hydrogen	−239.9	33.3	12.8
Helium	−267.9	5.3	2.3

Vapor vs. gas

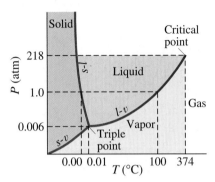

FIGURE 18–6 Phase diagram for water (note that the scales are not linear).

FIGURE 18–7 Phase diagram for carbon dioxide (CO_2).

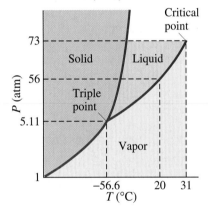

The behavior of a substance can be diagrammed not only on a *PV* diagram but also on a *PT* diagram. A *PT* diagram, often called a **phase diagram**, is particularly convenient for comparing the different phases of a substance. Figure 18–6 is the phase diagram for water. The curve labeled *l-v* represents those points where the liquid and vapor phases are in equilibrium—it is thus a graph of the boiling point versus pressure. Note that the curve correctly shows that at a pressure of 1 atm the boiling point of water is 100°C and that the boiling point is lowered for a decreased pressure. The curve *s-l* represents points where solid and liquid exist in equilibrium, and thus is a graph of the freezing point versus pressure. At 1 atm, the freezing point of water is, of course, 0°C, as shown. Notice also in Fig. 18–6 that at a pressure of 1 atm, the substance is in the liquid phase if the temperature is between 0°C and 100°C, but is in the solid or vapor phase if the temperature is below 0°C or above 100°C. The curve labeled *s-v* is the *sublimation point* versus pressure curve. **Sublimation** refers to the process whereby at low pressures a solid changes directly into the vapor phase without passing through the liquid phase. For water, this occurs at pressures less than 0.0060 atm. Carbon dioxide (CO_2), which in the solid phase is called dry ice, sublimates even at atmospheric pressure.

The intersection of the three curves (in Fig. 18–6) is the **triple point**, and it is only at this point that the three phases can exist together in equilibrium. Because the triple point corresponds to a unique value of temperature and pressure $(273.16 \text{ K}, 6.03 \times 10^{-3} \text{ atm}, \text{ for water})$, it is precisely reproducible and is often used as a point of reference (Section 17–10).

Notice that the *s-l* curve for water slopes upward to the left. This is true only of substances that *expand* upon freezing; for at a higher pressure, a lower temperature is needed to cause the liquid to freeze. More commonly, substances contract upon freezing and the *s-l* curve slopes upward to the right, as shown for CO_2 in Fig. 18–7.

The phase transitions we have been discussing are the common ones. Some substances, however, can exist in several forms in the solid phase. A transition from one of these phases to another occurs at a particular temperature and pressure, just like ordinary phase changes. For example, ice has been observed in at least eight different modifications at very high pressure. Ordinary helium has two distinct liquid phases, called helium I and II. They exist only at temperatures within a few degrees of absolute zero. Helium II exhibits very unusual properties referred to as **superfluidity**. It has essentially zero viscosity and exhibits strange properties such as actually climbing up the sides of an open container.

18–4 | Vapor Pressure and Humidity

Evaporation

If a glass of water is left out overnight, the water level will have dropped by morning. We say the water has evaporated, meaning that some of the water has changed to the vapor or gas phase.

This process of **evaporation** can be explained on the basis of kinetic theory. The molecules in a liquid move past one another with a variety of speeds that follow, approximately, the Maxwell distribution. There are strong attractive forces between these molecules, which is what keeps them close together in the liquid phase. A molecule in the upper regions of the liquid may, because of its speed, leave the liquid momentarily. But just as a rock thrown into the air returns to the Earth, so the attractive forces of the other molecules can pull the vagabond molecule back to the liquid surface—that is, if its velocity is not too large. A molecule with a high enough velocity, however, will escape from the liquid entirely, like a rocket escaping the Earth, and become part of the gas phase. Only those molecules that have kinetic energy above a particular value can escape to the gas phase. We have already seen that kinetic theory predicts that the relative number of molecules with kinetic energy above a particular value $(\text{such as } E_A \text{ in Fig. 18–3})$ increases with temperature. This is in accord with the well-known observation that the evaporation rate is greater at higher temperatures.

Because it is the fastest molecules that escape from the surface, the average speed of those remaining is less. When the average speed is less, the absolute temperature is less. Thus kinetic theory predicts that *evaporation is a cooling process.* You have no doubt noticed this effect when you stepped out of a warm shower and felt cold as the water on your body began to evaporate; and after working up a sweat on a hot day, even a slight breeze makes you feel cool through evaporation.

➡ PHYSICS APPLIED
Evaporation cools

Vapor Pressure

Air normally contains water vapor (water in the gas phase) and it comes mainly from evaporation. To look at this process in a little more detail, consider a closed container that is partially filled with water (it could just as well be any other liquid) and from which the air has been removed (Fig. 18–8). The fastest moving molecules quickly evaporate into the space above. As they move about, some of these molecules strike the liquid surface and again become part of the liquid phase: this is called **condensation**. The number of molecules in the vapor increases for a time, until a point is reached where the number returning to the liquid equals the number leaving in the same time interval. Equilibrium then exists, and the space is said to be *saturated.* The pressure of the vapor when it is saturated is called the **saturated vapor pressure** (or sometimes simply the vapor pressure).

The saturated vapor pressure does not depend on the volume of the container. If the volume above the liquid were reduced suddenly, the density of molecules in the vapor phase would be increased temporarily. More molecules would then be striking the liquid surface per second. There would be a net flow of molecules back to the liquid phase until equilibrium was again reached, and this would occur at the same value of the saturated vapor pressure, as long as the temperature had not changed.

The saturated vapor pressure of any substance depends on the temperature. At higher temperatures, more molecules have sufficient kinetic energy to break from the liquid surface into the vapor phase. Hence equilibrium will be reached at a higher pressure. The saturated vapor pressure of water at various temperatures is given in Table 18–2. Notice that even solids—for example, ice—have a measurable saturated vapor pressure.

In everyday situations, evaporation from a liquid takes place into the air above it rather than into a vacuum. This does not materially alter the discussion above relating to Fig. 18–8. Equilibrium will still be reached when there are sufficient molecules in the gas phase that the number reentering the liquid equals the number leaving. The concentration of particular molecules (such as water) in the gas phase is not affected by the presence of air, although collisions with air molecules may lengthen the time needed to reach equilibrium. Thus equilibrium occurs at the same value of the saturated vapor pressure as if air weren't there.

Of course, if the container is large or is not closed, all the liquid may evaporate before saturation is reached. And if the container is not sealed—as, for example, a room in your house—it is not likely that the air will become saturated with water vapor (unless it is raining outside).

Boiling

The saturated vapor pressure of a liquid increases with temperature. When the temperature is raised to the point where the saturated vapor pressure at that temperature equals the external pressure, **boiling** occurs (Fig. 18–9). As the boiling point is approached, tiny bubbles tend to form in the liquid, which indicate a change from the liquid to the gas phase. However, if the vapor pressure inside the bubbles is less than the external pressure, the bubbles immediately are crushed. As the temperature is increased, the saturated vapor pressure inside a bubble eventually becomes equal to or exceeds the external air pressure. The bubble will then not collapse but will increase in size and rise to the surface. Boiling has then begun. *A liquid boils when its saturated vapor pressure equals the external pressure.* This occurs for water under 1 atm (760 torr) of pressure at 100°C, as can be seen from Table 18–2.

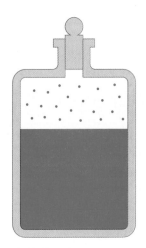

FIGURE 18–8 Vapor appears above a liquid in a closed container.

TABLE 18–2 Saturated Vapor Pressure of Water

Temp-erature (°C)	Saturated Vapor Pressure	
	torr (= mm-Hg)	Pa (= N/m²)
−50	0.030	4.0
−10	1.95	2.60×10^2
0	4.58	6.11×10^2
5	6.54	8.72×10^2
10	9.21	1.23×10^3
15	12.8	1.71×10^3
20	17.5	2.33×10^3
25	23.8	3.17×10^3
30	31.8	4.24×10^3
40	55.3	7.37×10^3
50	92.5	1.23×10^4
60	149	1.99×10^4
70	234	3.12×10^4
80	355	4.73×10^4
90	526	7.01×10^4
100	760	1.01×10^5
120	1489	1.99×10^5
150	3570	4.76×10^5

At boiling, saturated vapor pressure equals external pressure

FIGURE 18–9 Boiling: bubbles of water vapor float upward from the bottom of the pot (where the temperature is highest).

Relative humidity

The boiling point of a liquid clearly depends on the external pressure. At high elevations, the boiling point of water is somewhat less than at sea level since the air pressure is less. For example, on the summit of Mt. Everest (8850 m) the air pressure is about one-third of what it is at sea level, and from Table 18–2 we can see that water will boil at about 70°C. Cooking food by boiling takes longer at high elevations, since the temperature is less. Pressure cookers, however, reduce cooking time, because they build up a pressure as high as 2 atm, allowing higher boiling temperatures to be attained.

Partial Pressure and Humidity

When we refer to the weather as being dry or humid, we are referring to the water vapor content of the air. In a gas such as air, which is a mixture of several types of gases, the total pressure is the sum of the *partial pressures* of each gas present.[†] By **partial pressure**, we mean the pressure each gas would exert if it alone were present. The partial pressure of water in the air can be as low as zero and can vary up to a maximum equal to the saturated vapor pressure of water at the given temperature. Thus, at 20°C, the partial pressure of water cannot exceed 17.5 torr (see Table 18–2). The **relative humidity** is defined as the ratio of the partial pressure to the saturated vapor pressure at a given temperature. It is usually expressed as a percentage:

$$\text{Relative humidity} = \frac{\text{partial pressure of H}_2\text{O}}{\text{saturated vapor pressure of H}_2\text{O}} \times 100\%.$$

Thus, when the humidity is close to 100 percent, the air holds nearly all the water vapor it can.

> **EXAMPLE 18–6** **Relative humidity.** On a particular hot day, the temperature is 30°C and the partial pressure of water vapor in the air is 21.0 torr. What is the relative humidity?
>
> **SOLUTION** From Table 18–2, the saturated vapor pressure of water at 30°C is 31.8 torr. Hence the relative humidity is
>
> $$\frac{21.0 \text{ torr}}{31.8 \text{ torr}} \times 100\% = 66\%.$$

➡ **PHYSICS APPLIED**

Weather

Air is saturated with water vapor when the partial pressure of water in the air is equal to the saturated vapor pressure at that temperature. If the partial pressure of water exceeds the saturated vapor pressure, the air is said to be **supersaturated**. This situation can occur when a temperature decrease occurs. For example, suppose the temperature is 30°C and the partial pressure of water is 21 torr, which represents a humidity of 66 percent as we saw above. Suppose now that the temperature falls to, say, 20°C, as might happen at nightfall. From Table 18–2 we see that the saturated vapor pressure of water at 20°C is 17.5 torr. Hence the relative humidity would be greater than 100 percent, and the supersaturated air cannot hold this much water. The excess water condenses and appears as dew—or perhaps instead as fog or rain.

When air containing a given amount of water is cooled, a temperature is reached where the partial pressure of water equals the saturated vapor pressure. This is called the **dew point**. Measurement of the dew point is the most accurate means of determining the relative humidity. One method uses a polished metal surface in contact with air, which is gradually cooled down. The temperature at which moisture begins to appear on the surface is the dew point, and the partial pressure of water can then be obtained from saturated vapor pressure tables. If, for example, on a given day the temperature is 20°C, and the dew point is 5°C, then the partial pressure of water (Table 18–2) in the original air was 6.54 torr, whereas its saturated vapor pressure was 17.5 torr; hence the relative humidity was 6.54/17.5 = 37 percent.

[†] For example, 78 percent (by volume) of air molecules are nitrogen and 21 percent oxygen, with much smaller amounts of water vapor, argon, and other gases. At an air pressure of 1 atm, oxygen exerts a partial pressure of 0.21 atm and nitrogen 0.78 atm.

*18–5 Van der Waals Equation of State

In Section 18–3, we discussed how real gases deviate from ideal gas behavior, particularly at high densities or when near condensing to a liquid. We would like to understand these deviations using a microscopic (molecular) point of view. J. D. van der Waals (1837–1923) analyzed this problem and in 1873 arrived at an equation of state which fits real gases more accurately than the ideal gas law. His analysis is based on kinetic theory but takes into account: (1) the finite size of molecules (we previously neglected the actual volume of the molecules themselves, compared to the total volume of the container, and this assumption becomes poorer as the density increases and molecules become closer together); (2) the range of the forces between molecules may be greater than the size of the molecules (we previously assumed that intermolecular forces act only during collision, when the molecules are "in contact"). Let us now look at this analysis and derive the van der Waals equation of state.

Assume the molecules in a gas are spherical with radius r. If we assume these molecules behave like hard spheres, then two molecules collide and bounce off one another if the distance between their centers (Fig. 18–10) gets as small as $2r$. Thus the actual volume in which the molecules can move about is somewhat less than the volume V of the container holding the gas. The amount of "unavailable volume" depends on the number of molecules and on their size. Let b represent the "unavailable volume per mole" of gas. Then in the ideal gas law we replace V by $(V - nb)$, where n is the number of moles, and we obtain

$$P(V - nb) = nRT.$$

If we divide through by n

$$P\left(\frac{V}{n} - b\right) = RT. \tag{18–8}$$

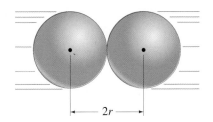

FIGURE 18–10 Molecules, of radius r, colliding.

This relation (sometimes called the **Clausius equation of state**) predicts that for a given temperature T and volume V, the pressure P will be greater than for an ideal gas. This makes sense since the reduced "available" volume means the number of collisions with the walls is increased.

Next we consider the effects of attractive forces between molecules, which are responsible for holding molecules in the liquid and solid states at lower temperatures. These forces are electrical in nature and although they act even when molecules are not touching, we assume their range is small—that is, they act mainly between nearest neighbors. Molecules at the edge of the gas, headed toward a wall of the container, are slowed down by a net force pulling them back into the gas. Thus these molecules will exert less force and less pressure on the wall than if there were no attractive forces. The reduced pressure will be proportional to the density of molecules in the layer of gas at the surface, and also to the density in the next layer, which exerts the inward force.[†] Therefore we expect the pressure to be reduced by a factor proportional to the density squared $(n/V)^2$, here written as moles per volume. If the pressure P is given by Eq. 18–8, then we should reduce this by an amount $a(n/V)^2$ where a is a proportionality constant. Thus we have

$$P = \frac{RT}{(V/n) - b} - \frac{a}{(V/n)^2}$$

or

$$\left(P + \frac{a}{(V/n)^2}\right)\left(\frac{V}{n} - b\right) = RT, \tag{18–9}$$

Van der Waals equation of state

which is the **van der Waals equation of state**.

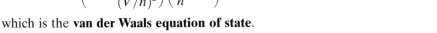

[†] This is similar to the gravitational force in which the force on mass m_1 due to mass m_2 is proportional to the product of their masses (Newton's law of universal gravitation, Chapter 6).

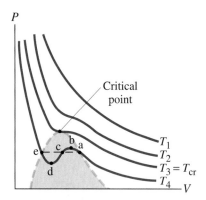

FIGURE 18–11 PV diagram for a van der Waals gas, shown for four different temperatures. For T_1, T_2, and T_3 (T_3 is chosen equal to the critical temperature), the curves fit experimental data very well for most gases. The curve labeled T_4, a temperature below the critical point, passes through the liquid–vapor region. The maximum (point b) and minimum (point d) would seem to be artifacts, since we usually see constant pressure, as indicated by the horizontal dashed line. However, for very pure supersaturated vapors or supercooled liquids, the sections ab and ed, respectively, have been observed. (The section bd would be unstable and has not been observed.)

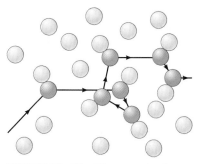

FIGURE 18–12 Zigzag path of a molecule colliding with other molecules.

FIGURE 18–13 Molecule at left moves to the right with speed $\overline{v}$. It collides with any molecule whose center is within the cylinder of radius $2r$.

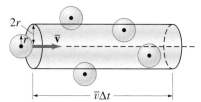

The constants a and b in the van der Waals equation are different for different gases and are determined by fitting to experimental data for each gas. For CO_2 gas, the best fit is obtained for $a = 3.6 \times 10^{-3}\,\mathrm{N \cdot m^4/mol^2}$ and $b = 4.2 \times 10^{-5}\,\mathrm{m^3/mol}$. Figure 18–11 shows a typical PV diagram for Eq. 18–9 (a "van der Waals gas") for four different temperatures, with detailed caption, and it should be compared to Fig. 18–5 for real gases.

Neither the van der Waals equation of state nor the many other equations of state that have been proposed are accurate for all gases under all conditions. Yet Eq. 18–9 is a very useful relation. And because it is quite accurate for many situations, its derivation gives us further insight into the nature of gases at the microscopic level. Note that at low densities $a/(V/n)^2 \ll P$ and $b \ll V/n$, so that the van der Waals equation reduces to the equation of state for an ideal gas, $PV = nRT$.

*18–6 Mean Free Path

If gas molecules were truly point particles, they would never collide with one another. Thus, when a person opened a perfume bottle, you would be able to smell it almost instantaneously across the room, since molecules travel hundreds of meters per second. In fact it takes some time before you detect an odor, and according to kinetic theory, this must be due to collisions between molecules of nonzero size.

If we were to follow the path of a particular molecule, we would expect to see it follow a zigzag path as shown in Fig. 18–12. Between each collision the molecule would move in a straight-line path. (Not quite true if we take account of the small intermolecular forces that act between collisions.) An important parameter for a given situation is the **mean free path**, which is defined as the average distance a molecule travels between collisions. We would expect that the greater the gas density, and the larger the molecules, the shorter the mean free path would be. We now determine the nature of this relationship for an ideal gas.

Suppose our gas is made up of molecules which are hard spheres of radius r. A collision will occur whenever the centers of two molecules come within a distance $2r$ of one another. Let us follow a molecule as it traces a straight-line path. In Fig. 18–13, the dashed line represents the path of our particle if it made no collisions. Also shown is a cylinder of radius $2r$. If the center of another molecule lies within this cylinder, a collision will occur. (Of course, when a collision occurs the particle's path would change direction, as would our imagined cylinder, but our result won't be altered by unbending a zigzag cylinder into a straight one for purposes of calculation.) Assume our molecule is an average one, moving at the mean speed $\overline{v}$ in the gas. For the moment, let us assume that the other molecules are not moving, and that the concentration of molecules (number per unit volume) is N/V. Then the number of molecules whose center lies within the cylinder of Fig. 18–13 is N/V times the volume of this cylinder, and this also represents the number of collisions that will occur. In a time Δt, our molecule travels a distance $\overline{v}\,\Delta t$, so the length of the cylinder is $\overline{v}\,\Delta t$ and its volume is $\pi(2r)^2\,\overline{v}\,\Delta t$. Hence the number of collisions that occur in a time Δt is $(N/V)\pi(2r)^2\overline{v}\,\Delta t$. We define the **mean free path**, l_{M}, as the average distance between collisions. This distance is equal to the distance traveled ($\overline{v}\,\Delta t$) in a time Δt divided by the number of collisions made in time Δt:

$$l_{\mathrm{M}} = \frac{\overline{v}\,\Delta t}{(N/V)\pi(2r)^2\overline{v}\,\Delta t} = \frac{1}{4\pi r^2(N/V)}. \tag{18–10a}$$

Thus we see that l_{M} is inversely proportional to the cross-sectional area ($= \pi r^2$) of the molecules and to their concentration (number/volume), N/V. However, Eq. 18–10a is not fully correct since we assumed the other molecules are all at rest. In

fact, they are moving, and the number of collisions in a time Δt must depend on the *relative* speed of the colliding molecules, rather than on $\bar{v}$. Hence the number of collisions per second is $(N/V)\pi(2r)^2 v_{rel}\, \Delta t$ (rather than $(N/V)\pi(2r)^2\bar{v}\, \Delta t$), where v_{rel} is the average relative speed of colliding molecules. A careful calculation shows that for a Maxwellian distribution of speeds $v_{rel} = \sqrt{2}\bar{v}$. Hence the mean free path is

$$l_M = \frac{1}{4\pi\sqrt{2}r^2(N/V)}.$$ **(18–10b)** *Mean free path*

EXAMPLE 18–7 ESTIMATE **Mean free path of air molecules at STP.**
Estimate the mean free path of air molecules at STP, standard temperature and pressure (0°C, 1 atm). The diameter of O_2 and N_2 molecules is about 3×10^{-10} m.

SOLUTION We saw in Example 17–8 that 1 mol of an ideal gas occupies a volume of 22.4×10^{-3} m^3 at STP. Hence

$$\frac{N}{V} = \frac{6.02 \times 10^{23}\ \text{molecules}}{22.4 \times 10^{-3}\ \text{m}^3} = 2.69 \times 10^{25}\ \text{molecules/m}^3.$$

Then

$$l_M = \frac{1}{4\pi\sqrt{2}\left(1.5 \times 10^{-10}\ \text{m}\right)^2\left(2.7 \times 10^{25}\ \text{m}^{-3}\right)} \approx 9 \times 10^{-8}\ \text{m}$$

which is about 300 times the diameter of a molecule.

At very low densities, such as in an evacuated vessel, the concept of mean free path loses meaning since collisions with walls of the container may occur more frequently than collisions with other molecules. For example, in a cubical box 20 cm on a side containing air at 10^{-7} torr, the mean free path is about 700 m, which means many more collisions are made with the walls than with other molecules. (Note, nonetheless, that the box contains over 10^{12} molecules.) If the concept of mean free path were to include any type of collision, it would be closer to 0.2 m than to the 700 m calculated from Eq. 18–10.

*18–7 Diffusion

If you carefully place a few drops of food coloring in a container of water as in Fig. 18–14, you will find that the color spreads throughout the water. The process may take some time (assuming you don't shake the glass), but eventually the color will become uniform. This mixing, known as **diffusion**, takes place because of the random movement of the molecules. Diffusion occurs in gases too. Common examples include perfume or smoke (or the odor of something cooking on the stove) diffusing in air, although convection (moving air currents) often plays a greater role in spreading odors than does diffusion. Diffusion depends on concentration, by which we mean the number of molecules or moles per unit volume. In general, *the diffusing substance moves from a region where its concentration is high to one where its concentration is low.*

Diffusion occurs from high toward low concentration

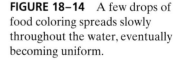

FIGURE 18–14 A few drops of food coloring spreads slowly throughout the water, eventually becoming uniform.

(a)　　　　　　　(b)　　　　　　　(c)

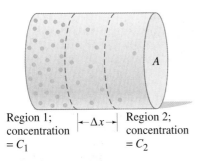

FIGURE 18–15 Diffusion occurs from a region of high concentration to one of lower concentration (only one type of molecule is shown).

Region 1; concentration $= C_1$ $\leftarrow \Delta x \rightarrow$ Region 2; concentration $= C_2$

Diffusion can be readily understood on the basis of kinetic theory and the random motion of molecules. Consider a tube of cross-sectional area A containing molecules in a higher concentration on the left than on the right, Fig. 18–15. We assume the molecules are in random motion. Yet there will be a net flow of molecules to the right. To see why this is true, let us consider the small section of tube of length Δx as shown. Molecules from both regions 1 and 2 cross into this central section as a result of their random motion. The more molecules there are in a region, the more will strike a given area or cross a boundary. Since there is a greater concentration of molecules in region 1 than in region 2, more molecules cross into the central section from region 1 than from region 2. There is, then, a net flow of molecules from left to right, from high concentration toward low concentration. The flow stops only when the concentrations become equal.

You might expect that the greater the difference in concentration, the greater the flow rate. This is indeed the case. In 1855, the physiologist Adolf Fick (1829–1901) determined experimentally that the rate of diffusion (J) is directly proportional to the change in concentration per unit distance $(C_1 - C_2)/\Delta x$ (which is called the **concentration gradient**), and to the cross-sectional area A (see Fig. 18–15):

$$J = DA \frac{C_1 - C_2}{\Delta x},$$

or, in terms of derivatives

$$J = DA \frac{dC}{dx}. \tag{18–11}$$

Diffusion equation (Fick's law)

D is a constant of proportionality called the **diffusion constant**. Equation 18–11 is known as the **diffusion equation**, or **Fick's law**. If the concentrations are given in mol/m^3, then J is the number of moles passing a given point per second; if the concentrations are given in kg/m^3, then J is the mass movement per second (kg/s).

The diffusion constant D will depend on the properties of the diffusing molecules and on those of the substance in which it is immersed (often water or air), and also on the temperature and the external pressure. The values of D for a variety of substances are given in Table 18–3.

➡ PHYSICS APPLIED

Diffusion time

TABLE 18–3 Diffusion Constants, D (20°C, 1 atm)

Diffusing Molecules	Medium	$D(m^2/s)$
H_2	Air	6.3×10^{-5}
O_2	Air	1.8×10^{-5}
O_2	Water	100×10^{-11}
Blood hemoglobin	Water	6.9×10^{-11}
Glycine (an amino acid)	Water	95×10^{-11}
DNA (mass 6×10^6 u)	Water	0.13×10^{-11}

EXAMPLE 18–8 ESTIMATE Diffusion of ammonia in air. To get an idea of the time required for diffusion, estimate how long it might take for ammonia (NH_3) to be detected 10 cm from a bottle after it is opened, assuming only diffusion.

SOLUTION This will be an order-of-magnitude calculation. The rate of diffusion J can be set equal to the number of molecules N diffusing across area A in a time t: $J = N/t$. Then $t = N/J$ and using Eq. 18–11,

$$t = \frac{N}{J} = \frac{N}{AD} \frac{\Delta x}{\Delta C}.$$

The average concentration (midway between bottle and nose) can be approximated by $\overline{C} \approx N/V$, where V is the volume over which the molecules move and is roughly of the order of $V \approx A \Delta x$ where Δx is 10 cm. We substitute $N = \overline{C} A \Delta x$ into the above equation:

$$t \approx \frac{(\overline{C} A \Delta x)\Delta x}{AD \Delta C} = \frac{\overline{C}}{\Delta C} \frac{(\Delta x)^2}{D}.$$

The concentration of ammonia is high near the bottle and low near the detecting nose, so $\overline{C} \approx \Delta C/2$ or $(\overline{C}/\Delta C) \approx \frac{1}{2}$. Since NH_3 molecules have a size somewhere between H_2 and O_2, from Table 18–3, we can estimate $D \approx 4 \times 10^{-5} m^2/s$. Then

$$t \approx \frac{1}{2} \frac{(0.10 \text{ m})^2}{(4 \times 10^{-5} \text{ m}^2/\text{s})} \approx 100 \text{ s}.$$

or about a minute or two. This seems rather long from experience, suggesting that air currents (convection) are more important than diffusion for transmitting odors.

Summary

According to the **kinetic theory** of gases, which is based on the idea that a gas is made up of molecules that are moving rapidly and at random, the average kinetic energy of the molecules is proportional to the Kelvin temperature T:

$$\overline{K} = \tfrac{1}{2}m\overline{v^2} = \tfrac{3}{2}kT$$

where k is Boltzmann's constant.

At any moment, there exists a wide distribution of molecular speeds within a gas. The **Maxwell distribution of speeds** is derived from simple kinetic theory assumptions, and is in good accord with experiment for gases at not too high a pressure.

The behavior of real gases at high pressure, and/or when near their liquefaction point, deviates from the ideal gas law. These deviations are due to the finite size of molecules and to the attractive forces between molecules.

Below the **critical temperature**, a gas can change to a liquid if sufficient pressure is applied; but if the temperature is higher than the critical temperature, no amount of pressure will cause a liquid surface to form.

The **triple point** of a substance is that unique temperature and pressure at which all three phases—solid, liquid, and gas—can coexist in equilibrium. Because of its precise reproducibility, the triple point of water is often taken as a standard reference point.

Evaporation of a liquid is the result of the fastest moving molecules escaping from the surface. Because the average molecular velocity is less after the fastest molecules escape, the temperature decreases when evaporation takes place.

Saturated vapor pressure refers to the pressure of the vapor above a liquid when the two phases are in equilibrium. The vapor pressure of a substance (such as water) depends strongly on temperature and is equal to atmospheric pressure at the boiling point.

Relative humidity of air at a given place is the ratio of the partial pressure of water vapor in the air to the saturated vapor pressure at that temperature; it is usually expressed as a percentage.

Questions

1. Why doesn't the size of different molecules enter into the gas laws?

2. When a gas is rapidly compressed (say, by pushing down a piston) its temperature increases. When a gas expands against a piston, it cools. Explain these changes in temperature using the kinetic theory, in particular noting what happens to the momentum of molecules when they strike the moving piston.

3. In Section 18–1 we assumed the gas molecules made perfectly elastic collisions with the walls of the container. This assumption is not necessary as long as the walls are at the same temperature as the gas. Why?

4. Explain in words how Charles's law follows from kinetic theory and the relation between average kinetic energy and the absolute temperature.

5. Explain in words how Gay-Lussac's law follows from kinetic theory.

6. As you go higher in the Earth's atmosphere, the ratio of N_2 molecules to O_2 molecules increases. Why?

7. Can you determine the temperature of a vacuum?

8. Is temperature a macroscopic or microscopic variable?

9. Discuss why the Maxwell distribution of speeds (Fig. 18–2) is not a symmetric curve.

10. Explain why the peak of the curve for 310 K in Fig. 18–3 is not as high as for 273 K. (Assume the total number of molecules is the same for both.)

11. Explain, using the Maxwell distribution of speeds, (a) why the Moon has very little atmosphere, (b) why hydrogen, if at one time in the Earth's atmosphere, would probably have escaped.

12. Explain why putting food in the freezer retards spoilage.

13. Would the average kinetic energy of molecules in an ideal gas correspond to $\overline{v}$, v_{rms}, v_p, or to some other value?

14. If the pressure in a gas is doubled while its volume is held constant, by what factor do (a) v_{rms} and (b) $\overline{v}$ change?

15. If a container of gas is at rest, the average velocity of molecules must be zero. Yet the average speed is not zero. Explain.

16. Draw, roughly, the Maxwellian distribution of *velocities*. Discuss what the values would be for mean, rms, and most probable velocities. Would the curve be symmetric?

17. What everyday observation would tell you that not all molecules in a material have the same speed?

18. We saw that the saturated vapor pressure of a liquid (say, water) does not depend on the external pressure. Yet the temperature of boiling does depend on the external pressure. Is there a contradiction? Explain.

19. Alcohol evaporates more quickly than water at room temperature. What can you infer about the molecular properties of one relative to the other?

20. Explain why a hot humid day is far more uncomfortable than a hot dry day at the same temperature.

21. Is it possible to boil water at room temperature (20°C) without heating it? Explain.

22. What exactly does it mean when we say that oxygen boils at −183°C?

23. A length of thin wire is placed over a block of ice (or an ice cube) at 0°C and weights are hung from the ends of the wire. It is found that the wire cuts its way through the ice cube, but leaves a solid block of ice behind it. This process is called *regelation*. Explain how this happens by inferring how the freezing point of water depends on pressure.

24. How do a gas and a vapor differ?

25. (*a*) At suitable temperatures and pressures, can ice be melted by applying pressure? (*b*) At suitable temperatures and pressures, can carbon dioxide be melted by applying pressure?

26. Why does dry ice not last long at room temperature?

27. Under what conditions can liquid CO_2 exist? Be specific. Can it exist as a liquid at normal room temperature?

* **28.** Name several ways to reduce the mean free path in a gas.

* **29.** Discuss why sound waves can travel in a gas only if their wavelength is somewhat larger than the mean free path.

Problems

Section 18–1

1. (I) (*a*) What is the average kinetic energy of an oxygen molecule at STP? (*b*) What is the total translational kinetic energy of 2.0 mol of O_2 molecules at 20°C?

2. (I) Calculate the rms speed of helium atoms near the surface of the sun at a temperature of about 6000 K.

3. (I) By what factor will the rms speed of gas molecules increase if the temperature is increased from 0°C to 100°C?

4. (I) A gas is at 20°C. To what temperature must it be raised to double the rms speed of its molecules?

5. (I) Twelve molecules have the following speeds, given in arbitrary units: 6, 2, 4, 6, 0, 4, 1, 8, 5, 3, 7, and 8. Calculate (*a*) the mean speed, and (*b*) the rms speed.

6. (II) The rms speed of molecules in a gas at 20.0°C is to be increased by 1.0%. To what temperature must it be raised?

7. (II) If the pressure in a gas is doubled while its volume is held constant, by what factor does v_{rms} change?

8. (II) Show that the rms speed of molecules in a gas is given by $v_{rms} = \sqrt{3P/\rho}$, where P is the pressure in the gas, and ρ is the gas density.

9. (II) Show that for a mixture of two gases at the same temperature, the ratio of their rms speeds is equal to the inverse ratio of the square roots of their molecular masses.

10. (II) What is the rms speed of nitrogen molecules contained in an 8.5-m^3 volume at 2.1 atm if the total amount of nitrogen is 1300 mol?

11. (II) Calculate (*a*) the rms speed of an oxygen molecule at 0°C and (*b*) determine how many times per second it would move back and forth across a 7.0-m-long room on the average, assuming it made very few collisions with other molecules.

12. (II) What is the average distance between nitrogen molecules at STP?

13. (II) The two isotopes of uranium, ^{235}U and ^{238}U (the superscripts refer to their atomic masses), can be separated by a gas diffusion process by combining them with fluorine to make the gaseous compound UF_6. Calculate the ratio of the rms speeds of these molecules for the two isotopes.

Section 18–2

14. (I) A group of 25 particles have the following speeds: two have speed 10 m/s, seven have 15 m/s, four have 20 m/s, three have 25 m/s, six have 30 m/s, one has 35 m/s, and two have 40 m/s. Determine (*a*) the average speed, (*b*) the rms speed, and (*c*) the most probable speed.

* **15.** (III) Starting from the Maxwell distribution of speeds, Eq. 18–6, show (*a*) $\int_0^\infty f(v)\,dv = N$, and (*b*)

$$\int_0^\infty v^2 f(v)\,dv/N = 3kT/m.$$

Section 18–3

16. (I) (*a*) At atmospheric pressure, in what phases can CO_2 exist? (*b*) For what range of pressures and temperatures can CO_2 be a liquid? Refer to Fig. 18–7.

17. (I) CO_2 exists in what phase when the pressure is 30 atm and the temperature is 30°C (Fig. 18–7)?

18. (I) Water is in which phase when the pressure is 0.01 atm and the temperature is (*a*) 90°C, (*b*) −20°C?

19. (II) You have a sample of water and are able to control temperature and pressure arbitrarily. (*a*) Using Fig. 18–6, describe the phase changes you would see if you started at a temperature of 100°C, a pressure of 220 atm, and decreased the pressure down to 0.004 atm while keeping the temperature fixed. (*b*) Repeat part (*a*) with the temperature at 0.0°C. Assume that you held the system at the starting conditions long enough for the system to stabilize before making further changes.

Section 18–4

20. (I) What is the dew point (approximately) if the humidity is 50 percent on a day when the temperature is 25°C?

21. (I) What is the air pressure at a place where water boils at 90°C?

22. (I) If the air pressure at a particular place in the mountains is 0.85 atm, estimate the temperature at which water boils.

23. (I) What is the temperature on a day when the partial pressure of water is 530 Pa and the relative humidity is 40 percent?

24. (I) What is the partial pressure of water on a day when the temperature is 25°C and the relative humidity is 35%?

25. (II) What is the approximate pressure inside a pressure cooker if the water is boiling at a temperature of 120°C? Assume no air escaped during the heating process, which started at 20°C.

26. (II) If the humidity in a room of volume 240 m^3 at 25°C is 80 percent, what mass of water can still evaporate from an open pan?

27. (II) An *autoclave* is a device used to sterilize laboratory instruments. It is essentially a high-pressure steam boiler and operates on the same principle as a pressure cooker. However, because hot steam under pressure is more effective in killing microorganisms than moist air at the same temperature and pressure, the air is removed and replaced by steam. Typically, the gauge pressure inside the autoclave is 1.0 atm. What is the temperature of the steam? Assume the steam is in equilibrium with boiling water.

28. (III) Air that is at its dew point of 5°C is drawn into a building where it is heated to 25°C. What will be the relative humidity at this temperature? Assume constant pressure of 1.0 atm. Take into account the expansion of the air.

* Section 18–5

* 29. (II) For oxygen gas, the van der Waals equation of state achieves its best fit for $a = 0.14 \, N \cdot m^4/mol^2$ and $b = 3.2 \times 10^{-5} \, m^3/mol$. Determine the pressure in 1.0 mole of the gas at 0°C if its volume is 0.40 L, calculated using (a) the van der Waals equation, (b) the ideal gas law.

* 30. (II) In the van der Waals equation of state, the constant b represents the amount of "unavailable volume" occupied by the molecules themselves. Thus V is replaced by $(V - nb)$, where n is the number of moles. For oxygen, b is about $3.2 \times 10^{-5} \, m^3/mol$. Estimate the diameter of an oxygen molecule.

* 31. (III) (a) From the van der Waals equation of state, show that the critical temperature and pressure are given by

$$T_{cr} = \frac{8a}{27bR}, \qquad P_{cr} = \frac{a}{27b^2}.$$

[Hint: Use the fact that the P versus V curve has an inflection point at the critical point so that the first and second derivatives are zero.] (b) Determine a and b for CO_2 from the measured values of $T_{cr} = 304 \, K$ and $P_{cr} = 72.8 \, atm$.

* 32. (III) (a) Show that a collision between two spherical molecules each of radius r, is equivalent to a collision between a point particle and a sphere of radius $2r$, and thus the center of one molecule cannot penetrate a volume equal to that of a sphere of radius $2r$ when two molecules collide. (b) Show that the total unavailable volume per mole is $b = 16\pi r^3 N_A/3$, where N_A is Avogadro's number. [Hint: In summing the total excluded volume, multiply by $\frac{1}{2}$ to avoid counting each pair of molecules twice.] Note that the unavailable volume is four times the actual molecular volume. (c) Estimate the diameter of a CO_2 molecule, for which $b = 4.2 \times 10^{-5} \, m^3/mol$.

* Section 18–6

* 33. (II) At about what pressure would the mean free path of air molecules be (a) 1.0 m and (b) equal to the diameter of air molecules, $\approx 3 \times 10^{-10} \, m$?

* 34. (II) (a) The mean free path of CO_2 molecules at STP is measured to be about $5.6 \times 10^{-8} \, m$. Estimate the diameter of a CO_2 molecule. (b) Do the same for He gas for which $l_M \approx 25 \times 10^{-8}$ at STP.

* 35. (II) A cubic box of volume $4.4 \times 10^{-3} \, m^3$ contains 70 marbles, each 1.5 cm in diameter. What is the mean free path for a marble (a) when the box is shaken vigorously and (b) when it is slightly shaken?

* 36. (II) A cubic box 1.20 m on a side is evacuated so the pressure of air inside is 10^{-6} torr. Estimate how many molecular collisions there are per each collision with a wall (0°C).

* 37. (II) A very small amount of hydrogen gas is released into the air. If the air is at 1.0 atm and 25°C, estimate the mean free path for a H_2 molecule. What assumptions did you make?

* 38. (II) Estimate the maximum allowable pressure in a 38-cm-long cathode ray tube if 98 percent of all electrons must hit the screen without first striking an air molecule.

* 39. (II) (a) Show that the number of collisions a molecule makes per second, called the *collision frequency*, f, is given by $f = \bar{v}/l_M$, and thus $f = 4\sqrt{2}\,\pi r^2 \bar{v} N/V$. (b) What is the collision frequency for N_2 molecules in air at $T = 20°C$ and $P = 1.0 \times 10^{-2} \, atm$?

* 40. (II) We saw in Example 18–7 that the mean free path of air molecules at STP, l_M, is about 9×10^{-8} m. Estimate the collision frequency f, the number of collisions per unit time.

* 41. (II) Suppose that a gas contains two types of molecules in concentrations n_1 and n_2. Their radii are r_1 and r_2. Use arguments similar to those leading to Eq. 18–10a to derive the following relation for the mean free path for type 1 molecules

$$l_{M1} = \frac{1}{4\pi r_1^2 n_1 + \pi(r_1 + r_2)^2 n_2}.$$

* 42. (III) At some instant, suppose we have N_0 identical molecules. Show that the number N of molecules that travel a distance x or more before the next collision is given by $N = N_0 e^{-x/l_M}$, where l_M is the mean free path. This is called the *survival equation*.

* Section 18–7

* 43. (I) Approximately how long would it take for the ammonia of Example 18–8 to be detected 1.5 m from the bottle after it is opened? What does this suggest about the relative importance of diffusion and convection for carrying odors?

* 44. (II) What is the time needed for a glycine molecule (see Table 18–3) to diffuse a distance of $15\mu m$ in water at 20°C if its concentration varies over that distance from 1.00 mol/m^3 to 0.40 mol/m^3? Compare this "speed" to its rms (thermal) speed. The molecular mass of glycine is about 75 u.

* 45. (II) Oxygen diffuses from the surface of insects to the interior through tiny tubes called tracheae. An average trachea is about 2 mm long and has cross-sectional area of $2 \times 10^{-9} \, m^2$. Assuming the concentration of oxygen inside is half what it is outside in the atmosphere, (a) show that the concentration of oxygen in the air (assume 21 percent is oxygen) at 20°C is about 8.7 mol/m^3, then (b) calculate the diffusion rate J, and (c) estimate the average time for a molecule to diffuse in. Assume the diffusion constant is $1 \times 10^{-5} \, m^2/s$.

General Problems

46. What is the rms speed of nitrogen molecules contained in a 12.8 m³ volume at 3.42 atm if the total amount of nitrogen is 1800 mol.

47. In outer space the density of matter is about one atom per cm³, mainly hydrogen atoms, and the temperature is about 2.7 K. Calculate the rms speed of these hydrogen atoms, and the pressure (in atmospheres).

48. Calculate approximately the total translational kinetic energy of all the molecules in an *E. coli* bacterium of mass 2.0×10^{-15} kg at 37°C. Assume 70 percent of the cell, by weight, is water, and the other molecules have an average molecular mass on the order of 10^5 u.

49. (a) Calculate the approximate rms speed of an amino acid, whose molecular mass is 89 u, in a living cell at 37°C. (b) What would be the rms speed of a protein of molecular mass 50,000 u at 37°C?

50. The escape speed from the Earth is 1.12×10^4 m/s, so that a gas molecule travelling away from Earth near the outer boundary of the Earth's atmosphere would, at this speed, be able to escape from the Earth's gravitational field and be lost to the atmosphere. At what temperature is the average speed of (a) oxygen molecules, and (b) helium atoms equal to 1.12×10^4 m/s?

51. The second postulate of kinetic theory is that the molecules are, on the average, far apart from one another. That is, their average separation is much greater than the diameter of each molecule. Is this assumption reasonable? To check, calculate the average volume occupied by one molecule of a gas at STP, and compare it to the size of the molecule (diameter of a typical gas molecule is about 0.2 nm). If the molecules were the diameter of ping-pong balls, say 4 cm, how far away, on average, would the next ping-pong ball be?

52. Estimate how many times per second a collision of a molecule of oxygen gas occurs with one of the walls of its container, which is a cube with sides 1.0 m. There are 2.0 mol of oxygen inside at 20°C. [*Hint:* Let $\bar{v}_x \approx \sqrt{\overline{v_x^2}}$.]

53. Consider a container of oxygen gas at a temperature of 20°C that is 0.50 m tall. Compare the gravitational potential energy of a molecule at the top of the container (assuming the potential energy is zero at the bottom) with the average kinetic energy of the molecules. Is it reasonable to neglect the potential energy?

54. In humid climates, people constantly *dehumidify* their cellars in order to prevent rot and mildew. If the cellar in a house (kept at 20°C) has 85 m² of floor space and a ceiling height of 2.8 m, what is the mass of water that must be removed from it in order to drop the humidity from 95 percent to a more reasonable 30 percent?

55. The temperature of an ideal gas is increased from 120°C to 290°C while the volume and the number of moles stay constant. By what factor does the pressure change? By what factor does v_{rms} change?

56. A scuba tank has a volume of 2800 cm³. For very deep dives, the tank is filled with 50 percent (by volume) pure oxygen and 50 percent pure helium. (a) How many molecules are there of each type in the tank if it is filled at 20°C to a gauge pressure of 10 atm? (b) What is the ratio of the average kinetic energies of the two types of molecule? (c) What is the ratio of the rms speeds of the two types of molecule?

57. A space vehicle returning from the Moon enters the atmosphere at a speed of about 40,000 km/h. Molecules (assume nitrogen) striking the nose of the vehicle with this speed correspond to what temperature? (Because of this high temperature, the nose of a space vehicle must be made of special materials; indeed, part of it does vaporize, and this is seen as a bright blaze upon reentry.)

58. At room temperature, it takes approximately 2.45×10^3 J to evaporate 1.00 g of water. Estimate the average speed of evaporating molecules. What multiple of v_{rms} (at 20°C) for water molecules is this? (Assume Eq. 18–4 holds.)

59. Calculate the total water vapor pressure in the air on the following two days: (a) a hot summer day, with the temperature 30°C and the relative humidity at 40%; (b) a cold winter day, with the temperature 5°C and the relative humidity at 80%.

* 60. At 300 K, an 8.50 mol sample of carbon dioxide occupies a volume of 0.200 m³. Calculate the gas pressure, first by assuming the ideal gas law, and then by using the van der Waals equation of state. (The values for *a* and *b* are given in the text following Eq. 18–9.) In this range of pressure and volume, the van der Waals equation is very accurate. What percentage error did you make in assuming ideal-gas-law behavior?

* 61. The density of atoms, mostly hydrogen, in interstellar space is about one per cubic centimeter. Estimate the mean free path of the hydrogen atoms, assuming an atomic diameter of 10^{-10} m.

* 62. Using the ideal gas law, find an expression for the mean free path l_M that involves pressure and temperature instead of (N/V). Use this expression to find the mean free path for nitrogen molecules at a pressure of 10 atm and 300 K.

When it is cold, warm clothes act as insulators to reduce heat loss from the body to the outside by conduction and convection.

Heat radiation from a campfire can warm you, and the fire can also transfer energy directly by heat conduction to a cooking pot.

Heat, like work, represents a transfer of energy. Heat is defined as a transfer of energy due to a difference of temperature. Work is a transfer of energy by mechanical means, not due to a temperature difference. The first law of thermodynamics links the two in a general statement of energy conservation: the heat Q added to a system minus the net work W done by the system equals the change in internal energy ΔU of the system: $\Delta U = Q - W$. Internal energy U is the sum total of all the energy of the molecules of the system.

Heat and the First Law of Thermodynamics

W hen a pot of cold water is placed on a hot burner of a stove, the temperature of the water increases. We say that heat flows from the hot burner to the cold water. When two objects at different temperatures are put in contact, heat spontaneously flows from the hotter one to the colder one. The spontaneous flow of heat is in the direction tending to equalize the temperature. If the two objects are kept in contact long enough for their temperatures to become equal, the two bodies are said to be in thermal equilibrium, and there is no further heat flow between them. For example, when the mercury in a fever thermometer is still rising, heat is flowing from the patient's mouth to the thermometer; when the mercury stops, the thermometer is then in equilibrium with the person's mouth, and they are at the same temperature.

Heat and temperature are often confused. They are very different concepts and we will make the clear distinction between them. We begin this Chapter by defining and using the concept of heat. We will also begin our discussion of thermodynamics, which is the name we give to the study of processes in which energy is transferred as heat and as work.

19–1 Heat as Energy Transfer

We use the term "heat" in everyday life as if we knew what we meant. But the term is often used inconsistently, so it is important for us to define heat clearly, and to clarify the phenomena and concepts related to heat.

We commonly speak of the "flow" of heat—heat flows from a stove burner to a pot of coffee, from the Sun to the Earth, from a person's mouth into a fever thermometer. Heat flows spontaneously from an object at higher temperature to one at lower temperature. Indeed, an eighteenth-century model of heat pictured heat flow as movement of a fluid substance called *caloric*. However, the caloric fluid was never able to be detected. In the nineteenth century, it was found that the various phenomena associated with heat could be described consistently without the need to use the fluid model. The new model viewed heat as being akin to work and energy, as we will discuss in a moment. First we note that a common unit for heat, still in use today, is named after caloric. It is called the **calorie** (cal) and is defined as *the amount of heat necessary to raise the temperature of 1 gram of water by 1 Celsius degree*. [To be precise, the particular temperature range from 14.5°C to 15.5°C is specified because the heat required is very slightly different at different temperatures. The difference is less than 1 percent over the range 0°C to 100°C and we will ignore it for most purposes.] More often used than the calorie is the **kilocalorie** (kcal), which is 1000 calories. Thus *1 kcal is the heat needed to raise 1 kg of water by 1 C°*. Sometimes a kilocalorie is called a **Calorie** (with a capital C), and it is by this unit that the energy value of food is specified (we might also call it the "dietary calorie"). In the British system of units, heat is measured in British thermal units (Btu). One Btu is defined as the heat needed to raise the temperature of 1 lb of water by 1 F°. It can be shown (Problem 4) that 1 Btu = 0.252 kcal = 1055 J.

The idea that heat is related to energy was pursued by a number of scientists in the 1800s, particularly by an English brewer, James Prescott Joule (1818–1889). Joule performed a number of experiments that were crucial for establishing our present-day view that heat, like work, represents a transfer of energy. One of Joule's experiments is shown (simplified) in Fig. 19–1. The falling weight causes the paddle wheel to turn and do work on the water: the friction between the water and the paddle wheel causes the temperature of the water to rise slightly (barely measurable, in fact, by Joule). Of course, the same temperature rise could also be obtained by heating the water on a hot stove. In this and a great many other experiments (some involving electrical energy), Joule determined that a given amount of work done was always equivalent to a particular amount of heat input. Quantitatively, 4.186 joules (J) of work was found to be equivalent to 1 calorie (cal) of heat. This is known as the **mechanical equivalent of heat**:

$$4.186 \text{ J} = 1 \text{ cal}$$
$$4.186 \times 10^3 \text{ J} = 1 \text{ kcal}.$$

As a result of these and other experiments, scientists came to interpret heat not as a substance, and not even as a form of energy. Rather, heat refers to a *transfer of energy*: when heat flows from a hot object to a cooler one, it is energy that is being transferred from the hot to the cold object. Thus, **heat** is *energy that is transferred from one body to another because of a difference in temperature*. In SI units, the unit for heat, as for any form of energy, is the joule. Nonetheless, calories and kcal are still sometimes used. Today the calorie is *defined* in terms of the joule (via the mechanical equivalent of heat, above), rather than in terms of the properties of water, as given previously. The latter is still handy to remember: 1 cal raises 1 g of water by 1 C°, or 1 kcal raises 1 kg of water by 1 C°.

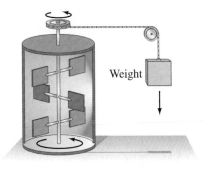

FIGURE 19–1 Joule's experiment on the mechanical equivalent of heat.

calorie (unit)

Kilocalorie (= dietary Calorie)

Btu

Mechanical equivalent of heat

Heat is energy transferred because of a ΔT

EXAMPLE 19–1 ESTIMATE Working off the extra calories. A young couple throw caution to the wind one afternoon, eating too much ice cream and cake. They realize that they each overate by about 500 Calories, and to compensate they want to do an equivalent amount of work climbing stairs or a mountain. How much vertical height must each person walk up? Each has a mass of 60 kg.

➡ **PHYSICS APPLIED**

Working off Calories

SOLUTION 500 Calories is 500 kcal, which in joules is

$$(500 \text{ kcal})(4.186 \times 10^3 \text{ J/kcal}) = 2.1 \times 10^6 \text{ J}.$$

The work done to climb a vertical height h is $W = mgh$. We want to solve for h given that $W = 2.1 \times 10^6$ J:

$$h = \frac{W}{mg} = \frac{2.1 \times 10^6 \text{ J}}{(60 \text{ kg})(9.80 \text{ m/s}^2)} = 3600 \text{ m}.$$

They need to climb a very high mountain (over 11,000 ft) or many flights of stairs. [The human body does not transform energy with 100 percent efficiency—more like 20 percent—just as no engine does. As we'll see in the next chapter, some energy is always "wasted," so the couple would actually have to climb only about $(0.2)(3600 \text{ m}) \approx 700 \text{ m}.$]

19-2 | Internal Energy

We introduce the concept of internal energy now since it will help clarify ideas about heat. The sum total of all the energy of all the molecules in an object is called its **thermal energy** or **internal energy**. (We will use the two terms interchangeably.) Occasionally, the term "heat content" of a body is used for this purpose, but it is not a good term because it can be confused with heat itself. Heat, as we have seen, is not the energy a body contains, but rather refers to the amount of energy transferred from one body to another at a different temperature.

Internal energy

Distinguishing Temperature, Heat, and Internal Energy

Using the kinetic theory, we can make a clear distinction between temperature, heat, and internal energy. Temperature (in kelvins) is a measure of the *average* kinetic energy of individual molecules. Thermal energy and internal energy refer to the *total* energy of all the molecules in the object. (Thus two equal-mass hot ingots of iron may have the same temperature, but two of them have twice as much thermal energy as one does.) Heat, finally, refers to a *transfer* of energy (such as thermal energy) from one object to another because of a difference in temperature.

Heat vs. internal energy vs. temperature

Notice that the direction of heat flow between two objects depends on their temperatures, not on how much internal energy each has. Thus, if 50 g of water at 30°C is placed in contact (or mixed) with 200 g of water at 25°C, heat flows *from* the water at 30°C *to* the water at 25°C even though the internal energy of the 25°C water is much greater because there is so much more of it.

Direction of heat flow depends on temperature

Internal Energy of an Ideal Gas

Let us calculate the internal energy of n moles of an ideal monatomic (one atom per molecule) gas. The internal energy, U, is the sum of the translational kinetic energies of all the atoms. This sum is just equal to the average kinetic energy per molecule times the total number of molecules, N:

$$U = N(\tfrac{1}{2}m\overline{v^2}).$$

Using Eq. 18–4, $\overline{K} = \tfrac{1}{2}m\overline{v^2} = \tfrac{3}{2}kT$, we can write this as

$$U = \tfrac{3}{2}NkT$$

or

$$U = \tfrac{3}{2}nRT, \qquad \text{[monatomic ideal gas]} \quad \textbf{(19-1)}$$

Internal energy of an ideal monatomic gas

where n is the number of moles. Thus, the internal energy of an ideal gas depends only on temperature and the number of moles of gas.

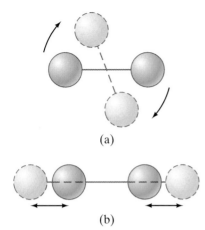

(a)

(b)

FIGURE 19–2 Molecules can have (a) rotational and (b) vibrational energy, as well as translational energy.

Relation between heat flow and temperature change

Specific heat

If the gas molecules contain more than one atom, then the rotational and vibrational energy of the molecules (Fig. 19–2) must also be taken into account. The internal energy will be greater at a given temperature than for a monatomic gas, but it will still be a function only of temperature for an ideal gas.

The internal energy of real gases also depends mainly on temperature, but where they deviate from ideal gas behavior, the internal energy also depends somewhat on pressure and volume.

The internal energy of liquids and solids is quite complicated, for it includes electrical potential energy associated with the forces (or "chemical" bonds) between atoms and molecules.

19–3 | Specific Heat

If heat is put into an object, its temperature rises. But how much does the temperature rise? That depends. As early as the eighteenth century, experimenters had recognized that the amount of heat Q required to change the temperature of a given material is proportional to the mass m of the material and to the temperature change ΔT. This remarkable simplicity in nature can be expressed in the equation

$$Q = mc\,\Delta T, \tag{19–2}$$

where c is a quantity characteristic of the material called its **specific heat**. Because $c = Q/m\Delta T$, specific heat is specified in units of J/kg·C° (the proper SI unit) or kcal/kg·C°. (The value of c in cal/gm·C° is the same as in kcal/kg·C°.) For water at 15°C and a constant pressure of 1 atm, $c = 1.00$ kcal/kg·C° or 4.19×10^3 J/kg·C°, since, by definition of the cal and the joule, it takes 1 kcal of heat to raise the temperature of 1 kg of water by 1 C°. Table 19–1 gives the values of specific heat for many substances at 20°C. The values of c depend to some extent on temperature (as well as slightly on pressure), but for temperature changes that are not too great, c can often be considered constant.[†]

TABLE 19–1 Specific Heats (at 1 atm constant pressure and 20°C unless otherwise stated)

	Specific Heat, c	
Substance	kcal/kg·C° (= cal/gm·C°)	J/kg·C°
Aluminum	0.22	900
Alcohol (ethyl)	0.58	2400
Copper	0.093	390
Glass	0.20	840
Iron or steel	0.11	450
Lead	0.031	130
Marble	0.21	860
Mercury	0.033	140
Silver	0.056	230
Wood	0.4	1700
Water		
Ice (−5°C)	0.50	2100
Liquid (15°C)	1.00	4186
Steam (110°C)	0.48	2010
Human body (average)	0.83	3470
Protein	0.4	1700

EXAMPLE 19–2 How heat depends on specific heat. (a) How much heat is required to raise the temperature of an empty 20-kg vat made of iron from 10°C to 90°C? (b) What if the vat is filled with 20 kg of water?

SOLUTION (a) From Table 19–1, the specific heat of iron is 450 J/kg·C°. The change in temperature is $(90°C − 10°C) = 80$ C°. Thus,

$$Q = mc\,\Delta T = (20\,\text{kg})(450\,\text{J/kg·C°})(80\,\text{C°}) = 7.2 \times 10^5\,\text{J} = 720\,\text{kJ}.$$

(b) The water alone would require

$$Q = mc\,\Delta T = (20\,\text{kg})(4186\,\text{J/kg·C°})(80\,\text{C°}) = 6.7 \times 10^6\,\text{J} = 6700\,\text{kJ}$$

or almost 10 times what an equal mass of iron requires. The total, for the vat plus the water, is 720 kJ + 6700 kJ = 7400 kJ.

If the iron vat in part (a) had been *cooled* from 90°C to 10°C, 720 kJ of heat would have flowed *out* of the iron. In other words, Eq. 19–2 is valid for heat flow either in or out, with a corresponding increase or decrease in temperature. We saw in part (b) of Example 19–2 that water requires almost 10 times as much heat as an equal mass of iron to make the same temperature change. Water has one of the highest specific heats of all substances, which makes it an ideal substance for hot-water space-heating systems and other uses that require a minimal drop in temperature for a given amount of heat transfer. It is the water content, too, that causes the apples rather than the crust in hot apple pie to burn our tongues, through heat transfer.

[†] To take into account the dependence of c on T, we can write Eq. 19–2 in differential form: $dQ = mc(T)\,dT$. Then the heat Q required to change the temperature from T_1 to T_2 is

$$Q = \int_{T_1}^{T_2} mc(T)\,dT,$$

where $c(T)$ is a function of temperature.

19–4 Calorimetry—Solving Problems

When different parts of an isolated system are at different temperatures, but are placed in thermal contact, heat will flow from the part at higher temperature to the part at lower temperature. If the system is completely isolated, no energy can flow into or out of it. Again, the *conservation of energy* plays an important role for us; the heat lost by one part of the system is equal to the heat gained by the other part:

heat lost = heat gained.

Energy conservation

Let us take an Example.

EXAMPLE 19–3 **The cup cools the tea.** If $200\,cm^3$ of tea at 95°C is poured into a 150-g glass cup initially at 25°C (Fig. 19–3), what will be the final temperature T of the mixture when equilibrium is reached, assuming no heat flows to the surroundings?

SOLUTION Since tea is mainly water, its specific heat is $4186\,J/kg\cdot C°$ (Table 19–1) and its mass m is its density times its volume $(V = 200\,cm^3 = 200 \times 10^{-6}\,m^3)$: $m = \rho V = (1.0 \times 10^3\,kg/m^3)(200 \times 10^{-6}\,m^3) = 0.20\,kg$. Applying conservation of energy, we can set

heat lost by tea = heat gained by cup

$$m_{tea}c_{tea}(95°C - T) = m_{cup}c_{cup}(T - 25°C)$$

where T is the as yet unknown final temperature. Putting in numbers and using Table 19–1, we solve for T, and find

$$(0.20\,kg)(4186\,J/kg\cdot C°)(95°C - T) = (0.15\,kg)(840\,J/kg\cdot C°)(T - 25°C)$$
$$79{,}400\,J - (836\,J/C°)T = (126\,J/C°)T - 6300\,J$$
$$T = 89°C.$$

The tea drops in temperature by 6 C° by coming into equilibrium with the cup.

Alternate Solution: We can set up this Example (and others) by a different approach. We can write that the total heat transferred into or out of the isolated system is zero:

$$\Sigma Q_i = 0.$$

Then each term is written as $Q = mc(T_i - T_f)$ where T_i and T_f are the initial and final temperatures. In the present Example:

$$\Sigma Q = m_{tea}c_{tea}(95°C - T) + m_{cup}c_{cup}(25°C - T) = 0.$$

Note that the second term is negative because T will be greater than 25°C. Solving the algebra gives the same result.

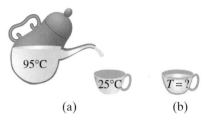

FIGURE 19–3 Example 19–3.

➡ **PROBLEM SOLVING**

Alternate approach

The exchange of energy, as exemplified in Example 19–3, is the basis for a technique known as **calorimetry**, which is the quantitative measurement of heat exchange. To make such measurements, a **calorimeter** is used; a simple water calorimeter is shown in Fig. 19–4. It is very important that the calorimeter be well insulated so that only a minimal amount of heat is exchanged with the outside. One important use of the calorimeter is in the determination of specific heats of substances. In the technique known as the "method of mixtures," a sample of the substance is heated to a high temperature, which is accurately measured, and then quickly placed in the cool water of the calorimeter. The heat lost by the sample will be gained by the water and the calorimeter. By measuring the final temperature of the mixture, the specific heat can be calculated, as illustrated in the following Example.

FIGURE 19–4
Simple water calorimeter.

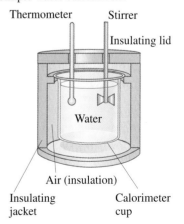

EXAMPLE 19–4 **Unknown specific heat determined by calorimetry.** An engineer wishes to determine the specific heat of a new alloy. A 0.150-kg sample of the alloy is heated to 540°C. It is then quickly placed in 400 g of water at 10.0°C, which is contained in a 200-g aluminum calorimeter cup. (We do not need to know the mass of the insulating jacket since we assume the air space between it and the cup insulates it well, so that its temperature does not change significantly.) The final temperature of the mixture is 30.5°C. Calculate the specific heat of the alloy.

SOLUTION Again we apply the conservation of energy and write that the heat lost equals the heat gained:

$$\begin{pmatrix} \text{heat lost} \\ \text{by sample} \end{pmatrix} = \begin{pmatrix} \text{heat gained} \\ \text{by water} \end{pmatrix} + \begin{pmatrix} \text{heat gained by} \\ \text{calorimeter cup} \end{pmatrix}$$

$$m_s c_s \, \Delta T_s = m_w c_w \, \Delta T_w + m_{cal} c_{cal} \, \Delta T_{cal}$$

where the subscripts s, w, and cal refer to the sample, water, and calorimeter, respectively. When we put in values and use Table 19–1, this equation becomes

$$(0.150 \, \text{kg})(c_s)(540°C - 30.5°C) = (0.40 \, \text{kg})(4186 \, \text{J/kg} \cdot \text{C°})(30.5°C - 10.0°C)$$
$$+ (0.20 \, \text{kg})(900 \, \text{J/kg} \cdot \text{C°})(30.5°C - 10.0°C)$$
$$76.4 \, c_s = (34{,}300 + 3700) \, \text{J/kg} \cdot \text{C°}$$
$$c_s = 500 \, \text{J/kg} \cdot \text{C°}.$$

In making this calculation, we have ignored any heat transferred to the thermometer and the stirrer (which is needed to quicken the heat transfer process and thus reduce heat loss to the outside). It can be taken into account by adding additional terms to the right side of the above equation and will result in a slight correction to the value of c_s (see Problem 14). It should be noted that the quantity $m_{cal} c_{cal}$ is often called the **water equivalent** of the calorimeter—that is, $m_{cal} c_{cal}$ is numerically equal to the mass of water (in kilograms) that would absorb the same amount of heat.

➤ **PHYSICS APPLIED**

Measuring Calorie content

A **bomb calorimeter** is used to measure the heat released when a substance burns. Important applications are the burning of foods to determine their Calorie content, and burning of seeds and other substances to determine their "energy content," or heat of combustion. A carefully weighed sample of the substance, together with an excess amount of oxygen at high pressure, is placed in a sealed container (the "bomb"). The bomb is placed in the water of the calorimeter and a fine wire passing into the bomb is then heated briefly, which causes the mixture to ignite.

19–5 | Latent Heat

When a material changes phase from solid to liquid, or from liquid to gas (see also Section 18–3), a certain amount of energy is involved in this **change of phase**. For example, let us trace what happens when a 1.0-kg block of ice at −40°C has heat added to it at a steady rate until all the ice has changed to water, then the (liquid) water is heated to 100°C and changed to steam above 100°C, all at 1 atm pressure. As shown in the graph of Fig. 19–5, as heat is added to the ice, its temperature rises at a rate of about 2 C°/kcal of heat added (since for ice, $c \approx 0.50 \, \text{kcal/kg} \cdot \text{C°}$). However, when 0°C is reached, the temperature stops increasing even though heat is still being added. Now as heat is added, the ice gradually changes to water in the liquid state without any change in temperature. After about 40 kcal have been added at 0°C, half the ice remains and half has changed to water. After about 80 kcal, or 330 kJ, has been added, all the ice has changed to water, still at 0°C. Further addition of heat causes the water's temperature to again increase, now at a rate of 1 C°/kcal. When 100°C is reached, the temperature again remains constant as the heat added changes the liquid water to vapor (steam). About 540 kcal (2260 kJ) is

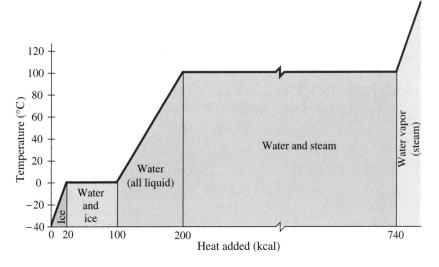

FIGURE 19–5 Temperature as a function of the heat added to bring 1.0 kg of ice at −40°C to steam above 100°C. Note the scale break between 200 and 740 kcal (the page isn't wide enough to fit it).

required to change the 1.0 kg of water completely to steam, after which the curve rises again, indicating that the temperature of the steam now rises as heat is added.

The heat required to change 1.0 kg of a substance from the solid to the liquid state is called the **heat of fusion**; it is denoted by L_F. The heat of fusion of water at 0°C is 79.7 kcal/kg or, in proper SI units, 333 kJ/kg $(= 3.33 \times 10^5 \text{ J/kg})$. The heat required to change a substance from the liquid to the vapor phase is called the **heat of vaporization**, L_V, and for water at 100°C it is 539 kcal/kg or 2260 kJ/kg. Other substances follow graphs similar to Fig. 19–5, although the melting-point and boiling-point temperatures are different, as are the specific heats and heats of fusion and vaporization. Values for the heats of fusion and vaporization, which are also called the **latent heats**, are given in Table 19–2 for a number of substances.

Heat of fusion

Heat of vaporization

The heats of vaporization and fusion also refer to the amount of heat *released* by a substance when it changes from a gas to a liquid, or from a liquid to a solid. Thus, steam releases 2260 kJ/kg when it changes to water, and water releases 333 kJ/kg when it becomes ice.

Of course, the heat involved in a change of phase depends not only on the latent heat, but also on the total mass of the substance. That is,

$$Q = mL,$$

where L is the latent heat of the particular process and substance, m is the mass of the substance, and Q is the heat required or given off during the phase change. For example, when 5.00 kg of water freezes at 0°C, $(5.00 \text{ kg})(3.33 \times 10^5 \text{ J/kg}) = 1.67 \times 10^6 \text{ J}$ of energy is released.

TABLE 19–2   Latent Heats (at 1 atm)

Substance	Melting Point (°C)	Heat of Fusion kcal/kg[†]	J/kg	Boiling Point (°C)	Heat of Vaporization kcal/kg[†]	J/kg
Oxygen	−218.8	3.3	0.14×10^5	−183	51	2.1×10^5
Nitrogen	−210.0	6.1	0.26×10^5	−195.8	48	2.00×10^5
Ethyl alcohol	−114	25	1.04×10^5	78	204	8.5×10^5
Ammonia	−77.8	8.0	0.33×10^5	−33.4	33	1.37×10^5
Water	0	79.7	3.33×10^5	100	539	22.6×10^5
Lead	327	5.9	0.25×10^5	1750	208	8.7×10^5
Silver	961	21	0.88×10^5	2193	558	23×10^5
Iron	1808	69.1	2.89×10^5	3023	1520	63.4×10^5
Tungsten	3410	44	1.84×10^5	5900	1150	48×10^5

[†] Numerical values in kcal/kg are the same in cal/g.

Calorimetry sometimes involves a change of state as the following Examples show. Indeed, latent heats are often measured using calorimetry.

EXAMPLE 19–5 Making ice.   How much energy does a refrigerator have to remove from 1.5 kg of water at 20°C to make ice at −12°C?

SOLUTION   Heat must flow out to reduce the water from 20°C to 0°C, to change it to ice, and then to lower the ice from 0°C to −12°C,

$$Q = mc_w(20°C − 0°C) + mL_F + mc_{ice}[0° − (−12°C)]$$
$$= (1.5\,kg)(4186\,J/kg·C°)(20\,C°) + (1.5\,kg)(3.33 × 10^5\,J/kg)$$
$$+ (1.5\,kg)(2100\,J/kg·C°)(12\,C°)$$
$$= 6.6 × 10^5\,J = 660\,kJ.$$

EXAMPLE 19–6 Will all the ice melt?   At a party, a 0.50-kg chunk of ice at −10°C is placed in 3.0 kg of "iced" tea at 20°C. At what temperature and in what phase will the final mixture be? The tea can be considered as water.

➥ **PROBLEM SOLVING**

First determine (or estimate) the final state

SOLUTION   In this situation, before we can write down an equation, we must first check to see if the final state will be all ice, a mixture of ice and water at 0°C, or all water. To bring the 3.0 kg of water at 20°C down to 0°C would require an energy release of

$$m_w c_w(20°C − 0°C) = (3.0\,kg)(4186\,J/kg·C°)(20\,C°) = 250\,kJ.$$

On the other hand, to raise the ice from −10°C to 0°C would require

$$m_{ice} c_{ice}[0°C − (−10°C)] = (0.50\,kg)(2100\,J/kg·C°)(10\,C°) = 10.5\,kJ,$$

and to change the ice to water at 0°C would require

$$m_{ice} L_F = (0.50\,kg)(333\,kJ/kg) = 167\,kJ,$$

for a total of 10.5 kJ + 167 kJ = 177 kJ. This is not enough energy to bring the 3.0 kg of water at 20°C down to 0°C, so we know that the mixture must end up all water, somewhere between 0°C and 20°C. Now we can determine the final temperature T by applying the conservation of energy and writing

Then determine the final temperature

$$\begin{pmatrix} \text{heat to raise} \\ \text{0.50 kg of ice} \\ \text{from −10°C} \\ \text{to 0°C} \end{pmatrix} + \begin{pmatrix} \text{heat to change} \\ \text{0.50 kg} \\ \text{of ice} \\ \text{to water} \end{pmatrix} + \begin{pmatrix} \text{heat to raise} \\ \text{0.50 kg of water} \\ \text{from 0°C} \\ \text{to } T \end{pmatrix} = \begin{pmatrix} \text{heat lost by} \\ \text{3.0 kg of} \\ \text{water cooling} \\ \text{from 20°C to } T \end{pmatrix}$$

so

$$10.5\,kJ + 167\,kJ + (0.50\,kg)(4186\,J/kg·C°)(T)$$
$$= (3.0\,kg)(4186\,J/kg·C°)(20°C − T).$$

Solving for T gives $T = 5.1°C$.

We can make use of kinetic theory to see why energy is needed to melt or vaporize a substance. At the melting point, the latent heat of fusion does not increase the kinetic energy (and the temperature) of the molecules in the solid, but instead is used to overcome the potential energy associated with the forces between the molecules. That is, work must be done against these attractive forces to break the molecules loose from their relatively fixed positions in the solid so they can freely roll over one another in the liquid phase. Similarly, energy is required for molecules held close together in the liquid phase to escape into the gaseous phase. This process is a more violent reorganization of the molecules than is melting (the average distance between the molecules is greatly increased) and hence the heat of vaporization is generally much greater than the heat of fusion for a given substance.

492   CHAPTER 19   Heat and the First Law of Thermodynamics

1. Be sure you have enough information to apply energy conservation. Ask yourself: is the system isolated (or very nearly so, enough to get a good estimate)? Do we know or can we calculate all significant sources of heat energy flow?

2. Apply conservation of energy. One approach is to write

$$\text{heat gained} = \text{heat lost.}$$

For each substance in the system, a heat (energy) term will appear on either the left or right of this equation.

3. If no phase changes occur, each term in the energy conservation equation (above) will have the form

$$Q(\text{gain}) = mc(T_f - T_i) \quad \text{or} \quad Q(\text{lost}) = mc(T_i - T_f)$$

where T_i and T_f are the initial and final temperatures of the substance, and m and c are its mass and specific heat.

4. If phase changes do or might occur, there may be terms in the energy conservation equation of the form $Q = mL$, where L is the latent heat. But *before* applying energy conservation, determine (or estimate) in which phase the final state will be, as we did in Example 19–6 by calculating the different contributing values for heat Q.

5. Be sure each term appears on the correct side of the energy equation (heat gained or heat lost) and that each ΔT is positive.

6. Note that when the system reaches thermal equilibrium, the final temperature of each substance will have the *same* value. There is only one T_f.

7. Solve your energy equation for the unknown.

The latent heat to change a liquid to a gas is needed not only at the boiling point. Water can change from the liquid to the gas phase even at room temperature. This process is called **evaporation** (see also Section 18–4). The value of the heat of vaporization increases slightly with a decrease in temperature: at 20°C, for example, it is 2450 kJ/kg (585 kcal/kg) compared to 2260 kJ/kg (539 kcal/kg) at 100°C. When water evaporates, it cools, since the energy required (the latent heat of vaporization) comes from the water itself. So its internal energy, and therefore its temperature, must drop.[†]

19–6 | The First Law of Thermodynamics

Up to now in this chapter we have discussed internal energy and heat. But work too is often involved in thermodynamic processes.

In Chapter 8 we saw that work is done when energy is transferred from one body to another by mechanical means. In Section 19–1 we saw that heat is a transfer of energy from one body to a second body at a lower temperature. Thus, heat is much like work. To distinguish them, *heat* is defined as a *transfer of energy due to a difference in temperature*, whereas work is a transfer of energy that is not due to a temperature difference.

Heat distinguished from work

In discussing thermodynamics, we shall often refer to particular systems. A **system** is any object or set of objects that we wish to consider. Everything else in the universe we will refer to as its "environment." There are several categories of systems. A **closed system** is one for which no mass enters or leaves (but energy may be exchanged with the environment). In an **open system**, mass may enter or leave (as well as energy). Many (idealized) systems we study in physics are closed systems. But many systems, including plants and animals, are open systems since they exchange materials (food, oxygen, waste products) with the environment. A closed system is said to be **isolated** if no energy in any form passes across its boundaries; otherwise it is not isolated.

Open and closed systems

[†] According to kinetic theory, evaporation is a cooling process because it is the fastest-moving molecules that escape from the surface. Hence the average speed of the remaining molecules is less, so by Eq. 18–4 the temperature is less.

In Section 19–2, we defined the internal energy of a system as the sum total of all the energy of the molecules of the system. We would expect that the internal energy of a system would be increased if work were done on the system, or if heat were added to it. Similarly the internal energy would be decreased if heat flowed out of the system or if work were done by the system on something else.

Thus, from conservation of energy, it is reasonable to propose an important law: the change in internal energy of a closed system, ΔU, will be equal to the heat added to the system minus the work done by the system; in equation form:

$$\Delta U = Q - W, \tag{19–3}$$

where Q is the net heat *added* to the system and W is the net work done *by* the system. We must be careful and consistent in following the sign conventions for Q and W. Because W in Eq. 19–3 is the work done *by* the system, then if work is done *on* the system, W will be negative and U will increase. [Of course, we could have defined W as the work done *on* the system, in which case there would be a plus sign in Eq. 19–3; but it is conventional to define W and Q as we have done.] Similarly, Q is positive for heat added to the system, so if heat leaves the system, Q is negative. Equation 19–3 is known as the **first law of thermodynamics**. It is one of the great laws of physics, and its validity rests on experiments (such as Joule's) in which no exceptions have been seen. Since Q and W represent energy transferred into or out of the system, the internal energy changes accordingly. Thus, the first law of thermodynamics is a great and broad statement of the *law of conservation of energy*. It is worth noting that the conservation of energy law was not formulated until the nineteenth century, for it depended on the interpretation of heat as a transfer of energy.

Heat added is +
Heat lost is −
Work on system is −
Work by system is +

First law of thermodynamics is conservation of energy

Equation 19–3 applies to a closed system. It also applies to an open system if we take into account the change in internal energy due to the increase or decrease in the amount of matter. For an isolated system, no work is done and no heat enters or leaves the system, so $W = Q = 0$, and hence $\Delta U = 0$.

A given system, in a particular state, can be said to have a certain amount of internal energy, U. This cannot be said of heat or work. A system in a given state does not "have" a certain amount of heat or work. Rather, when work is done on a system (such as compressing a gas), or when heat is added or removed from a system, the state of the system *changes*. Thus, work and heat are involved in *thermodynamic processes* that can change the system from one state to another; they are not characteristic of the state itself, as are pressure P, volume V, temperature T, internal energy U, and the mass m or number of moles n. Because U is a *state variable*, which we saw in Chapter 17 is a variable that depends only on the state of the system and not on how the system arrived in that state, we can write

Internal energy is a property of a system; work and heat are not

$$\Delta U = U_2 - U_1 = Q - W,$$

where U_1 and U_2 represent the internal energy of the system in states 1 and 2, and Q and W are the heat added to the system and work done by the system in going from state 1 to state 2.

It is sometimes useful to write the first law of thermodynamics in differential form:

$$dU = dQ - dW.$$

Here, dU represents an infinitesimal change in internal energy when an infinitesimal amount[†] of heat dQ is added to the system, and the system does an infinitesimal amount of work dW.

[†] The differential form of the first law is often written

$$dU = dQ - dW,$$

where the bars on the differential sign (d) are used to remind us that W and Q are not functions of the state variables (such as P, V, T, n). Internal energy, U, is a function of the state variables, and so dU represents the differential (called an *exact differential*) of some function U. The differentials dW and dQ are not exact differentials (they are not the differential of some mathematical function); they thus only represent infinitesimal amounts. This issue won't really be of concern in this book.

EXAMPLE 19–7 **Using the first law.** (*a*) An amount of heat equal to 2500 J is added to a system, and 1800 J of work is done on the system. What is the change in internal energy of the system?

SOLUTION We use the first law of thermodynamics, Eq. 19–3. The heat added to the system is $Q = 2500$ J. The work done by the system W is -1800 J, with a minus sign because 1800 J done *on* the system equals -1800 J done *by* the system. Hence

$$\Delta U = 2500\,\text{J} - (-1800\,\text{J}) = 2500\,\text{J} + 1800\,\text{J} = 4300\,\text{J}.$$

You may have intuitively thought that the 2500 J and the 1800 J would need to be added together, since both refer to energy transferred into the system. You would have been right. We did this exercise in detail to emphasize the importance of keeping careful track of signs.

19–7 | Applying the First Law of Thermodynamics; Calculating the Work

Let us analyze some simple processes in the light of the first law of thermodynamics.

Isothermal Processes ($\Delta T = 0$)

First we consider an idealized process that is carried out at constant temperature. Such a process is called an **isothermal** process (from the Greek meaning "same temperature"). If the system is an ideal gas, then $PV = nRT$ (Eq. 17–3), so for a fixed amount of gas kept at constant temperature, $PV = $ constant. Thus the process follows a curve like AB on the PV diagram shown in Fig. 19–6, which is a curve for $PV = $ constant. Each point on the curve, such as point A, represents the state of the system at a given moment—that is, its pressure P and volume V. At a lower temperature, another isothermal process would be represented by a curve like A′B′ in Fig. 19–6 (the product $PV = nRT = $ constant is less when T is less). The curves shown in Fig. 19–6 are referred to as *isotherms*.

   Let us assume that the gas is enclosed in a container fitted with a movable piston, Fig. 19–7, and that the gas is in contact with a **heat reservoir** (a body whose mass is so large that, ideally, its temperature does not change significantly when heat is exchanged with our system). We also assume that the process of compression (volume decreases) or expansion (volume increases) is done **quasistatically** ("almost statically"), by which we mean extremely slowly, so that all of the gas stays in equilibrium at the same constant temperature. If the gas is initially in a state represented by point A in Fig. 19–6 and an amount of heat Q is added to the system, the system will move to another point, B, on the diagram. If the temperature is to remain constant, the gas must expand and do an amount of work W on the environment (it exerts a force on the piston and moves it through a distance). The temperature and mass are kept constant so, from Eq. 19–1, the internal energy does not change: $\Delta U = \frac{3}{2}nR\Delta T = 0$. Hence, by the first law of thermodynamics, Eq. 19–3, $\Delta U = Q - W = 0$, so $W = Q$: the work done by the gas in an isothermal process equals the heat added to the gas.

Adiabatic Processes ($Q = 0$)

An **adiabatic** process is one in which no heat is allowed to flow into or out of the system: $Q = 0$. This situation can occur if the system is extremely well insulated, or the process happens so quickly that heat—which flows slowly—has no time to flow in or out. The very rapid expansion of gases in an internal combustion engine

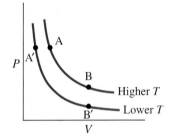

FIGURE 19–6 PV diagram for an ideal gas undergoing isothermal processes at two different temperatures.

Heat reservoir

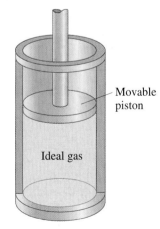

FIGURE 19–7 An ideal gas in a cylinder fitted with a movable piston.

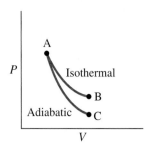

FIGURE 19–8 *PV* diagram for adiabatic (AC) and isothermal (AB) processes on an ideal gas.

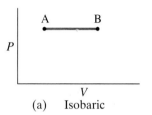

(a) Isobaric

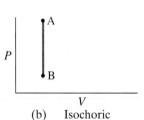

(b) Isochoric

FIGURE 19–9 (a) Isobaric ("same pressure") process. (b) Isochoric ("same volume") process.

FIGURE 19–10 The work done by a gas when its volume increases by $dV = A \, dl$ is $dW = P \, dV$.

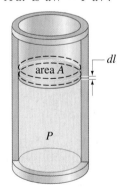

is one example of a process that is very nearly adiabatic. A slow adiabatic expansion of an ideal gas follows a curve like that labeled AC in Fig. 19–8. Since $Q = 0$, we have from Eq. 19–3 that $\Delta U = -W$. That is, the internal energy decreases if the gas expands; hence the temperature decreases as well (because $\Delta U = \frac{3}{2} nR\Delta T$). Thus an adiabatic *PV* curve is steeper than an isotherm: in Fig. 19–8 the product $PV (= nRT)$ is less at point C than at point B (curve AB is for an isothermal process, for which $\Delta U = 0$ and $\Delta T = 0$). In an adiabatic compression (going from C to A, for example), work is done *on* the gas, and hence the internal energy increases and the temperature rises. In a diesel engine, the rapid adiabatic compression reduces the volume by a factor of 15 or more; the temperature rise is so great that the air-fuel mixture ignites spontaneously.

CONCEPTUAL EXAMPLE 19–8 **Simple adiabatic process.** Here is an adiabatic process that you can do with just a rubber band. Hold the rubber band loosely with two hands and gauge the temperature with your lips. Stretch the rubber band suddenly and again touch it lightly to your lips. You should notice an increase in temperature. Explain clearly why the temperature increases.

RESPONSE Stretching the rubber band *suddenly* makes the process adiabatic because there is no time for heat to enter or leave the system, so $Q = 0$. You do work on the system, representing an energy input, so by the first law of thermodynamics the internal energy increases, corresponding to an increase in temperature (Eq. 19–1); so the rubber band heats up.

Isobaric and Isochoric Processes

Isothermal and adiabatic processes are just two possible processes that can occur. Two other simple thermodynamic processes are illustrated on the *PV* diagram of Fig. 19–9: (a) an **isobaric** process is one in which the pressure is kept constant, so the process is represented by a horizontal straight line on the *PV* diagram, Fig. 19–9a; (b) an **isochoric** or isovolumetric process is one in which the volume does not change (Fig. 19–9b). In these, and in all other processes, the first law of thermodynamics holds.

Work Done in Volume Changes

We often want to calculate the work done in a process. Suppose we have a gas confined to a cylindrical container fitted with a movable piston (Fig. 19–10). We must always be careful to define exactly what our system is. In this case we choose our system to be the gas; so the container's walls and the piston are parts of the environment. Now let us calculate the work done by the gas when it expands quasistatically, so that P and T are defined for the system at all instants.[†] The gas expands against the piston, whose area is A. The gas exerts a force $F = PA$ on the piston, where P is the pressure in the gas. The work done by the gas to move the piston an infinitesimal displacement $d\boldsymbol{l}$ is

$$dW = \mathbf{F} \cdot d\boldsymbol{l} = PA \, dl = P \, dV \qquad (19\text{–}4)$$

since the infinitesimal increase in volume is $dV = A \, dl$. If the gas were *compressed* so that $d\boldsymbol{l}$ pointed into the gas, the volume would decrease and $dV < 0$. The work done by the gas in this case would then be negative, which is equivalent to saying that positive work was done *on* the gas, not by it. For a finite change in volume from V_A to V_B, the work W done by the gas will be

$$W = \int dW = \int_{V_A}^{V_B} P \, dV. \qquad (19\text{–}5)$$

Equations 19–4 and 19–5 are valid for the work done in any volume change—by a gas, a liquid, or a solid—as long as it is done quasistatically.

[†] If the gas expanded or were compressed quickly, there would be turbulence, and different parts would be at different pressure (and temperature).

In order to integrate Eq. 19–5, we need to know how the pressure varies during the process, and this depends on the type of process. Let us first consider a quasistatic isothermal expansion of an ideal gas. This process is represented by the curve between points A and B on the *PV* diagram of Fig. 19–11. The work done by the gas in this process, according to Eq. 19–5, is just the area between the *PV* curve and the *V* axis, and is shown shaded in Fig. 19–11. We can do the integral in Eq. 19–5 for an ideal gas by using the ideal gas law, $P = nRT/V$. The work done is

$$W = \int_{V_A}^{V_B} P \, dV = nRT \int_{V_A}^{V_B} \frac{dV}{V} = nRT \ln \frac{V_B}{V_A}. \quad \begin{bmatrix} \text{isothermal process;} \\ \text{ideal gas} \end{bmatrix} \quad \textbf{(19–6)}$$

Let us next consider a different way of taking an ideal gas between the same states A and B. This time, let us lower the pressure in the gas from P_A to P_B, as indicated by the line AD in Fig. 19–12. (In this *isochoric* process, heat must be allowed to flow out of the gas so its temperature drops.) Then let the gas expand from V_A to V_B at constant pressure ($= P_B$), which is indicated by the line DB in Fig. 19–12. (In this *isobaric* process, heat must be added to the gas to raise its temperature.) No work is done in the isochoric process AD, since $dV = 0$:

$$W = 0. \qquad \text{[isochoric process; ideal gas]}$$

In the isobaric process DB the pressure remains constant, so

$$W = \int_{V_A}^{V_B} P \, dV = P_B(V_B - V_A) = P \, \Delta V. \quad \begin{bmatrix} \text{isobaric process;} \\ \text{ideal gas} \end{bmatrix} \quad \textbf{(19–7a)}$$

The work done is again represented by the area between the curve (ADB) on the *PV* diagram, and the *V* axis as indicated by the shading in Fig. 19–12. Using the ideal gas law, we can also write

$$W = P_B(V_B - V_A) = nRT_B\left(1 - \frac{V_A}{V_B}\right). \quad \begin{bmatrix} \text{isobaric process;} \\ \text{ideal gas} \end{bmatrix} \quad \textbf{(19–7b)}$$

As can be seen from the shaded areas in Figs. 19–11 and 19–12, or by putting in numbers in Eqs. 19–6 and 19–7 (try it for $V_B = 2V_A$), the work done in these two processes is different. This is a general result. *The work done in taking a system from one state to another depends not only on the initial and final states but also on the type of process (or "path").*

This result reemphasizes the fact that work cannot be considered a property of a system. The same is true of heat. The heat input required to change the gas from state A to state B depends on the process; for the isothermal process of Fig. 19–11, the heat input turns out to be greater than for the process ADB of Fig. 19–12. In general, *the amount of heat added or removed in taking a system from one state to another depends not only on the initial and final states but also on the path or process.*

EXAMPLE 19–9 **First law in isobaric and isochoric processes.** An ideal gas is slowly compressed at a constant pressure of 2.0 atm from 10.0 L to 2.0 L, path B to D in Fig. 19–12. (In this process, some heat flows out and the temperature drops.) Heat is then added to the gas, holding the volume constant, and the pressure and temperature are allowed to rise until the temperature reaches its original value, process D to A in Fig. 19–12. Calculate (*a*) the total work done by the gas in the process BDA, and (*b*) the total heat flow into the gas.

SOLUTION (*a*) Work is done only in the first part, the compression (BD)

$$W = P \, \Delta V = (2.0 \times 10^5 \, \text{N/m}^2)(2.0 \times 10^{-3} \, \text{m}^3 - 10.0 \times 10^{-3} \, \text{m}^3) = -1.6 \times 10^3 \, \text{J}.$$

From D to A no work is done ($\Delta V = 0$); so the total work done by the gas is $-1.6 \times 10^3 \, \text{J}$, where the minus means that $+1.6 \times 10^3 \, \text{J}$ of work is done *on* the gas. (*b*) Since the temperature at the beginning and at the end of the process is the same, there is no change in internal energy of our ideal gas: $\Delta U = 0$. From the first law of thermodynamics we have

$$0 = \Delta U = Q - W,$$

so $Q = W = -1.60 \times 10^3 \, \text{J}$. Since Q is negative, 1600 J of heat flows out of the gas for the whole process, BDA.

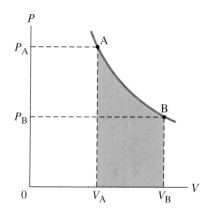

FIGURE 19–11 Work done by an ideal gas in an isothermal process equals the area under the *PV* curve. Shaded area equals the work done by the gas when it expands from V_A to V_B.

FIGURE 19–12 Process ADB consists of an isochoric (AD) and an isobaric (DB) process.

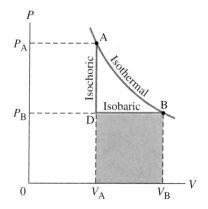

EXAMPLE 19–10 **Work done in an engine.** In an engine, 0.25 moles of gas in the cylinder expands rapidly and adiabatically against the piston. In the process, the temperature drops from 1150 K to 400 K. How much work does the gas do? Assume the gas is ideal.

SOLUTION Instead of calculating the work directly, we can more simply use the first law of thermodynamics, if we can determine ΔU since we know $Q = 0$ because the process is adiabatic. We determine ΔU from Eq. 19–1 for the internal energy of an ideal monatomic gas:

$$\Delta U = U_f - U_i = \tfrac{3}{2} nR(T_f - T_i)$$
$$= \tfrac{3}{2}(0.25 \text{ mol})(8.315 \text{ J/mol·K})(400 \text{ K} - 1150 \text{ K}) = -2300 \text{ J}.$$

Then, from Eq. 19–3, $W = Q - \Delta U = 0 - (-2300 \text{ J}) = 2300 \text{ J}.$

Free Expansion

One type of adiabatic process is a so-called **free expansion** in which a gas is allowed to expand in volume adiabatically without doing any work. The apparatus to accomplish a free expansion is shown in Fig. 19–13. It consists of two well-insulated compartments (to ensure no heat flow in or out) connected by a valve or stopcock. One compartment is filled with gas, the other is empty. When the valve is opened, the gas expands to fill both containers. No heat flows in or out ($Q = 0$), and no work is done because the gas does not move any other object. Thus $Q = W = 0$ and by the first law of thermodynamics, $\Delta U = 0$. *The internal energy of a gas does not change in a free expansion.* For an ideal gas, $\Delta T = 0$ also, since U depends only on T (Section 19–2). Experimentally, the free expansion has been used to determine if the internal energy of *real gases* depends only on T. The experiments are very difficult to do accurately, but it has been found that the temperature of a real gas drops very slightly in a free expansion. Thus the internal energy of real gases does depend, a little, on pressure or volume as well as on temperature.

Note, incidentally, that a free expansion could not be plotted on a *PV* diagram, because the process is rapid, not quasistatic. The intermediate states are not equilibrium states, and hence the pressure (and even the volume at some instants) is not clearly defined.

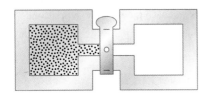

FIGURE 19–13 Free expansion.

For a free expansion
$Q = 0, W = 0, \Delta U = 0$

19–8	Molar Specific Heats for Gases, and the Equipartition of Energy

In Section 19–3 we discussed the concept of specific heat and applied it to solids and liquids. Much more than for solids and liquids, the values of the specific heat for gases depends on how the process is carried out. Two important processes are those in which either the volume or the pressure is kept constant. Although for solids and liquids it matters little, Table 19–3 shows that the specific heats of gases at constant volume (c_V) and at constant pressure (c_P) are quite different.

Molar Specific Heats for Gases

The difference in specific heats for gases is nicely explained in terms of the first law of thermodynamics and kinetic theory. Indeed, the values of the specific heats can be calculated using the kinetic theory, and the results are in close agreement with experiment. Our discussion is simplified if we use **molar specific heats**, C_V and C_P, which are defined as the heat required to raise 1 mol of the gas by 1 C° at constant volume and at constant pressure, respectively. That is, in analogy to Eq. 19–2, the heat Q needed to raise the temperature of n moles of gas by ΔT is

$$Q = nC_V \Delta T \qquad \text{[volume constant]} \qquad \textbf{(19–8a)}$$
$$Q = nC_P \Delta T. \qquad \text{[pressure constant]} \qquad \textbf{(19–8b)}$$

Molar specific heats

TABLE 19–3   Specific heats of gases at 15°C

Gas	Specific heats (kcal/kg·K)		Molar specific heats (cal/mol·K)		$C_P - C_V$ (cal/mol·K)	$\gamma = \dfrac{C_P}{C_V}$
	c_V	c_P	C_V	C_P		
Monatomic						
He	0.75	1.15	2.98	4.97	1.99	1.67
Ne	0.148	0.246	2.98	4.97	1.99	1.67
Diatomic						
N_2	0.177	0.248	4.96	6.95	1.99	1.40
O_2	0.155	0.218	5.03	7.03	2.00	1.40
Triatomic						
CO_2	0.153	0.199	6.80	8.83	2.03	1.30
H_2O (100°C)	0.350	0.482	6.20	8.20	2.00	1.32
Polyatomic						
C_2H_6	0.343	0.412	10.30	12.35	2.05	1.20

It is clear from the definition of molar specific heat (or by comparing Eqs. 19–2 and 19–8) that

$$C_V = Mc_V$$
$$C_P = Mc_P,$$

where M is the molecular mass of the gas ($M = m/n$ in grams/mol). The values for molar specific heats are included in Table 19–3, and we see that the values are nearly the same for different gases that have the same number of atoms per molecule.

Let us now make use of the kinetic theory and see first why the specific heats of gases are higher for constant-pressure processes than for constant-volume processes. Let us imagine that an ideal gas is slowly heated via these two different processes—first at constant volume, and then at constant pressure. In both of these processes, we let the temperature increase by the same amount, ΔT. In the process done at constant volume, no work is done since $\Delta V = 0$. Thus, according to the first law of thermodynamics, the heat added (which we now denote by Q_V) all goes into increasing the internal energy of the gas:

$$Q_V = \Delta U.$$

In the process carried out at constant pressure, work is done, and hence the heat added, Q_P, must not only increase the internal energy but also is used to do the work $W = P\Delta V$. Thus, more heat must be added in this process than in the first process at constant volume. For the process at constant pressure, we have from the first law of thermodynamics

$$Q_P = \Delta U + P\Delta V.$$

Since ΔU is the same in the two processes (ΔT was chosen to be the same), we can combine the two above equations:

$$Q_P - Q_V = P\Delta V.$$

From the ideal gas law, $V = nRT/P$, so for a process at constant pressure we have $\Delta V = nR\Delta T/P$. Putting this into the above equation and using Eqs. 19–8, we find

$$nC_P\,\Delta T - nC_V\,\Delta T = P\left(\frac{nR\Delta T}{P}\right)$$

or, after cancellations,

$$C_P - C_V = R. \tag{19–9}$$

Since the gas constant $R = 8.315\,\text{J/mol·K} = 1.99\,\text{cal/mol·K}$, our prediction is that C_P will be larger than C_V by about 1.99 cal/mol·K. Indeed, this is very close to what is obtained experimentally, as can be seen in the next to last column in Table 19–3.

Let us now calculate the molar specific heat of a monatomic gas using the kinetic theory model of gases. First we consider a process carried out at constant volume. Since no work is done in this process the first law of thermodynamics tells us that if heat Q is added to the gas, the internal energy of the gas changes by

$$\Delta U = Q.$$

For an ideal monatomic gas, the internal energy, U, is the total kinetic energy of all the molecules,

$$U = N\left(\tfrac{1}{2}m\overline{v^2}\right) = \tfrac{3}{2}nRT$$

as we saw in Section 19–2. Then, using Eq. 19–8a, we can write $\Delta U = Q$ in the form

$$\Delta U = \tfrac{3}{2}nR\,\Delta T = nC_V\,\Delta T \qquad\qquad \textbf{(19–10)}$$

or

$$C_V = \tfrac{3}{2}R. \qquad\qquad \textbf{(19–11)}$$

Since $R = 8.315\,\text{J/mol·K} = 1.99\,\text{cal/mol·K}$, kinetic theory predicts that $C_V = 2.98\,\text{cal/mol·K}$ for an ideal monatomic gas. This is very close to the experimental values for monatomic gases such as helium and neon (Table 19–3). From Eq. 19–9, C_P is predicted to be about $4.97\,\text{cal/mol·K}$, also in agreement with experiment.

Equipartition of Energy

The measured molar specific heats for more complex gases (Table 19–3), such as diatomic (two atoms) and triatomic (three atoms) gases, increase with the increased number of atoms per molecule. We can explain this by assuming that the internal energy includes not only translational kinetic energy but other forms of energy as well. Take, for example, a diatomic gas. As shown in Fig. 19–14 the two atoms can rotate about two different axes (but rotation about a third axis passing through the two atoms would give rise to very little energy since the moment of inertia is so small). The molecules can have rotational as well as translational kinetic energy. It is useful to introduce the idea of **degrees of freedom**, by which we mean the number of independent ways molecules can possess energy. For example, a monatomic gas is said to have three degrees of freedom, since an atom can have velocity along the x axis, the y axis, and the z axis. These are considered to be three independent motions because a change in any one of the components would not affect any of the others. A diatomic molecule has the same three degrees of freedom associated with translational kinetic energy plus two more degrees of freedom associated with rotational kinetic energy, for a total of five degrees of freedom. A quick look at Table 19–3 indicates that the C_V for diatomic gases is about $\tfrac{5}{3}$ times as great as for a monatomic gas—that is, in the same ratio as their degrees of freedom. This led nineteenth-century physicists to an important idea, the **principle of equipartition of energy**. This principle states that energy is shared equally among the active degrees of freedom, and in particular each active degree of freedom of a molecule has on the average an energy equal to $\tfrac{1}{2}kT$. Thus, the average energy for a molecule of a monatomic gas would be $\tfrac{3}{2}kT$ (which we already knew) and of a diatomic gas $\tfrac{5}{2}kT$. Hence the internal energy of a diatomic gas would be $U = N\left(\tfrac{5}{2}kT\right) = \tfrac{5}{2}nRT$, where n is the number of moles. Using the same argument we did for monatomic gases, we see that for diatomic gases the molar specific heat at constant volume would be $\tfrac{5}{2}R = 4.97\,\text{cal/mol·K}$, in accordance with measured values. More complex molecules have even more degrees of freedom and thus greater molar specific heats.

The situation was complicated, however, by measurements that showed that for diatomic gases at very low temperatures, C_V has a value of only $\tfrac{3}{2}R$, as if it had only three degrees of freedom. And at very high temperatures, C_V was about $\tfrac{7}{2}R$, as if there were seven degrees of freedom. The explanation is that at low temperatures, nearly all molecules have only translational kinetic energy. That is, no energy goes into rotational energy, so only three degrees of freedom are "active." At very high temperatures, on the other hand, all five degrees of freedom are active plus

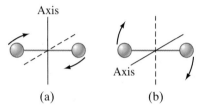

Axis

Axis

(a)   (b)

FIGURE 19–14 A diatomic molecule can rotate about two different axes.

Degrees of freedom

Equipartition of energy

two additional ones. We can interpret the two new degrees of freedom as being associated with the two atoms vibrating as if they were connected by a spring, as shown in Fig. 19–15. One degree of freedom comes from the kinetic energy of the vibrational motion, and the second comes from the potential energy of vibrational motion $(\frac{1}{2}kx^2)$. At room temperature, these two degrees of freedom are apparently not active. See Fig. 19–16. Just why fewer degrees of freedom are "active" at lower temperatures was eventually explained by Einstein using the quantum theory. [According to quantum theory, energy does not take on continuous values but is quantized—it can have only certain values, and there is a certain minimum energy. The minimum rotational and vibrational energies are higher than for simple translational kinetic energy, so at lower temperatures and lower translational kinetic energy, there is not enough energy to excite the rotational or vibrational kinetic energy.] Calculations based on kinetic theory and the principle of equipartition of energy (as modified by the quantum theory) give numerical results in accord with experiment.

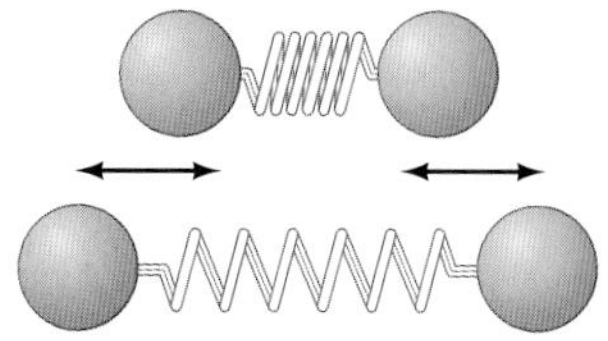

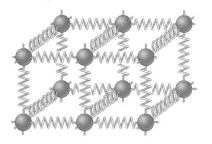

FIGURE 19–15 A diatomic molecule can vibrate, as if the two atoms were connected by a spring. Of course they are not connected by a spring, but rather exert forces on each other that are electrical in nature, but of a form that resembles a spring force.

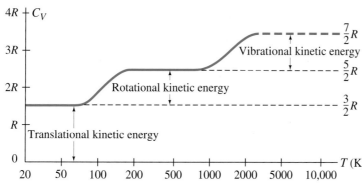

FIGURE 19–16 Molar specific heat C_V as a function of temperature for hydrogen molecules (H_2). As the temperature is increased, some of the translational kinetic energy can be transferred in collisions into rotational kinetic energy and, at still higher temperature, into vibrational kinetic energy. [Note: H_2 dissociates into two atoms at about 3200 K, so the last part of the curve is shown dashed.]

Solids

The principle of equipartition of energy can be applied to solids as well. The molar specific heat of any solid, at high temperature, is close to $3R$ (6.0 cal/mol·K), Fig. 19–17. This is called the Dulong and Petit value after the scientists who first measured it in 1819. (Note that Table 19–1 gave the specific heats per kilogram, not per mole.) At high temperatures, each atom apparently has six degrees of freedom, although some are not active at low temperatures. Each atom in a crystalline solid can vibrate about its equilibrium position as if it were connected by springs to each of its neighbors (Fig. 19–18). Thus it can have three degrees of freedom for kinetic energy and three more associated with potential energy of vibration in each of the x, y, and z directions, which is in accord with the measured values.

FIGURE 19–17 Molar heat capacities of solids as a function of temperature.

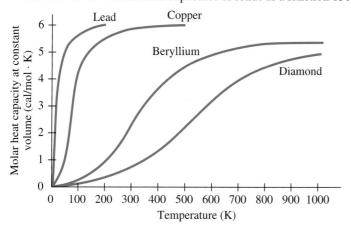

FIGURE 19–18 The atoms in a crystalline solid can vibrate about their equilibrium positions as if they were connected to their neighbors by springs. (The forces between atoms are actually electrical in nature.)

19–9 Adiabatic Expansion of a Gas

The PV curve for the quasistatic (slow) adiabatic expansion ($Q = 0$) of an ideal gas was shown in Fig. 19–8 (curve AC). It is somewhat steeper than for an isothermal process ($\Delta T = 0$), which indicates that for the same change in volume the change in pressure will be greater. Hence the temperature of the gas must drop during an adiabatic expansion. Conversely, the temperature rises during an adiabatic compression.

We can derive the relation between the pressure P and the volume V of an ideal gas that is allowed to slowly expand adiabatically. We begin with the first law of thermodynamics, written in differential form:

$$dU = dQ - dW = -dW = -P\, dV,$$

since $dQ = 0$ for an adiabatic process. Equation 19–10 gives us a relation between ΔU and C_V, which is valid for any ideal gas process since U is a function only of T for an ideal gas. We write this in differential form:

$$dU = nC_V\, dT.$$

When we combine these last two equations, we obtain

$$nC_V\, dT + P\, dV = 0.$$

We next take the differential of the ideal gas law, $PV = nRT$, allowing P, V, and T to vary:

$$P\, dV + V\, dP = nR\, dT.$$

We solve for dT in this relation and substitute it into the previous relation and get

$$nC_V\left(\frac{P\, dV + V\, dP}{nR}\right) + P\, dV = 0$$

or

$$(C_V + R)P\, dV + C_V V\, dP = 0.$$

We note from Eq. 19–9 that $C_V + R = C_P$, so we have

$$C_P P\, dV + C_V V\, dP = 0,$$

or

$$\frac{C_P}{C_V} P\, dV + V\, dP = 0.$$

We define

$$\gamma = \frac{C_P}{C_V} \tag{19–12}$$

so that our last equation becomes

$$\frac{dP}{P} + \gamma\frac{dV}{V} = 0.$$

This is integrated to become

$$\ln P + \gamma \ln V = \text{constant}.$$

This simplifies (using the rules for addition and multiplication of logarithms) to

$$PV^{\gamma} = \text{constant}. \qquad \begin{bmatrix}\text{quasistatic adiabatic} \\ \text{process; ideal gas}\end{bmatrix} \tag{19–13}$$

This is the relation between P and V for a quasistatic adiabatic expansion or contraction. We will find it very useful when we discuss heat engines in the next chapter. Table 19–3 gives values of γ for some real gases. Figure 19–8 compares an adiabatic expansion (Eq. 19–13) in curve AC to an isothermal expansion

(PV = constant) in curve AB. It is important to remember that the ideal gas law, $PV = nRT$, continues to hold even for an adiabatic expansion (PV^γ = constant); clearly PV is not constant, meaning T is not constant.

EXAMPLE 19–11 **Adiabatic vs. isothermal expansion.** An ideal monatomic gas is allowed to expand slowly until its pressure is reduced to exactly half its original value. By what factor does the volume change if the process is (*a*) adiabatic; (*b*) isothermal?

SOLUTION (*a*) From Eq. 19–13, $P_1 V_1^\gamma = P_2 V_2^\gamma$; hence

$$\frac{V_2}{V_1} = \left(\frac{P_1}{P_2}\right)^{1/\gamma} = (2)^{3/5} = 1.52$$

since $\gamma = C_P / C_V = (5/2)/(3/2) = 5/3$.
(*b*) Equation 19–13 does not apply for an isothermal process. But the ideal gas law, $PV = nRT$, always applies for an ideal gas. Hence, since $T_1 = T_2$, then $P_1 V_1 = P_2 V_2$ and

$$\frac{V_2}{V_1} = \frac{P_1}{P_2} = 2.0.$$

19–10 Heat Transfer: Conduction, Convection, Radiation

Heat is transferred from one place or body to another in three different ways: by *conduction*, *convection*, and *radiation*. We now discuss each of these in turn; but in practical situations, any two or all three may be operating at the same time. We start with conduction.

Three methods of heat transfer

Conduction

When a metal poker is put in a hot fire, or a silver spoon is placed in a hot bowl of soup, the exposed end of the poker or spoon soon becomes hot as well, even though it is not directly in contact with the source of heat. We say that heat has been conducted from the hot end to the cold end.

Heat **conduction** in many materials can be visualized as the result of molecular collisions. As one end of the object is heated, the molecules there move faster and faster. As they collide with their slower-moving neighbors, they transfer some of their energy to these molecules whose speeds thus increase. These in turn transfer some of their energy by collision with molecules still farther along the object. Thus the energy of thermal motion is transferred by molecular collision along the object. In metals, according to modern theory, it is collisions of free electrons within the metal with each other and with the metal lattice atoms that are visualized as being mainly responsible for conduction.

Heat conduction takes place only if there is a difference in temperature. Indeed, it is found experimentally that the rate of heat flow through a substance is proportional to the difference in temperature between its ends. The rate of heat flow also depends on the size and shape of the object. To investigate this quantitatively, let us consider the heat flow through a uniform object, as illustrated in Fig. 19–19. It is found experimentally that the heat flow ΔQ per time interval Δt is given by

$$\frac{\Delta Q}{\Delta t} = kA\frac{T_1 - T_2}{l}, \tag{19–14a}$$

where A is the cross-sectional area of the object, l is the distance between the two ends, which are at temperatures T_1 and T_2, and k is a proportionality constant called the **thermal conductivity**, which is characteristic of the material.

FIGURE 19–19 Heat conduction between areas at temperatures T_1 and T_2. If T_1 is greater than T_2, the heat flows to the right; the rate is given by Eq. 19–14a.

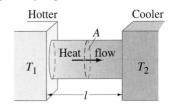

Rate of heat flow by conduction

TABLE 19–4
Thermal Conductivities

Substance	Thermal Conductivity, k	
	kcal/(s·m·C°)	J/(s·m·C°)
Silver	10×10^{-2}	420
Copper	9.2×10^{-2}	380
Aluminum	5.0×10^{-2}	200
Steel	1.1×10^{-2}	40
Ice	5×10^{-4}	2
Glass	2.0×10^{-4}	0.84
Brick	2.0×10^{-4}	0.84
Concrete	2.0×10^{-4}	0.84
Water	1.4×10^{-4}	0.56
Human tissue	0.5×10^{-4}	0.2
Wood	0.3×10^{-4}	0.1
Fiberglass	0.12×10^{-4}	0.048
Cork	0.1×10^{-4}	0.042
Wool	0.1×10^{-4}	0.040
Goose down	0.06×10^{-4}	0.025
Polyurethane	0.06×10^{-4}	0.024
Air	0.055×10^{-4}	0.023

FIGURE 19–20 Example 19–12.

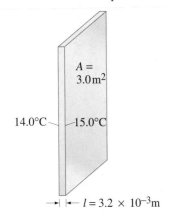

$A = 3.0\,\mathrm{m}^2$

14.0°C | 15.0°C

$l = 3.2 \times 10^{-3}\,\mathrm{m}$

Wind can cause much greater heat loss

➡ PHYSICS APPLIED

Thermal windows

In some cases (such as when k or A cannot be considered constant) we need to consider the limit of an infinitesimally thin slab of thickness dx. Then Eq. 19–14a becomes

$$\frac{dQ}{dt} = -kA\frac{dT}{dx},\qquad\qquad(19\text{–}14\mathrm{b})$$

where dT/dx is the temperature gradient and the negative sign is included since the heat flow is in the direction opposite to the temperature gradient.[†]

The thermal conductivities, k, for a variety of substances are given in Table 19–4. Substances for which k is large conduct heat rapidly and are said to be good **conductors**. Most metals fall in this category, although there is a wide range even among them as you may observe by holding the ends of a silver spoon and a stainless-steel spoon immersed in the same hot cup of soup. Substances for which k is small, such as fiberglass and down, are poor conductors of heat and are therefore good **insulators**. The relative magnitudes of k can explain simple phenomena such as why a tile floor feels much colder on the feet than a rug-covered floor at the same temperature. Tile is a better conductor of heat than the rug. Heat transferred from your foot to the rug is not conducted away rapidly, so the rug's surface quickly heats up to the temperature of your foot. But the tile conducts the heat away rapidly and thus can take more heat from your foot quickly, so your foot's surface temperature drops.

EXAMPLE 19–12 **Heat loss through windows.** A major source of heat loss from a house is through the windows. Calculate the rate of heat flow through a glass window $2.0\,\mathrm{m} \times 1.5\,\mathrm{m}$ in area and 3.2 mm thick, if the temperatures at the inner and outer surfaces are 15.0°C and 14.0°C, respectively (Fig. 19–20).

SOLUTION Since $A = (2.0\,\mathrm{m})(1.5\,\mathrm{m}) = 3.0\,\mathrm{m}^2$, $l = 3.2 \times 10^{-3}\,\mathrm{m}$, and using Table 19–4 to get k, we have from Eq. 19–14a

$$\frac{\Delta Q}{\Delta t} = kA\frac{T_1 - T_2}{l} = \frac{(0.84\,\mathrm{J/s \cdot m \cdot C°})(3.0\,\mathrm{m}^2)(15.0°\mathrm{C} - 14.0°\mathrm{C})}{(3.2 \times 10^{-3}\,\mathrm{m})}$$

$$= 790\,\mathrm{J/s}.$$

This is equivalent to $(790\,\mathrm{J/s})/(4.19 \times 10^3\,\mathrm{J/kcal}) = 0.19\,\mathrm{kcal/s}$, or $(0.19\,\mathrm{kcal/s}) \cdot (3600\,\mathrm{s/h}) = 680\,\mathrm{kcal/h}$.

You might notice in this Example that 15°C is not very warm for the living room of a house. The room itself may indeed be much warmer, and the outside might be colder than 14°C. But the temperatures of 15°C and 14°C were specified as those at the window surfaces, and there is usually a considerable drop in temperature of the air in the vicinity of the window both on the inside and the outside. That is, the layer of air on either side of the window acts as an insulator, and normally the major part of the temperature drop between the inside and outside of the house takes place across the air layer. If there is a heavy wind, the air outside a window will constantly be replaced with cold air; the temperature gradient across the glass will be greater and there will be a much greater rate of heat loss. Increasing the width of the air layer, such as using two panes of glass separated by an air gap, will reduce the heat loss more than simply increasing the glass thickness, since the thermal conductivity of air is much less than for glass.

[†] This is quite similar to the relations describing diffusion (Chapter 18) and the flow of fluids through a tube (Chapter 13). In those cases, the flow of matter was found to be proportional to the concentration gradient or to the pressure gradient. This close similarity is one reason we speak of the "flow" of heat. Yet we must keep in mind that no substance is flowing in this case—it is energy that is being transferred.

The insulating properties of clothing come from the insulating properties of air. Without clothes, our bodies would heat the air in contact with the skin and would soon become reasonably comfortable because air is a very good insulator. But since air moves—there are breezes and drafts, and people themselves move about—the warm air would be replaced by cold air, thus increasing the temperature difference and the heat loss from the body. Clothes keep us warm by holding air so it cannot move readily. It is not the cloth that insulates us, but the air that the cloth traps. Down is a very good insulator because even a small amount of it fluffs up and traps a great amount of air. On this basis, can you see one reason why drapes in front of a window reduce heat loss from a house?

For practical purposes, the thermal properties of building materials, particularly when considered as insulation, are commonly specified by R-values (or "thermal resistance"), defined for a given thickness l of material as:

$$R = \frac{l}{k}.$$

The R-value of a given piece of material thus combines the thickness l and the thermal conductivity k in one number. In the United States, R-values are given in British units (although often not stated at all!), as $\text{ft}^2 \cdot \text{h} \cdot \text{F}°/\text{Btu}$. Table 19–5 gives R-values for some common building materials: note that R-values increase directly with material thickness. For example, 2 inches of fiberglass has $R = 6\,\text{ft}^2 \cdot \text{h} \cdot \text{F}°/\text{Btu}$, half that for 4 inches.

Convection

Although liquids and gases are generally not very good conductors of heat, they can transfer heat quite rapidly by convection. **Convection** is the process whereby heat is transferred by the mass movement of molecules from one place to another. Whereas conduction involves molecules (and/or electrons) moving only over small distances and colliding, convection involves the movement of molecules over large distances.

A forced-air furnace, in which air is heated and then blown by a fan into a room, is an example of *forced convection. Natural convection* occurs as well, and one familiar example is that hot air rises. For instance, the air above a radiator (or other type of heater) expands as it is heated, and hence its density decreases; because its density is less, it rises, just as a log submerged in water floats upward because its density is less than that of water. Warm or cold ocean currents, such as the balmy Gulf Stream, represent natural convection on a large scale. Wind is another example of convection, and weather in general is a result of convective air currents.

When a pot of water is heated (Fig. 19–21), convection currents are set up as the heated water at the bottom of the pot rises because of its reduced density and is replaced by cooler water from above. This principle is used in many heating systems, such as the hot-water radiator system shown in Fig. 19–22. Water is heated in the furnace and as its temperature increases, it expands and rises, as shown. This causes the water to circulate in the system. Hot water then enters the radiators, heat is transferred by conduction to the air, and the cooled water returns to the furnace. Thus, the water circulates because of convection; pumps are sometimes used to improve circulation. The air throughout the room also becomes heated as a result of convection. The air heated by the radiators rises and is replaced by cooler air, resulting in convective air currents, as shown.

Other types of furnaces also depend on convection. Hot-air furnaces with registers (openings) near the floor often do not have fans but depend on natural convection, which can be appreciable. In other systems, a fan is used. In either case, it is important that cold air can return to the furnace so that convective currents circulate throughout the room if the room is to be uniformly heated.

TABLE 19–5 R-values

Material	Thickness	R-value ($\text{ft}^2 \cdot \text{h} \cdot \text{F}°/\text{Btu}$)
Glass	$\frac{1}{8}$ inch	1
Brick	$3\frac{1}{2}$ inch	0.6–1
Plywood	$\frac{1}{2}$ inch	0.6
Fiberglass insulation	4 inches	12

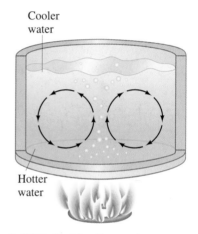

FIGURE 19–21 Convection currents in a pot of water being heated on a stove.

FIGURE 19–22 Convection plays a role in heating a house. The circular green arrows show the convective air currents in the rooms.

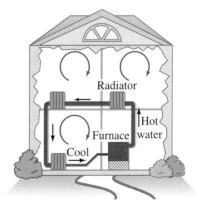

Radiation

Convection and conduction require the presence of matter as a medium to carry the heat from the hotter to the colder region. But a third type of heat transfer occurs without any medium at all. All life on Earth depends on the transfer of energy from the Sun, and this energy is transferred to the Earth over empty (or nearly empty) space. This form of energy transfer is heat—since the Sun's temperature is much higher (6000 K) than Earth's—and is referred to as **radiation**. The warmth we receive from a fire is mainly radiant energy (most of the air heated by a fire rises by convection up the chimney and does not reach us).

As we shall see in later chapters, radiation involves electromagnetic waves. Suffice it to say for now that radiation from the Sun consists of visible light plus many other wavelengths that the eye is not sensitive to, including infrared (IR) radiation which is mainly responsible for heating the Earth.

The rate at which an object radiates energy has been found to be proportional to the fourth power of the Kelvin temperature, T. That is, a body at 2000 K as compared to one at 1000 K radiates energy at a rate $2^4 = 16$ times greater. The rate of radiation is also proportional to the area A of the emitting object, so the rate at which energy leaves the object, $\Delta Q / \Delta t$, is

Radiation $\propto T^4$

$$\frac{\Delta Q}{\Delta t} = e\sigma A T^4. \tag{19–15}$$

This is called the **Stefan-Boltzmann equation**, and σ is a universal constant called the **Stefan-Boltzmann constant** which has the value

Stefan-Boltzmann constant

$$\sigma = 5.67 \times 10^{-8}\,\text{W/m}^2\cdot\text{K}^4.$$

The factor e, called the **emissivity**, is a number between 0 and 1 that is characteristic of the material. Very black surfaces, such as charcoal, have emissivity close to 1, whereas bright shiny surfaces have e close to zero and thus emit correspondingly less radiation. The value of e depends somewhat on the temperature of the body.

Not only do light shiny surfaces emit less radiation, but they absorb little of the radiation that falls upon them (most is reflected). Black and very dark objects, on the other hand, absorb nearly all the radiation that falls on them—which is why light-colored clothing is usually preferable to dark clothing on a hot day. Thus, **a good absorber is also a good emitter**.

Good absorber is good emitter

Any object not only emits energy by radiation, but also absorbs energy radiated by other bodies. If an object of emissivity e and area A is at a temperature T_1, it radiates energy at a rate $e\sigma A T_1^4$. If the object is surrounded by an environment at temperature T_2 and high emissivity (≈ 1), the rate the surroundings radiate energy is proportional to T_2^4, and the rate that energy is absorbed by the object is proportional to T_2^4. The *net* rate of radiant heat flow from the object is given by the equation

Net flow rate of heat radiation

$$\frac{\Delta Q}{\Delta t} = e\sigma A\big(T_1^4 - T_2^4\big), \tag{19–16}$$

where A is the surface area of the object, T_1 its temperature and e its emissivity (at temperature T_1), and T_2 is the temperature of the surroundings. Notice in this equation that the rate of heat absorption by an object was taken to be $e\sigma A T_2^4$. That is, the proportionality constant is the same for both emission and absorption. This must be true to correspond with the experimental fact that equilibrium between object and surroundings is reached when they come to the same temperature. That is, $\Delta Q / \Delta t$ must equal zero when $T_1 = T_2$, so the coefficients of emission and absorption terms must be the same. This confirms the idea that a good emitter is a good absorber.

Because both the object and its surroundings radiate energy, there is a net transfer of energy from one to the other unless everything is at the same tempera-

ture. From Eq. 19–16 it is clear that if $T_1 > T_2$, the net flow of heat is from the body to the surroundings, so the body cools. But if $T_1 < T_2$, the net heat flow is from the surroundings into the body, and its temperature rises. If different parts of the surroundings are at different temperatures, Eq. 19–16 becomes more complicated.

EXAMPLE 19–13 **ESTIMATE** **Two teapots.** A ceramic teapot ($e = 0.70$) and a shiny one ($e = 0.10$) each hold 0.75 L of tea at 95°C. (a) Estimate the rate of heat loss from each, and (b) estimate the temperature drop after 30 min for each. Consider only radiation and assume the surroundings are at 20°C.

SOLUTION (a) A teapot that holds 0.75 L can be approximated by a cube 10 cm on a side, with five sides exposed, so its surface area would be about $5 \times 10^{-2}\,\text{m}^2$. The rate of heat loss would be about

$$\frac{\Delta Q}{\Delta t} = e\sigma A(T_1^4 - T_2^4)$$
$$= e(5.67 \times 10^{-8}\,\text{W/m}^2\cdot\text{K}^4)(5 \times 10^{-2}\,\text{m}^2)\big[(368\,\text{K})^4 - (293\,\text{K})^4\big]$$
$$= e(30)\,\text{W},$$

or about 20 W for the ceramic pot ($e = 0.70$) and 3 W for the shiny one ($e = 0.10$). (b) To estimate the temperature drop, we use the concept of specific heat and ignore the contribution of the pots compared to that of the 0.75 L of water. Then, using Eq. 19–2,

$$\frac{\Delta T}{\Delta t} = \frac{\Delta Q/\Delta t}{mc} = \frac{e(30)\,\text{J/s}}{(0.75\,\text{kg})(4.19 \times 10^3\,\text{J/kg}\cdot\text{C}^\circ)} = e(0.010)\,\text{C}^\circ/\text{s},$$

which for 30 min (1800 s) represents about 12 C° for the ceramic pot and about 2 C° for the shiny one. The shiny one clearly has an edge, at least as far as radiation is concerned. However, convection and conduction could play a greater role than radiation.

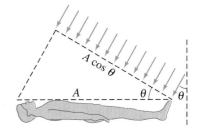

FIGURE 19–23 Radiant energy striking a body at an angle θ.

Heating of an object by radiation from the Sun cannot be calculated using Eq. 19–16 since this equation assumes a uniform temperature, T_2, of the environment surrounding the object, whereas the Sun is essentially a point source. Hence the Sun must be treated as a separate source of energy. Heating by the Sun is calculated using the fact that about 1350 J of energy strikes the atmosphere of the Earth from the Sun per second per square meter of area at right angles to the Sun's rays. This number, 1350 W/m², is called the **solar constant**. The atmosphere may absorb as much as 70 percent of this energy before it reaches the ground, depending on the cloud cover. On a clear day, about 1000 W/m² reaches the Earth's surface. An object of emissivity e with area A facing the Sun absorbs heat at a rate, in watts, of about

$$\frac{\Delta Q}{\Delta t} = (1000\,\text{W/m}^2)eA\cos\theta,$$

where θ is the angle between the Sun's rays and a line perpendicular to the area A (Fig. 19–23). That is, $A\cos\theta$ is the "effective" area, at right angles to the Sun's rays. The explanations for the seasons and the polar ice caps (Fig. 19–24), and why the Sun heats the Earth more at midday than at sunrise or sunset, are also related to this $\cos\theta$ factor.

Notice that if a person wears light-colored clothing, e is much smaller, so the energy absorbed is less.

An interesting application of thermal radiation to diagnostic medicine is **thermography**. A special instrument, the thermograph, scans the body, measuring the intensity of radiation from many points and forming a picture that resembles an X-ray (Fig. 19–25). Areas where metabolic activity is high, such as in tumors, can often be detected on a thermogram as a result of their higher temperature and consequent increased radiation.

Radiation from the Sun

FIGURE 19–24 June sun makes an angle of about 23° with the equator. Thus (a) θ in the southern United States is near 0° (direct summer sun), whereas (b) in the southern hemisphere, θ is 50° or 60°, and less heat can be absorbed—hence it is winter. (c) At the poles, there is never strong direct sun; $\cos\theta$ varies from about $\frac{1}{2}$ in summer to 0 in winter; with so little heating, ice can form.

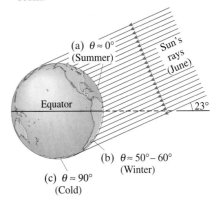

FIGURE 19–25 Thermograms of a healthy person's arms and hands (a) before smoking, (b) after smoking a cigarette, showing temperature decrease due to impaired blood circulation associated with smoking. The thermograms have been color-coded according to temperature; the scale on the right goes from blue (cold) to white (hot).

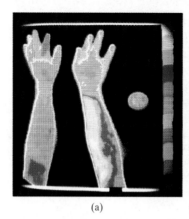 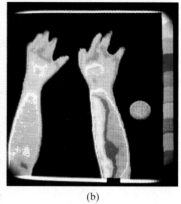

(a)   (b)

➡ P H Y S I C S   A P P L I E D

Astronomy—size of a star

EXAMPLE 19–14 **ESTIMATE** | **Star radius.** The giant star Betelgeuse emits radiant energy at a rate 10^4 times greater than our Sun, whereas its surface temperature is only half (2900 K) that of our Sun. Estimate the radius of Betelgeuse, assuming $e = 1$. The Sun's radius is $r_S = 7 \times 10^8$ m.

SOLUTION We assume both Betelgeuse and the Sun are spherical, with surface area $4\pi r^2$. We solve Eq. 19–15 for A:

$$4\pi r^2 = A = \frac{(\Delta Q/\Delta t)}{e\sigma T^4}.$$

Then

$$\frac{r_B^2}{r_S^2} = \frac{(\Delta Q/\Delta t)_B}{(\Delta Q/\Delta t)_S} \cdot \frac{T_S^4}{T_B^4} = (10^4)(2^4) = 16 \times 10^4.$$

Hence $r_B = \sqrt{16 \times 10^4}\, r_S = (400)(7 \times 10^8\,\text{m}) \approx 3 \times 10^{11}$ m. If Betelgeuse were our Sun, it would envelop us (Earth is 1.5×10^{11} m from the Sun).

Summary

Thermal energy, or **internal energy**, U, refers to the total energy of all the molecules in a body. For an ideal monatomic gas

$$U = \tfrac{3}{2}NkT = \tfrac{3}{2}nRT.$$

Heat refers to the transfer of energy from one body to another because of a difference of temperature. Heat is thus measured in energy units, such as joules.

Heat and thermal energy are also sometimes specified in calories or kilocalories, where

$$1\,\text{cal} = 4.186\,\text{J}$$

is the amount of heat needed to raise the temperature of 1 g of water by 1 C°.

The **specific heat**, c, of a substance is defined as the energy (or heat) required to change the temperature of unit mass of substance by 1 degree; as an equation,

$$Q = mc\,\Delta T,$$

where Q is the heat absorbed or given off, ΔT is the temperature increase or decrease, and m is the mass of the substance.

When heat flows within an isolated system, conservation of energy tells us that the heat gained by one part of the system is equal to the heat lost by the other part of the system; this is the basis of **calorimetry**, which is the quantitative measurement of heat exchange.

Exchange of energy occurs, without a change in temperature, whenever a substance changes phase. The **heat of fusion** is the heat required to melt 1 kg of a solid into the liquid phase; it is also equal to the heat given off when the substance changes from liquid to solid. The **heat of vaporization** is the energy required to change 1 kg of a substance from the liquid to the vapor phase; it is also the energy given off when the substance changes from vapor to liquid.

The **first law of thermodynamics** states that the change in internal energy, ΔU, of a system is equal to the heat added to the system, Q, minus the work, W, done by the system:

$$\Delta U = Q - W.$$

This important law is a broad restatement of the conservation of energy and is found to hold for all processes.

Two simple thermodynamic processes are **isothermal**, which is a process carried out at constant temperature, and **adiabatic**, a process in which no heat is exchanged.

The work done by (or on) a gas to change its volume by dV is $dW = P\,dV$, where P is the pressure.

Work and heat are not functions of the state of a system (as are P, V, T, n, and U) but depend on the type of process that takes a system from one state to another.

The **molar specific heat** of an ideal gas at constant volume, C_V, and at constant pressure, C_P, are related by $C_P - C_V = R$, where R is the gas constant. For a monatomic ideal gas, $C_V = \frac{3}{2}R$.

For ideal gases made up of diatomic or more complex molecules, C_V is equal to $\frac{1}{2}R$ times the number of **degrees of freedom** of the molecule. Unless the temperature is very high, some of the degrees of freedom may not be active and so do not contribute. According to the **principle of equipartition of energy**, energy is shared equally among the active degrees of freedom in an amount $\frac{1}{2}kT$ per molecule on average.

When an ideal gas expands (or contracts) adiabatically ($Q = 0$), the relation $PV^\gamma = $ constant holds, where $\gamma = C_P/C_V$.

Heat is transferred from one place (or body) to another in three different ways. In **conduction**, energy is transferred from higher-kinetic energy molecules or electrons to lower-kinetic energy neighbors when they collide.

Convection is the transfer of energy by the mass movement of molecules over considerable distances.

Radiation, which does not require the presence of matter, is energy transfer by electromagnetic waves, such as from the Sun. All bodies radiate energy in an amount that is proportional to the fourth power of their Kelvin temperature (T^4) and to their surface area. The energy radiated (or absorbed) also depends on the nature of the surface (dark surfaces absorb and radiate more than bright shiny ones), which is characterized by the emissivity, e.

Questions

1. Is heat really involved in the Joule experiment, Fig. 19–1?

2. What happens to the work done when a jar of orange juice is vigorously shaken?

3. When a hot object warms a cooler object, does temperature flow between them? Are the temperature changes of the two objects equal?

4. (a) If two objects of different temperature are placed in contact, will heat naturally flow from the object with higher internal energy to the object with lower internal energy? (b) Is it possible for heat to flow even if the internal energies of the two objects are the same? Explain.

5. In warm regions where tropical plants grow but the temperature may drop below freezing a few times in the winter, the destruction of sensitive plants due to freezing can be reduced by watering them in the evening. Explain.

6. The specific heat of water is quite large. Explain why this fact makes water particularly good for heating systems (that is, hot-water radiators).

7. Why does water in a canteen stay cooler if the cloth jacket surrounding the canteen is kept moist?

8. Explain why burns caused by steam on the skin are often more severe than burns caused by water at 100°C.

9. Explain, using the concepts of latent heat and internal energy, why water cools (its temperature drops) when it evaporates.

10. Will potatoes cook faster if the water is boiling faster?

11. The temperature very high in the Earth's atmosphere can be 700°C. Yet an animal there would freeze to death rather than roast. Explain.

12. What happens to the internal energy of water vapor in the air that condenses on the outside of a cold glass of water? Is work done or heat exchanged? Explain.

13. Use the conservation of energy to explain why the temperature of a gas increases when it is compressed—say, by pushing down on a cylinder—whereas the temperature decreases when the gas expands.

14. In an isothermal process, 3700 J of work is done by an ideal gas. Is this enough information to tell how much heat has been added to the system? If so, how much?

15. One liter of air is cooled at constant pressure until its volume is halved. Then it is allowed to expand isothermally back to its original volume. Draw the process on a PV diagram.

16. Is it possible for the temperature of a system to remain constant even though heat flows into or out of it? If so, give examples.

17. Discuss how the first law of thermodynamics can apply to metabolism in humans. In particular, note that a person does work W, but very little heat Q is added to the body (rather, it tends to flow out). Why then doesn't the internal energy drop drastically in time?

18. Explain in words why C_P is greater than C_V.

19. Explain why the temperature of a gas increases when it is adiabatically compressed.

20. An ideal monatomic gas is allowed to expand slowly to twice its volume (1) isothermally; (2) adiabatically; (3) isobarically. Plot each on a PV diagram. In which process is ΔU the greatest, and in which is ΔU the least? In which is W the greatest and the least? In which is Q the greatest and the least?

21. Ceiling fans are sometimes reversible, so that they drive the air down in one season and pull it up in another season. Which way should you set the fan for summer? For winter?

22. Down sleeping bags and parkas are often specified as so many inches or centimeters of *loft*, the actual thickness of the garment when it is fluffed up. Explain.

23. Microprocessor chips nowadays have a "heat sink" glued on top that looks like a series of fins. Why is it shaped like that?

24. Sea breezes are often encountered on sunny days at the shore of a large body of water. Explain in light of the fact that the temperature of the land rises more rapidly than that of the nearby water.

25. The Earth cools off at night much more quickly when the weather is clear than when cloudy. Why?

26. Explain why air-temperature readings are always taken with the thermometer in the shade.

27. A premature baby in an incubator can be dangerously cooled even when the air temperature in the incubator is warm. Explain.

28. Heat loss occurs through windows by the following processes: (1) ventilation around edges; (2) through the frame, particularly if it is metal; (3) through the glass panes and (4) radiation. (a) For the first three, what is (are) the mechanism(s): conduction, convection, or radiation? (b) Heavy curtains reduce which of these heat losses? Explain in detail.

29. A piece of wood lying in the Sun absorbs more heat than a piece of shiny metal. Yet the wood feels less hot than the metal when you pick it up. Explain.

30. Explain why cities situated on the ocean tend to have less extreme temperatures than inland cities at the same latitude.

31. Early in the day, after the sun has reached the slope of a mountain, there tends to be a gentle upward movement of air. Later, after a slope goes into shadow, there is a gentle downdraft. Explain.

32. In the Northern Hemisphere the amount of heat required to heat a room where the windows face north is much higher than that required where the windows face south. Explain.

| Problems

Section 19–1

1. (I) How much heat (joules) is required to raise the temperature of 30.0 kg of water from 15°C to 95°C?

2. (I) To what temperature will 7700 J of heat raise 3.0 kg of water initially at 10.0°C?

3. (II) When a 3.0-g bullet, traveling with a speed of 400 m/s, passes through a tree, its speed is reduced to 200 m/s. How much heat Q is produced and shared by the bullet and the tree?

4. (II) A British thermal unit (Btu) is a unit of heat in the British system of units. One Btu is defined as the heat needed to raise 1 pound of water by 1 F°. Show that

$$1 \text{ Btu} = 0.252 \text{ kcal} = 1055 \text{ J}.$$

5. (II) A water heater can generate 7200 kcal/h. How much water can it heat from 15°C to 50°C per hour?

6. (II) A small immersion heater is rated at 350 W. Estimate how long it will take to heat a cup of soup (assume this is 250 mL of water) from 20°C to 60°C.

7. (II) How many kilocalories of heat are generated when the brakes are used to bring a 1000-kg car to rest from a speed of 95 km/h?

Sections 19–3 and 19–4

8. (I) What is the specific heat of a metal substance if 135 kJ of heat is needed to raise 5.1 kg of the metal from 20°C to 30°C?

9. (I) An automobile cooling system holds 16 L of water. How much heat does it absorb if its temperature rises from 20°C to 90°C?

10. (II) A 35-g glass thermometer reads 21.6°C before it is placed in 135 mL of water. When the water and thermometer come to equilibrium, the thermometer reads 39.2°C. What was the original temperature of the water?

11. (II) The 1.20-kg head of a hammer has a speed of 6.5 m/s just before it strikes a nail (Fig. 19–26) and is brought to rest. Estimate the temperature rise of a 14-g iron nail generated by ten such hammer blows done in quick succession. Assume the nail absorbs all the energy.

12. (II) What will be the equilibrium temperature when a 245-g block of copper at 300°C is placed in a 150-g aluminum calorimeter cup containing 820 g of water at 12.0°C?

13. (II) A hot iron horseshoe (mass = 0.40 kg) which has just been forged, is dropped into 1.35 L of water in a 0.30-kg iron pot initially at 20°C. If the final equilibrium temperature is 25°C, estimate the initial temperature of the hot horseshoe.

14. (II) When 215 g of a substance is heated to 330°C and then plunged into a 100-g aluminum calorimeter cup containing 150 g of water at 12.5°C, the final temperature, as registered by a 17-g glass thermometer, is 35.0°C. What is the specific heat of the substance?

15. (II) How long does it take a 750-W coffeepot to bring to a boil 0.75 L of water initially at 8.0°C? Assume that the part of the pot which is heated with the water is made of 360 g of aluminum, and that no water boils away.

16. (II) Estimate the Calorie content of 100 g of Mullaney's fudge brownie from the following measurements. A 10-g sample of the brownie is allowed to dry before putting it in a bomb calorimeter. The aluminum bomb has a mass of 0.615 kg and is placed in 2.00 kg of water contained in an aluminum calorimeter cup of mass 0.524 kg. The initial temperature of the mixture is 15.0°C and its temperature after ignition is 36.0°C.

17. (II) (a) Show that if the specific heat varies as a function of temperature, $c(T)$, the heat needed to raise the temperature of a substance from T_1 to T_2 is given by

$$Q = \int_{T_1}^{T_2} mc(T)\, dT.$$

(b) Suppose $c(T) = c_0(1 + aT)$ for some substance, where $a = 2.0 \times 10^{-3} \text{ C}^{-1}$ and T is the Celsius temperature. Determine the heat required to raise the temperature from T_1 to T_2. (c) What is the mean value of c over the range T_1 to T_2 for part (b), expressed in terms of c_0, the specific heat at 0°C?

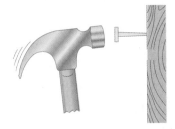

FIGURE 19–26
Problem 11.

18. (I) How much heat is needed to melt 15.50 kg of silver that is initially at 20°C?

19. (I) During exercise, a person may give off 180 kcal of heat in 30 min by evaporation of water from the skin. How much water has been lost?

20. (I) If 2.80×10^5 J of energy is supplied to a flask of liquid oxygen at −183°C, how much oxygen will evaporate?

21. (II) What will be the final result when equal amounts of ice at 0°C and steam at 100°C are mixed together?

22. (II) A 40-g ice cube at its melting point is dropped into an insulated container of liquid nitrogen. How much nitrogen evaporates if it is at its boiling point of 77 K and has a latent heat of vaporization of 200 kJ/kg? Assume for simplicity that the specific heat of ice is a constant and is equal to its value near its melting point.

23. (II) A cube of ice is taken from the freezer at −8.5°C and placed in a 75-g aluminum calorimeter filled with 300 g of water at room temperature of 20°C. The final situation is observed to be all water at 17°C. What was the mass of the ice cube?

24. (II) An iron boiler of mass 230 kg contains 760 kg of water at 20°C. A heater supplies energy at the rate of 52,000 kJ/h. How long does it take for the water (a) to reach the boiling point, and (b) to all have changed to steam?

25. (II) On a hot day's race, a bicyclist consumes 8.0 L of water over the span of four hours. Making the approximation that all of the cyclist's energy goes into evaporating this water as sweat, how much energy in kcal did the rider use during the ride? (Since the efficiency of the rider is only about 20 percent, most of the energy consumed does go to heat, so our approximation is not far off.)

26. (II) What mass of steam at 100°C must be added to 1.00 kg of ice at 0°C to yield liquid water at 20°C?

27. (II) The specific heat of mercury is 138 J/kg·C°. Determine the latent heat of fusion of mercury using the following calorimeter data: 1.00 kg of solid Hg at its melting point of −39.0°C is placed in a 0.620-kg aluminum calorimeter with 0.430 kg of water at 12.80°C; the resulting equilibrium temperature is 5.06°C.

28. (II) A 54.0-kg ice-skater moving at 4.8 m/s glides to a stop. Assuming the ice is at 0°C and that 50 percent of the heat generated by friction is absorbed by the ice, how much ice melts?

29. (II) At a crime scene, the forensic investigator notes that the 8.2-g lead bullet that was stopped in a door-frame apparently melted completely on impact. Assuming the bullet was fired at room temperature (20°C), what does the investigator calculate the *minimum* muzzle velocity of the gun was?

Sections 19–6 and 19–7

30. (I) In Example 19–9, if the heat lost from the gas in the process BD is 2.78×10^3 J, what is the change in internal energy of the gas?

31. (I) An ideal gas expands isothermally, performing 5.00×10^3 J of work in the process. Calculate (a) the change in internal energy of the gas, and (b) the heat absorbed during this expansion.

32. (I) 1.0 L of air initially at 6.0 atm of (absolute) pressure is allowed to expand isothermally until the pressure is 1.0 atm. It is then compressed at constant pressure to its initial volume and lastly is brought back to its original pressure by heating at constant volume. Draw the process on a PV diagram, including numbers and labels for the axes.

33. (I) Sketch a PV diagram of the following process: 2.0 L of ideal gas at atmospheric pressure are cooled at constant pressure to a volume of 1.0 L, and then expanded isothermally back to 2.0 L, whereupon the pressure is increased at constant volume until the original pressure is reached.

34. (I) A gas is enclosed in a cylinder fitted with a light frictionless piston and maintained at atmospheric pressure. When 1400 kcal of heat is added to the gas, the volume is observed to increase slowly from 12.0 m³ to 18.2 m³. Calculate (a) the work done by the gas and (b) the change in internal energy of the gas.

35. (II) An ideal gas has its pressure cut in half slowly, while being kept in a rigid wall container. In the process, 1300 kJ of heat left the gas. (a) How much work was done during this process? (b) What was the change in internal energy of the gas during this process?

36. (II) In an engine, an ideal gas is compressed adiabatically to half its volume. In doing so, 2350 J of work is done on the gas. (a) How much heat flows into or out of the gas? (b) What is the change in internal energy of the gas? (c) Does its temperature rise or fall?

37. (II) An ideal gas expands at a constant pressure of 5.0 atm from 400 mL to 710 mL. Heat then flows out of the gas, at constant volume, and the pressure and temperature are allowed to drop until the temperature reaches its original value. Calculate (a) the total work done by the gas in the process, and (b) the total heat flow into the gas.

38. (II) Consider the following two-step process. Heat is allowed to flow out of an ideal gas at constant volume so that its pressure drops from 2.2 atm to 1.5 atm. Then the gas expands at constant pressure, from a volume of 6.8 L to 10.0 L, where the temperature reaches its original value. See Fig. 19–27. Calculate (a) the total work done by the gas in the process, (b) the change in internal energy of the gas in the process, and (c) the total heat flow into or out of the gas.

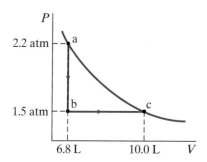

FIGURE 19–27 Problem 38.

39. (II) Suppose 2.00 mol of an ideal gas of volume $V_1 = 3.50$ m³ at $T_1 = 300$ K is allowed to expand isothermally to $V_2 = 7.00$ m³ at $T_2 = 300$ K. Determine (a) the work done by the gas, (b) the heat added to the gas, and (c) the change in internal energy of the gas.

40. (II) Determine (a) the work done and (b) the change in internal energy of 1.00 kg of water when it is all boiled to steam at 100°C. Assume a constant pressure of 1.00 atm.

41. (II) How much work is done to slowly compress, isothermally, 2.50 L of nitrogen at 0°C and 1 atm to 1.50 L at 0°C?

42. (II) The PV diagram in Fig. 19–28 shows two possible states of a system containing 1.45 moles of a monatomic ideal gas. ($P_1 = P_2 = 450$ N/m^2, $V_1 = 2.00$ m^3, $V_2 = 8.00$ m^3). (a) Draw the process which depicts an isobaric expansion from state 1 to state 2 and label this process (A). (b) Find the work done by the gas and the change in internal energy of the gas in process (A). (c) Draw the process which depicts an isothermal expansion from state 1 to the volume V_2 followed by an isochoric increase in temperature to state 2 and label this two-step process (B). (d) Find the change in internal energy of the gas for the two-step process (B).

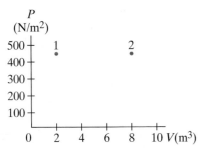

FIGURE 19–28 Problem 42.

43. (II) Show that Eqs. 19–4 and 19–5 are valid for any shape volume that changes. To do so, draw an arbitrary closed curve to represent the boundary of the volume; then draw a slightly larger curve to represent an increase in volume; choose a small section of the original boundary, of area ΔA, and show that $dW = P \, \Delta A \, dl = P \, dV$ for this section; then integrate over the whole boundary, and finally integrate over a finite volume.

44. (II) When a gas is taken from a to c along the curved path in Fig. 19–29, the work done by the gas is $W = -35$ J and the heat added to the gas is $Q = -63$ J. Along path abc, the work done is $W = -48$ J. (a) What is Q for path abc? (b) If $P_c = \frac{1}{2}P_b$, what is W for path cda? (c) What is Q for path cda? (d) What is $U_a - U_c$? (e) If $U_d - U_c = 5$ J, what is Q for path da?

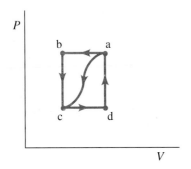

FIGURE 19–29 Problems 44, 45, and 46.

45. (III) In the process of taking a gas from state a to state c along the curved path shown in Fig. 19–29, 80 J of heat leave the system and 55 J of work are done *on* the system. (a) Determine the change in internal energy, $U_a - U_c$. (b) When the gas is taken along the path cda, the work done by the gas is $W = 38$ J. How much heat Q is added to the gas in the process cda? (c) If $P_a = 2.5P_d$, how much work is done by the gas in the process abc? (d) What is Q for path abc? (e) If $U_a - U_b = 10$ J, what is Q for the process bc? Here is a summary of what is given:

$$Q_{a \to c} = -80 \text{ J} \qquad U_a - U_b = 10 \text{ J}$$
$$W_{a \to c} = -55 \text{ J} \qquad P_a = 2.5P_d$$
$$W_{cda} = 38 \text{ J}$$

46. (III) Suppose a gas is taken clockwise around the rectangular cycle shown in Fig. 19–29, starting at b, then to a, to d, to c, and returning to b. Using the values given in Problem 45. (a) describe each leg of the process, and then calculate (b) the net work done during the cycle, (c) the net heat flow during the cycle, and (d) the total internal energy change during the cycle. (e) What percentage of the *intake* heat was turned into usable work: i.e., how efficient is this "rectangular" cycle (give as a percentage)?

* 47. (III) Determine the work done by 1.00 mol of a van der Waals gas (Section 18–5) when it expands from volume V_1 to V_2 isothermally.

Section 19–8

48. (I) What is the internal energy of 4.50 mol of an ideal diatomic gas at 600 K, assuming all degrees of freedom are active?

49. (I) If a heater supplies 1.8×10^6 J/h to a room 6.5 m $\times$ 4.6 m $\times$ 3.0 m containing air at 20°C and 1.0 atm, by how much will the temperature rise in one hour, assuming no heat losses to the outside?

50. (I) Show that if the molecules of a gas have n degrees of freedom, then theory predicts $C_V = (n/2)R$ and $C_P = [(n + 2)/2]R$.

51. (I) Estimate the molar specific heat and specific heat at both constant pressure and constant volume for hydrogen gas (H_2) at room temperature.

52. (II) Show that the work done by n moles of an ideal gas when it expands adiabatically is $W = nC_V(T_1 - T_2)$, where T_1 and T_2 are the initial and final temperatures, and C_V is the molar specific heat at constant volume.

53. (II) A certain gas has specific heat $c_V = 0.0356$ kcal/kg·C°, which changes little over a wide temperature range. What is the atomic mass of this gas? What gas is it?

54. (II) The specific heat at constant volume of a particular gas is 0.182 kcal/kg·K at room temperature, and its molecular mass is 34. (a) What is its specific heat at constant pressure? (b) What do you think is the molecular structure of this gas?

55. (II) An audience of 2500 fills a concert hall of volume 30,000 m^3. If there were no ventilation, by how much would the temperature rise over a period of 2.0 h due to the metabolism of the people (70 W/person)?

56. (II) A sample of 770 mol of nitrogen gas is maintained at a constant pressure of 1.00 atm in a flexible container. The gas is heated from 40°C to 180°C. Calculate (*a*) the heat added to the gas, (*b*) the work done by the gas, and (*c*) the change in internal energy.

57. (II) One mole of N_2 gas at 0°C is heated to 100°C at constant pressure (1.00 atm). Determine (*a*) the change in internal energy, (*b*) the work the gas does, and (*c*) the heat added to it.

58. (III) A 1.00-mol sample of an ideal diatomic gas at a pressure of 1.00 atm and temperature of 490 K undergoes a process in which its pressure increases linearly with temperature. The final temperature and pressure are 720 K and 1.60 atm. Determine (*a*) the change in internal energy, (*b*) the work done by the gas, and (*c*) the heat added to the gas. (Assume 5 active degrees of freedom.)

Section 19–9

59. (I) A 1.00-mol sample of an ideal diatomic gas, originally at 1.00 atm and 20°C, expands adiabatically to twice its volume. What are the final pressure and temperature for the gas? (Assume no molecular vibration.)

60. (II) Show, using Eqs. 19–4 and 19–13, that the work done by a gas that slowly expands adiabatically from pressure P_1 and volume V_1, to P_2 and V_2, is given by $W = (P_1V_1 - P_2V_2)/(\gamma - 1)$.

61. (II) 1.0 mol of an ideal monatomic gas at 300 K and 3.0 atm expands adiabatically to a final pressure of 1.0 atm. How much work does the gas do in the expansion?

62. (II) An ideal gas at 400 K is expanded adiabatically to 4.2 times its original volume. Determine its resulting temperature if the gas is (*a*) monatomic; (*b*) diatomic (no vibrations); (*c*) diatomic (molecules do vibrate).

63. (II) A 4.65-mol sample of an ideal diatomic gas expands adiabatically from a volume of $0.1210\,m^3$ to $0.750\,m^3$. Initially the pressure was 1.00 atm. Determine: (*a*) the initial and final temperatures; (*b*) the change in internal energy; (*c*) the heat lost by the gas; (*d*) the work done *on* the gas. (Assume no molecular vibration.)

64. (II) An ideal monatomic gas, consisting of 2.6 mol of volume $0.084\,m^3$, expands adiabatically. The initial and final temperatures are 25°C and −68°C. What is the final volume of the gas?

65. (III) A 1.00-mol sample of an ideal monatomic gas, originally at a pressure of 1.00 atm, undergoes a three-step process: (1) it is expanded adiabatically from $T_1 = 550\,K$ to $T_2 = 389\,K$; (2) it is compressed at constant pressure until its temperature reaches T_3; (3) it then returns to its original pressure and temperature by a constant-volume process. (*a*) Plot these processes on a *PV* diagram. (*b*) Determine T_3. (*c*) Calculate the change in internal energy, the work done by the gas, and the heat added to the gas for each process, and (*d*) for the complete cycle.

Section 19–10

66. (I) Calculate the rate of heat flow by conduction in Example 19–12 assuming there are strong gusty winds and that the external temperature is −5°C.

67. (I) (*a*) How much power is radiated by a tungsten sphere (emissivity $e = 0.35$) of radius 18.0 cm at a temperature of 25°C? (*b*) If the sphere is enclosed in a room whose walls are kept at −5°C, what is the *net* flow rate of energy out of the sphere?

68. (I) Over what distance must there be heat flow by conduction from the blood capillaries beneath the skin to the surface if the temperature difference is 0.50 C°? Assume 200 W must be tranferred through the whole body's surface area of $1.5\,m^2$.

69. (I) What is the rate of energy absorption from the Sun by a person lying flat on the beach on a clear day if the Sun makes a 30° angle with the vertical? Assume that $e = 0.70$, the area of the body exposed to the Sun is $0.80\,m^2$, and that $1000\,W/m^2$ reaches the Earth's surface.

70. (II) Two rooms, each a cube 4.0 m on a side, share a 15-cm thick brick wall. Because of a number of 100-W light-bulbs in one room, the air is at 30°C, while in the other room it is at 10°C. How many of the 100-W light-bulbs are needed to maintain the temperature difference across the wall?

71. (II) How long does it take the Sun to melt a block of ice at 0°C with an area of $1.0\,m^2$ and thickness 1.6 cm? Assume that the Sun's rays make an angle of 30° with the normal to the area, and that the emissivity of ice is 0.050.

72. (II) A copper rod and an aluminum rod of the same length and cross-sectional area are attached end to end (Fig. 19–30). The copper end is placed in a furnace which is maintained at a constant temperature of 250°C. The aluminum end is placed in an ice bath held at constant temperature of 0.0°C. Calculate the temperature at the point where the two rods are joined.

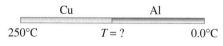

FIGURE 19–30 Problem 72.

73. (II) (*a*) Estimate, using the solar constant, the rate at which the whole Earth receives energy from the Sun. (*b*) Assume the Earth radiates an equal amount back into space (that is, the Earth is in equilibrium). Then, assuming the Earth is a perfect emitter ($e = 1.0$), estimate its average surface temperature.

74. (II) Suppose the insulating qualities of the wall of a house come mainly from a 4.0-inch layer of brick and an R-19 layer of insulation, as shown in Fig. 19–31. What is the total rate of heat loss through such a wall, if its total area is $240\ \text{ft}^2$ and the temperature difference across it is $10\ \text{F}°$?

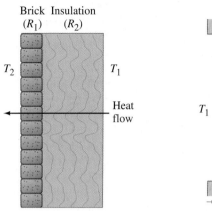

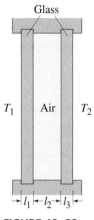

FIGURE 19–31 Two layers of insulation of a wall of a building. Problem 74.

FIGURE 19–32 Problem 75.

75. (II) A double-glazed window is one with two panes of glass separated by an air space, Fig. 19–32. (a) Show that the rate of heat flow by conduction is given by

$$\frac{\Delta Q}{\Delta t} = \frac{A(T_2 - T_1)}{l_1/k_1 + l_2/k_2 + l_3/k_3},$$

where k_1, k_2, and k_3 are the thermal conductivities for glass, air, and glass, respectively. (b) Generalize this expression for any number of materials placed next to one another.

76. (III) Approximately how long should it take $8.50\ \text{kg}$ of ice at $0°\text{C}$ to melt when it is placed in a carefully sealed Styrofoam icebox of dimensions $25\ \text{cm} \times 35\ \text{cm} \times 50\ \text{cm}$ whose walls are $1.5\ \text{cm}$ thick? Assume that the conductivity of Styrofoam is double that of air and that the outside temperature is $30°\text{C}$.

77. (III) A house thermostat is normally set to $22°\text{C}$, but at night it is turned down to $12°\text{C}$ for $7.0\ \text{h}$. Estimate how much more heat would be needed (state as a percentage of daily usage) if the thermostat were not turned down at night. Assume that the outside temperature averages $0°\text{C}$ for the $7.0\ \text{h}$ at night and $8°\text{C}$ for the remainder of the day, and that the heat loss from the house is proportional to the difference in temperature inside and out. To obtain an estimate from the data, you will have to make other simplifying assumptions; state what these are.

78. (III) A cylindrical pipe has inner radius R_1 and outer radius R_2. The interior of the pipe carries hot water at temperature T_1. The temperature outside is $T_2\ (< T_1)$. (a) Show that the rate of heat loss for a length L of pipe is

$$\frac{dQ}{dt} = \frac{2\pi k(T_1 - T_2)L}{\ln(R_2/R_1)},$$

where k is the thermal conductivity of the pipe. (b) Suppose the pipe is steel with $R_1 = 3.0\ \text{cm}$, $R_2 = 4.0\ \text{cm}$, and $T_2 = 30°\text{C}$. If the pipe holds still water at $T_1 = 71°\text{C}$, what will be the initial rate of change of its temperature? (c) Suppose water at $71°\text{C}$ enters the pipe and moves at a speed of $8.0\ \text{cm/s}$. What will be its temperature drop per centimeter of travel?

General Problems

79. To get an idea of how much thermal energy is contained in the world's oceans, estimate the heat liberated when a cube, $1\ \text{km}$ on a side, of ocean water is cooled by $1\ \text{K}$. (Approximate the ocean water by pure water for this estimate.)

80. A 15-g lead bullet is tested by firing it into a fixed block of wood with a mass of $0.92\ \text{kg}$. If the block and imbedded bullet together absorb all the heat energy generated and, after thermal equilibrium has been reached, the system has a temperature rise measured as $0.020\ \text{C}°$, estimate the entering speed of the bullet.

81. The specific heat of mercury is $0.033\ \text{kcal/kg·C}°$. When $1.00\ \text{kg}$ of solid mercury at its melting point of $-39°\text{C}$ is placed in a 0.50-kg aluminum calorimeter filled with $1.20\ \text{kg}$ of water at $20.0°\text{C}$, the final temperature of the mixture is found to be $16.5°\text{C}$. What is the heat of fusion of mercury in kcal/kg?

82. A scuba diver releases a 3.00-cm-diameter (spherical) bubble of air from a depth of $14.0\ \text{m}$. Assume the temperature is constant at $298\ \text{K}$, and that the air behaves as a perfect gas. (a) How large is the bubble when it reaches the surface? (b) Sketch a PV diagram for the process. (c) Apply the first law of thermodynamics to the bubble, and find the work done by the air in rising to the surface, the change in its internal energy, and the heat added or removed from the air in the bubble as it rises. Take the density of water to be $1000\ \text{kg/m}^3$.

83. A reciprocating compressor is a device that compresses air by a back-and-forth straight-line motion, like a piston in a cylinder. Consider a reciprocating compressor running at $400\ \text{rpm}$. During a compression stroke, $1.00\ \text{mol}$ of air is compressed. The initial temperature of the air is $390\ \text{K}$, the engine of the compressor is supplying $10\ \text{kW}$ of power to compress the air, and heat is being removed at the rate of $1.5\ \text{kW}$. Calculate the temperature change per compression stroke.

84. Suppose $2.0\ \text{mol}$ of neon (an ideal monatomic gas) at STP are compressed slowly and isothermally to $\frac{1}{5}$ the original volume. The gas is then allowed to expand quickly and adiabatically back to its original volume. Find the highest and lowest temperatures and pressures attained by the gas, and show on a PV diagram where these values occur.

85. At very low temperatures, the molar specific heat of many substances varies as the cube of the absolute temperature:

$$C = k\frac{T^3}{T_0^3},$$

which is sometimes called Debye's law. For rock salt, $T_0 = 281\ \text{K}$ and $k = 1940\ \text{J/mol·K}$. Determine the heat needed to raise $3.5\ \text{mol}$ of salt from $22.0\ \text{K}$ to $55.0\ \text{K}$.

86. (a) Find the total power radiated into space by the Sun, assuming it to be a perfect emitter at $T = 5500$ K. The Sun's radius is 7.0×10^8 m. (b) From this, determine the power per unit area arriving at the Earth, 1.5×10^{11} m away (Fig. 19–33).

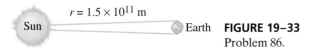

$r = 1.5 \times 10^{11}$ m

Sun   Earth **FIGURE 19–33**
                    Problem 86.

87. During light activity, a 70-kg person may generate 200 kcal/h. Assuming that 20 percent of this goes into useful work and the other 80 percent is converted to heat, calculate the temperature rise of the body after 1.00 h if none of this heat were transferred to the environment.

88. The *heat capacity*, C, of an object is defined as the amount of heat needed to raise its temperature by 1 C°. Thus, to raise the temperature by ΔT requires heat Q given by

$$Q = C \, \Delta T.$$

(a) Write the heat capacity C in terms of the specific heat, c, of the material. (b) What is the heat capacity of 1.0 kg of water? (c) Of 25 kg of water?

89. A very long 2.00-cm-diameter lead rod absorbs 410 kJ of heat. By how much does its length change? What would happen if the rod were only 2.0 cm long?

90. A mountain climber wears down clothing 3.5 cm thick with total surface area 1.9 m². The temperature at the surface of the clothing is −20°C and at the skin is 34°C. Determine the rate of heat flow by conduction through the clothing (a) assuming it is dry and that the thermal conductivity, k, is that of down, and (b) assuming the clothing is wet, so that k is that of water and the jacket has matted down to 0.50 cm thickness.

91. A marathon runner has an average metabolism rate for the race of about 1000 kcal/h. If the runner has a mass of 65.0 kg, how much water would the runner lose to evaporation from the skin for a race that lasts 2.5 h?

92. Estimate the rate at which heat can be conducted from the interior of the body to the surface. Assume that the thickness of tissue is 4.0 cm, that the skin is at 34°C and the interior at 37°C, and that the surface area is 1.5 m². Compare this to the measured value of about 230 W that must be dissipated by a person working lightly. This clearly shows the necessity of convective cooling by the blood.

93. *Newton's law of cooling* states that for small temperature differences, if a body at a temperature T_1 is in surroundings at a temperature T_2, the body cools at a rate given by

$$\frac{\Delta Q}{\Delta t} = K(T_1 - T_2)$$

where K is a constant. It includes the effects of conduction, convection, and radiation. That this linear relationship should hold is obvious if only conduction is considered. Show that it is also approximately true for radiation by showing that Eq. 19–16 reduces to

$$\Delta Q / \Delta t = 4\sigma e A T_2^3 (T_1 - T_2)$$
$$= \text{constant} \times (T_1 - T_2)$$

if $(T_1 - T_2)$ is small.

94. A house has well-insulated walls 15.5 cm thick (assume conductivity of air) and area 410 m², a roof of wood 6.5 cm thick and area 280 m², and plain glass windows 0.65 cm thick and total area 33 m². (a) Assuming that the heat loss is only by conduction, calculate the rate at which heat must be supplied to this house to maintain its temperature at 23°C if the outside temperature is −10°C. (b) If the house is initially at 10°C, estimate how much heat must be supplied to raise the temperature to 23°C within 30 min. Assume that only the air needs to be heated and that its volume is 750 m³. Take the specific heat of air to be 0.24 kcal/kg·C°.

95. A leaf of area 40 cm² and mass 4.5×10^{-4} kg directly faces the Sun on a clear day. The leaf has an emissivity of 0.85 and a specific heat of 0.80 kcal/kg·K. (a) Estimate the rate of rise of the leaf's temperature. (b) Calculate the temperature the leaf would reach if it lost all its heat by radiation (the surroundings are at 20°C). (c) In what other ways can the heat be dissipated by the leaf?

96. An iron meteorite melts when it enters the Earth's atmosphere. If its initial temperature, outside of the atmosphere, was −125°C, calculate the minimum velocity the meteorite must have had before it entered Earth's atmosphere.

97. Write a formula for the density of a gas when it is allowed to expand (a) as a function of temperature when the pressure is kept constant and (b) as a function of pressure when temperature is kept constant.

98. If 2.0 moles of an ideal monatomic gas expand adiabatically, performing 7500 J of work in the process, what is the change in temperature of the gas?

99. When 5.30×10^4 J of heat are added to a gas enclosed in a cylinder fitted with a light frictionless piston maintained at atmospheric pressure, the volume is observed to increase from 2.2 m³ to 4.1 m³. Calculate (a) the work done by the gas, and (b) the change in internal energy of the gas. (c) Graph this process on a PV diagram.

100. A diesel engine accomplishes ignition without a spark plug by an adiabatic compression of air to a temperature above the ignition temperature of the diesel fuel, which is injected into the cylinder at the peak of the compression. Suppose air is taken into the cylinder at 300 K and volume V_1 and is compressed adiabatically to 560°C ($\approx$ 1000°F) and volume V_2. Assuming that the air behaves as an ideal gas whose ratio of C_P to C_V is 1.4, calculate the compression ratio V_1/V_2 of the engine.

101. An ice sheet forms on a lake. The air above the sheet is at −15°C, whereas the water is at 0°C. Assume that the heat of fusion of the water freezing on the lower surface is conducted through the sheet to the air above. How much time will it take to form a sheet of ice 25 cm thick?

102. The temperature within the Earth's crust increases about 1.0 C° for each 30 m of depth. The thermal conductivity of the crust is 0.80 W/C°·m. (a) Determine the heat transferred from the interior to the surface for the entire Earth in one day. (b) Compare this heat to the amount of energy incident on the Earth in one day due to radiation from the Sun.

103. A 100-W lightbulb generates 95 W of heat, which is dissipated through a glass bulb that has a radius of 3.0 cm and is 1.0 mm thick. What is the difference in temperature between the inner and outer surfaces of the glass?

Two uses for a heat engine: a modern coal-burning power plant, and an old steam locomotive. Both produce steam which does work—on turbines to generate electricity, and on a piston that moves linkage to turn locomotive wheels. The efficiency of any engine—no matter how carefully engineered—is limited by nature as described in the second law of thermodynamics. This great law is best stated in terms of a quantity called entropy, which is unlike any other. Entropy is *not* conserved, but instead is constrained always to increase in any real process. Entropy is a measure of disorder. The second law of thermodynamics tells us that as time moves forward, the disorder in the universe increases.

We also discuss many practical matters such as heat engines, heat pumps, and refrigeration.

CHAPTER 20

Second Law of Thermodynamics

I n this final chapter on heat and thermodynamics, we discuss the famous second law of thermodynamics, and the quantity "entropy" that arose from this fundamental law and is its quintessential expression. We also discuss heat engines—the engines that transform heat into work in power plants, trains, and motor vehicles—because they first showed us that a new law was needed. Finally, we briefly discuss the third law of thermodynamics.

20–1 The Second Law of Thermodynamics— Introduction

The first law of thermodynamics states that energy is conserved. There are, however, many processes we can imagine that conserve energy but are not observed to occur in nature. For example, when a hot object is placed in contact with a cold object, heat flows from the hotter one to the colder one, never spontaneously the reverse. If heat were to leave the colder object and pass to the hotter one, energy could still be conserved. Yet it doesn't happen spontaneously.[†] As a second example, consider what happens when you drop a rock and it hits the ground. The initial

---

[†] By spontaneously, we mean by itself without input of work of some sort. (A refrigerator does move heat from a cold environment to a warmer one, but only by doing work.)

potential energy of the rock changes to kinetic energy as the rock falls, and when the rock hits the ground this energy in turn is transformed into internal energy (thermal energy) of the rock and the ground in the vicinity of the impact; the molecules move faster and the temperature rises slightly. But have you seen the reverse happen—a rock at rest on the ground suddenly rise up in the air because the thermal energy of molecules is transformed into kinetic energy of the rock as a whole? Energy could be conserved in this process, yet we never see it happen.

The first law of thermodynamics, conservation of energy, would not be violated if either of these processes occurred in reverse. To explain this lack of reversibility, scientists in the latter half of the nineteenth century came to formulate a new principle known as the **second law of thermodynamics**. This law is a statement about which processes occur in nature and which do not. It can be stated in a variety of ways, all of which are equivalent. One statement, due to R. J. E. Clausius (1822–1888), is that

heat flows naturally from a hot object to a cold object; heat will not flow spontaneously from a cold object to a hot object.

SECOND LAW OF THERMODYNAMICS (Clausius statement)

Since this statement applies to one particular process, it is not obvious how it applies to other processes. A more general statement is needed that will include other possible processes in a more obvious way.

The development of a general statement of the second law was based partly on the study of heat engines. A **heat engine** is any device that changes thermal energy into mechanical work, such as steam engines and automobile engines. We now examine heat engines, both from a practical point of view and to show their importance in developing the second law of thermodynamics.

20–2 Heat Engines

It is easy to produce thermal energy by doing work—for example, by simply rubbing your hands together briskly, or indeed by any frictional process. But to get work from thermal energy is more difficult, and the invention of a practical device to do this came only about 1700 with the development of the steam engine.

The basic idea behind any heat engine is that mechanical energy can be obtained from thermal energy only when heat is allowed to flow from a high temperature to a low temperature. In the process, some of the heat can then be transformed to mechanical work, as diagrammed schematically in Fig. 20–1. That is, a heat input $|Q_H|$ at a high temperature T_H is partly transformed into work $|W|$ and partly exhausted as heat $|Q_L|$ at a lower temperature T_L. By conservation of energy, $|Q_H| = |W| + |Q_L|$. The high and low temperatures, T_H and T_L, are called the **operating temperatures** of the engine. We will be interested only in engines that run in a repeating *cycle* (that is, the system returns repeatedly to its starting point) and thus can run continuously.

We use absolute value signs around Q_L, Q_H and W because we are interested only in the magnitudes, and thus avoid worrying about the sign conventions established in Section 19–6. These quantities are thus positive, and arrows on diagrams (as in Fig. 20–1) tell us the direction of energy transfers.

The temperatures T_H and T_L may or may not be precisely constant, depending on the cycle.

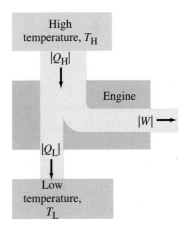

FIGURE 20–1 Schematic diagram of energy transfers for a heat engine.

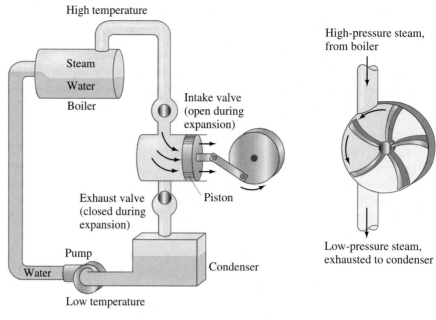

FIGURE 20–2 Steam engines.

(a) Reciprocating type

(b) Turbine (boiler and condenser not shown)

Steam Engine and Internal Combustion Engine

➥ PHYSICS APPLIED

Engines

The operation of two practical engines, the steam engine and the internal combustion engine (used in most automobiles), is illustrated in Figs. 20–2 and 20–3. Steam engines are of two main types, each making use of steam heated by combustion of coal, oil, or gas (or nuclear energy). In the so-called reciprocating type, Fig. 20–2a, the heated steam passes through the open intake valve and expands against a piston, forcing it to move. When the piston reaches the end of its stroke and begins to return to its original position, the exhaust valve opens and the piston forces the gases out. In a steam turbine, Fig. 20–2b, everything is essentially the same, except that the reciprocating piston is replaced by a rotating turbine that resembles a paddlewheel with many sets of blades. Most of our electricity today is generated using

FIGURE 20–3 Four-cycle internal combustion engine: (a) the gasoline–air mixture flows into the cylinder as the piston moves down; (b) the piston moves upward and compresses the gas; (c) firing of the spark plug ignites the gasoline–air mixture, raising it to a high temperature; (d) the gases, now at high temperature and pressure, expand against the piston in this, the power stroke; (e) the burned gases are pushed out to the exhaust pipe; the intake valve then opens, and the whole cycle repeats.

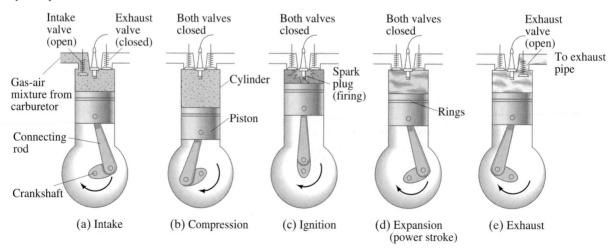

(a) Intake

(b) Compression

(c) Ignition

(d) Expansion (power stroke)

(e) Exhaust

steam turbines.[†] The material that is heated and cooled, steam in this case, is called the **working substance**. In a steam engine, the high temperature is obtained by burning coal, oil, or other fuel to heat the steam. In an internal combustion engine, the high temperature is achieved by burning the gasoline–air mixture in the cylinder itself (ignited by the spark plug), Fig. 20–3.

Why a ΔT is needed to drive a heat engine

To see why a *temperature difference* is required to run an engine, let us examine the steam engine. In the reciprocating engine, for example (Fig. 20–2a), suppose there were no condenser or pump, and that the steam was at the same temperature throughout the system. This would mean that the pressure of the gas being exhausted would be the same as that on intake. Thus, although work would be done by the gas *on* the piston when it expanded, an equal amount of work would have to be done *by* the piston to force the steam out the exhaust; hence, no net work would be done. In a real engine, the exhausted gas is cooled to a lower temperature and condensed so that the exhaust pressure is less than the intake pressure. Thus, although the piston must do work on the gas to expel it on the exhaust stroke, it is less than the work done by the gas on the piston during the intake. So a net amount of work can be obtained—but only if there is a difference of temperature. Similarly, in the gas turbine if the gas were not cooled, the pressure on each side of the blades would be the same; by cooling the gas on the exhaust side, the pressure on the front side of the blade is greater and hence the turbine turns.

The **efficiency**, *e*, of any heat engine can be defined as the ratio of the work it does, $|W|$, to the heat input at the high temperature, $|Q_H|$ (Fig. 20–1):

$$e = \frac{|W|}{|Q_H|}.$$

This is a sensible definition since $|W|$ is the output (what you get from the engine), whereas $|Q_H|$ is what you put in and pay for in fuel that burns. Because energy is conserved, the heat input $|Q_H|$ must equal the work done plus the heat that flows out at the low temperature $(|Q_L|)$:

$$|Q_H| = |W| + |Q_L|.$$

Thus $|W| = |Q_H| - |Q_L|$, and the efficiency of an engine is

$$e = \frac{|W|}{|Q_H|} = \frac{|Q_H| - |Q_L|}{|Q_H|} = 1 - \frac{|Q_L|}{|Q_H|}. \qquad \textbf{(20–1)}$$

Efficiency of any heat engine

EXAMPLE 20–1 **Car efficiency.** An automobile engine has an efficiency of 20 percent and produces an average of 23,000 J of mechanical work per second during operation. How much heat is discharged from this engine per second?

SOLUTION The output heat is $|Q_L|$. We are given $e = 0.20$, so from Eq. 20–1 we find that

$$\frac{|Q_L|}{|Q_H|} = 1 - e = 0.80.$$

We also know that $e = |W|/|Q_H|$, by definition, so in one second

$$|Q_H| = \frac{|W|}{e} = \frac{23,000 \text{ J}}{0.20} = 1.15 \times 10^5 \text{ J}.$$

Thus

$$|Q_L| = 0.80 |Q_H| = (0.80)(1.15 \times 10^5 \text{ J}) = 9.2 \times 10^4 \text{ J}.$$

The engine discharges $9.2 \times 10^4 \text{ J/s} = 92,000$ watts.

[†] Even nuclear power plants utilize steam turbines; the nuclear fuel—uranium—merely serves as fuel to heat the steam.

It is clear from Eq. 20–1 that the efficiency of an engine will be greater if $|Q_L|$ can be made small. However, from experience with a wide variety of systems, it has not been found possible to reduce $|Q_L|$ to zero. If $|Q_L|$ could be reduced to zero we would have a 100 percent efficient engine, as diagrammed in Fig. 20–4. That such a perfect engine (running continuously in a cycle) is not possible is another way of expressing the second law of thermodynamics. This can be stated formally as follows:

No device is possible whose sole effect is to transform a given amount of heat completely into work.

This is known as the **Kelvin-Planck statement of the second law of thermodynamics**. Said another way, *there can be no perfect (100 percent efficient) heat engine* such as that diagrammed in Fig. 20–4.

If the second law were not true, so that a perfect engine could be built, some rather remarkable things could happen. For example, if the engine of a ship did not need a low-temperature reservoir to exhaust heat into, the ship could sail across the ocean using the vast resources of the internal energy of the ocean water. Indeed, we would have no fuel problems at all!

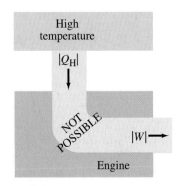

FIGURE 20–4 Schematic diagram of a hypothetical perfect heat engine in which all the heat input would be used to do work.

20–3 | Reversible and Irreversible Processes; Carnot Engine

In the early nineteenth century, the French scientist N. L. Sadi Carnot (1796–1832) studied in detail the process of transforming heat into mechanical energy. His aim had been to determine how to increase the efficiency of heat engines, but his studies soon led him to investigate the foundations of thermodynamics itself.

As an aid in this pursuit, Carnot, in 1824, invented (on paper) an idealized type of engine which we now call the "Carnot engine." The importance of the Carnot engine is not really as a practical engine, but rather as an aid to the understanding of heat engines in general, and also because Carnot and his engine contributed to the establishment and understanding of the second law of thermodynamics.

Reversible and Irreversible Processes

Reversible process

The Carnot engine involves *reversible processes*, so before we discuss it we must discuss what is meant by reversible and irreversible processes. A **reversible process** is one that is carried out infinitely slowly, so that the process can be considered as a series of equilibrium states, and the whole process could be done in reverse with no change in magnitude of the work done or heat exchanged. For example, a gas contained in a cylinder fitted with a tight, movable, but frictionless piston could be compressed isothermally in a reversible way if done infinitely slowly. Not all very slow (quasistatic) processes are reversible, however. If there is friction present, for example (as between the movable piston and cylinder just mentioned), the work done in one direction (going from some state A to state B) will not be the negative of the work done in the reverse direction (state B to state A). Such a process would not be considered reversible. Of course a perfectly reversible process is not possible in reality since it would require an infinite time; reversible processes can be approached arbitrarily closely, however, and they are very important theoretically.

Irreversible process

All real processes are **irreversible**: they are not done infinitely slowly. There could be turbulence in the gas, friction would be present, and so on; any process could not be done precisely in reverse since the heat lost to friction would not reverse itself, the turbulence would be different, and so on. For any given volume there would not be a well-defined pressure P and temperature T since the system would not always be in an equilibrium state. Thus a real, irreversible, process cannot be plotted on a PV diagram (except insofar as it may approach an ideal

reversible process). But a reversible process (since it is a quasistatic series of equilibrium states) always can be plotted on a *PV* diagram; and a reversible process, when done in reverse, retraces the same path on a *PV* diagram. Although all real processes are irreversible, reversible processes are conceptually important, just as the concept of an ideal gas is.

Carnot's Engine

Now let us look at Carnot's idealized engine. The Carnot engine makes use of a **reversible cycle**, by which we mean a series of reversible processes that take a given substance (the *working substance*) from an initial equilibrium state through many other equilibrium states and returns it again to the same initial state. In particular, the Carnot engine utilizes the **Carnot cycle**, which is illustrated in Fig. 20–5, with the working substance assumed to be an ideal gas. Let us take point a as the initial state. The gas is first expanded isothermally, and reversibly, path ab, at temperature T_H. To do so, we can imagine the gas to be in contact with a heat reservoir at a constant temperature T_H which delivers heat $|Q_H|$ to our working substance. Next the gas is expanded adiabatically and reversibly, path bc; no heat is exchanged and the temperature of the gas is reduced to T_L. The third step is a reversible isothermal compression, path cd, in contact with a heat reservoir at a constant low temperature, T_L, during which heat $|Q_L|$ flows out of the working substance. Finally, the gas is compressed adiabatically, path da, back to its original state. Thus a Carnot cycle consists of two isothermal and two adiabatic processes.

Carnot cycle

It is easy to see that the net work done in one cycle by a Carnot engine (or any other type of engine using a reversible cycle) is equal to the area enclosed by the curve representing the cycle on the *PV* diagram, the curve abcd in Fig. 20–5. (See Section 19–7.)

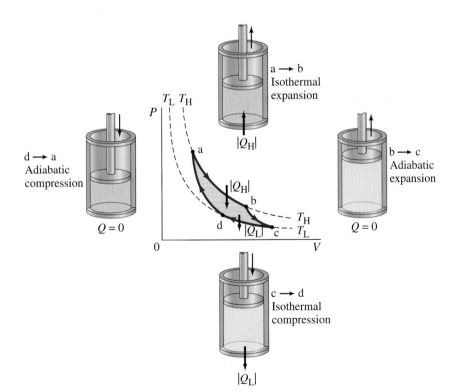

FIGURE 20–5 The Carnot cycle. Heat engines work in a cycle, and the cycle for the Carnot engine begins at point a on this *PV* diagram. (1) The gas is first expanded isothermally, with the addition of heat $|Q_H|$, along the path ab at temperature T_H. (2) Next the gas expands adiabatically from b to c—no heat is exchanged, but the temperature drops to T_L. (3) The gas is then compressed at constant temperature T_L, path c to d, and heat $|Q_L|$ flows out. (4) Finally, the gas is compressed adiabatically, path da, back to its original state. No Carnot engine actually exists, but as a theoretical engine it played an important role in the development of thermodynamics.

Carnot Efficiency and the Second Law of Thermodynamics

The efficiency of a Carnot engine, like any heat engine, is given by Eq. 20–1:

$$e = 1 - \frac{|Q_L|}{|Q_H|}.$$

For a Carnot engine, however, we can show that the efficiency depends only on the temperatures of the heat reservoirs, T_H and T_L. In the first isothermal process, ab in Fig. 20–5, the work done by the gas is (see Eq. 19–6)

$$W_{ab} = nRT_H \ln \frac{V_b}{V_a},$$

where n is the number of moles of the ideal gas used as working substance. Since the internal energy of an ideal gas does not change when the temperature remains constant, the first law of thermodynamics tells us that the heat added to the gas equals the work done by the gas:

$$|Q_H| = nRT_H \ln \frac{V_b}{V_a}.$$

Similarly, the heat lost by the gas in the isothermal process cd is

$$|Q_L| = nRT_L \ln \frac{V_c}{V_d}.$$

Since paths bc and da are adiabatic, we have (see Eq. 19–13)

$$P_b V_b^\gamma = P_c V_c^\gamma \quad \text{and} \quad P_d V_d^\gamma = P_a V_a^\gamma.$$

Also, from the ideal gas law,

$$\frac{P_b V_b}{T_H} = \frac{P_c V_c}{T_L} \quad \text{and} \quad \frac{P_d V_d}{T_L} = \frac{P_a V_a}{T_H}.$$

When we divide these last equations, term by term, into the corresponding set of equations on the line above, we obtain

$$T_H V_b^{\gamma-1} = T_L V_c^{\gamma-1} \quad \text{and} \quad T_L V_d^{\gamma-1} = T_H V_a^{\gamma-1}.$$

Next we divide the equation on the left by the one on the right and get

$$\left(\frac{V_b}{V_a}\right)^{\gamma-1} = \left(\frac{V_c}{V_d}\right)^{\gamma-1}.$$

Hence

$$\frac{V_b}{V_a} = \frac{V_c}{V_d}$$

or

$$\ln \frac{V_b}{V_a} = \ln \frac{V_c}{V_d}.$$

Using this result in our equations for $|Q_H|$ and $|Q_L|$ above we have

$$\frac{|Q_L|}{|Q_H|} = \frac{T_L}{T_H}. \qquad \text{[Carnot cycle]} \quad \textbf{(20–2)}$$

Hence the efficiency of a reversible Carnot engine is

Carnot (ideal) efficiency

$$e_{\text{ideal}} = 1 - \frac{|Q_L|}{|Q_H|} = 1 - \frac{T_L}{T_H}. \qquad \left[\begin{array}{l}\text{Carnot efficiency;} \\ \text{Kelvin temperatures}\end{array}\right] \quad \textbf{(20–3)}$$

The temperatures T_L and T_H are the absolute or Kelvin temperatures as measured on the ideal gas temperature scale. Thus the efficiency of a Carnot engine depends only on the temperatures T_L and T_H.

We could imagine other possible reversible cycles that could be used for an ideal reversible engine. According to a theorem stated by Carnot:

All reversible engines operating between the same two constant temperatures T_H and T_L have the same efficiency. Any irreversible engine operating between the same two fixed temperatures will have an efficiency less than this.

Carnot's theorem

This is known as **Carnot's theorem**.[†] It tells us that Eq. 20–3, $e = 1 - (T_L/T_H)$, applies to any ideal reversible engine with fixed input and exhaust temperatures, T_H and T_L, and that this equation represents a maximum possible efficiency for a real (i.e., irreversible) engine.

In practice, the efficiency of real engines is always less than the Carnot efficiency. Well-designed engines reach perhaps 60 to 80 percent of Carnot efficiency.

EXAMPLE 20–2 **Steam engine efficiency.** A steam engine operates between 500°C and 270°C. What is the maximum possible efficiency of this engine?

SOLUTION We must first change the temperature to kelvins. Thus, $T_H = 773$ K and $T_L = 543$ K. Then from Eq. 20–3,

$$e_{ideal} = 1 - \frac{543}{773} = 0.30.$$

Thus, the maximum (or Carnot) efficiency is 30 percent. Realistically, an engine might attain 0.70 of this value or 21 percent. Note in this Example that the exhaust temperature is still rather high, 270°C. Steam engines are often arranged in series so that the exhaust of one engine is used as intake by a second or third engine.

EXAMPLE 20–3 **A phony claim?** An engine manufacturer makes the following claims: The heat input per second of the engine is 9.0 kJ at 475 K. The heat output per second is 4.0 kJ at 325 K. Do you believe these claims?

SOLUTION The efficiency of the engine is (Eq. 20–1)

$$e = \frac{|Q_H| - |Q_L|}{|Q_H|} = \frac{9.0\,\text{kJ} - 4.0\,\text{kJ}}{9.0\,\text{kJ}} = 0.56.$$

However, the maximum possible efficiency is given by the Carnot efficiency, Eq. 20–3:

$$e_{ideal} = \frac{T_H - T_L}{T_H} = \frac{475\,\text{K} - 325\,\text{K}}{475\,\text{K}} = 0.32.$$

So the manufacturer's claims violate the second law of thermodynamics and cannot be believed.

It is clear from Eq. 20–3 that a 100 percent efficient engine is not possible. Only if the exhaust temperature, T_L, were at absolute zero would 100 percent efficiency be obtainable. But reaching absolute zero is a practical (as well as theoretical) impossibility.[‡] Thus we can state, as we already did in Section 20–2, that **no device is possible whose sole effect is to transform a given amount of heat completely into work**. As we saw in Section 20–2, this is known as the *Kelvin-Planck statement of the second law of thermodynamics*. It tells us that there can be no perfect (100 percent efficient) heat engine such as the one diagrammed in Fig. 20–4.

SECOND LAW OF THERMODYNAMICS (Kelvin-Planck statement)

[†] Carnot's theorem can be shown to follow directly from either the Clausius or Kelvin-Planck statements of the second law of thermodynamics.

[‡] This result is known as the *third law of thermodynamics*, as discussed in Section 20–10.

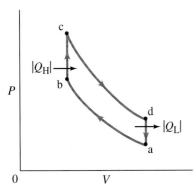

FIGURE 20–6 The Otto cycle.

*The Otto Cycle

The operation of an automobile internal combustion engine (see Fig. 20–3) can be approximated by a reversible cycle known as the *Otto cycle*, whose PV diagram is shown in Fig. 20–6. Unlike the Carnot cycle, the input and exhaust temperatures of the Otto cycle are *not* constant. Paths ab and cd are adiabatic, and paths bc and da are at constant volume. The gas (gasoline-air mixture) enters the cylinder at point a and is compressed adiabatically (compression stroke) to point b. At b ignition occurs (spark plug) and the burning of the gas adds heat $|Q_H|$ to the system at constant volume (approximately in a real engine). The temperature and pressure rise, and then in the power stroke, cd, the gas expands adiabatically. In the exhaust stroke, da, heat $|Q_L|$ is ejected to the environment (in a real engine, the gas leaves the engine and is replaced by a new mixture of air and fuel).

EXAMPLE 20–4 **The Otto cycle.** (*a*) For an ideal gas as working substance, show that the efficiency of an Otto cycle engine is

$$e = 1 - \left(\frac{V_a}{V_b}\right)^{1-\gamma}$$

where γ is the ratio of specific heats (Sections 19–8 and 19–9) and V_a/V_b is the *compression ratio*. (*b*) Calculate the efficiency for a compression ratio $V_a/V_b = 8.0$ assuming a diatomic gas like O_2 and N_2.

SOLUTION The heat exchanges take place at constant volume in the ideal Otto cycle, so from Eq. 19–8a:

$$|Q_H| = nC_V(T_c - T_b) \quad \text{and} \quad |Q_L| = nC_V(T_d - T_a).$$

Then from Eq. 20–1,

$$e = 1 - \frac{|Q_L|}{|Q_H|} = 1 - \left[\frac{T_d - T_a}{T_c - T_b}\right].$$

To get this in terms of the compression ratio, V_a/V_b, we use the result from Section 19–9, Eq. 19–13, $PV^\gamma = $ constant during the adiabatic processes ab and cd. Thus

$$P_a V_a^\gamma = P_b V_b^\gamma \quad \text{and} \quad P_c V_c^\gamma = P_d V_d^\gamma.$$

We use the ideal gas law, $P = nRT/V$, and substitute P into these two equations

$$T_a V_a^{\gamma-1} = T_b V_b^{\gamma-1} \quad \text{and} \quad T_c V_c^{\gamma-1} = T_d V_d^{\gamma-1}.$$

Then the efficiency (see above) is

$$e = 1 - \left[\frac{T_d - T_a}{T_c - T_b}\right] = 1 - \left[\frac{T_c(V_c/V_d)^{\gamma-1} - T_b(V_b/V_a)^{\gamma-1}}{T_c - T_b}\right].$$

But processes bc and da are at constant volume, so $V_c = V_b$ and $V_d = V_a$. Hence $V_c/V_d = V_b/V_a$ and

$$e = 1 - \left[\frac{(V_b/V_a)^{\gamma-1}(T_c - T_b)}{T_c - T_b}\right] = 1 - \left(\frac{V_b}{V_a}\right)^{\gamma-1} = 1 - \left(\frac{V_a}{V_b}\right)^{1-\gamma}.$$

(*b*) For diatomic molecules (Section 19–8), $\gamma = C_P/C_V = 1.4$ so

$$e = 1 - (8.0)^{1-\gamma} = 1 - (8.0)^{-0.4} = 0.56.$$

Real engines do not reach this high efficiency because they do not follow perfectly the Otto cycle, plus there is friction, turbulence, heat loss and incomplete combustion of the gases.

(a)	(b)	(c)

FIGURE 20–7 (a) An array of mirrors focuses sunlight on a boiler to produce steam at a solar energy installation. (b) A fossil-fuel steam plant. (c) Large cooling towers at an electric generating plant.

* Thermal Pollution

Much of the energy we utilize in everyday life—from motor vehicles to most of the electricity produced by power plants—makes use of a heat engine. Electricity produced by falling water at dams, by windmills, or by solar cells (Fig. 20–7a) does not involve a heat engine. But over 90 percent of the electric energy produced in the U.S. is done at fossil-fuel steam plants (Fig. 20–7b), and they make use of a heat engine (essentially steam engines). In electric power plants, the steam drives the turbines and generators whose output is electric energy. Even nuclear power plants use nuclear fuel to run a steam engine. The heat output $|Q_L|$ from every heat engine, from power plants to cars, is referred to as **thermal pollution** because this heat $|Q_L|$ must be absorbed by the environment—such as by water from rivers or lakes or via large cooling towers (Fig. 20–7c). This heat raises the temperature of the cooling water, altering the natural ecology of aquatic life (largely because warmer water holds less oxygen); or, in the case of air cooling towers, the output heat raises the temperature of the atmosphere which affects the weather. Air pollution—by which we mean the chemicals released in the burning of fossil fuels in cars, power plants, and industrial furnaces—gives rise to smog and other problems $\left(CO_2\right.$ buildup in the atmosphere, causing the greenhouse effect and global warming$\left.\right)$ that can be controlled to some extent and hopefully more so in the coming years. But thermal pollution is unavoidable. Engineers can try to design and build engines that are more efficient, but they cannot surpass the Carnot efficiency and must live with T_L being at best the ambient temperature of water or air. The second law of thermodynamics tells us the limit imposed by nature. What we can do, in the light of the second law of thermodynamics, is use less energy and conserve our fuel resources.

➥ **PHYSICS APPLIED**

*Heat engines
and thermal pollution*

20–4 | Refrigerators, Air Conditioners, and Heat Pumps

The operating principle of refrigerators, air conditioners, and heat pumps is just the reverse of a heat engine. Each operates to transfer heat out of a cool environment into a warm environment. As diagrammed in Fig. 20–8, by doing work $|W|$, heat is taken from a low-temperature region, T_L (inside a refrigerator, say), and a greater amount of heat is exhausted at a high temperature, T_H (the room). You can often feel this heated air blowing out beneath a refrigerator. The

FIGURE 20–8 Schematic diagram of energy transfers for a refrigerator or air conditioner.

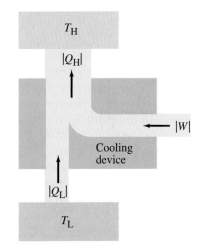

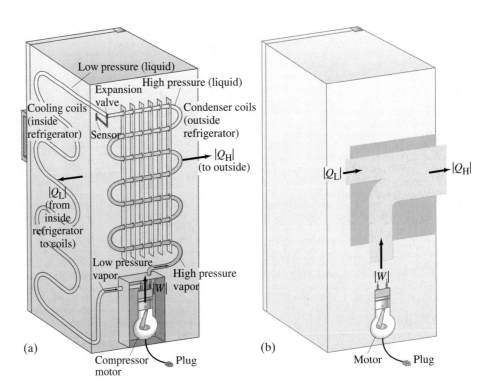

FIGURE 20–9 (a) Typical refrigerator system. The compressor motor forces a gas at high pressure through a heat exchanger (condenser) on the outside walls of the refrigerator where $|Q_H|$ is given off and the gas cools to become liquid. The liquid passes from a high-pressure region, via a valve, to low-pressure tubes on the inside walls of the refrigerator; the liquid evaporates at this lower pressure and thus absorbs heat ($|Q_L|$) from the inside of the refrigerator. The fluid returns to the compressor where the cycle begins again. (b) Schematic diagram, like Fig. 20–8.

work $|W|$ is usually done by a compressor motor which compresses a fluid, as illustrated in Fig. 20–9.

A perfect refrigerator—one in which no work is required to take heat from the low-temperature region to the high-temperature region—is not possible. This is **the Clausius statement of the second law of thermodynamics**, already mentioned in Section 20–1, which can be stated formally as

> **No device is possible whose sole effect is to transfer heat from one system at one temperature into a second system at a higher temperature.**

To make heat flow from a low-temperature body (or system) to one at a higher temperature, work must be done. Thus, *there can be no perfect refrigerator.*

The **coefficient of performance (CP)** of a refrigerator is defined as the heat $|Q_L|$ removed from the low-temperature area (inside a refrigerator) divided by the work $|W|$ done to remove the heat (Fig. 20–8 or 20–9b):

$$CP = \frac{|Q_L|}{|W|}. \qquad \left[\begin{array}{c}\text{Refrigerator and}\\ \text{air conditioner}\end{array}\right] \quad \textbf{(20–4a)}$$

This makes sense since the more heat, $|Q_L|$, that can be removed from the inside of the refrigerator for a given amount of work, the better (more efficient) the refrigerator. Energy is conserved, so from the first law of thermodynamics we can write (see Fig. 20–8 or 20–9b) $|Q_L| + |W| = |Q_H|$, or $|W| = |Q_H| - |Q_L|$. Then Eq. 20–4a becomes

$$CP = \frac{|Q_L|}{|W|} = \frac{|Q_L|}{|Q_H| - |Q_L|}. \qquad \textbf{(20–4b)}$$

For an ideal refrigerator (not a perfect one, which is impossible), the best one could do would be

$$CP_{\text{ideal}} = \frac{T_L}{T_H - T_L}, \qquad \textbf{(20–4c)}$$

as for an ideal (Carnot) engine (Eq. 20–3).

An air conditioner works very much like a refrigerator, although the actual construction details are different because an air conditioner takes heat $|Q_L|$ from

SECOND LAW OF THERMODYNAMICS
(Clausius statement)

*Coefficient of
performance (CP)*

CP for

refrigerator

and

air conditioner

➡ **P H Y S I C S   A P P L I E D**

Air conditioner

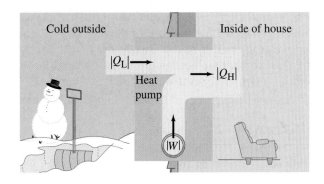

FIGURE 20–10 A heat pump "pumps" heat from the cold outside to the warm inside of a house.

inside a room or building at a low temperature, and deposits heat $|Q_H|$ outside to the environment at a higher temperature. Equations 20–4 also describe the coefficient of performance for an air conditioner.

Heat naturally flows from high temperature to low temperature. Refrigerators and air conditioners do work to accomplish the opposite: to make heat flow from cold to hot. We might say they "pump" heat from cold areas to hotter areas, against the natural tendency of heat to flow from hot to cold, just as water can be pumped uphill, against the natural tendency to flow downhill. The term **heat pump** is usually reserved for a device that can heat a house in winter by taking heat $|Q_L|$ from the outside at low temperature and delivering heat $|Q_H|$ to the warmer inside of the house, by doing work $|W|$; see Fig. 20–10. As in a refrigerator, there is an indoor and an outdoor heat exchanger (coils of the refrigerator) and a compressor. The operating principle is like that for a refrigerator or air conditioner; but the objective of a heat pump is to heat (deliver $|Q_H|$), rather than to cool (remove $|Q_L|$). Thus, the coefficient of performance of a heat pump is defined differently than for an air conditioner because it is the heat $|Q_H|$ delivered to the inside of the house that is important now:

$$\text{CP} = \frac{|Q_H|}{|W|}.$$

[heat pump] **(20–5)**

CP for heat pump

Most heat pumps can, however, be "turned around" and used as air conditioners in the summer.

EXAMPLE 20–5 **Heat pump.** A heat pump has a coefficient of performance of 3.0 and is rated to do work at 1500 W. (*a*) How much heat can it add to a room per second? (*b*) If the heat pump were turned around to act as an air conditioner in the summer, what would you expect its coefficient of performance to be, assuming all else stays the same?

SOLUTION (*a*) We use Eq. 20–5 for the heat pump and, since our device does 1500 J of work per second, it can pour heat into the room at a rate of

$$|Q_H| = \text{CP} \times |W| = 3.0 \times 1500\,\text{J} = 4500\,\text{J}$$

per second, or at a rate of 4500 W.

(*b*) If our device is turned around in summer, it can take heat $|Q_L|$ from inside the house and do 1500 J of work per second, and then dumps $|Q_H|$ = 4500 J per second to the hot outside. Energy is conserved, so $|Q_L| + |W| = |Q_H|$ (see Fig. 20–10, but reverse inside and outside of house), so

$$|Q_L| = |Q_H| - |W| = 4500\,\text{J} - 1500\,\text{J} = 3000\,\text{J}.$$

The coefficient of performance as an air conditioner would thus be (Eq. 20–4a)

$$\text{CP} = \frac{|Q_L|}{|W|} = \frac{3000\,\text{J}}{1500\,\text{J}} = 2.0.$$

A good heat pump can be a money saver and an energy saver. Compare, for example, our heat pump in this Example to, say, a 1500-watt electric heater. We plug the latter into the wall, it draws 1500 watts of electricity and delivers 1500 watts of heat to the room. Our heat pump when plugged into the wall also draws 1500 W of electricity (which is what we pay for), but it delivers 4500 W of heat!

20–5 | Entropy

We have seen several aspects of the second law of thermodynamics; but we have not yet arrived at a general statement of it. Both the Clausius and Kelvin-Planck statements deal with rather specific situations. Yet, as mentioned at the beginning of this chapter, there are a great many processes that simply are not observed in nature, even though they would not violate the first law of thermodynamics if they did occur. To cover all these processes, a more general statement of the second law of thermodynamics is needed. This general statement can be made in terms of a quantity, introduced by Clausius in the 1860s, called *entropy*, which we now discuss.

In our study of the Carnot cycle we found (Eq. 20–2) that $|Q_L|/|Q_H| = T_L/T_H$. We rewrite this as

$$\frac{|Q_H|}{T_H} = \frac{|Q_L|}{T_L}.$$

In this relation, both $|Q_H|$ and $|Q_L|$ are positive since they are absolute values. Let us now remove the absolute value signs and recall our original convention as used in the first law (see Section 19–6), that Q is positive when it represents a heat flow into the system (as Q_H) and negative for a heat flow out of the system (as Q_L). Then this relation becomes

$$\frac{Q_H}{T_H} + \frac{Q_L}{T_L} = 0. \qquad \text{[Carnot cycle]} \quad \textbf{(20–6)}$$

Now consider *any* reversible cycle, as represented by the smooth (oval-shaped) curve in Fig. 20–11. Any reversible cycle can be approximated as a series of Carnot cycles. Figure 20–11 shows only six—the isotherms (dashed lines) are connected by adiabatic paths for each—and the approximation becomes better and better if we increase the number of Carnot cycles. Equation 20–6 is valid for each of these cycles, so we can write

$$\Sigma \frac{Q}{T} = 0 \qquad \text{[Carnot cycles]} \quad \textbf{(20–7)}$$

for the sum of all these cycles. But note that the heat output Q_L of one cycle is approximately equal to the negative of the heat input, Q_H, of the cycle below it (actual equality in the limit of an infinite number of infinitely thin Carnot cycles). Hence the heat flows on the inner paths of all these Carnot cycles cancel out, so the net heat transferred, and the work done, is the same for the series of Carnot cycles as for the original cycle. Hence, in the limit of infinitely many Carnot cycles, Eq. 20–7 applies to any reversible cycle. In this case Eq. 20–7 becomes

$$\oint \frac{dQ}{T} = 0, \qquad \text{[reversible cycle]} \quad \textbf{(20–8)}$$

where dQ represents an infinitesimal heat flow.[†] The symbol $\oint$ means take the integral around a closed path; the integral can be started at any point on the path such

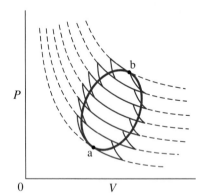

FIGURE 20–11 Any reversible cycle can be approximated as a series of Carnot cycles. (The dashed lines represent isotherms.)

[†] dQ is often written $đQ$: see footnote at end of Section 19–6.

as at a or b in Fig. 20–11, and proceed in either direction. Let us divide the cycle of Fig. 20–11 into two parts as indicated in Fig. 20–12. Then we rewrite Eq. 20–8 as

$$\int_{a}^{b}{}_{I}\frac{dQ}{T} + \int_{b}^{a}{}_{II}\frac{dQ}{T} = 0.$$

The first term is the integral from point a to point b along path I in Fig. 20–12, and the second term is the integral from b back to a along path II. (Path I plus path II is the whole cycle.) If one path is taken in reverse, say path II, dQ at each point becomes $-dQ$, since the path is reversible. We can therefore write

$$\int_{a}^{b}{}_{I}\frac{dQ}{T} = \int_{a}^{b}{}_{II}\frac{dQ}{T}. \qquad \text{[reversible paths]} \quad \textbf{(20–9)}$$

Since our cycle is arbitrary, Eq. 20–9 tells us that the integral of dQ/T between any two equilibrium states, a and b, does not depend on the path of the process. We therefore define a new quantity, called the **entropy**, S, by the relation

$$dS = \frac{dQ}{T}. \qquad \textbf{(20–10)}$$

From Eq. 20–8, we have that

$$\oint dS = 0. \qquad \text{[reversible cycle]} \quad \textbf{(20–11)}$$

The quantity ΔS, where

$$\Delta S = S_b - S_a = \int_{a}^{b} dS = \int_{a}^{b}\frac{dQ}{T}, \qquad \text{[reversible processes]} \quad \textbf{(20–12)}$$

is seen from Eq. 20–9 to be *independent of the path between the two points a and b*. This is an important result. It tells us that the difference in entropy, $S_b - S_a$, between two equilibrium states of a system does not depend on how you get from one state to the other. Thus *entropy is a state variable*—its value depends only on the state of the system, and not on the process or the past history of how it got there.[†] This is in clear distinction to Q and W which are *not* state variables; their values do depend on the processes undertaken.

20–6 Entropy and the Second Law of Thermodynamics

We have defined a new quantity, S, the entropy, which can be used to describe the state of the system, along with P, T, V, U, and n. But what does this rather abstract quantity have to do with the second law of thermodynamics? To answer this, let us take some examples in which we calculate the entropy changes during particular processes. But note first that Eq. 20–12 can be applied only to reversible processes. How then do we calculate $\Delta S = S_b - S_a$ for a real process that is irreversible? What we can do is this: we figure out some other *reversible* process that takes the system between the same two states, and calculate ΔS for this reversible process. This will equal ΔS for the irreversible process since ΔS depends only on the initial and final states of the system.

[†]Equation 20–12 says nothing about the absolute value of S; it only gives the change in S. This is much like potential energy (Chapter 8). However, one form of the so-called *third law of thermodynamics* (see also Section 20–10) states that as $T \to 0, S \to 0$.

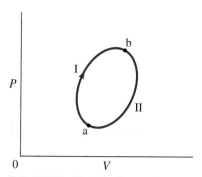

FIGURE 20–12 The integral, $\oint dS$, of the entropy for a reversible cycle is zero. Hence the difference in entropy between states a and b, $S_b - S_a = \int_{a}^{b} dS$, is the same for path I as for path II.

Entropy defined

Entropy is a state variable

How to determine ΔS (using a reversible process)

EXAMPLE 20–6 **Entropy change in melting.** A 1.00-kg piece of ice at 0°C melts very slowly to water at 0°C. Assume the ice is in contact with a heat reservoir whose temperature is only infinitesimally greater than 0°C. Determine the entropy change of (*a*) the ice cube and (*b*) the heat reservoir.

SOLUTION (*a*) The process is carried out at a constant temperature $T = 273$ K and is done reversibly, so we can use Eq. 20–12:

$$\Delta S_{\text{ice}} = \int \frac{dQ}{T} = \frac{1}{T} \int dQ = \frac{Q}{T}.$$

Since the heat needed to melt the ice is $Q = mL$, where the heat of fusion $L = 79.7$ kcal/kg $= 3.33 \times 10^5$ J/kg, we have

$$\Delta S_{\text{ice}} = \frac{mL}{T} = \frac{(1.00 \text{ kg})(79.7 \text{ kcal/kg})}{273 \text{ K}} = 0.292 \text{ kcal/K},$$

or 1220 J/K.
(*b*) Heat $Q = mL$ is *removed* from the heat reservoir, so (since $T = 273$ K and is constant)

$$\Delta S_{\text{res}} = -\frac{Q}{T} = -0.292 \text{ kcal/K}.$$

Note that the *total* entropy change, $\Delta S_{\text{ice}} + \Delta S_{\text{res}}$, is zero.

EXAMPLE 20–7 **ESTIMATE** **Entropy change when mixing water.** A sample of 50.0 kg of water at 20.0°C is mixed with 50.0 kg of water at 24.0°C. Estimate the change in entropy without using calculus.

SOLUTION The final temperature of the mixture will be 22.0°C, since we started with equal amounts of water. A quantity of heat,

$$Q = mc\,\Delta T = (50.0 \text{ kg})(1.00 \text{ kcal/kg}\cdot\text{K})(2.0 \text{ C}°) = 100 \text{ kcal},$$

flows out of the hot water as it cools down from 24°C to 22°C, and this heat flows into the cold water as it warms from 20°C to 22°C. The total change in entropy, ΔS, will be the sum of the changes in entropy of the hot water, ΔS_{H}, and that of the cold water, ΔS_{C}:

$$\Delta S = \Delta S_{\text{H}} + \Delta S_{\text{C}}.$$

We estimate entropy changes by writing $\Delta S = Q/T_{\text{av}}$, where T_{av} is an "average" temperature for each process, which ought to give a reasonable estimate since the temperature change is small. For the hot water we use an average temperature of 23°C (296 K), and for the cold water an average temperature of 21°C (294 K). Thus

$$\Delta S_{\text{H}} \approx -\frac{100 \text{ kcal}}{296 \text{ K}} = -0.338 \text{ kcal/K}$$

and

$$\Delta S_{\text{C}} \approx \frac{100 \text{ kcal}}{294 \text{ K}} = 0.340 \text{ kcal/K}.$$

Note that the entropy of the hot water (S_{H}) decreases since heat flows out of the hot water. But the entropy of the cold water (S_{C}) increases by a greater amount. The total change in entropy is

$$\Delta S = \Delta S_{\text{H}} + \Delta S_{\text{C}} \approx -0.338 \text{ kcal/K} + 0.340 \text{ kcal/K}$$

$$\approx +0.002 \text{ kcal/K}.$$

We see that although the entropy of one part of the system decreased, the entropy of the other part increased by a greater amount so that the net change in entropy of the whole system is positive.

We can now easily show in general that for an isolated system of two bodies, the flow of heat from the higher-temperature (T_H) body to the lower-temperature (T_L) body always results in an increase in the total entropy. The two bodies eventually come to some intermediate temperature, T_M. The heat lost by the hotter body $(Q_H = -Q$ where Q is positive$)$ is equal to the heat gained by the colder one $(Q_L = Q)$, so the total change in entropy is

$$\Delta S = \Delta S_H + \Delta S_L = -\frac{Q}{T_{HM}} + \frac{Q}{T_{LM}},$$

where T_{HM} is some intermediate temperature between T_H and T_M for the hot body as it cools from T_H to T_M, and T_{LM} is the counterpart for the cold body; since the temperature of the hot body is, at all times during the process, greater than that of the cold body, then $T_{HM} > T_{LM}$. Hence

$$\Delta S = Q\left(\frac{1}{T_{LM}} - \frac{1}{T_{HM}}\right) > 0.$$

One body decreases in entropy, while the other gains in entropy, but the *total* change is positive.

EXAMPLE 20–8 **Entropy changes in a free expansion.** Consider the *adiabatic free expansion* of n moles of an ideal gas from volume V_1 to volume V_2, where $V_2 > V_1$, as was discussed in Section 19–7, Fig. 19–13. Calculate the change in entropy (a) of the gas and (b) of the surrounding environment. (c) Evaluate ΔS for 1.00 mole, with $V_2 = 2.00\ V_1$.

SOLUTION As we saw in Section 19–7, the gas is initially in a closed container of volume V_1, and, with the opening of a valve, adiabatically it expands into a previously empty container; the total volume of the two containers is V_2. The whole apparatus is thermally insulated from the surroundings, so no heat flows into the gas, $Q = 0$. The gas does no work, $W = 0$, so there is no change in internal energy, $\Delta U = 0$, and the temperature of the initial and final states is the same, $T_2 = T_1 = T$.

(a) The process takes place very quickly, and so is irreversible. So we cannot apply Eq. 20–12 to this process. Instead we must think of a reversible process that will take the gas from volume V_1 to V_2 at the same temperature, and use Eq. 20–12 on this reversible process to get ΔS. A reversible isothermal process will do the trick; in such a process, the internal energy does not change, so (from the first law)

$$dQ = dW = P\,dV.$$

Then

$$\Delta S_{gas} = \int \frac{dQ}{T} = \frac{1}{T}\int_{V_1}^{V_2} P\,dV;$$

from the ideal gas law $P = nRT/V$, so

$$\Delta S_{gas} = \frac{nRT}{T}\int_{V_1}^{V_2}\frac{dV}{V} = nR\ln\frac{V_2}{V_1}.$$

Since $V_2 > V_1$, $\Delta S_{gas} > 0$.

(b) Since no heat is transferred to the surrounding environment, there is no change of the state of the environment due to this process. Hence $\Delta S_{env} = 0$. Note that the total change in entropy, $\Delta S_{gas} + \Delta S_{env}$ is greater than zero.

(c) Since $n = 1.00$ and $V_2 = 2.00\ V_1$, then $\Delta S_{gas} = R\ln 2.00 = 5.76\ \text{J/K}$.

EXAMPLE 20–9 **Heat conduction.** A red-hot 2.0-kg piece of iron at temperature $T_1 = 880$ K is thrown into a huge lake whose temperature is $T_2 = 280$ K. Assume the lake is so large that its temperature rise is insignificant. Determine the change in entropy (*a*) of the iron and (*b*) of the surrounding environment (the lake).

SOLUTION (*a*) The process is irreversible, but the same entropy change will occur for a reversible process, and we use the concept of specific heat, Eq. 19–2. We assume the specific heat of the iron is constant at $c = 0.11$ kcal/kg·K. Then $dQ = mc\, dT$ and in a quasistatic reversible process

$$\Delta S_{\text{iron}} = \int \frac{dQ}{T} = mc \int_{T_1}^{T_2} \frac{dT}{T} = mc \ln \frac{T_2}{T_1} = -mc \ln \frac{T_1}{T_2}.$$

Putting in numbers, we find

$$\Delta S_{\text{iron}} = -(2.0\,\text{kg})(0.11\,\text{kcal/kg}\cdot\text{K}) \ln \frac{880\,\text{K}}{280\,\text{K}} = -0.25\,\text{kcal/K}$$

or $\Delta S = -1100$ J/K.

(*b*) The initial and final temperatures of the lake are the same, $T = 280$ K. The lake receives from the iron an amount of heat

$$Q = mc(T_2 - T_1) = (2.0\,\text{kg})(0.11\,\text{kcal/kg}\cdot\text{K})(880\,\text{K} - 280\,\text{K}) = 130\,\text{kcal}.$$

Strictly speaking, this is an irreversible process (the lake heats up locally before equilibrium is reached), but is equivalent to a reversible isothermal transfer of heat $Q = 130$ kcal at $T = 280$ K. Hence

$$\Delta S_{\text{env}} = \frac{130\,\text{kcal}}{280\,\text{K}} = 0.46\,\text{kcal/K}$$

or 1900 J/K. Thus, although the entropy of the iron actually decreases, the *total* change in entropy of iron plus environment is positive: 0.46 kcal/K − 0.25 kcal/K = +0.21 kcal/K, or 800 J/K.

In each of these examples, the entropy of our system plus that of the environment (or surroundings) either stayed constant or increased. For any *reversible* process, such as that in Example 20–6, the total entropy change is zero. This can be seen in general as follows: any reversible process can be considered as a series of quasistatic isothermal transfers of heat ΔQ between a system and the environment, which differ in temperature only by an infinitesimal amount. Hence the change in entropy of either the system or environment is $\Delta Q/T$ and that of the other is $-\Delta Q/T$, so the total is

$$\Delta S = \Delta S_{\text{syst}} + \Delta S_{\text{env}} = 0. \qquad \text{[any reversible process]}$$

In Examples 20–7, 20–8, and 20–9, we found that the total entropy of system plus environment increases. Indeed, it has been found that for all real (irreversible) processes, the total entropy increases. No exceptions have been found. We can thus make the *general statement of the* **second law of thermodynamics** as follows:

The entropy of an isolated system never decreases. It either stays constant (reversible processes) or increases (irreversible processes).

Since all real processes are irreversible, we can equally well state the second law as:

SECOND LAW OF
THERMODYNAMICS
(general statement)

The total entropy of any system plus that of its environment increases as a result of any natural process:

$$\Delta S = \Delta S_{\text{syst}} + \Delta S_{\text{env}} > 0. \qquad\qquad \textbf{(20–13)}$$

Although the entropy of one part of the universe may decrease in any process (see the Examples above), the entropy of some other part of the universe always increases by a greater amount, so the total entropy always increases.

Now that we finally have a quantitative general statement of the second law of thermodynamics, we can see that it is an unusual law. It differs considerably from other laws of physics, which are typically equalities (such as $F = ma$) or conservation laws (such as for energy and momentum). The second law introduces a new quantity, the entropy, S, but does not tell us it is conserved. Quite the opposite. Entropy always increases in time.

Entropy increases in all real processes

The second law of thermodynamics summarizes, very succinctly, which processes are observed in nature, and which are not. Or, said another way, it tells us about the *direction* processes go. For the reverse of any of the processes in the last few Examples, the entropy would decrease; and we never observe them. For example, we never observe heat flowing spontaneously from a cold body to a hot body, the reverse of Example 20–9. Nor do we ever observe a gas spontaneously compressing itself into a smaller volume, the reverse of Example 20–8 (gases always expand to fill their containers). Nor do we see thermal energy transform into kinetic energy of a rock so the rock rises spontaneously from the ground. Any of these processes would be consistent with the first law of thermodynamics (conservation of energy). It is the second law of thermodynamics they are not consistent with, and this is why we need the second law. If you were to see a movie run backward, you would probably realize it immediately because you would see odd occurrences—such as rocks rising spontaneously from the ground, or air rushing in from the atmosphere to fill an empty balloon (the reverse of free expansion). When watching a movie or video, we are tipped off to a reversal of time by observing whether entropy is increasing or decreasing. Hence entropy has been called **time's arrow**, for it tells us in which direction time is going.

Time's arrow

Is the general statement of the second law of thermodynamics—"the principle of entropy increase"—consistent with the Clausius and Kelvin-Planck statements? Yes, it is. This is easy to see, for if a process occurred in which heat flowed spontaneously out of a low-temperature (T_L) reservoir (so its entropy decreased), and all of it flowed into a high-temperature (T_H) reservoir (which increased in entropy), in violation of the Clausius statement, then the total change in entropy $\Delta S = Q/T_H - Q/T_L$ would be less than zero, since $T_L < T_H$. Thus the principle of entropy increase implies the Clausius statement. Can you show the equivalence of the entropy principle to the Kelvin-Planck statement?

Equivalence of second law statements

20–7 | Order to Disorder

The concept of entropy, as we have discussed it so far, may seem rather abstract. But we can relate it to the more ordinary concepts of *order* and *disorder*. In fact, the entropy of a system can be considered a *measure of the disorder of the system*. Then the second law of thermodynamics can be stated simply as:

Natural processes tend to move toward a state of greater disorder.

SECOND LAW OF THERMODYNAMICS (general statement)

Exactly what we mean by disorder may not always be clear, so we now consider a few examples. Some of these will show us how this very general statement of the second law actually applies beyond what we usually consider as thermodynamics.

Let us look at some very simple processes. First, a jar containing separate layers of salt and pepper is more orderly than when the salt and pepper are all mixed up. Shaking a jar containing separate layers results in a mixture, and no amount of shaking brings the layers back again. The natural process is from a state of relative order (layers) to one of relative disorder (a mixture), not the reverse. That is, disorder increases. Second, a solid coffee cup is a more "orderly"

(a)

(b)

(c)

FIGURE 20–13 Have you ever observed this process, a broken cup spontaneously reassembling and rising up to a table?

object than the pieces of a broken cup. Cups break when they fall, but they do not spontaneously mend themselves (Fig. 20–13). Again, the normal course of events is an increase of disorder.

Now let us consider some processes for which we have actually calculated the entropy change, and see that an increase in entropy results in an increase in disorder (or vice versa). When ice melts to water at 0°C, the entropy of the water increases, as we saw in Example 20–6. Intuitively, we can think of solid water, ice, as being more ordered than the less orderly fluid state which can flow all over the place. This change from order to disorder can be seen more clearly from the molecular point of view: the orderly arrangement of water molecules in an ice crystal has changed to the disorderly and somewhat random motion of the molecules in the fluid state.

When a hot object is put in contact with a cold object, heat flows from the high temperature to the low until the two objects reach the same intermediate temperature. At the beginning of the process we can distinguish two classes of molecules—those with a high average kinetic energy and those with a low average kinetic energy. After the process, all the molecules are in one class with the same average kinetic energy, and we no longer have the more orderly arrangement of molecules in two classes. Order has gone to disorder. Furthermore, note that the separate hot and cold objects could serve as the hot- and cold-temperature regions of a heat engine, and thus could be used to obtain useful work. But once the two objects are put in contact and reach the same temperature, no work can be obtained. Disorder has increased, since a system that has the ability to perform work must surely be considered to have a higher order than a system no longer able to do work.

An interesting example of the increase in entropy relates to the theory of biological evolution and to the growth of organisms. Clearly, a human being is a highly ordered organism. The process of evolution from the early macromolecules and simple forms of life to *Homo sapiens* represents increasing order. So, too, the development of an individual from a single cell to a grown person is a process of increasing order. Do these processes violate the second law of thermodynamics? No, they do not. In the processes of evolution and growth, and even during the life of an individual, waste products are eliminated. These small molecules that remain as a result of metabolism are simple molecules without much order in comparison to the macromolecules of life such as DNA and proteins. Thus they represent relatively higher disorder or entropy. Indeed, the total entropy of the molecules cast aside by organisms during the processes of evolution and growth is greater than the decrease in entropy associated with the order of the growing individual or evolving species.

20–8 | Energy Availability; Heat Death

In the process of heat conduction from a hot body to a cold one, we have seen that entropy increases and that order goes to disorder. The separate hot and cold objects could serve as the high- and low-temperature regions for a heat engine and thus could be used to obtain useful work. But after the two objects are put in contact with each other and reach the same uniform temperature, no work can be obtained from them. With regard to being able to do useful work, order has gone to disorder in this process.

The same can be said about a falling rock that comes to rest upon striking the ground. Just before hitting the ground, all the kinetic energy of the rock could have been used to do useful work. But once the rock's mechanical kinetic energy becomes thermal energy, this is no longer possible.

Both these examples illustrate another important aspect of the second law of thermodynamics—*in any natural process, some energy becomes unavailable to do useful work*. In any process, no energy is ever lost (it is always conserved). Rather, it becomes less useful—it can do less useful work. As time goes on, **energy is degraded**, in a sense; it goes from more orderly forms (such as mechanical) eventually to the least orderly form, internal or thermal energy. Entropy is a factor here because the amount of energy that becomes unavailable to do work is proportional to the change in entropy during any process.[†]

Degradation of energy

A natural outcome of this is the prediction that as time goes on, the universe will approach a state of maximum disorder. Matter will become a uniform mixture, heat will have flowed from high-temperature regions to low-temperature regions until the whole universe is at one temperature. No work can then be done. All the energy of the universe will have become degraded to thermal energy. All change will cease. This, the so-called **heat death** of the universe, has been much discussed by philosophers. This final state seems an inevitable consequence of the second law of thermodynamics, although it lies very far in the future. Yet it is based on the assumption that the universe is finite, which cosmologists are not fully sure of. The answers are not all in yet; and that makes science interesting.

"Heat death"

*⎡20–9⎤ Statistical Interpretation of Entropy and the Second Law

The ideas of entropy and disorder are made clearer with the use of a statistical or probabilistic analysis of the molecular state of a system. This statistical approach, which was first applied toward the end of the nineteenth century by Ludwig Boltzmann (1844–1906), makes a distinction between the "macrostate" and the "microstate" of a system. The **microstate** of a system would be specified when the position and velocity of every particle (or molecule) is given. The **macrostate** of a system is specified by giving the macroscopic properties of the system—the temperature, pressure, number of moles, and so on. In reality, we can know only the macrostate of a system. There are generally far too many molecules in a system to be able to know the velocity and position of every one at a given moment. Nonetheless, it is important to recognize that a great many different microstates can correspond to the *same* macrostate.

Microstates and macrostates

Let us take a simple example. Suppose you repeatedly shake four coins in your hand and drop them on the table. Specifying the number of heads and the number of tails that appear on a given throw is the macrostate of this system. Specifying each coin as being a head or a tail is the microstate of the system. In the following table we see the number of microstates that correspond to each macrostate:

Macrostate	Possible microstates (H = heads, T = tails)	Number of microstates
4 heads	H H H H	1
3 heads, 1 tail	H H H T, H H T H, H T H H, T H H H	4
2 heads, 2 tails	H H T T, H T H T, T H H T, H T T H, T H T H, T T H H	6
1 head, 3 tails	T T T H, T T H T, T H T T, H T T T	4
4 tails	T T T T	1

[†] It can be shown that the amount of energy that becomes unavailable to do useful work is equal to $T_L \Delta S$, where T_L is the lowest available temperature and ΔS is the total increase in entropy during the process.

A basic principle behind the statistical approach is that *each microstate is equally probable*. Thus the number of microstates that give the same macrostate corresponds to the relative probability of that macrostate occurring. The macrostate of two heads and two tails is the most probable one in our case of tossing four coins. Out of the total of 16 possible microstates, six correspond to two heads and two tails, so the probability of throwing two heads and two tails is 6 out of 16, or 38 percent. The probability of throwing one head and three tails is 4 out of 16, or 25 percent. The probability of four heads is only 1 in 16, or 6 percent. Of course if you threw the coins 16 times, you might not find that two heads and two tails appears exactly 6 times, or four tails exactly once. These are only probabilities or averages. But if you made 1600 throws, very nearly 38 percent of them would be two heads and two tails. The greater the number of tries, the closer are the percentages to the calculated probabilities.

If we consider tossing more coins, say 100, the relative probability of throwing all heads (or all tails) is greatly reduced. There is only one microstate corresponding to all heads. For 99 heads and 1 tail, there are 100 microstates since each of the coins could be the one tail. The relative probabilities for other macrostates are given in Table 20–1. There are a total of about 10^{30} microstates possible.[†] Thus the relative probability of finding all heads is 1 in 10^{30}, an incredibly unlikely event! The probability of obtaining 50 heads and 50 tails (see Table 20–1) is $1.0 \times 10^{29}/10^{30} = 0.10$ or 10 percent. The probability of obtaining between 45 and 55 heads is 90 percent.

Thus we see that as the number of coins increases, the probability of obtaining an orderly arrangement (all heads or all tails) becomes extremely unlikely. The least orderly arrangement (half heads, half tails) is the most probable; and the probability of being within a certain percentage (say 5 percent) of the most probable arrangement greatly increases as the number of coins increases. These same ideas can be applied to the molecules of a system. For example the most probable state of a gas (say the air in a room) is one in which the molecules take up the whole space and move about randomly; this corresponds to the Maxwellian distribution, Fig. 20–14a (see Section 18–2). On the other hand, the very orderly arrangement of all the molecules located in one corner of the room and all moving with the same velocity (Fig. 20–14b) is extremely unlikely.

From these examples, it is clear that probability is directly related to disorder and hence to entropy. That is, the most probable state is the one with greatest entropy or greatest disorder and randomness. Boltzmann showed that, consistent with Clausius's definition $(dS = dQ/T)$ that the entropy of a system in a given (macro) state can be written

$$S = k \ln \mathcal{W}, \tag{20-14}$$

where k is Boltzmann's constant $(k = R/N_A = 1.38 \times 10^{-23} \, \text{J/K})$ and $\mathcal{W}$ is the number of microstates corresponding to the given macrostate. That is, $\mathcal{W}$ is

[†]Each coin has two possibilities, heads or tails. Then the possible number of microstates is $2 \times 2 \times 2 \times \ldots = 2^{100} = 1.27 \times 10^{30}$ (using a calculator or logarithms).

TABLE 20–1 Probabilities of various macrostates for 100 coin tosses

Macrostate		Number of microstates
Heads	**Tails**	$\mathcal{W}$
100	0	1
99	1	1.0×10^2
90	10	1.7×10^{13}
80	20	5.4×10^{20}
60	40	1.4×10^{28}
55	45	6.1×10^{28}
50	50	1.0×10^{29}
45	55	6.1×10^{28}
40	60	1.4×10^{28}
20	80	5.4×10^{20}
10	90	1.7×10^{13}
1	99	1.0×10^2
0	100	1

FIGURE 20–14 (a) Most probable distribution of molecular speeds in a gas (Maxwellian, or random); (b) orderly, but highly unlikely, distribution of speeds in which all molecules have nearly the same speed.

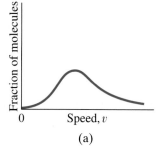

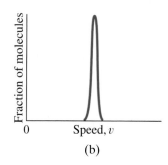

proportional to the probability of occurrence of that state. $\mathcal{W}$ is called the **thermodynamic probability**, or, sometimes, the **disorder parameter**.

| EXAMPLE 20–10 | Free expansion—statistical determination of entropy.

Use Eq. 20–14 to determine the change in entropy for the adiabatic *free expansion* of a gas, a calculation we did macroscopically in Example 20–8. Assume $\mathcal{W}$, the number of microstates for each macrostate, is the number of possible positions.

SOLUTION Assume the number of moles is $n = 1$ and then the number of molecules is $N = nN_A = N_A$, and assume the volume doubles, just as in Example 20–8. Because the volume doubles, the number of possible positions for each molecule doubles. Since there are N_A molecules, the total number of microstates increases by a factor of 2^{N_A} when the volume doubles. That is

$$\frac{\mathcal{W}_2}{\mathcal{W}_1} = 2^{N_A}.$$

The change in entropy is, from Eq. 20–14,

$$\Delta S = S_2 - S_1 = k\left(\ln \mathcal{W}_2 - \ln \mathcal{W}_1\right) = k \ln \frac{\mathcal{W}_2}{\mathcal{W}_1} = k \ln 2^{N_A} = kN_A \ln 2 = R \ln 2$$

which is the same result we got in Example 20–8.

In terms of probability, the second law of thermodynamics—which tells us that entropy increases in any process—reduces to the statement that those processes occur which are most probable. The second law thus becomes a trivial statement. However there is an additional element now. The second law in terms of probability does not *forbid* a decrease in entropy. Rather, it says the probability is extremely low. It is not impossible that salt and pepper should separate spontaneously into layers, or that a broken tea cup should mend itself. It is even possible that a lake should freeze over on a hot summer day (that is, heat flow out of the cold lake into the warmer surroundings). But the probability for such events occurring is extremely small. In our coin examples above, we saw that increasing the number of coins from 4 to 100 reduced drastically the probability of large deviations from the average, or most probable, arrangement. In ordinary systems we are dealing with incredibly large numbers of molecules: in 1 mol alone there are 6×10^{23} molecules. Hence the probability of deviation far from the average is incredibly tiny. For example, it has been calculated that the probability that a stone resting on the ground should transform 1 cal of thermal energy into mechanical energy and rise up into the air is much less likely than the probability that a group of monkeys typing randomly would by chance produce the complete works of Shakespeare.

*20–10 Thermodynamic Temperature Scale; Absolute Zero, and the Third Law of Thermodynamics

In Section 20–3 we discussed the Carnot engine and other (ideal) reversible engines. We saw that the efficiency of any reversible engine operating between two heat reservoirs depends only on the temperatures of these two reservoirs; and the efficiency does not depend on the working substance—it could be helium, water, or something else, and the efficiency would be the same. The efficiency is given by

$$e = 1 - \frac{|Q_L|}{|Q_H|},$$

where $|Q_H|$ is the heat absorbed from the high-temperature reservoir and $|Q_L|$ is the heat exhausted to the low-temperature reservoir. Since $|Q_L/Q_H|$ is the same for any reversible engine operating between the same two temperatures, Kelvin

suggested using this fact to define an absolute temperature scale. That is, the ratio of the temperatures of the two reservoirs, T_H and T_L, is defined as the ratio of the heats, $|Q_H|$ and $|Q_L|$, exchanged with them by a Carnot or other reversible engine:

$$\frac{T_H}{T_L} = \frac{|Q_H|}{|Q_L|}.$$ **(20–15)**

This is the basis for the **Kelvin** or **thermodynamic temperature scale**.

In Section 20–3, we saw that this same relation, $T_H/T_L = |Q_H/Q_L|$, holds for a Carnot engine (Eq. 20–2) when the temperatures are based on the ideal gas temperature scale (Section 17–10) which we have been using up to now. Indeed, to complete the definition of the thermodynamic scale, we assign the value $T_{tp} = 273.16$ K to the triple point of water so that

$$T = (273.16 \text{ K})\left(\frac{|Q|}{|Q_{tp}|}\right),$$

where $|Q|$ and $|Q_{tp}|$ are the heats exchanged by a Carnot engine with reservoirs at temperatures T and T_{tp}. This corresponds precisely to the definition of the ideal gas scale. Thus, the thermodynamic scale is identical to the ideal gas scale over the range of validity of the latter (below about 1 K, no substance is a gas, so the ideal gas scale cannot be used). The thermodynamic scale is now considered the standard scale since it can be used over the entire range of possible temperatures, and it is also independent of the substance used. From a practical point of view, it is especially useful at very low temperatures.

Very low temperatures are difficult to obtain experimentally. In fact, it is found experimentally that the closer the temperature is to absolute zero, the more difficult it is to reduce the temperature further. And it is generally accepted that it is not possible to actually reach absolute zero in any finite number of processes. This last statement is one way to state[†] the **third law of thermodynamics**. Since the maximum efficiency that any heat engine can have is the Carnot efficiency

Third law of thermodynamics

$$e = 1 - \frac{T_L}{T_H},$$

and since T_L can never be zero, we see that a 100-percent efficient heat engine is not possible.

[†] See also the statement in the footnote on page 529.

PROBLEM SOLVING	**Thermodynamics**

1. Define the system you are dealing with; be careful to distinguish the system under study from its surroundings.

2. Be careful of signs associated with work and heat. In the first law work done *by* the system is positive; work done *on* the system is negative. Heat added to the system is positive, but heat removed from it is negative. With heat engines, we usually consider heat and work as positive and write energy conservation equations with + and − signs taking into account directions.

3. Watch the units used for work and heat; work is most often expressed in joules, and heat in calories or kilocalories. Be consistent: choose only one unit for use throughout a given problem.

4. Temperatures must generally be expressed in kelvins; temperature **differences** may be expressed in C° or K.

5. Efficiency (or coefficient of performance) is a ratio of two energy transfers: useful output divided by required input. Efficiency (but *not* coefficient of performance) is always less than 1 in value, and hence is often stated as a percentage.

6. The entropy of a system increases when heat is added to the system, and decreases when heat is removed. Because entropy is a state variable, the change in entropy ΔS for an irreversible process can be determined by calculating ΔS for a reversible process between the same two states.

Summary

A **heat engine** is a device for changing thermal energy, by means of heat flow, into useful work.

The **efficiency** of a heat engine is defined as the ratio of the work W done by the engine to the heat input $|Q_H|$. Because of conservation of energy, the work output equals $|Q_H| - |Q_L|$, where $|Q_L|$ is the heat exhausted to the environment; hence the efficiency

$$e = \frac{|W|}{|Q_H|} = 1 - \frac{|Q_L|}{|Q_H|}.$$

Carnot's (idealized) engine consists of two isothermal and two adiabatic processes in a reversible cycle. For a **Carnot engine**, or any reversible engine operating between two temperatures, T_H and T_L, the efficiency is

$$e_{ideal} = 1 - \frac{T_L}{T_H}.$$

Irreversible (real) engines always have an efficiency less than this.

All heat engines give rise to thermal pollution because they exhaust heat to the environment.

The operation of **refrigerators** and **air conditioners** is the reverse of that of a heat engine: work is done to extract heat from a cool region and exhaust it to a region at a higher temperature. A **heat pump** does work to bring heat from a cold exterior into a warmer interior.

The **second law of thermodynamics** can be stated in several equivalent ways:

(a) heat flows spontaneously from a hot object to a cold one, but not the reverse;

(b) there can be no 100 percent efficient heat engine—that is, one that can change a given amount of heat completely into work;

(c) natural processes tend to move toward a state of greater disorder or greater **entropy**.

Statement (c) is the most general statement of the second law of thermodynamics, and can be restated as: the total entropy, S, of any system plus that of its environment increases as a result of any natural process:

$$\Delta S > 0.$$

Entropy, which is a state variable, is a quantitative measure of the disorder of a system. The change in entropy of a system during a reversible process is given by $\Delta S = \int dQ/T$.

The second law of thermodynamics tells us in which direction processes tend to proceed; hence entropy is called "time's arrow."

As time goes on, energy is degraded to less useful forms—that is, it is less available to do useful work.

Questions

1. Is it possible to cool down a room on a hot summer day by leaving the refrigerator door open? Explain.

2. Can mechanical energy ever be transformed completely into heat or internal energy? Can the reverse happen? In each case, if your answer is no, explain why not; if yes, give examples.

3. Would a definition of heat engine efficiency as $e = |W|/|Q_L|$ be a useful one? Explain.

4. What plays the role of high-temperature and low-temperature reservoirs in (a) an internal combustion engine, (b) a steam engine? Are they, strictly speaking, heat reservoirs?

5. Which will give the greater improvement in the efficiency of a Carnot engine, a 10°C increase in the high-temperature reservoir, or a 10°C decrease in the low-temperature reservoir?

6. The oceans contain a tremendous amount of thermal energy. Why, in general, is it not possible to put this energy to useful work?

7. Discuss the factors that keep real engines from reaching Carnot efficiency.

8. The expansion valve in a refrigeration system, Fig. 20–9, is crucial for cooling the fluid. Explain how the cooling occurs.

9. Describe a process in nature that is nearly reversible.

10. (a) Describe how heat could be added to a system reversibly. (b) Could you use a stove burner to add heat to a system reversibly? Explain.

11. Powdered milk is very slowly (quasistatically) added to water while being stirred. Is this a reversible process?

12. Two identical systems are taken from state a to state b by two different *irreversible* processes. Will the change in entropy for the system be the same for each process? For the environment? Answer carefully and completely.

13. It can be said that the *total change in entropy during a process is a measure of the irreversibility of the process.* Discuss why this is valid, starting with the fact that $\Delta S = 0$ for a reversible process.

14. Use arguments, other than the principle of entropy increase, to show that for an adiabatic process, $\Delta S = 0$ if it is done reversibly and $\Delta S > 0$ if done irreversibly.

15. A gas is allowed to expand (a) adiabatically and (b) isothermally. In each process, does the entropy increase, decrease, or stay the same?

16. Entropy is often called "time's arrow" because it tells us in which direction natural processes occur. If a movie film were run backward, name some processes that you might see that would tell you that time was "running backward."

17. Give three examples, other than those mentioned in this chapter, of naturally occurring processes in which order goes to disorder. Discuss the observability of the reverse process.

18. Which do you think has the greater entropy, 1 kg of solid iron or 1 kg of liquid iron? Why?

19. What happens if you remove the lid of a bottle containing chlorine gas? Does the reverse process ever happen? Why or why not?

20. Think up several processes (other than those already mentioned in the text) that would obey the first law of thermodynamics, but, if they actually occurred, would violate the second law.

21. Describe how a free expansion, for which the entropy increases, can be considered as a process in which order goes to disorder. [*Hint*: consider a stack of papers dropped into a large box versus dropped all over the floor.]

22. Suppose you collect a lot of papers strewn all over the floor and put them in a neat stack; does this violate the second law of thermodynamics? Explain.

23. The first law of thermodynamics is sometimes whimsically stated as, "You can't get something for nothing," and the second law as, "You can't even break even." Explain how these statements could be equivalent to the formal statements.

24. Give three examples of naturally occurring processes that illustrate the degradation of usable energy into internal energy.

25. Living organisms convert relatively simple food molecules into a complex structure as they grow. Is this a violation of the second law of thermodynamics? Explain.

Problems

Section 20–2

1. (I) A heat engine exhausts 8500 J of heat while performing 2700 J of useful work. What is the efficiency of this engine?

2. (I) A heat engine does 8200 J of work in each cycle while absorbing 18.0 kcal of heat from a high-temperature reservoir. What is the efficiency of this engine?

3. (I) A typical power plant puts out 500 MW of electric power. Estimate the heat discharged per second, assuming that the plant has an efficiency of 38 percent.

4. (II) A four-cylinder gasoline engine has an efficiency of 0.25 and delivers 180 J of work per cycle per cylinder. The engine fires at 25 cycles per second. (*a*) Determine the work done per second. (*b*) What is the total heat input per second from the gasoline? (*c*) If the energy content of gasoline is 130 MJ per gallon, how long does one gallon last?

5. (II) The burning of gasoline in a car releases about 3.0×10^4 kcal/gal. If a car averages 38 km/gal when driving 90 km/h, which requires 20 hp, what is the efficiency of the engine under those conditions?

6. (II) Figure 20–15 is a *PV* diagram for a reversible heat engine in which 1.0 mol of argon, a nearly ideal monatomic gas, is initially at STP (point a). Points b and c are on an isotherm at $T = 423$ K. Process ab is at constant volume, process ac at constant pressure. (*a*) Is the path of the cycle carried out clockwise or counterclockwise? (*b*) What is the efficiency of this engine?

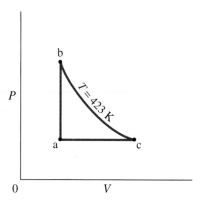

FIGURE 20–15 Problem 6.

7. (II) A 38-percent-efficient power plant puts out 810 MW (megawatts) of power (electrical). Cooling towers are used to take away the exhaust heat. If the air temperature is allowed to rise 7.5 C°, what volume of air (km^3) is heated per day? Will the local climate be heated significantly? If the heated air were to form a layer 200 m thick, how large an area would it cover for 24 h of operation? (The heat capacity of air is about 7.0 cal/mol·C° at constant pressure.)

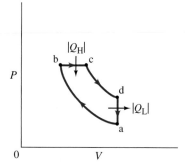

FIGURE 20–16
Problem 8.

8. (III) The operation of a *Diesel engine* can be idealized by the cycle shown in Fig. 20–16. Air is drawn into the cylinder during the intake stroke (not part of the idealized cycle). The air is compressed adiabatically, path ab. At point b diesel oil is injected into the cylinder which immediately burns since the temperature is very high. Combustion is slow, and during the first part of the power stroke, the gas expands at (nearly) constant pressure, path bc. After burning, the rest of the power stroke is adiabatic, path cd. Path da corresponds to the exhaust stroke. (*a*) Show that, for a quasistatic reversible engine undergoing this cycle using an ideal gas, the ideal efficiency is

$$e = 1 - \frac{(V_a/V_c)^{-\gamma} - (V_a/V_b)^{-\gamma}}{\gamma[(V_a/V_c)^{-1} - (V_a/V_b)^{-1}]},$$

where V_a/V_b is the "compression ratio", V_a/V_c is the "expansion ratio", and γ is defined by Eq. 19–12. (*b*) If $V_a/V_b = 15$ and $V_a/V_c = 5.0$, calculate the efficiency assuming the gas is diatomic (like N_2 and O_2) and ideal.

Section 20–3

9. (I) What is the maximum efficiency of a heat engine whose operating temperatures are 530°C and 305°C?

10. (I) The exhaust temperature of a heat engine is 220°C. What must be the high temperature if the Carnot efficiency is to be 36 percent?

11. (II) (a) Show that the work done by a Carnot engine is equal to the area enclosed by the Carnot cycle on a PV diagram, Fig. 20–5. (See Section 19–7.) (b) Generalize this to any reversible cycle.

12. (II) A heat engine exhausts its heat at 360°C and has a Carnot efficiency of 35 percent. What exhaust temperature would enable it to achieve a Carnot efficiency of 50 percent?

13. (II) A nuclear power plant operates at 75 percent of its maximum theoretical (Carnot) efficiency between temperatures of 660°C and 360°C. If the plant produces electric energy at the rate of 1.1 GW, how much exhaust heat is discharged per hour?

14. (II) An engine that operates at half its theoretical (Carnot) efficiency operates between 525°C and 280°C while producing work at the rate of 850 kW. How much heat is wasted per hour?

15. (II) Assume that a hiker needs 4000 kcal of energy to supply a day's worth of metabolism. Estimate the maximum height the person can climb in one day, using only this amount of energy. As a rough prediction, treat the person as an isolated heat engine, operating between the internal temperature of 37°C (98.6°F) and the ambient air temperature of 20°C.

16. (II) A Carnot engine performs work at the rate of 570 kW while using 1350 kcal of heat per second. If the temperature of the heat source is 580°C, at what temperature is the waste heat exhausted?

17. (II) A heat engine utilizes a heat source at 580°C and has a Carnot efficiency of 29 percent. To increase the efficiency to 35 percent, what must be the temperature of the heat source?

18. (II) At a steam power plant, steam engines work in pairs, the heat output of the first one being the approximate heat input of the second. The operating temperatures of the first are 680°C and 430°C, and of the second 415°C and 280°C. If the heat of combustion of coal is 2.8×10^7 J/kg, at what rate must coal be burned if the plant is to put out 900 MW of power? Assume the efficiency of the engines is 65 percent of the ideal (Carnot) efficiency.

19. (II) Water is used to cool the power plant in Problem 18. If the water temperature is allowed to increase by no more than 5.5 C°, estimate how much water must pass through the plant per hour.

20. (II) Show that if two different adiabatic paths intersected at a single point on a PV diagram, they could be connected by an isotherm to form a cycle, and that an engine run on this cycle would violate the second law of thermodynamics. What do you conclude about the crossing of adiabatic lines?

21. (III) A Carnot cycle, shown in Fig. 20–5, has the following conditions: $V_a = 6.0$ L, $V_b = 15.0$ L, $T_H = 470°C$, and $T_L = 290°C$. The gas used in the cycle is 0.50 mole of a diatomic gas, $\gamma = 1.4$. Calculate (a) the pressures at a and b; (b) the volumes at c and d. (c) What is the work done along process ab? (d) What is the heat lost along process cd? (e) Calculate the net work done for the whole cycle. (f) What is the efficiency of the cycle, using the definition $e = |W|/|Q_H|$? Show that this is the same as given by Eq. 20–3.

22. (III) One mole of monatomic gas undergoes a Carnot cycle with $T_H = 350°C$ and $T_L = 210°C$. The initial pressure is 10 atm. During the isothermal expansion, the volume doubles. (a) Find the values of the pressure and volume at the points a, b, c, and d (see Fig. 20–5). (b) Determine $Q, W,$ and ΔU for each segment of the cycle. (c) Calculate the efficiency of the cycle.

Section 20–4

23. (I) The low temperature of a freezer cooling coil is −15°C and the discharge temperature is 30°C. What is the maximum theoretical coefficient of performance?

24. (I) If an ideal refrigerator keeps its contents at −15°C when the house temperature is 22°C, what is its coefficient of performance?

25. (I) A restaurant refrigerator has a coefficient of performance of 5.0. If the temperature in the kitchen outside the refrigerator is 29°C, what is the lowest temperature that could be obtained inside the refrigerator if it were ideal?

26. (II) A heat pump is used to keep a house warm at 22°C. How much work is required of the pump to deliver 2800 J of heat into the house if the outdoor temperature is (a) 0°C, (b) −15°C? Assume ideal (Carnot) behavior.

27. (II) An ideal engine has an efficiency of 35 percent. If it were run backward as a heat pump, what would be its coefficient of performance?

28. (II) The engine of Example 20–2 is run in reverse. How long would it take to freeze a tray of a dozen 40-g compartments of liquid water at room temperature (20°C) into a dozen ice cubes at the freezing point, assuming that it takes 450 W of input electric power to run it? Assume ideal (Carnot) behavior.

29. (II) A "Carnot" refrigerator (reverse of a Carnot engine) absorbs heat from the freezer compartment at a temperature of −17°C and exhausts it into the room at 25°C. (a) How much work must be done by the refrigerator to change 0.50 kg of water at 25°C into ice at −17°C? (b) If the compressor output is 200 W, what minimum time is needed to take 0.50 kg of 25°C water and freeze it at 0°C?

30. (II) (a) Given that the coefficient of performance of a refrigerator is defined (Eq. 20–4a) as

$$CP = \frac{|Q_L|}{|W|},$$

show that for an ideal (Carnot) refrigerator,

$$CP_{ideal} = \frac{T_L}{T_H - T_L}.$$

(b) Write the CP in terms of the efficiency e of the reversible heat engine obtained by running the refrigerator backward. (c) What is the coefficient of performance for an ideal refrigerator that maintains a freezer compartment at −16°C when the condenser's temperature is 22°C?

31. (II) What volume of water at 0°C can a freezer make into ice cubes in one hour, if the coefficient of performance of the cooling unit is 7.0 and the power input is 1.0 kilowatt?

32. (II) A central heat pump operating as an air conditioner draws 36,000 Btu per hour from a building and operates between the temperatures of 24°C and 38°C. (a) If its coefficient of performance is 24% of that of a Carnot air conditioner, what is the effective coefficient of performance? (b) What is the power required of the compressor motor? (c) What is the power in terms of hp?

Sections 20–5 and 20–6

33. (I) A 10.0-kg box having an initial speed of 3.0 m/s slides along a rough table and comes to rest. Estimate the total change in entropy of the universe. Assume all objects are at room temperature (293 K).

34. (I) What is the change in entropy of 1.00 m^3 of water at 0°C when it is frozen to ice at 0°C?

35. (II) If the water in Problem 34 were frozen by being in contact with a great deal of ice at −10°C, what would be the total change in entropy of the process?

36. (II) If 1.00 kg of water at 100°C is changed by a reversible process to steam at 100°C, determine the change in entropy of (a) the water, (b) the surroundings, and (c) the universe as a whole. (d) How would your answers differ if the process were irreversible?

37. (II) An aluminum rod conducts 7.50 cal/s from a heat source maintained at 240°C to a large body of water at 27°C. Calculate the rate at which entropy increases in this process.

38. (II) A 3.8-kg piece of aluminum at 30°C is placed in 1.0 kg of water in a Styrofoam container at room temperature (20°C). Estimate the net change in entropy of the system.

39. (II) What is the total change in entropy when 2.5 kg of water at 0°C is frozen to ice at 0°C by being in contact with 450 kg of ice at −15°C?

40. (II) When 2.0 kg of water at 20°C is mixed with 3.0 kg of water at 80°C in a well-insulated container, what is the change in entropy of the system?

41. (II) Show that the principle of entropy increase is equivalent to the Kelvin-Planck statement of the second law of thermodynamics.

42. (II) The temperature of 2.0 mol of an ideal diatomic gas goes from 25°C to 45°C at a constant volume. What is the change in entropy?

43. (II) (a) Calculate the change in entropy of 1.00 kg of water when it is heated from 0°C to 100°C. (b) Does the entropy of the surroundings change? If so, by how much?

44. (II) A 150-g insulated aluminum cup at 20°C is filled with 240 g of water at 100°C. (a) What is the final temperature of the mixture? (b) What is the total change in entropy as a result of the mixing process?

45. (II) Two samples of an ideal gas are initially at the same temperature and pressure; they are each compressed reversibly from a volume V to volume $V/2$, one isothermally, the other adiabatically. (a) In which sample is the final pressure greater? (b) Determine the change in entropy of the gas for each process. (c) What is the entropy change of the environment for each process?

46. (II) One mole of nitrogen (N_2) gas and one mole of argon (Ar) gas are in separate, equal-sized, insulated containers at the same temperature. The containers are then connected and the gases (assumed ideal) allowed to mix. (a) What is the change in entropy of the system and (b) of the environment? (c) Repeat part (a) but assume one container is twice as large as the other.

47. (II) (a) Why would you expect the total entropy change in a Carnot cycle to be zero? (b) Do a calculation to show that it is zero.

48. (III) The specific heat per mole of potassium at low temperatures is given by $C_V = aT + bT^3$, where $a = 2.08$ mJ/mol·K^2 and $b = 2.57$ mJ/mol·K^4. Calculate the entropy change of 0.25 mol of potassium when its temperature is lowered from 3.0 K to 1.0 K.

Section 20–8

49. (III) A general theorem states that the amount of energy that becomes unavailable to do useful work in any process is equal to $T_L \Delta S$, where T_L is the lowest temperature available and ΔS is the total change in entropy during the process. Show that this is valid in the specific cases of (a) a falling rock that comes to rest when it hits the ground; (b) the free adiabatic expansion of an ideal gas; and (c) the conduction of heat, Q, from a high-temperature (T_H) reservoir to a low-temperature (T_L) reservoir. [*Hint*: In part (c) compare to a Carnot engine.]

50. (III) Determine the work available in a 5.0 kg block of copper at 420 K if the surroundings are at 290 K. Use results of Problem 49.

* Section 20–9

* 51. (II) Suppose you repeatedly shake six coins in your hand and drop them on the table. Construct a table showing the number of microstates that correspond to each macrostate. What is the probability of obtaining (a) three heads and three tails and (b) six heads?

* 52. (II) Calculate the relative probabilities, when you throw two dice, of obtaining (a) a 7, (b) an 11, (c) a 5.

* 53. (II) Rank the following five-card hands in order of increasing probability: (a) four aces and a king; (b) six of hearts, eight of diamonds, queen of clubs, three of hearts, jack of spades; (c) two jacks, two queens, and an ace; and (d) any hand having no two equal-value cards. Explain your ranking using microstates and macrostates.

* 54. (II) (a) Suppose you have four coins, all with tails up. You now rearrange them so two heads and two tails are up. What was the change in entropy of the coins? (b) Suppose your system is the 100 coins of Table 20–1; what is the change in entropy of the coins if they are mixed randomly initially, 50 heads and 50 tails, and you arrange them so all 100 are heads? (c) Compare these entropy changes to ordinary thermodynamic entropy changes, such as Examples 20–6, 7, 8 and 9.

General Problems

55. It is not necessary that a heat engine's hot environment be hotter than ambient temperature. Liquid nitrogen (90 K) is about as cheap as bottled water. What would be the efficiency of an engine that made use of heat transferred from air at room temperature (293 K) to the liquid nitrogen "fuel"?

56. It has been suggested that a heat engine could be developed that made use of the temperature difference between water at the surface of the ocean and that several hundred meters deep. In the tropics, the temperatures may be 27°C and 4°C, respectively. What is the maximum efficiency such an engine could have? Why might such an engine be feasible in spite of the low efficiency? Can you imagine any adverse environmental effects that might occur?

57. Two 1100-kg cars are traveling 95 km/h in opposite directions when they collide and are brought to rest. Estimate the change in entropy of the universe as a result of this collision. Assume $T = 20°C$.

58. A 120-g insulated aluminum cup at 15°C is filled with 210 g of water at 50°C. After a few minutes, equilibrium is reached. Determine (a) the final temperature, and (b) the total change in entropy.

59. (a) What is the coefficient of performance of an ideal heat pump that extracts heat from 6°C air outside and deposits heat inside your house at 24°C? (b) If this heat pump operates on 1000 W of electrical energy, what is the maximum heat it can deliver into your house each hour?

60. An inventor claims to have designed and built an engine that produces 1.50 MW (megawatts) of usable work while taking in 3.00 MW of thermal energy at 425 K, and rejecting 1.50 MW of thermal energy at 215 K. Is there anything fishy about his claim? Explain.

61. Suppose a power plant delivers energy at 900 MW using steam turbines. The steam goes into the turbines superheated at 600 K and deposits its unused heat in river water at 285 K. Assume that the turbine operates as an ideal Carnot engine. (a) If the river flow rate is 37 m³/s, calculate the average temperature increase of the river water downstream from the power plant. (b) What is the entropy increase per kilogram of the downstream river water in J/kg·K?

62. A 100-hp car engine operates at about 15 percent efficiency. Assume the engine's water temperature of 85°C is its cold-temperature (exhaust) reservoir and 500°C is its thermal "intake" temperature (the temperature of the exploding gas/air mixture). (a) Calculate its efficiency relative to its maximum possible (Carnot) efficiency. (b) Estimate how much power (in watts) goes into moving the car, and how much heat, in joules and in kcal, is exhausted to the air in 1.0 h.

63. A falling rock has kinetic energy K just before striking the ground and coming to rest. What is the total change in entropy of the rock plus environment as a result of this collision?

64. An aluminum can, with negligible heat capacity, is filled with 500 g of water at 0°C and then is brought into thermal contact with a similar can filled with 500 g of water at 50°C. Find the change in entropy of the system if no heat is allowed to exchange with the surroundings.

65. (II) Thermodynamic processes can be represented not only on PV and PT diagrams; another useful one is a TS (temperature-entropy) diagram. (a) Draw a TS diagram for a Carnot cycle. (b) What does the area within the curve represent?

66. (II) A real heat engine working between heat reservoirs at 400 K and 850 K produces 600 J of work per cycle for a heat input of 1600 J. (a) Compare the efficiency of this real engine to that of a Carnot engine. (b) Calculate the total entropy change of the universe for each cycle of the real engine. (c) Calculate the total entropy change of the universe for a Carnot engine operating between the same two temperatures. (d) Show that the difference in work done by these two engines per cycle is $T_L \Delta S$, where T_L is the temperature of the low-temperature reservoir (400 K) and ΔS is the entropy increase per cycle of the real engine. (See also Problem 49 and Section 20–8.)

67. The *Stirling cycle*, shown in Fig. 20–17, is useful to describe external combustion engines as well as solar-power systems. Find the efficiency of the cycle in terms of the parameters shown, assuming a monatomic gas as the working substance. The processes ab and cd are isothermal whereas bc and da are at constant volume. How does it compare to the Carnot efficiency?

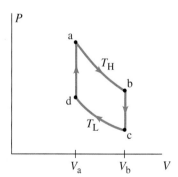

FIGURE 20–17 Problem 67.

(Problems continue on the next page.)

68. One mole of an ideal monatomic gas at STP first undergoes an isothermal expansion so that the volume at b is 2.5 times the volume at a (Fig. 20–18). Next, heat is extracted at a constant volume so that the pressure drops. The gas is then compressed adiabatically back to the original state. (*a*) Calculate the pressures at b and c. (*b*) Determine the temperature at c. (*c*) Determine the work done, heat input or extracted, and the change in entropy for each process. (*d*) What is the efficiency of this cycle?

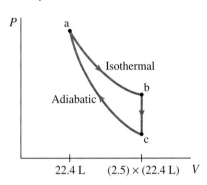

FIGURE 20–18 Problem 68.

69. As a mechanical engineer, you must design a cooling unit for a new freezer. The enclosure has an inner surface area of $6.0\,\text{m}^2$ and is bounded by walls that are $10\,\text{cm}$ thick, with a thermal conductivity of $0.050\,\text{W/m}\cdot\text{K}$. The inside must be kept at $-10\,\text{C}°$ in a room that is at $20°\text{C}$. The motor for the cooling unit must run no more than 15% of the time. What is the minimum power requirement of the cooling motor?

70. A gas turbine operates under the *Brayton cycle*, which is depicted in the *PV* diagram of Fig. 20–19. In process ab the air-fuel mixture undergoes an adiabatic compression. This is followed, in process bc, with an isobaric (constant pressure) heating, by combustion. Process cd is an adiabatic expansion with expulsion of the products to the atmosphere. The return step, da, takes place at constant pressure. If the working gas behaves like an ideal gas, show that the efficiency of the Brayton Cycle is

$$e = 1 - \left(\frac{P_b}{P_a}\right)^{\frac{1-\gamma}{\gamma}}.$$

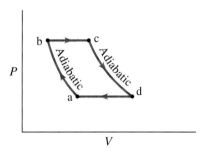

FIGURE 20–19 Problem 70.

*71. Estimate the probability that a bridge player will be dealt (*a*) all four aces (among 13 cards), and (*b*) all 13 cards of one suit.

APPENDIX

Mathematical Formulas

A–1 Quadratic Formula

If
$$ax^2 + bx + c = 0$$
then
$$x = \frac{-b \pm \sqrt{b^2 - 4ac}}{2a}$$

A–2 Binomial Expansion

$$(1 \pm x)^n = 1 \pm nx + \frac{n(n-1)}{2!} x^2 \pm \frac{n(n-1)(n-2)}{3!} x^3 + \cdots$$

$$(x + y)^n = x^n \left(1 + \frac{y}{x}\right)^n = x^n \left(1 + n\frac{y}{x} + \frac{n(n-1)}{2!} \frac{y^2}{x^2} + \cdots\right)$$

A–3 Other Expansions

$$e^x = 1 + x + \frac{x^2}{2!} + \frac{x^3}{3!} + \cdots$$

$$\ln(1 + x) = x - \frac{x^2}{2} + \frac{x^3}{3} - \frac{x^4}{4} + \cdots$$

$$\sin\theta = \theta - \frac{\theta^3}{3!} + \frac{\theta^5}{5!} - \cdots$$

$$\cos\theta = 1 - \frac{\theta^2}{2!} + \frac{\theta^4}{4!} - \cdots$$

$$\tan\theta = \theta + \frac{\theta^3}{3} + \frac{2}{15}\theta^5 + \cdots \qquad |\theta| < \frac{\pi}{2}$$

In general:
$$f(x) = f(0) + \left(\frac{df}{dx}\right)_0 x + \left(\frac{d^2f}{dx^2}\right)_0 \frac{x^2}{2!} + \cdots$$

A–4 Areas and Volumes

Object	Surface area	Volume
Circle, radius r	πr^2	—
Sphere, radius r	$4\pi r^2$	$\frac{4}{3}\pi r^3$
Right circular cylinder, radius r, height h	$2\pi r^2 + 2\pi rh$	$\pi r^2 h$
Right circular cone, radius r, height h	$\pi r^2 + \pi r \sqrt{r^2 + h^2}$	$\frac{1}{3}\pi r^2 h$

A–5 Plane Geometry

1.

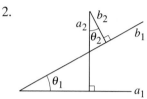

FIGURE A–1

If line a_1 is parallel to line a_2, then $\theta_1 = \theta_2$.

2.

FIGURE A–2

If $a_1 \perp a_2$ and $b_1 \perp b_2$, then $\theta_1 = \theta_2$.

3. The sum of the angles in any plane triangle is 180°.

4. *Pythagorean theorem:*

In any right triangle (one angle = 90°) of sides a, b, and c:

$$a^2 + b^2 = c^2$$

where c is the length of the hypotenuse (opposite the 90° angle).

FIGURE A–3

A–6 Trigonometric Functions and Identities

(See Fig. A–4.)

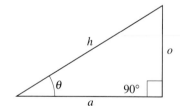

FIGURE A–4

$$\sin \theta = \frac{o}{h} \qquad \csc \theta = \frac{1}{\sin \theta} = \frac{h}{o}$$

$$\cos \theta = \frac{a}{h} \qquad \sec \theta = \frac{1}{\cos \theta} = \frac{h}{a}$$

$$\tan \theta = \frac{o}{a} = \frac{\sin \theta}{\cos \theta} \qquad \cot \theta = \frac{1}{\tan \theta} = \frac{a}{o}$$

$$a^2 + o^2 = h^2 \qquad \text{[Pythagorean theorem]}.$$

Figure A–5 shows the signs (+ or −) that cosine, sine, and tangent take on for angles θ in the four quadrants (0° to 360°). Note that angles are measured counterclockwise from the x axis as shown; negative angles are measured from *below* the x axis, clockwise: for example, −30° = +330°, and so on.

FIGURE A–5

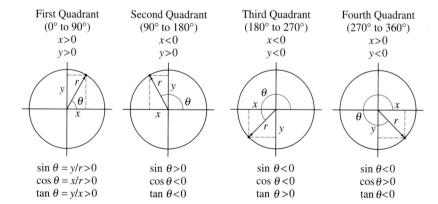

First Quadrant (0° to 90°) $x>0$ $y>0$	Second Quadrant (90° to 180°) $x<0$ $y>0$	Third Quadrant (180° to 270°) $x<0$ $y<0$	Fourth Quadrant (270° to 360°) $x>0$ $y<0$
$\sin \theta = y/r > 0$ $\cos \theta = x/r > 0$ $\tan \theta = y/x > 0$	$\sin \theta > 0$ $\cos \theta < 0$ $\tan \theta < 0$	$\sin \theta < 0$ $\cos \theta < 0$ $\tan \theta > 0$	$\sin \theta < 0$ $\cos \theta > 0$ $\tan \theta < 0$

The following are some useful identities among the trigonometric functions:

$$\sin^2\theta + \cos^2\theta = 1, \quad \sec^2\theta - \tan^2\theta = 1, \quad \csc^2\theta - \cot^2\theta = 1$$

$$\sin 2\theta = 2\sin\theta\cos\theta$$

$$\cos 2\theta = \cos^2\theta - \sin^2\theta = 2\cos^2\theta - 1 = 1 - 2\sin^2\theta$$

$$\tan 2\theta = \frac{2\tan\theta}{1 - \tan^2\theta}$$

$$\sin(A \pm B) = \sin A\cos B \pm \cos A\sin B$$

$$\cos(A \pm B) = \cos A\cos B \mp \sin A\sin B$$

$$\tan(A \pm B) = \frac{\tan A \pm \tan B}{1 \mp \tan A\tan B}$$

$$\sin(180° - \theta) = \sin\theta$$

$$\cos(180° - \theta) = -\cos\theta$$

$$\sin(90° - \theta) = \cos\theta$$

$$\cos(90° - \theta) = \sin\theta$$

$$\cos(-\theta) = \cos\theta$$

$$\sin(-\theta) = -\sin\theta$$

$$\tan(-\theta) = -\tan\theta$$

$$\sin\tfrac{1}{2}\theta = \sqrt{\frac{1 - \cos\theta}{2}}, \quad \cos\tfrac{1}{2}\theta = \sqrt{\frac{1 + \cos\theta}{2}}, \quad \tan\tfrac{1}{2}\theta = \sqrt{\frac{1 - \cos\theta}{1 + \cos\theta}}$$

$$\sin A \pm \sin B = 2\sin\left(\frac{A \pm B}{2}\right)\cos\left(\frac{A \mp B}{2}\right).$$

For any triangle (see Fig. A–6):

$$\frac{\sin\alpha}{a} = \frac{\sin\beta}{b} = \frac{\sin\gamma}{c} \qquad \text{[Law of sines]}$$

$$c^2 = a^2 + b^2 - 2ab\cos\gamma. \qquad \text{[Law of cosines]}$$

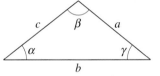

FIGURE A–6

A–7 Logarithms

The following identities apply to common logs (base 10), natural logs (base e) which are often abbreviated ln, or logs to any other base.

$$\log(ab) = \log a + \log b$$

$$\log\left(\frac{a}{b}\right) = \log a - \log b$$

$$\log a^n = n\log a.$$

A–8 Vectors

Vector addition is covered in Sections 3–2 to 3–5.
Vector multiplication is covered in Sections 3–3, 7–2 and 11–1.

Derivatives and Integrals

B–1 Derivatives: General Rules

(See also Section 2–3.)

$$\frac{dx}{dx} = 1$$

$$\frac{d}{dx}\left[af(x)\right] = a\frac{df}{dx} \qquad (a = \text{constant})$$

$$\frac{d}{dx}\left[f(x) + g(x)\right] = \frac{df}{dx} + \frac{dg}{dx}$$

$$\frac{d}{dx}\left[f(x)g(x)\right] = \frac{df}{dx}g + f\frac{dg}{dx}$$

$$\frac{d}{dx}\left[f(y)\right] = \frac{df}{dy}\frac{dy}{dx} \qquad [\text{chain rule}]$$

$$\frac{dx}{dy} = \frac{1}{\left(\dfrac{dy}{dx}\right)} \qquad \text{if } \frac{dy}{dx} \neq 0.$$

B–2 Derivatives: Particular Functions

$$\frac{da}{dx} = 0 \qquad (a = \text{constant})$$

$$\frac{d}{dx}x^n = nx^{n-1}$$

$$\frac{d}{dx}\sin ax = a\cos ax$$

$$\frac{d}{dx}\cos ax = -a\sin ax$$

$$\frac{d}{dx}\tan ax = a\sec^2 ax$$

$$\frac{d}{dx}\ln ax = \frac{1}{x}$$

$$\frac{d}{dx}e^{ax} = ae^{ax}$$

B–3 Indefinite Integrals: General Rules

(See also Section 7–3.)

$$\int dx = x$$

$$\int af(x)\,dx = a\int f(x)\,dx \qquad (a = \text{constant})$$

$$\int [f(x) + g(x)]\,dx = \int f(x)\,dx + \int g(x)\,dx$$

$$\int u\,dv = uv - \int v\,du \qquad \text{[integration by parts]}$$

B–4 Indefinite Integrals: Particular Functions

(An arbitrary constant can be added to the right side of each equation.)

$$\int a\,dx = ax \qquad (a = \text{constant})$$

$$\int x^m\,dx = \frac{1}{m+1}x^{m+1} \qquad (m \neq -1)$$

$$\int \sin ax\,dx = -\frac{1}{a}\cos ax$$

$$\int \cos ax\,dx = \frac{1}{a}\sin ax$$

$$\int \tan ax\,dx = \frac{1}{a}\ln|\sec ax|$$

$$\int \frac{1}{x}\,dx = \ln x$$

$$\int e^{ax}\,dx = \frac{1}{a}e^{ax}$$

$$\int \frac{dx}{x^2 + a^2} = \frac{1}{a}\tan^{-1}\frac{x}{a}$$

$$\int \frac{dx}{x^2 - a^2} = \frac{1}{2a}\ln\left(\frac{x-a}{x+a}\right) \qquad (x^2 > a^2)$$

$$= -\frac{1}{2a}\ln\left(\frac{a+x}{a-x}\right) \qquad (x^2 < a^2)$$

$$\int \frac{dx}{\sqrt{x^2 \pm a^2}} = \ln\left(x + \sqrt{x^2 \pm a^2}\right)$$

$$\int \frac{dx}{(x^2 \pm a^2)^{\frac{3}{2}}} = \frac{\pm x}{a^2\sqrt{x^2 \pm a^2}}$$

$$\int \frac{x\,dx}{(x^2 \pm a^2)^{\frac{3}{2}}} = \frac{-1}{\sqrt{x^2 \pm a^2}}$$

$$\int \sin^2 ax\,dx = \frac{x}{2} - \frac{\sin 2ax}{4a}$$

$$\int xe^{-ax}\,dx = -\frac{e^{-ax}}{a^2}(ax + 1)$$

$$\int x^2 e^{-ax}\,dx = -\frac{e^{-ax}}{a^3}(a^2x^2 + 2ax + 2)$$

B–5 A few Definite Integrals

$$\int_0^\infty x^n e^{-ax}\,dx = \frac{n!}{a^{n+1}}$$

$$\int_0^\infty e^{-ax^2}\,dx = \sqrt{\frac{\pi}{4a}}$$

$$\int_0^\infty xe^{-ax^2}\,dx = \frac{1}{2a}$$

$$\int_0^\infty x^2 e^{-ax^2}\,dx = \sqrt{\frac{\pi}{16a^3}}$$

$$\int_0^\infty x^3 e^{-ax^2}\,dx = \frac{1}{2a^2}$$

$$\int_0^\infty x^{2n} e^{-ax^2}\,dx = \frac{1 \cdot 3 \cdot 5 \cdots (2n-1)}{2^{n+1}a^n}\sqrt{\frac{\pi}{a}}$$

Gravitational Force due to a Spherical Mass Distribution

In Chapter 6 (Section 6–1), we stated that the gravitational force exerted by or on a uniform sphere acts as if all the mass of the sphere were concentrated at its center, if the other mass is outside the sphere. In other words, the gravitational force that a uniform sphere exerts on a particle outside it is

$$F = G\frac{mM}{r^2},$$ [m outside sphere of mass M]

where m is the mass of the particle, M the mass of the sphere, and r the distance of m from the center of the sphere. Now we will derive this result. We will use the concepts of infinitesimally small quantities and integration.

First we consider a very thin, uniform spherical shell (like a thin-walled basketball) of mass M whose thickness t is small compared to its radius R (Fig. C–1). The force on a particle of mass m at a distance r from the center of the shell can be calculated as the vector sum of the forces due to all the particles of the shell. We imagine the shell divided up into thin (infinitesimal) circular strips so that all points on a strip are equidistant from our particle m. One of these circular strips, labeled AB, is shown in Fig. C–1. It is $Rd\theta$ wide, t thick, and has a radius $R\sin\theta$. The force on our particle m due to a tiny piece of the strip at point A is represented by the vector $\mathbf{F}_A$ shown. The force due to a tiny piece of the strip at point B, which is diametrically opposite A, is the force $\mathbf{F}_B$. We take the two pieces at A and B to be of equal mass, so $F_A = F_B$. The horizontal components of $\mathbf{F}_A$ and $\mathbf{F}_B$ are each equal to

$$F_A\cos\phi$$

and point toward the center of the shell. The vertical components of $\mathbf{F}_A$ and $\mathbf{F}_B$ are of equal magnitude and point in opposite directions, and so cancel. Since for every point on the strip there is a corresponding point diametrically opposite (as with A and B), we see that the net force due to the entire strip points toward the center of the shell. Its magnitude will be

$$dF = G\frac{m\,dM}{l^2}\cos\phi,$$

where dM is the mass of the entire circular strip and l is the distance from all

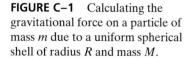

FIGURE C–1 Calculating the gravitational force on a particle of mass m due to a uniform spherical shell of radius R and mass M.

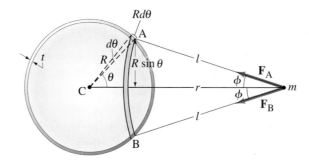

points on the strip to m, as shown. We write dM in terms of the density ρ; by density we mean the mass per unit volume (Section 13–1). Hence, $dM = \rho \, dV$, where dV is the volume of the strip and equals $(2\pi R \sin\theta)(t)(R \, d\theta)$. Then the force dF due to the circular strip shown is

$$dF = G\frac{m\rho 2\pi R^2 t \sin\theta \, d\theta}{l^2}\cos\phi. \qquad (C\text{–}1)$$

To get the total force F that the entire shell exerts on the particle m, we must integrate over all the circular strips: that is, from $\theta = 0°$ to $\theta = 180°$. But our expression for dF contains l and ϕ, which are functions of θ. From Fig. C–1 we can see that

$$l\cos\phi = r - R\cos\theta.$$

Furthermore, we can write the law of cosines for triangle CmA:

$$\cos\theta = \frac{r^2 + R^2 - l^2}{2rR}. \qquad (C\text{–}2)$$

With these two expressions we can reduce our three variables (l, θ, ϕ) to only one, which we take to be l. We do two things with Eq. C–2: (1) We put it into the equation for $l\cos\phi$ above:

$$\cos\phi = \frac{1}{l}(r - R\cos\theta) = \frac{r^2 + l^2 - R^2}{2rl};$$

and (2) we take the differential of both sides of Eq. C–2 (because $\sin\theta \, d\theta$ appears in the expression for dF, Eq. C–1):

$$-\sin\theta \, d\theta = -\frac{2l \, dl}{2rR} \qquad \text{or} \qquad \sin\theta \, d\theta = \frac{l \, dl}{rR},$$

since r and R are considered constants when summing over the strips. Now we insert these into Eq. C–1 for dF and find

$$dF = Gm\rho\pi t\frac{R}{r^2}\left(1 + \frac{r^2 - R^2}{l^2}\right)dl.$$

Now we integrate to get the net force on our thin shell of radius R. To integrate over all the strips ($\theta = 0°$ to $180°$), we must go from $l = r - R$ to $l = r + R$ (see Fig. C–1). Thus,

$$F = Gm\rho\pi t\frac{R}{r^2}\left[l - \frac{r^2 - R^2}{l}\right]_{l=r-R}^{l=r+R}$$

$$= Gm\rho\pi t\frac{R}{r^2}(4R).$$

The volume V of the spherical shell is its area $(4\pi R^2)$ times the thickness t. Hence the mass $M = \rho V = \rho 4\pi R^2 t$, and finally

$$F = G\frac{mM}{r^2}. \qquad \left[\begin{array}{c}\text{particle of mass } m \text{ outside a} \\ \text{thin uniform spherical shell of mass } M\end{array}\right]$$

This result gives us the force a thin shell exerts on a particle of mass m a distance r from the center of the shell, and *outside* the shell. We see that the force is the same as that between m and a particle of mass M at the center of the shell. In other words, for purposes of calculating the gravitational force exerted on or by a uniform spherical shell, we can consider all its mass concentrated at its center.

What we have derived for a shell holds also for a solid sphere, since a solid sphere can be considered as made up of many concentric shells, from $R = 0$ to $R = R_0$, where R_0 is the radius of the solid sphere. Why? Because if each shell has

mass dM, we write for each shell, $dF = Gm\,dM/r^2$, where r is the distance from the center C to mass m and is the same for all shells. Then the total force equals the sum or integral over dM, which gives the total mass M. Thus the result

$$F = G\frac{mM}{r^2} \qquad \begin{bmatrix} \text{particle of mass } m \text{ outside} \\ \text{solid sphere of mass } M \end{bmatrix} \quad \text{(C–3)}$$

is valid for a solid sphere of mass M even if the density varies with distance from the center. (It is not valid if the density varies within each shell—that is, depends not only on R.) Thus the gravitational force exerted on or by spherical objects, including nearly spherical objects like the Earth, Sun, and Moon, can be considered to act as if the objects were point particles.

This result, Eq. C–3, is true only if the mass m is outside the sphere. Let us next consider a point mass m that is located inside the spherical shell of Fig. C–1. Here, r would be less than R, and the integration over l would be from $l = R - r$ to $l = R + r$, so

$$\left[l - \frac{r^2 - R^2}{l} \right]_{R-r}^{R+r} = 0.$$

Thus the force on any mass inside the shell would be zero. This result has particular importance for the electrostatic force, which is also an inverse square law. For the gravitational situation, we see that at points within a solid sphere, say 1000 km below the earth's surface, only the mass up to that radius contributes to the net force. The outer shells beyond the point in question contribute no net gravitational effect.

The results we have obtained here can also be reached using the gravitational analog of Gauss's law for electrostatics (Chapter 22).

APPENDIX D

Selected Isotopes

(1) Atomic Number Z	(2) Element	(3) Symbol	(4) Mass Number A	(5) Atomic Mass[†]	(6) % Abundance (or Radioactive Decay Mode)	(7) Half-life (if radioactive)
0	(Neutron)	n	1	1.008665	β^-	10.4 min
1	Hydrogen	H	1	1.007825	99.985%	
	Deuterium	D	2	2.014102	0.015%	
	Tritium	T	3	3.016049	β^-	12.33 yr
2	Helium	He	3	3.016029	0.000137%	
			4	4.002603	99.999863%	
3	Lithium	Li	6	6.015122	7.5%	
			7	7.016004	92.5%	
4	Beryllium	Be	7	7.016929	EC, γ	53.12 days
			9	9.012182	100%	
5	Boron	B	10	10.012937	19.9%	
			11	11.009305	80.1%	
6	Carbon	C	11	11.011434	β^+, EC	20.39 min
			12	12.000000	98.90%	
			13	13.003355	1.10%	
			14	14.003242	β^-	5730 yr
7	Nitrogen	N	13	13.005739	β^+	9.965 min
			14	14.003074	99.63%	
			15	15.000108	0.37%	
8	Oxygen	O	15	15.003065	β^+, EC	122.24 s
			16	15.994915	99.76%	
			18	17.999160	0.20%	
9	Fluorine	F	19	18.998403	100%	
10	Neon	Ne	20	19.992440	90.48%	
			22	21.991386	9.25%	
11	Sodium	Na	22	21.994437	β^+, EC, γ	2.6019 yr
			23	22.989770	100%	
			24	23.990963	β^-, γ	14.9590 h
12	Magnesium	Mg	24	23.985042	78.99%	
13	Aluminum	Al	27	26.981538	100%	

[†]The masses given in column (5) are those for the neutral atom, including the Z electrons.

(1) Atomic Number Z	(2) Element	(3) Symbol	(4) Mass Number A	(5) Atomic Mass†	(6) % Abundance (or Radioactive Decay Mode)	(7) Half-life (if radioactive)
14	Silicon	Si	28	27.976927	92.23%	
			31	30.975363	β^-, γ	157.3 min
15	Phosphorus	P	31	30.973762	100%	
			32	31.973907	β^-	14.262 days
16	Sulfur	S	32	31.972071	95.02%	
			35	34.969032	β^-	87.32 days
17	Chlorine	Cl	35	34.968853	75.77%	
			37	36.965903	24.23%	
18	Argon	Ar	40	39.962383	99.600%	
19	Potassium	K	39	38.963707	93.2581%	
			40	39.963999	0.0117%	
					β^-, EC, γ, β^+	1.28×10^9 yr
20	Calcium	Ca	40	39.962591	96.941%	
21	Scandium	Sc	45	44.955910	100%	
22	Titanium	Ti	48	47.947947	73.8%	
23	Vanadium	V	51	50.943964	99.750%	
24	Chromium	Cr	52	51.940512	83.79%	
25	Manganese	Mn	55	54.938049	100%	
26	Iron	Fe	56	55.934942	91.72%	
27	Cobalt	Co	59	58.933200	100%	
			60	59.933822	β^-, γ	5.2714 yr
28	Nickel	Ni	58	57.935348	68.077%	
			60	59.930791	26.233%	
29	Copper	Cu	63	62.929601	69.17%	
			65	64.927794	30.83%	
30	Zinc	Zn	64	63.929147	48.6%	
			66	65.926037	27.9%	
31	Gallium	Ga	69	68.925581	60.108%	
32	Germanium	Ge	72	71.922076	27.66%	
			74	73.921178	35.94%	
33	Arsenic	As	75	74.921596	100%	
34	Selenium	Se	80	79.916522	49.61%	
35	Bromine	Br	79	78.918338	50.69%	
36	Krypton	Kr	84	83.911507	57.0%	
37	Rubidium	Rb	85	84.911789	72.17%	
38	Strontium	Sr	86	85.909262	9.86%	
			88	87.905614	82.58%	
			90	89.907737	β^-	28.79 yr
39	Yttrium	Y	89	88.905848	100%	
40	Zirconium	Zr	90	89.904704	51.45%	
41	Niobium	Nb	93	92.906377	100%	
42	Molybdenum	Mo	98	97.905408	24.13%	

† The masses given in column (5) are those for the neutral atom, including the Z electrons.

(1) Atomic Number Z	(2) Element	(3) Symbol	(4) Mass Number A	(5) Atomic Mass†	(6) % Abundance (or Radioactive Decay Mode)	(7) Half-life (if radioactive)
43	Technetium	Tc	98	97.907216	β^-, γ	4.2×10^6 yr
44	Ruthenium	Ru	102	101.904349	31.6%	
45	Rhodium	Rh	103	102.905504	100%	
46	Palladium	Pd	106	105.903483	27.33%	
47	Silver	Ag	107	106.905093	51.839%	
			109	108.904756	48.161%	
48	Cadmium	Cd	114	113.903358	28.73%	
49	Indium	In	115	114.903878	95.7%; β^-, γ	4.41×10^{14} yr
50	Tin	Sn	120	119.902197	32.59%	
51	Antimony	Sb	121	120.903818	57.36%	
52	Tellurium	Te	130	129.906223	33.80%	7.9×10^{20} yr
53	Iodine	I	127	126.904468	100%	
			131	130.906124	β^-, γ	8.0207 days
54	Xenon	Xe	132	131.904154	26.9%	
			136	135.907220	8.9%	
55	Cesium	Cs	133	132.905446	100%	
56	Barium	Ba	137	136.905821	11.23%	
			138	137.905241	71.70%	
57	Lanthanum	La	139	138.906348	99.9098%	
58	Cerium	Ce	140	139.905434	88.48%	
59	Praseodymium	Pr	141	140.907647	100%	
60	Neodymium	Nd	142	141.907718	27.13%	
61	Promethium	Pm	145	144.912744	EC, γ, α	17.7 yr
62	Samarium	Sm	152	151.919728	26.7%	
63	Europium	Eu	153	152.921226	52.2%	
64	Gadolinium	Gd	158	157.924101	24.84%	
65	Terbium	Tb	159	158.925343	100%	
66	Dysprosium	Dy	164	163.929171	28.2%	
67	Holmium	Ho	165	164.930319	100%	
68	Erbium	Er	166	165.930290	33.6%	
69	Thulium	Tm	169	168.934211	100%	
70	Ytterbium	Yb	174	173.938858	31.8%	
71	Lutecium	Lu	175	174.940767	97.4%	
72	Hafnium	Hf	180	179.946549	35.100%	
73	Tantalum	Ta	181	180.947996	99.988%	
74	Tungsten (wolfram)	W	184	183.950933	30.67%	
75	Rhenium	Re	187	186.955751	62.60%; β^-	4.35×10^{10} yr
76	Osmium	Os	191	190.960927	β^-, γ	15.4 days
			192	191.961479	41.0%	
77	Iridium	Ir	191	190.960591	37.3%	
			193	192.962923	62.7%	
78	Platinum	Pt	195	194.964774	33.8%	

†The masses given in column (5) are those for the neutral atom, including the Z electrons.

(1) Atomic Number Z	(2) Element	(3) Symbol	(4) Mass Number A	(5) Atomic Mass†	(6) % Abundance (or Radioactive Decay Mode)	(7) Half-life (if radioactive)
79	Gold	Au	197	196.966551	100%	
80	Mercury	Hg	199	198.968262	16.87%	
			202	201.970625	29.86%	
81	Thallium	Tl	205	204.974412	70.476%	
82	Lead	Pb	206	205.974449	24.1%	
			207	206.975880	22.1%	
			208	207.976635	52.4%	
			210	209.984173	β^-, γ, α	22.3 yr
			211	210.988731	β^-, γ	36.1 min
			212	211.991887	β^-, γ	10.64 h
			214	213.999798	β^-, γ	26.8 min
83	Bismuth	Bi	209	208.980383	100%	
			211	210.987258	α, γ, β^-	2.14 min
84	Polonium	Po	210	209.982857	α, γ	138.376 days
			214	213.995185	α, γ	164.3 μs
85	Astatine	At	218	218.008681	α, β^-	1.5 s
86	Radon	Rn	222	222.017570	α, γ	3.8235 days
87	Francium	Fr	223	223.019731	β^-, γ, α	21.8 min
88	Radium	Ra	226	226.025402	α, γ	1600 yr
89	Actinium	Ac	227	227.027746	β^-, γ, α	21.773 yr
90	Thorium	Th	228	228.028731	α, γ	1.9116 yr
			232	232.038050	100%; α, γ	1.405×10^{10} yr
91	Protactinium	Pa	231	231.035878	α, γ	3.276×10^4 yr
92	Uranium	U	232	232.037146	α, γ	68.9 yr
			233	233.039628	α, γ	1.592×10^5 yr
			235	235.043923	0.720%, α, γ	7.038×10^8 yr
			236	236.045561	α, γ	2.342×10^7 yr
			238	238.050782	99.2745%; α, γ	4.468×10^9 yr
			239	239.054287	β^-, γ	23.45 min
93	Neptunium	Np	237	237.048166	α, γ	2.144×10^6 yr
			239	239.052931	β^-, γ	2.3565 d
94	Plutonium	Pu	239	239.052157	α, γ	24,110 yr
			244	244.064197	α	8.08×10^7 yr
95	Americium	Am	243	243.061373	α, γ	7370 yr
96	Curium	Cm	247	247.070346	α, γ	1.56×10^7 yr
97	Berkelium	Bk	247	247.070298	α, γ	1380 yr
98	Californium	Cf	251	251.079580	α, γ	898 yr
99	Einsteinium	Es	252	252.082972	α, EC, γ	472 d
100	Fermium	Fm	257	257.095099	α, γ	101 d
101	Mendelevium	Md	258	258.098425	α, γ	51.5 d
102	Nobelium	No	259	259.10102	α, EC	58 min
103	Lawrencium	Lr	262	262.10969	α, EC, fission	216 min

†The masses given in column (5) are those for the neutral atom, including the Z electrons.

(1) Atomic Number Z	(2) Element	(3) Symbol	(4) Mass Number A	(5) Atomic Mass†	(6) % Abundance (or Radioactive Decay Mode)	(7) Half-life (if radioactive)
104	Rutherfordium	Rf	261	261.10875	α	65 s
105	Dubnium	Db	262	262.11415	α, fission, EC	34 s
106	Seaborgium	Sg	266	266.12193	α, fission	21 s
107	Bohrium	Bh	264	264.12473	α	0.44 s
108	Hassium	Hs	269	269.13411	α	9 s
109	Meitnerium	Mt	268	268.13882	α	0.07 s
110			271	271.14608	α	0.06 s
111			272	272.15348	α	1.5 ms
112			277	277	α	0.24 ms
114			289	289	α	20 s
116			289	289	α	0.6 ms
118			293	293	α	0.1 ms

†The masses given in column (5) are those for the neutral atom, including the Z electrons.

Answers to Odd-Numbered Problems

CHAPTER 1

1. (a) 1×10^{10} yr; (b) 3×10^{17} s.
3. (a) 1.156×10^3; (b) 2.18×10^1;
   (c) 6.8×10^{-3}; (d) 2.7635×10^1;
   (e) 2.19×10^{-1}; (f) 2.2×10^1.
5. 7.7%.
7. (a) 4%; (b) 0.4%; (c) 0.07%.
9. 1.0×10^5 s.
11. 9%.
13. (a) 0.286 6 m; (b) 0.000 085 V;
    (c) 0.000 760 kg;
    (d) 0.000 000 000 060 0 s;
    (e) 0.000 000 000 000 022 5 m;
    (f) 2,500,000,000 volts.
15. 1.8 m.
17. (a) 0.111 yd^2; (b) 10.76 ft^2.
19. (a) 3.9×10^{-9} in;
    (b) 1.0×10^8 atoms.
21. (a) 0.621 mi/h;
    (b) 1 m/s = 3.28 ft/s; (c) 0.278 m/s.
23. (a) 9.46×10^{15} m;
    (b) 6.31×10^4 AU; (c) 7.20 AU/h.
25. (a) 10^3; (b) 10^4; (c) 10^{-2}; (d) 10^9.
27. ≈20%.
29. 1×10^5 cm^3.
31. (a) ≈600 dentists.
33. ≈3×10^8 kg/yr.
35. 51 km.
37. $A = [L/T^4] = $ m/s^4,
    $B = [L/T^2] = $ m/s^2.
39. (a) 0.10 nm; (b) 1.0×10^5 fm;
    (c) 1.0×10^{10} Å; (d) 9.5×10^{25} Å.
41. (a) 3.16×10^7 s; (b) 3.16×10^{16} ns;
    (c) 3.17×10^{-8} yr.
43. (a) 1,000 drivers.
45. 1×10^{11} gal/yr.
47. 9 cm.
49. 4×10^5 t.
51. ≈4 yr.
53. 1.9×10^2 m.
55. (a) 3%, 3%; (b) 0.7%, 0.2%.

CHAPTER 2

1. 5.0 h.
3. 61 m.
5. 0.78 cm/s (toward +x).
7. ≈300 m/s.
9. (a) 10.1 m/s;
   (b) +3.4 m/s, away from trainer.
11. (a) 0.28 m/s; (b) 1.2 m/s;
    (c) 0.28 m/s; (d) 1.6 m/s;
    (e) −1.0 m/s.
13. (a) 13.4 m/s;
    (b) +4.5 m/s, away from master.
15. 24 s.
17. 55 km/h, 0.
19. 6.73 m/s.
21. 5.2 s
23. −7.0 m/s^2, 0.72.
25. (a) 4.7 m/s^2; (b) 2.2 m/s^2;
    (c) 0.3 m/s^2; (d) 1.6 m/s^2.
27. $v = (6.0$ m/s$) + (17$ m/s$^2)t$,
    $a = 17$ m/s^2.
29. 1.5 m/s^2, 99 m.
31. 1.7×10^2 m.
33. 4.41 m/s^2, $t = 2.61$ s.
35. 55.0 m.
37. (a) 2.3×10^2 m; (b) 31 s;
    (c) 15 m, 13 m.
39. (a) 103 m; (b) 64 m.
41. 31 m/s.
43. (b) 3.45 s.
45. 32 m/s (110 km/h).
47. 2.83 s.
49. (a) 8.81 s; (b) 86.3 m/s.
51. 1.44 s.
53. 15 m/s.
55. 5.44 s.
59. 0.035 s.
61. 1.8 m above the top of the window.
63. 52 m.
65. 19.8 m/s, 20.0 m.
67. (a) $v = (g/k)(1 - e^{-kt})$;
    (b) $v_{\text{term}} = g/k$.
69. $6h_{\text{Earth}}$.
71. 1.3 m.
73. (b) $H_{50} = 9.8$ m; (c) $H_{100} = 39$ m.
75. (a) 1.3 m; (b) 6.1 m/s; (c) 1.2 s.
77. (a) 3.88 s; (b) 73.9 m; (c) 38.0 m/s,
    48.4 m/s.
79. (a) 52 min; (b) 31 min.
81. (a) $v_0 = 26$ m/s; (b) 35 m; (c) 1.2 s;
    (d) 4.1 s.
83. (a) 4.80 s; (b) 37.0 m/s; (c) 75.2 m.
85. She should decide to stop!
87. $\Delta v_{0\text{down}} = 0.8$ m/s, $\Delta v_{0\text{up}} = 0.9$ m/s.
89. 29.0 m.

CHAPTER 3

1. 263 km, 13° S of W.
3. $\mathbf{V}_{\text{wrong}} = \mathbf{V}_2 - \mathbf{V}_1$.
5. 13.6 m, 18° N of E,

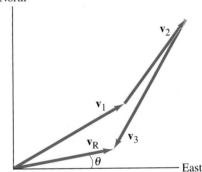

7. (a)

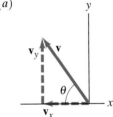

   (b) $V_x = -11.7$, $V_y = 8.16$;
   (c) 14.3, 34.9° above $-x$-axis.
9. (a) $V_N = 476$ km/h,
   $V_W = 421$ km/h;
   (b) $d_N = 1.43 \times 10^3$ km,
   $d_W = 1.26 \times 10^3$ km.
11. (a) 4.2, 45° below +x-axis;
    (b) 5.1, 79° below +x-axis.
13. (a) 53.7, 1.40° above −x-axis;
    (b) 53.7, 1.40° below +x-axis.
15. (a) 94.5, 11.8° below −x-axis;
    (b) 150, 35.3° below +x-axis.
17. (a) $A_x = \pm 82.9$; (b) 166.6, 12.1°
    above −x-axis.
19. $(7.60$ m/s$)\mathbf{i} - (4.00$ m/s$)\mathbf{k}$; 8.59 m/s.
21. (a) Unknown; (b) 4.11 m/s^2, 33.2°
    north of east; (c) unknown.

23. (a) $\mathbf{v} = (4.0\,\text{m/s}^2)t\mathbf{i} + (3.0\,\text{m/s}^2)t\mathbf{j}$;
(b) $(5.0\,\text{m/s}^2)t$;
(c) $\mathbf{r} = (2.0\,\text{m/s}^2)t^2\mathbf{i} + (1.5\,\text{m/s}^2)t^2\mathbf{j}$;
(d) $\mathbf{v} = (8.0\,\text{m/s})\mathbf{i} + (6.0\,\text{m/s})\mathbf{j}$,
$|\mathbf{v}| = 10.0\,\text{m/s}$,
$\mathbf{r} = (8.0\,\text{m})\mathbf{i} + (6.0\,\text{m})\mathbf{j}$.

25. (a) $-(18.0\,\text{m/s})\sin(3.0\,\text{s}^{-1})t\mathbf{i}$
$+ (18.0\,\text{m/s})\cos(3.0\,\text{s}^{-1})t\mathbf{j}$;
(b) $-(54.0\,\text{m/s}^2)\cos(3.0\,\text{s}^{-1})t\mathbf{i}$
$- (54.0\,\text{m/s}^2)\sin(3.0\,\text{s}^{-1})t\mathbf{j}$;
(c) circle; (d) $a = (9.0\,\text{s}^{-2})r$, $180°$.

27. 44 m, 6.3 m.

29. 38° and 52°.

31. 1.95 s.

33. 22 m.

35.

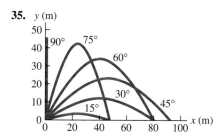

37. 5.71 s.

39. (a) 65.7 m; (b) 7.32 s; (c) 267 m;
(d) 42.2 m/s, 30.1° above the horizontal.

43. Unsuccessful, 34.7 m.

45. (a) $\mathbf{v}_0 = 3.42\,\text{m/s}$, 47.5° above the
horizontal; (b) 5.32 m above the
water; (c) $\mathbf{v}_f = 10.5\,\text{m/s}$, 77° below
the horizontal.

47. $\theta = \tan^{-1}(gt/v_0)$ below the
horizontal.

49. $\theta = \frac{1}{2}\tan^{-1}(-\cot\phi)$.

51. $7.29g$ up.

53. $5.9 \times 10^{-3}\,\text{m/s}^2$ toward the Sun.

55. $0.94g$.

59. 2.7 m/s, 22° from the river bank.

61. 23.1 s.

63. 1.41 m/s.

65. (a) 1.82 m/s; (b) 3.22 m/s.

67. (a) 60 m; (b) 75 s.

69. 58 km/h, 31°, 58 km/h opposite
to $\mathbf{v}_{12}$.

71. 0.0889 m/s².

73. $D_x = 60\,\text{m}$, $D_y = -35\,\text{m}$,
$D_z = -12\,\text{m}$; 70 m; $\theta_h = 30°$ from
the x-axis toward the $-y$-axis,
$\theta_v = 9.8°$ below the horizontal.

75. 7.0 m/s.

77. $\pm 28.5°$, ± 25.2.

79. 170 km/h, 41.5° N of E.

81. 1.6 m/s².

83. 2.7 s, 1.9 m/s.

85. (a) $Dv/(v^2 - u^2)$;
(b) $D/(v^2 - u^2)^{1/2}$.

87. 54.6° below the horizontal.

89. (a) 464 m/s; (b) 355 m/s.

91. Row at an angle of 23° upstream
and run 243 m in a total time of
20.7 min.

93. 1.8×10^3 rev/day.

CHAPTER 4

3. 6.9×10^2 N.

5. (a) 5.7×10^2 N; (b) 99 N;
(c) 2.1×10^2 N; (d) 0.

7. 107 N.

9. -9.3×10^5 N, 25% of the weight of
the train.

11. $m > 1.9$ kg.

13. 2.1×10^2 N.

15. $-1.40\,\text{m/s}^2$ (down).

17. a (downward) $\geq 1.2\,\text{m/s}^2$.

19. $a_{\text{max}} = 0.557\,\text{m/s}^2$.

21. (a) 2.2 m/s²; (b) 18 m/s; (c) 93 s.

23. 3.0×10^3 N downward.

25. (a) 1.4×10^2 N; (b) 14.5 m/s.

27. Southwesterly direction.

29.

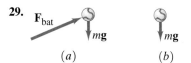

31. (a) 1.13 m/s², 52.2° below $-x$-axis;
(b) 0.814 m/s², 42.3° above $+x$-axis.

33. (a) $m_1g - F_T = m_1a$, $F_T - m_2g$
$= m_2a$.

35. (a) lower bucket = 34 N, upper
bucket = 68N; (b) lower bucket
= 40 N, upper bucket = 80 N.

37. 1.4×10^3 N.

39. $F_B = 6890\,\text{N}$, $F_A + F_B = 8860\,\text{N}$.

41. (a) 2.2 m up the plane; (b) 2.2 s.

43. $\frac{5}{2}(F_0/m)t_0^2$.

47. (a)

47. (a)

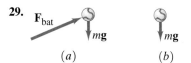

(b) $a = m_2g/(m_1 + m_2)$,
$F_T = m_1m_2g/(m_1 + m_2)$.

49. $a = [m_2 + m_C(\ell_2/\ell)]g/$
$(m_1 + m_2 + m_C)$.

51. 1.74 m/s², $F_{T1} = 22.6\,\text{N}$,
$F_{T2} = 20.9\,\text{N}$.

53. $(m + M)g\tan\theta$.

55. $F_{T1} = [4m_1m_2m_3/$
$(m_1m_3 + m_2m_3 + 4m_1m_2)]g$,
$F_{T3} = [8m_1m_2m_3/$
$(m_1m_3 + m_2m_3 + 4m_1m_2)]g$,
$a_1 = [(m_1m_3 - 3m_2m_3 + 4m_1m_2)/$
$(m_1m_3 + m_2m_3 + 4m_1m_2)]g$,
$a_2 = [(-3m_1m_3 + m_2m_3 + 4m_1m_2)/$
$(m_1m_3 + m_2m_3 + 4m_1m_2)]g$,
$a_3 = [(m_1m_3 + m_2m_3 - 4m_1m_2)/$
$(m_1m_3 + m_2m_3 + 4m_1m_2)]g$.

57. $v = \{[2m_2\ell_2 + m_C(\ell_2^2/\ell)]g/$
$(m_1 + m_2 + m_C)\}^{1/2}$.

59. 2.0×10^{-2} N.

61. 4.3 N.

63. 1.5×10^4 N.

65. 1.2 s, no change.

67. (a) 2.45 m/s² (up the incline);
(b) 0.50 kg; (c) 7.35 N, 4.9 N.

69. 1.3×10^2 N.

71. 8.8°.

73. 82 m/s (300 km/h).

75. (a) $F = \frac{1}{2}Mg$;
(b) $F_{T1} = F_{T2} = \frac{1}{2}Mg$, $F_{T3} = \frac{3}{2}Mg$,
$F_{T4} = Mg$.

77. -8.3×10^2 N.

79. (a) 0.606 m/s²; (b) 150 kN.

CHAPTER 5

1. 35 N, no force.

3. (a)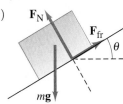

(b)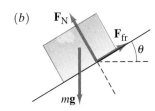

(c)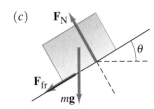

5. 0.20.

7. 69 N, $\mu_k = 0.54$.

9. 8.0 kg.

11. 1.3 m.

13. 1.3×10^3 N.

15. (a) 0.58; (b) 5.7 m/s; (c) 15 m/s.

17. (a) 1.4 m/s^2; (b) 5.4×10^2 N; (c) 1.41 m/s^2, 2.1×10^2 N.

19. (a) 86 cm up the plane; (b) 1.5 s.

21. (a) 2.8 m/s^2; (b) 2.1 N.

23. $a = \{\sin\theta - [(\mu_1 m_1 + \mu_2 m_2)/(m_1 + m_2)]\cos\theta\}g$,
$F_T = [m_1 m_2(\mu_2 - \mu_1)/(m_1 + m_2)]g\cos\theta$.

25. (a) $\mu_k = (v_0^2/2gd\cos\theta) - \tan\theta$; (b) $\mu_s \geq \tan\theta$.

27. (a) 2.0 m/s^2 up the plane; (b) 5.4 m/s^2 up the plane.

29. $\mu_k = 0.40$.

31. (a) $c = 14$ kg/m; (b) 5.7×10^2 N.

33. $F_{min} = (m + M)g(\sin\theta + \mu\cos\theta)/(\cos\theta - \mu_s\sin\theta)$.

35. $v_{max} = 21$ m/s, independent of the mass.

37. (a) 0.25 m/s^2 toward the center; (b) 6.3 N toward the center.

39. Yes, $v_{top,\,min} = (gR)^{1/2}$.

41. 0.34.

43. 2.1×10^2 N.

45. 5.91°, 14.3 N.

47. (a) 5.8×10^3 N; (b) 4.1×10^2 N; (c) 31 m/s.

49. $F_T = 2\pi m R f^2$.

51. 66 km/h $< v <$ 123 km/h.

53. (a) $(1.6 \text{ m/s}^2)\mathbf{i}$;
(b) $(0.98 \text{ m/s}^2)\mathbf{i} - (1.7 \text{ m/s}^2)\mathbf{j}$;
(c) $-(4.9 \text{ m/s}^2)\mathbf{i} - (1.6 \text{ m/s}^2)\mathbf{j}$.

55. (a) 9.0 m/s^2; (b) 15 m/s^2.

57. $\tau = m/b$.

59. (a) $v = (mg/b) + [v_0 - (mg/b)]e^{-bt/m}$;
(b) $v = -(mg/b) + [v_0 + (mg/b)]e^{-bt/m}$, $v \geq 0$.

61. (b) 1.8°.

63. 10 m.

65. $\mu_s = 0.41$.

67. 2.3.

69. 101 N, $\mu_k = 0.719$.

71. (b) Will slide.

73. Emerges with a speed of 13 m/s.

75. 27.6 m/s, 0.439 rev/s.

77. $\Sigma F_{tan} = 3.3 \times 10^3$ N, $\Sigma F_R = 2.0 \times 10^3$ N.

79. (a) $F_{NC} > F_{NB} > F_{NA}$;
(b) heaviest at C, lightest at A;
(c) $v_{Amax} = (gR)^{1/2}$.

81. (a) 1.23 m/s; (b) 3.01 m/s.

83. $\phi = 31°$.

85. (a) $r = v^2/g\cos\theta$; (b) 92 m.

87. (a) 59 s; (b) greater normal force.

89. 29.2 m/s.

91. 302 m, 735 m.

93. $g(1 - \mu_s\tan\phi)/4\pi^2 f^2(\tan\phi + \mu_s)$
$< r < g(1 + \mu_s\tan\phi)/4\pi^2 f^2(\tan\phi - \mu_s)$.

CHAPTER 6

1. 1.52×10^3 N.

3. 1.6 m/s^2.

5. $g_h = 0.91 g_{surface}$.

7. 1.9×10^{-8} N toward center of square.

9. $Gm^2\{(2/x_0^2) + [3x_0/(x_0^2 + y_0^2)^{3/2}]\}\mathbf{i}$
$+ Gm^2\{[3y_0/(x_0^2 + y_0^2)^{3/2}] + (4/y_0^2)\}\mathbf{j}$.

11. 1.26.

13. 3.46×10^8 m from Earth's center.

15. (b) g decreases with an increase in height; (c) 9.493 m/s^2.

19. 7.56×10^3 m/s.

21. 2.0 h.

23. (a) 56 kg; (b) 56 kg; (c) 75 kg; (d) 38 kg; (e) 0.

25. (a) 22 N (toward the Moon); (b) -1.7×10^2 N (away from the Moon).

27. (a) Gravitational force provides required centripetal acceleration; (b) 9.6×10^{26} kg.

29. 7.9×10^3 m/s.

31. $v = (Gm/L)^{1/2}$.

33. 0.0587 days (1.41 h).

35. 1.6×10^2 yr.

37. 2×10^8 yr.

39. $r_{Europa} = 6.71 \times 10^5$ km, $r_{Ganymede} = 1.07 \times 10^6$ km, $r_{Callisto} = 1.88 \times 10^6$ km.

41. 9.0 Earth-days.

43. (a) 2.1×10^2 A.U. $(3.1 \times 10^{13}$ m); (b) 4.2×10^2 A.U.; (c) 4.2×10^2.

45. (a) 5.9×10^{-3} N/kg; (b) not significant.

47. 2.7×10^3 km.

49. 6.7×10^{12} m/s^2.

51. 4.4×10^7 m/s^2.

53. $G' = 1 \times 10^{-4}$ N$\cdot$m^2/kg$^2 \approx 10^6 G$.

55. 5 h 35 min, 19 h 50 min.

57. (a) 10 h; (b) 6.5 km; (c) 4.2×10^{-3} m/s^2.

59. 5.4×10^{12} m, in the Solar System, Pluto.

61. $2.3 g_{Earth}$.

63. $m_P = g_P r^2/G$.

67. 7.9×10^3 m/s.

CHAPTER 7

1. 6.86×10^3 J.

3. 1.27×10^4 J.

5. 8.1×10^3 J.

7. 1 J $= 1 \times 10^7$ erg $= 0.738$ ft$\cdot$lb.

9. 1.0×10^4 J.

13. (a) 3.6×10^2 N; (b) -1.3×10^3 J; (c) -4.6×10^3 J; (d) 5.9×10^3 J; (e) 0.

15. $W_{FN} = W_{mg} = 0$,
$W_{FP} = -W_{fr} = 2.0 \times 10^2$ J.

21. (a) -16.1; (b) -238; (c) -3.9.

23. $\mathbf{C} = -1.3\mathbf{i} + 1.8\mathbf{j}$.

25. $\theta_x = 42.7°, \theta_y = 63.8°, \theta_z = 121°$.

27. 95°, $-35°$ from x-axis.

31. 0.089 J.

33. 2.3×10^3 J.

35. 2.7×10^3 J.

37. $(kX^2/2) + (aX^4/4) + (bX^5/5)$.

39. (a) 5.0×10^{10} J.

41. (a) $\sqrt{3}$; (b) $\frac{1}{4}$.

43. -5.02×10^5 J.

45. 3.0×10^2 N in the direction of the motion of the ball.

47. 24 m/s (87 km/h or 54 mi/h), the mass cancels.

49. (a) 72 J; (b) -35 J; (c) 37 J.

51. 10.2 m/s.

53. $\mu_k = F/2mg$.

55. (a) 6.5×10^2 J; (b) -4.9×10^2 J; (c) 0; (d) 4.0 m/s.

57. (a) 1.66×10^5 J; (b) 21.0 m/s; (c) 2.13 m.

59. $v_p = 2.0 \times 10^7$ m/s, $v_{pc} = 2.0 \times 10^7$ m/s; $v_e = 2.9 \times 10^8$ m/s, $v_{ec} = 8.4 \times 10^8$ m/s.

61. 1.74×10^3 J.

63. (a) 15 J; (b) 4.2×10^2 J; (c) -1.8×10^2 J; (d) -2.5×10^2 J; (e) 0; (f) 10 J.

65. (a) 12 J; (b) 10 J; (c) -2.1 J.

67. 86 kJ, $\theta = 42°$.

69. $(A/k)e^{-(0.10\,\text{m})k}$.

71. 1.5 N.

73. 5.0×10^3 N/m.

75. (a) 6.6°; (b) 10.3°.

CHAPTER 8

1. 0.924 m.

3. 2.2×10^3 J.

5. (a) 51.7 J; (b) 15.1 J; (c) 51.7 J.

7. (a) Conservative; (b) $\frac{1}{2}kx^2 - \frac{1}{4}ax^4 - \frac{1}{5}bx^5 +$ constant.

9. (a) $\frac{1}{2}k(x^2 - x_0^2)$; (b) same.

11. 45.4 m/s.

13. 6.5 m/s.

15. (a) 1.0×10^2 N/m; (b) 22 m/s^2.

17. (a) 8.03 m/s; (b) 3.44 m.

19. (a) $v_{\max} = [v_0^2 + (kx_0^2/m)]^{1/2}$; (b) $x_{\max} = [x_0^2 + (mv_0^2/k)]^{1/2}$.

21. (a) 2.29 m/s; (b) 1.98 m/s; (c) 1.98 m/s; (d) $F_{Ta} = 0.87$ N, $F_{Tb} = 0.80$ N, $F_{Tc} = 0.80$ N; (e) $v_a = 2.59$ m/s, $v_b = 2.31$ m/s, $v_c = 2.31$ m/s.

23. $k = 12Mg/h$.

25. 4.5×10^6 J.

27. (a) 22 m/s; (b) 2.9×10^2 m.

29. 13 m/s.

31. 0.23.

33. 0.40.

35. (a) 0.13 m; (b) 0.77; (c) 0.46 m/s.

37. (a) $K = GM_E m_S/2r_S$; (b) $U = -GM_E m_S/r_S$; (c) $-\frac{1}{2}$.

39. (a) 6.2×10^5 m/s; (b) 4.2×10^4 m/s, $v_{esc}/v_{orbit} = \sqrt{2}$.

45. (a) 1.07×10^4 m/s; (b) 1.17×10^4 m/s; (c) 1.12×10^4 m/s.

47. (a) $dv_{esc}/dr = -\frac{1}{2}(2GM_E/r^3)^{1/2}$ $= -v_{esc}/2r$; (b) 1.09×10^4 m/s.

49. $GmM_E/12r_E$.

51. 1.1×10^4 m/s.

55. 5.4×10^2 N.

57. (a) 1.0×10^3 J; (b) 1.0×10^3 W.

59. 2.1×10^4 W, 28 hp.

61. 4.8×10^2 W.

63. 1.2×10^3 W.

65. 1.8×10^6 W.

67. (a) -25 W; (b) $+4.3 \times 10^3$ W; (c) $+1.5 \times 10^3$ W.

69. (a) 80 J; (b) 60 J; (c) 80 J; (d) 5.7 m/s at $x = 0$; (e) 32 m/s^2 at $x = \pm x_0$.

71. $a^2/4b$.

73. 8.0 m/s.

75. 32.5 hp.

77. (a) 28 m/s; (b) 1.2×10^2 m.

79. (a) $(2gL)^{1/2}$; (b) $(1.2gL)^{1/2}$.

81. (a) 1.1×10^6 J; (b) 60 W (0.081 hp); (c) 4.0×10^2 W (0.54 hp).

83. (a) 40 m/s; (b) 2.6×10^5 W.

87. (a) 29°; (b) 6.4×10^2 N; (c) 9.2×10^2 N.

89. (a) $-\dfrac{U_0}{r}\left(\dfrac{r_0}{r} + 1\right)e^{-r/r_0}$; (b) 0.030; (c) $F(r) = -C/r^2$, 0.11.

91. 6.7 hp.

93. (a) 2.8 m; (b) 1.5 m; (c) 1.5 m.

95. 76 hp.

97. (a) 5.00×10^3 m/s; (b) 2.89×10^3 m/s.

CHAPTER 9

1. 6.0×10^7 N, up.

3. (a) 0.36 kg·m/s; (b) 0.12 kg·m/s.

5. $(26 \text{ N·s})\mathbf{i} - (28 \text{ N·s})\mathbf{j}$.

7. (a) $(8h/g)^{1/2}$; (b) $(2gh)^{1/2}$; (c) $-(8m^2gh)^{1/2}$ (up); (d) mg (down), a surprising result.

9. 3.4×10^4 kg.

11. 4.4×10^3 m/s.

13. -0.667 m/s (opposite to the direction of the package).

15. 2, lesser kinetic energy has greater mass.

17. $\frac{3}{2}v_0\mathbf{i} - v_0\mathbf{j}$.

19. 1.1×10^{-22} kg·m/s, 36° from the direction opposite to the electron's.

21. (a) $(100 \text{ m/s})\mathbf{i} + (50 \text{ m/s})\mathbf{j}$; (b) 3.3×10^5 J.

23. 130 N, not large enough.

25. 1.1×10^3 N.

27. (a) $2mv/\Delta t$; (b) $2mv/t$.

29. (a)

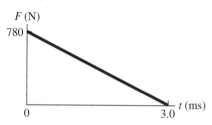

(b) 1.2 N·s; (c) 1.2 N·s; (d) 3.9 g.

31. (a) $(0.84 \text{ N}) + (1.2 \text{ N/s})t$; (b) 18.5 N; (c) $(0.12 \text{ kg/s})\{[(49 \text{ m}^2/\text{s}^2) - (1.18 \text{ m}^2/\text{s}^3)t]^{1/2} + (9.80 \text{ m/s}^2)t\}$, 18.3 N.

33. $v_1' = -1.40$ m/s (rebound), $v_2' = 2.80$ m/s.

35. (a) 2.7 m/s; (b) 0.84 kg.

37. 3.2×10^3 m/s.

39. (a) 1.00; (b) 0.89; (c) 0.29; (d) 0.019.

41. (a) 0.32 m; (b) -3.1 m/s (rebound), 4.9 m/s; (c) Yes.

43. (a) $+M/(m + M)$; (b) 0.964.

45. 141°.

47. (b) $e = (h'/h)^{1/2}$.

49. (a) $v_1' = v_2' = 1.9$ m/s; (b) $v_1' = -1.6$ m/s, $v_2' = 7.9$ m/s; (c) $v_1' = 0$, $v_2' = 5.2$ m/s; (d) $v_1' = 3.1$ m/s, $v_2' = 0$; (e) $v_1' = -4.0$ m/s, $v_2' = 12$ m/s; result for (c) is reasonable, result for (d) is not reasonable, result for (e) is not reasonable.

51. 61° from first eagles's direction, 6.8 m/s.

53. (a) 30°; (b) $v_2' = v/\sqrt{3}$, $v_1' = v/\sqrt{3}$; (c) $\frac{2}{3}$.

55. $\theta'_1 = 76°$, $v'_n = 5.1 \times 10^5$ m/s, $v'_{He} = 1.8 \times 10^5$ m/s.

59. 6.5×10^{-11} m from the carbon atom.

61. 0.030 nm above center of H triangle.

63. $x_{CM} = 1.10$ m (East), $y_{CM} = -1.10$ m (South).

65. $x_{CM} = 0$, $y_{CM} = 2r/\pi$.

67. $x_{CM} = 0$, $y_{CM} = 0$, $z_{CM} = 3h/4$ above the point.

69. (a) 4.66×10^6 m.

71. (a) $x_{CM} = 4.6$ m; (b) 4.3 m; (c) 4.6 m.

73. $mv/(M + m)$ up, balloon will also stop.

75. 55 m.

77. 0.899 hp.

79. (a) 2.3×10^3 N; (b) 2.8×10^4 N; (c) 1.1×10^4 hp.

81. A "scratch shot".

83. 1.4×10^4 N, 43.3°.

85. 5.1×10^2 m/s.

87. $m_2 = 4.00m$.

89. 50%.

91. (a) No; (b) $v_1/v_2 = -m_2/m_1$; (c) m_2/m_1; (d) does not move; (e) center of mass will move.

93. 8.29 m/s.

95. (a) 2.5×10^{-13} m/s; (b) 1.7×10^{-17}; (c) 0.19 J.

97. $m \leq M/3$.

99. 29.6 km/s.

101. (a) 2.3 N·s; (b) 4.5×10^2 N.

103. (a) Inelastic collision; (b) 0.10 s; (c) -1.4×10^5 N.

105. 0.28 m, 1.1 m.

CHAPTER 10

1. (a) $\pi/6$ rad $= 0.524$ rad; (b) $19\pi/60 = 0.995$ rad; (c) $\pi/2 = 1.571$ rad; (d) $2\pi = 6.283$ rad; (e) $7\pi/3 = 7.330$ rad.

3. 2.3×10^3 m.

5. (a) 0.105 rad/s; (b) 1.75×10^{-3} rad/s; (c) 1.45×10^{-4} rad/s; (d) zero.

7. (a) 464 m/s; (b) 185 m/s; (c) 355 m/s.

9. (a) 262 rad/s; (b) 46 m/s, 1.2×10^4 m/s² radial.

11. 7.4 cm.

13. (a) 1.75×10^{-4} rad/s²; (b) $a_R = 1.17 \times 10^{-2}$ m/s², $a_{tan} = 7.44 \times 10^{-4}$ m/s².

15. (a) 0.58 rad/s2; (b) 12 s.

17. (a) $(1.67 \text{ rad/s}^4)t^3 - (1.75 \text{ rad/s}^3)t^2$; (b) $(0.418 \text{ rad/s}^4)t^4 - (0.583 \text{ rad/s}^3)t^3$; (c) 6.4 rad/s, 2.0 rad.

19. (a) ω_1 is in the $-x$-direction, ω_2 is in the $+z$-direction; (b) $\omega = 61.0$ rad/s, 35.0° above $-x$-axis; (c) $-(1.75 \times 10^3 \text{ rad/s}^2)\mathbf{j}$.

21. (a) 35 m·N; (b) 30 m·N.

23. 1.2 m·N (clockwise).

25. 3.5×10^2 N, 2.0×10^3 N.

27. 53 m·N.

29. (a) 3.5 kg·m²; (b) 0.024 m·N.

31. 2.25×10^3 kg·m², 8.8×10^3 m·N.

33. 9.5×10^4 m·N.

35. 10 m/s.

37. (a)

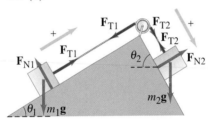

(b) $F_{T1} = 47$ N, $F_{T2} = 75$ N; (c) 7.0 m·N, 1.7 kg·m².

39. Thin hoop (through center): $k = R_0$;

Thin hoop (through diameter): $k = [(R_0^2/2) + (w^2/12)]^{1/2}$;

Solid cylinder (through center): $k = R/\sqrt{2}$;

Hollow cylinder (through center): $k = [(R_1^2 + R_2^2)/2]^{1/2}$;

Uniform sphere (through center): $k = (2r_0^2/5)^{1/2}$;

Rod (through center): $k = \ell/\sqrt{12}$;

Rod (through end): $k = \ell/\sqrt{3}$;

Plate (through center): $k = [(\ell^2 + w^2)/12]^{1/2}$.

41. (a) 4.18 rad/s²; (b) 8.37 m/s²; (c) 421 m/s²; (d) 3.07×10^3 N; (e) 1.14°.

43. (a) $I_a = Ms^2/12$; (b) $I_b = Ms^2/12$.

45. (a) $5.30MR_0^2$; (b) -15%.

47. (a) $9MR_0^2/16$; (b) $MR_0^2/4$; (c) $5MR_0^2/4$.

51. (b) $M\ell^2/12$, $Mw^2/12$.

53. 0.38 rev/s.

55. (a) As moment of inertia increases, angular velocity must decrease; (b) 1.6.

57. (a) 7.1×10^{33} kg·m²/s; (b) 2.7×10^{40} kg·m²/s.

59. 0.45 rad/s, 0.80 rad/s.

61. 2.33×10^4 J.

63. 5×10^9, loss of gravitational potential energy.

65. 1.4 m/s.

67. (a) 2.5 kg·m²; (b) 0.58 kg·m²; (c) 0.35 s; (d) -72 J; (e) rotating.

69. 12.4 m/s.

71. 1.4×10^2 J.

73. (a) 4.48 m/s; (b) 1.21 J; (c) $\mu_s \geq 0.197$.

75. $v = [10g(R_0 - r_0)/7]^{1/2}$.

77. (a) 4.5×10^5 J; (b) 0.18 (18%); (c) 1.71 m/s²; (d) 6.4%.

79. (a) $12v_0^2/49 \mu_k g$; (b) $v = 5v_0/7$, $\omega = 5v_0/7R$.

81. (a) 4.5 m/s², 19 rad/s²; (b) 5.8 m/s; (c) 15.3 J; (d) 1.4 J; (e) $K = 16.7$ J, $\Delta E = 0$; (f) $a = 4.5$ m/s², $v = 5.8$ m/s, 14.1 J.

83. $\theta_{Sun} = 9.30 \times 10^{-3}$ rad (0.53°), $\theta_{Moon} = 9.06 \times 10^{-3}$ rad (0.52°).

85. $\omega_1/\omega_2 = R_2/R_1$.

87. $\ell/2$, $\ell/2$.

89. (a) $-(I_W/I_P)\omega_W$ (down); (b) $-(I_W/2I_P)\omega_W$ (down); (c) $(I_W/I_P)\omega_W$ (up); (d) 0.

91. (a) $\omega_R/\omega_F = N_F/N_R$; (b) 4.0; (c) 1.5.

93. (a) 1.5×10^2 rad/s²; (b) 1.2×10^3 N.

95. (a) 0.070 rad/s²; (b) 40 rpm.

97. 7.9 N.

99. (b) 2.2×10^3 rad/s; (c) 24 min.

101. (a) 2.9 m; (b) 3.6 m.

103. (a) 1.2 rad/s; (b) 2.0×10^3 J, 1.2×10^3 J, loss of 8.0×10^2 J, decrease of 40%.

105. (a) 1.7 m/s; (b) 0.84 m/s.

107. (a) $h_{min} = 2.7R_0$; (b) $h_{min} = 2.7R_0 - 1.7r_0$.

109. (a) 0.84 m/s; (b) 0.96.

CHAPTER 11

7. (a) $-7.0\mathbf{i} - 14.0\mathbf{j} + 19.3\mathbf{k}$; (b) 164°.

11. $-(30.3 \text{ m·N})\mathbf{k}$ (in $-z$-direction).

13. $(18 \text{ m·kN})\mathbf{i} \pm (14 \text{ m·kN})\mathbf{j} \mp (19 \text{ m·kN})\mathbf{k}$.

19. $(55\mathbf{i} - 90\mathbf{j} + 42\mathbf{k}) \, \text{kg} \cdot \text{m}^2/\text{s}$.

21. (a) $[(7m/9) + (M/6)]\ell^2\omega^2$;
(b) $[(14m/9) + (M/3)]\ell^2\omega$.

23. $2.30 \, \text{m/s}^2$.

25. (a) $L = [R_0M_1 + R_0M_2 + (I/R_0)]v$;
(b) $a = M_2g/[M_1 + M_2 + (I/R_0^2)]$.

27. Rod rotates at 7.8 rad/s about the center of mass, which moves with constant velocity of 0.21 m/s.

31. $F_1 = [(d + r\cos\phi)/2d]m_1r\omega^2\sin\phi$,
$F_2 = [(d - r\cos\phi)/2d]m_1r\omega^2\sin\phi$.

33. $16 \, \text{N}, -7.5 \, \text{N}$.

35. $3m^2v^2/g(3m + 4M)(m + M)$.

37. $(1 - 4.7 \times 10^{-13})\omega_E$.

39. (a) 14 m/s; (b) 6.8 rad/s.

41. $1.02 \times 10^{-3} \, \text{kg} \cdot \text{m}^2$.

43. 2.2 rad/s (0.35 rev/s).

45. $\tan^{-1}(r\omega^2/g)$.

47. (a) g, along a radial line; (b) $0.998g$, 0.0988° south from a radial line; (c) $0.997g$, along a radial line.

49. North or south direction.

51. (a) South; (b) $\omega D^2 \sin\lambda/v_0$; (c) 0.46 m.

53. (a) $(-9.0\mathbf{i} + 12\mathbf{j} - 8.0\mathbf{k}) \, \text{kg} \cdot \text{m}^2/\text{s}$;
(b) $(9.0\mathbf{j} - 6.0\mathbf{k}) \, \text{m} \cdot \text{N}$.

55. (a) Turn in the direction of the lean;
(b) $\Delta L = 0.98 \, \text{kg} \cdot \text{m}^2/\text{s}$,
$\Delta L = 0.18L_0$.

57. (a) $1.8 \times 10^3 \, \text{kg} \cdot \text{m}^2/\text{s}^2$;
(b) $1.8 \times 10^3 \, \text{m} \cdot \text{N}$; (c) $2.1 \times 10^3 \, \text{W}$.

59. $v_{\text{CM}} = (3g\ell/4)^{1/2}$.

61. $(19 \, \text{m/s})(1 - \cos\theta)^{1/2}$.

63. (a) $2.3 \times 10^4 \, \text{rev/s}$;
(b) $5.7 \times 10^3 \, \text{rev/s}$.

CHAPTER 12

1. 379 N, 141°.

3. $1.6 \times 10^3 \, \text{m} \cdot \text{N}$.

5. 6.52 kg.

7. 2.84 m from the adult.

9. 0.32 m.

11. $F_{T1} = 3.4 \times 10^3 \, \text{N}$,
$F_{T2} = 3.9 \times 10^3 \, \text{N}$.

13. $F_1 = -2.94 \times 10^3 \, \text{N}$ (down),
$F_2 = 1.47 \times 10^4 \, \text{N}$.

15. Top hinge: $F_{Ax} = 55.2 \, \text{N}$,
$F_{Ay} = 63.7 \, \text{N}$; bottom hinge:
$F_{Bx} = -55.2 \, \text{N}, F_{By} = 63.7 \, \text{N}$.

17. (a)

(b) $1.5 \times 10^4 \, \text{N}$; (c) $6.7 \times 10^3 \, \text{N}$.

19. $F_T = 1.4 \times 10^3 \, \text{N}$ (up),
$F_{\text{bone}} = 2.1 \times 10^3 \, \text{N}$ (down).

21. $2.7 \times 10^3 \, \text{N}$.

23. 89.5 cm from the feet.

25. $F_1 = 5.8 \times 10^3 \, \text{N}, F_2 = 5.6 \times 10^3 \, \text{N}$.

27. (a) $2.1 \times 10^2 \, \text{N}$; (b) $2.0 \times 10^3 \, \text{N}$.

29. $7.1 \times 10^2 \, \text{N}$.

31. $F_T = 2.5 \times 10^2 \, \text{N}$,
$F_{AH} = 2.5 \times 10^2 \, \text{N}$,
$F_{AV} = 2.0 \times 10^2 \, \text{N}$.

33. (a) 1.00 N; (b) 1.25 N.

35. $\theta_{\text{max}} = 40°$, same.

37. (a) $F_T = 182 \, \text{N}$; (b) $F_{N1} = 352 \, \text{N}$,
$F_{N2} = 236 \, \text{N}$; (c) $F_B = 298 \, \text{N}, 52.4°$.

39. $1.0 \times 10^2 \, \text{N}$.

41. (a) $1.2 \times 10^5 \, \text{N/m}^2$; (b) 2.4×10^{-6}.

43. (a) $1.3 \times 10^5 \, \text{N/m}^2$; (b) 6.5×10^{-7}; (c) 0.0062 mm.

45. $9.6 \times 10^6 \, \text{N/m}^2$.

47. $9.0 \times 10^7 \, \text{N/m}^2, 9.0 \times 10^2 \, \text{atm}$.

49. $2.2 \times 10^7 \, \text{N}$.

51. (a) $1.1 \times 10^2 \, \text{m} \cdot \text{N}$; (b) wall; (c) all three.

53. $3.9 \times 10^2 \, \text{N}$, thicker strings, maximum strength is exceeded.

55. (a) $4.4 \times 10^{-5} \, \text{m}^2$; (b) 2.7 mm.

57. 1.2 cm.

61. (a) $F_T = 129 \, \text{kN}$;
$F_A = 141 \, \text{kN}, 23.5°$;
(b) $F_{DE} = 64.7 \, \text{kN}$ (tension),
$F_{CE} = 32.3 \, \text{kN}$ (compression),
$F_{CD} = 64.7 \, \text{kN}$ (compression),
$F_{BD} = 64.7 \, \text{kN}$ (tension),
$F_{BC} = 64.7 \, \text{kN}$ (tension),
$F_{AC} = 97.0 \, \text{kN}$ (compression),
$F_{AB} = 64.7 \, \text{kN}$ (compression).

63. (a) $4.8 \times 10^{-2} \, \text{m}^2$; (b) $6.8 \times 10^{-2} \, \text{m}^2$.

65. $F_{AB} = 5.44 \times 10^4 \, \text{N}$ (compression),
$F_{ACx} = 2.72 \times 10^4 \, \text{N}$ (tension),
$F_{BC} = 5.44 \times 10^4 \, \text{N}$ (tension),
$F_{BD} = 5.44 \times 10^4 \, \text{N}$ (compression),
$F_{CD} = 5.44 \times 10^4 \, \text{N}$ (tension),
$F_{CE} = 2.72 \times 10^4 \, \text{N}$ (tension),
$F_{DE} = 5.44 \times 10^4 \, \text{N}$ (compression).

67. 12 m.

69. $M_C = 0.191 \, \text{kg}, M_D = 0.0544 \, \text{kg}$,
$M_A = 0.245 \, \text{kg}$.

71. (a) $Mg[h/(2R - h)]^{1/2}$;
(b) $Mg[h(2R - h)]^{1/2}/(R - h)$.

73. $\theta_{\text{max}} = 29°$.

75. 6, 2.0 m apart.

77. 3.8.

79. $5.0 \times 10^5 \, \text{N}, 3.2 \, \text{m}$.

81. (a) 600 N; (b) $F_A = 0, F_B = 1200 \, \text{N}$;
(c) $F_A = 150 \, \text{N}, F_B = 1050 \, \text{N}$;
(d) $F_A = 750 \, \text{N}, F_B = 450 \, \text{N}$.

83. $6.5 \times 10^2 \, \text{N}$.

85. 0.67 m.

87. Right end is safe, left end is not safe, 0.10 m.

89. (a) $F_L = 3.3 \times 10^2 \, \text{N}$ up,
$F_R = 2.3 \times 10^2 \, \text{N}$ down;
(b) 65 cm from right hand;
(c) 123 cm from right hand.

91. $\theta \geq 40°$.

93. (b) beyond the table;
(c) $D = L\sum_{i=1}^{n}\frac{1}{2i}$; (d) 32 bricks.

95. $F_{TB} = 134 \, \text{N}, F_{TA} = 300 \, \text{N}$.

97. $2.6w$, 31° above horizontal.

CHAPTER 13

1. $3 \times 10^{11} \, \text{kg}$.

3. $4.3 \times 10^2 \, \text{kg}$.

5. 0.8477.

7. (a) $3 \times 10^7 \, \text{N/m}^2$;
(b) $2 \times 10^5 \, \text{N/m}^2$.

9. 1.1 m.

11. $8.28 \times 10^3 \, \text{kg}$.

13. $1.2 \times 10^5 \, \text{N/m}^2$,
$2.3 \times 10^7 \, \text{N}$ (down),
$1.2 \times 10^5 \, \text{N/m}^2$.

15. $6.54 \times 10^2 \, \text{kg/m}^3$.

17. $3.36 \times 10^4 \, \text{N/m}^2$ (0.331 atm).

19. (a) $1.41 \times 10^5 \, \text{Pa}$; (b) $9.8 \times 10^4 \, \text{Pa}$.

21. (a) 0.34 kg; (b) $1.5 \times 10^4 \, \text{N}$ (up).

23. (c) $\geq 0.38h$, no.

27. $4.70 \times 10^3 \, \text{kg/m}^3$.

29. $8.5 \times 10^2 \, \text{kg}$.

31. Copper.

33. (a) $1.14 \times 10^6 \, \text{N}$; (b) $4.0 \times 10^5 \, \text{N}$.

35. (b) Above the center of gravity.

37. 0.88.

39. $7.9 \times 10^2 \, \text{kg}$.

43. 4.1 m/s.

45. 9.5 m/s.

47. $1.5 \times 10^5 \, \text{N/m}^2 = 1.5 \, \text{atm}$.

49. $4.11 \times 10^{-3} \, \text{m}^3/\text{s}$.

51. $1.7 \times 10^6 \, \text{N}$.

59. (a) $2[h_1(h_2 - h_1)]^{1/2}$;
  (b) $h_1' = h_2 - h_1$.

61. $0.072\ \text{Pa} \cdot \text{s}$.

63. $4.0 \times 10^3\ \text{Pa}$.

65. 11 cm.

67. (a) Laminar; (b) 3200, turbulent.

69. 1.9 m.

71. $9.1 \times 10^{-3}\ \text{N}$.

73. (a) $\gamma = F/4\pi r$; (b) 0.024 N/m.

75. (a) 0.88 m; (b) 0.55 m; (c) 0.24 m.

77. $1.5 \times 10^2\ \text{N} \le F \le 2.2 \times 10^2\ \text{N}$.

79. 0.051 atm.

81. 0.63 N.

83. 5 km.

85. $5.3 \times 10^{18}\ \text{kg}$.

87. 2.6 m.

89. 39 people.

91. 37 N, not float.

93. $d = D[v_0^2/(v_0^2 + 2gy)]^{1/4}$.

95. (a) 3.2 m/s; (b) 19 s.

97. $1.9 \times 10^2\ \text{m/s}$.

CHAPTER 14

1. 0.60 m.

3. 1.15 Hz.

5. (a) 2.4 N/m; (b) 12 Hz.

7. (a) $0.866\ x_{\max}$; (b) $0.500\ x_{\max}$.

9. $0.866\ A$.

11. $[(k_1 + k_2)/m]^{1/2}/2\pi$.

13. (a) 8/7 s, 0.875 Hz; (b) 3.3 m,
  -10.4 m/s; (c) $+18$ m/s, -57 m/s^2.

15. 3.6 Hz.

19. (a) $y = -(0.220\ \text{m}) \sin[(37.1\ \text{s}^{-1})t]$;
  (b) maximum extensions at 0.0423 s,
  0.211 s, 0.381 s, …; minimum
  extensions at 0.127 s, 0.296 s,
  0.465 s, ….

21. $f = (3k/M)^{1/2}/2\pi$.

25. (a) $x = (12.0\ \text{cm}) \cos[(25.6\ \text{s}^{-1})t + 1.89\ \text{rad}]$;
  (b) $t_{\max} = 0.294$ s, 0.539 s, 0.784 s, …;
  $t_{\min} = 0.171$ s, 0.416 s, 0.661 s, …;
  (c) -3.77 cm; (d) $+13.1$ N (up);
  (e) 3.07 m/s, 0.110 s.

27. (a) 0.650 m; (b) 1.34 Hz; (c) 29.8 J;
  (d) $K = 25.0$ J, $U = 4.8$ J.

29. 9.37 m/s.

31. $A_1 = 2.24A_2$.

33. (a) $4.2 \times 10^2\ \text{N/m}$; (b) 3.3 kg.

35. 352.6 m/s.

39. 0.9929 m.

41. (a) 0.248 m; (b) 2.01 s.

43. (a) $-12°$; (b) $+1.9°$; (c) $-13°$.

45. $\frac{1}{3}$.

47. 1.08 s.

49. 0.31 g.

51. (a) 1.6 s.

53. 3.5 s.

55. (a) 0.727 s; (b) 0.0755;
  (c) $x = (0.189\ \text{m})e^{-(0.108/\text{s})t}$
  $\sin[(8.64\ \text{s}^{-1})t]$.

57. (a) 8.3×10^{-4}%; (b) 39 periods.

59. (a) 5.03 Hz; (b) $0.0634\ \text{s}^{-1}$;
  (c) 110 oscillations.

61.

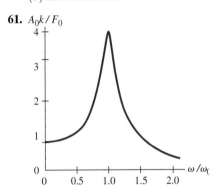

65. (a) 198 s; (b) $8.7 \times 10^{-6}\ \text{W}$;
  (c) $8.8 \times 10^{-4}\ \text{Hz}$ on either side of f_0.

69. (a) 0.63 Hz; (b) 0.65 m/s; (c) 0.077 J.

71. 151 N/m, 20.3 m.

73. 0.11 m.

75. 3.6 Hz.

77. (a) 1.1 Hz; (b) 13 J.

79. (a) 90 N/m; (b) 8.9 cm.

81. $k = \rho_{\text{water}} g A$.

83. Water will oscillate with SHM,
  $k = 2\rho g A$, the density and the cross
  section.

85. $T = 2\pi(ma/2k\ \Delta a)^{1/2}$.

87. (a) 1.64 s; (b) 0.67 m.

CHAPTER 15

1. 2.3 m/s.

3. 1.26 m.

5. 0.72 m.

7. 2.7 N.

9. (a) 75 m/s; (b) $7.8 \times 10^3\ \text{N}$.

11. (a) $1.3 \times 10^3\ \text{km}$;
  (b) cannot be determined.

13. (a) 0.25; (b) 0.50.

17. (a) 0.30 W; (b) 0.25 cm.

19. $D = D_M \sin[2\pi(x/\lambda + t/T) + \phi]$.

21. (a) 41 m/s; (b) $6.4 \times 10^4\ \text{m/s}^2$;
  (c) 41 m/s, $8.2 \times 10^3\ \text{m/s}^2$.

23. (a, c)

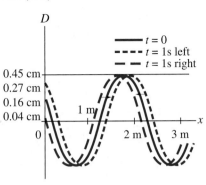

  (b) $D = (0.45\ \text{m}) \cos[(3.0\ \text{m}^{-1})x - (6.0\ \text{s}^{-1})t + 1.2]$.
  (d) $D = (0.45\ \text{m}) \cos[(3.0\ \text{m}^{-1})x + (6.0\ \text{s}^{-1})t + 1.2]$.

25. $D = -(0.020\ \text{cm}) \cos[(8.01\ \text{m}^{-1})x - (2.76 \times 10^3\ \text{s}^{-1})t]$.

27. The function is a solution.

31. (a) $v_2/v_1 = (\mu_1/\mu_2)^{1/2}$;
  (b) $\lambda_2/\lambda_1 = v_2/v_1 = (\mu_1/\mu_2)^{1/2}$;
  (c) lighter cord.

33. (c) $A_T = [2k_1/(k_2 + k_1)]A = [2v_2/(v_1 + v_2)]A$.

35. (b) $2D_M \cos(\frac{1}{2}\phi)$, purely sinusoidal;
  (d) $D = \sqrt{2}\ D_M \sin(kx - \omega t + \pi/4)$.

37. 440 Hz, 880 Hz, 1320 Hz, 1760 Hz.

39. $f_n = n(0.50\ \text{Hz}), n = 1, 2, 3, …$;
  $T_n = (2.0\ \text{s})/n, n = 1, 2, 3, …$.

41. 70 Hz.

45. 4.

47. (a) $D_2 = (4.2\ \text{cm}) \sin[(0.71\ \text{cm}^{-1})x + (47\ \text{s}^{-1})t + 2.1]$;
  (b) $D_{\text{resultant}} = (8.4\ \text{cm})$
  $\sin[(0.71\ \text{cm}^{-1})x + 2.1]$
  $\cos[(47\ \text{s}^{-1})t]$.

49. 308 Hz.

51. (a)

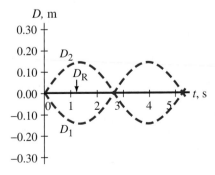

(b)

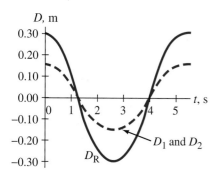

53. 5.4 km/s.

55. 29°.

57. 24°.

59. Speed will be greater in the less dense rod by a factor of $\sqrt{2}$.

61. (a) 0.050 m; (b) 2.3.

63. 0.99 m.

65. (a) solid curves,

(c) dashed curves;

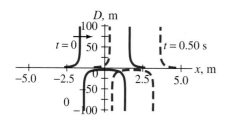

(b)
$D = (4.0\,\text{m}^3)/\{[x - (3.0\,\text{m/s})t]^2 - 2.0\,\text{m}^2\};$

(d)

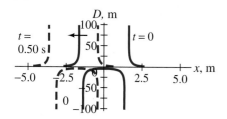

$D = (4.0\,\text{m}^3)/\{[x + (3.0\,\text{m/s})t]^2 - 2.0\,\text{m}^2\}.$

67. (a) 784 Hz, 1176 Hz, 880 Hz,

1320 Hz; (b) 1.26; (c) 1.12; (d) 0.794.

69. $\lambda_n = 4L/(2n - 1), n = 1, 2, 3, \ldots.$

71. $y = (3.5\,\text{cm})\cos\big[(1.05\,\text{cm}^{-1})x$

$- (1.39\,\text{s}^{-1})t\big].$

73.

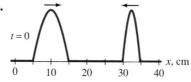
$t = 0$

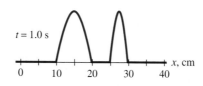
$t = 1.0\,\text{s}$

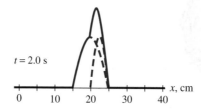
$t = 2.0\,\text{s}$

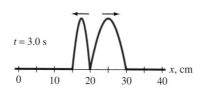
$t = 3.0\,\text{s}$

CHAPTER 16

1. 2.6×10^2 m.

3. 5.4×10^2 m.

5. 1200 m, 300 m.

7. (a) 1.1×10^{-8} m; (b) 1.1×10^{-10} m.

9. (a) $\Delta P = (4 \times 10^{-5}\,\text{Pa})$
$\sin\big[(0.949\,\text{m}^{-1})x$
$- (315\,\text{s}^{-1})t\big];$
(b) $\Delta P = (4 \times 10^{-3}\,\text{Pa})$
$\sin\big[(94.9\,\text{m}^{-1})x$
$- (3.15 \times 10^4\,\text{s}^{-1})t\big].$

11. (a) 49 dB; (b) 3.2×10^{-10} W/m².

13. 150 Hz to 20,000 Hz.

15. (a) 9; (b) 9.5 dB.

17. (a) Higher frequency is greater by a factor of 2; (b) 4.

19. (a) 5.0×10^{-13} W; (b) 6.3×10^4 yr.

21. 87 dB.

23. (a) 5.10×10^{-5} m; (b) 29.8 Pa.

25. (a) 1.5×10^3 W; (b) 3.4×10^2 m.

27. (b) 190 dB.

29. (a) 570 Hz; (b) 860 Hz.

31. 8.6 mm < L < 8.6 m.

33. (a) 110 Hz, 330 Hz, 550 Hz, 770 Hz; (b) 220 Hz, 440 Hz, 660 Hz, 880 Hz.

35. (a) 0.656 m; (b) 262 Hz, 1.31 m; (c) 1.31 m, 262 Hz.

37. -2.6%.

39. (a) 0.578 m; (b) 869 Hz.

41. 215 m/s.

43. $0.64, 0.20, -2$ dB, -7 dB.

45. 28.5 kHz.

47. (a) 130.5 Hz, or 133.5 Hz; (b) $\pm 2.3\%$.

49. (a) 343 Hz; (b) 1030 Hz, 1715 Hz.

53. 346 Hz.

57. (a) 1690 Hz; (b) 1410 Hz.

59. 30,890 Hz.

61. 120 Hz.

63. 91 Hz.

65. 90 beats/min.

67. (a) 570 Hz; (b) 570 Hz; (c) 570 Hz; (d) 570 Hz; (e) 594 Hz; (f) 595 Hz.

71. (a) 120; (b) 0.96°.

73. (a) 37°; (b) 1.7.

75. 0.278 s.

77. 55 m.

79. 410 km/h (255 mi/h).

81. 1, 0.444, 0.198, 0.0878, 0.0389.

83. 18.1 W.

85. 15 W.

87. 2.3 Hz.

89. $\Delta P_M/\Delta P_{M0} = D_M/D_{M0} = 10^6.$

91. 50 dB.

93. 17.5 m/s.

95. 2.3 kHz.

97. 550 Hz.

99. (a) 2.8×10^3 Hz; (b) 1.80 m; (c) 0.12 m.

101. (a) 2.2×10^{-7} m; (b) 5.4×10^{-5} m.

CHAPTER 17

1. 0.548.

3. (a) 20°C; (b) ≈ 3300°F.

5. 104.0°F.

7. -40°F $= -40$°C.

9. $\Delta L_{\text{Invar}} = 2.0 \times 10^{-6}$ m,
$\Delta L_{\text{steel}} = 1.2 \times 10^{-4}$ m,
$\Delta L_{\text{marble}} = 2.5 \times 10^{-5}$ m.

11. -69°C.

13. 5.1 mL.

15. 0.06 cm³.

19. -40 min.

21. -2.8×10^{-3} (0.28%).

23. 3.5×10^7 N/m².

25. (a) 27°C; (b) 4.3×10^3 N.

27. -459.7°F.

29. 1.07 m³.

31. 1.43 kg/m³.

33. (a) 14.8 m³; (b) 1.81 atm.

35. 1.80×10^3 atm.

37. 37°C.

39. 3.43 atm.

41. 0.588 kg/m³, water vapor is not an ideal gas.

43. 2.69×10^{25} molecules/m³.

45. 4.9×10^{22} molecules.

47. 7.7×10^3 N.

49. (a) 71.2 torr; (b) 157°C.

51. (a) 0.19 K; (b) 0.051%.

53. (a) Low; (b) 0.017%.

55. 1/6.

57. 5.1×10^{27} molecules, 8.4×10^3 mol.

59. 11 L, not advisable.

61. (a) 9.3×10^2 kg; (b) 1.0×10^2 kg.

63. 1.1×10^{44} molecules.

65. 3.3×10^{-7} cm.

67. 1.1×10^3 m.

69. 15 h.

71. 0.66×10^3 kg/m³.

73. ± 0.11 C°.

77. 3.6 m.

CHAPTER 18

1. (a) 5.65×10^{-21} J; (b) 7.3×10^3 J.

3. 1.17.

5. (a) 4.5; (b) 5.2.

7. $\sqrt{2}$.

11. (a) 461 m/s; (b) 19 s⁻¹.

13. 1.00429.

17. Vapor.

19. (a) Gas, liquid, vapor;
(b) gas, liquid, solid, vapor.

21. 0.69 atm.

23. 11°C.

25. 1.96 atm.

27. 120°C.

29. (a) 5.3×10^6 Pa; (b) 5.7×10^6 Pa.

31. (b) $b = 4.28 \times 10^{-5}$ m³/mol,
$a = 0.365$ N·m⁴/mol².

33. (a) 10^{-7} atm; (b) 300 atm.

35. (a) 6.3 cm; (b) 0.58 cm.

37. 2×10^{-7} m.

39. (b) 4.7×10^7 s⁻¹.

43. 7.8 h.

45. (b) 4×10^{-11} mol/s; (c) 0.7 s.

47. 2.6×10^2 m/s,
4×10^{-17} N/m² ≈ 4×10^{-22} atm.

49. (a) 2.9×10^2 m/s; (b) 12 m/s.

51. Reasonable, 70 cm.

53. $mgh = 4.3 \times 10^{-5}(\frac{1}{2}mv_{rms}^2)$,
reasonable.

55. $P_2/P_1 = 1.43$, $T_2/T_1 = 1.20$.

57. 1.4×10^5 K.

59. (a) 1.7×10^3 Pa; (b) 7.0×10^2 Pa.

61. 2×10^{13} m.

CHAPTER 19

1. 1.0×10^7 J.

3. 1.8×10^2 J.

5. 2.1×10^2 kg/h.

7. 83 kcal.

9. 4.7×10^6 J.

11. 40 C°.

13. 186°C.

15. 7.1 min.

17. (b) $mc_0[(T_2 - T_1) + a(T_2^2 - T_1^2)/2]$;
(c) $c_{mean} = c_0[1 + \frac{1}{2}a(T_2 + T_1)]$.

19. 0.334 kg (0.334 L).

21. $\frac{2}{3}m$ steam and $\frac{4}{3}m$ water at 100°C.

23. 9.4 g.

25. 4.7×10^3 kcal.

27. 1.22×10^4 J/kg.

29. 360 m/s.

31. (a) 0; (b) 5.00×10^3 J.

33.

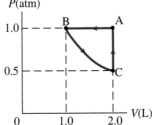

35. (a) 0; (b) −1300 kJ.

37. (a) 1.6×10^2 J; (b) $+1.6 \times 10^2$ J.

39. $W = 3.46 \times 10^3$ J, $\Delta U = 0$,
$Q = +3.46 \times 10^3$ J (into the gas).

41. +129 J.

45. (a) +25 J; (b) +63 J; (c) −95 J;
(d) −120 J; (e) −15 J.

47. $W = RT \ln\left(\dfrac{V_2 - b}{V_1 - b}\right) + a\left(\dfrac{1}{V_2} - \dfrac{1}{V_1}\right)$.

49. 22°C/h.

51. 4.98 cal/mol·K, 2.49 kcal/kg·K;
6.97 cal/mol·K, 3.48 kcal/kg·K.

53. 83.7 g/mol, krypton.

55. 46 C°.

57. (a) 2.08×10^3 J; (b) 8.32×10^2 J;
(c) 2.91×10^3 J.

59. 0.379 atm, −51°C.

61. 1.33×10^3 J.

63. (a) $T_1 = 317$ K, $T_2 = 153$ K;
(b) -1.59×10^4 J; (c) -1.59×10^4 J;
(d) $Q = 0$.

65. (a)

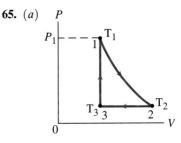

(b) 231 K;
(c) $Q_{1\to2} = 0$,
$\Delta U_{1\to2} = -2.01 \times 10^3$ J,
$W_{1\to2} = +2.01 \times 10^3$ J;
$W_{2\to3} = -1.31 \times 10^3$ J,
$\Delta U_{2\to3} = -1.97 \times 10^3$ J,
$Q_{2\to3} = -3.28 \times 10^3$ J;
$W_{3\to1} = 0$,
$\Delta U_{3\to1} = +3.98 \times 10^3$ J,
$Q_{3\to1} = +3.98 \times 10^3$ J;
(d) $W_{cycle} = +0.70 \times 10^3$ J,
$Q_{cycle} = +0.70 \times 10^3$ J,
$\Delta U_{cycle} = 0$.

67. (a) 64 W; (b) 22 W.

69. 4.8×10^2 W.

71. 31 h.

73. (a) 1.7×10^{17} W; (b) 278K (5°C).

75. (b) $\Delta Q/\Delta t = A(T_2 - T_1)/\Sigma(\ell_i/k_i)$.

77. 22%.

79. 4×10^{15} J.

81. 2.8 kcal/kg.

83. 30 C°.

85. 682 J.

87. 2.8 C°.

89. 2.58 cm, rod vaporizes.

91. 4.3 kg.

95. (a) 2.3 C°/s; (b) 84°C;
(c) convection, conduction,
evaporation.

97. (a) $\rho = m/V = (mP/nR)/T$;
(b) $\rho = (m/nRT)P$.

99. (a) 1.9×10^5 J; (b) -1.4×10^5 J;

(c)

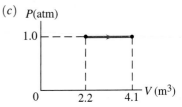

101. 3.2×10^5 s = 3.7 d.

103. 10 C°.

CHAPTER 20

1. 24%.

3. 816 MW.

5. 18%.

7. 13 km^3/day, 63 km^2.

9. 28.0%.

13. 1.2×10^{13} J/h.

15. 1.4×10^3 m/day.

17. 660°C.

19. 3.7×10^8 kg/h.

21. (a) $P_a = 5.15 \times 10^5$ Pa,
$P_b = 2.06 \times 10^5$ Pa;
(b) $V_c = 30.0$ L, $V_d = 12.0$ L;
(c) 2.83×10^3 J; (d) -2.14×10^3 J;
(e) 0.69×10^3 J; (f) 24%.

23. 5.7.

25. -21°C.

27. 2.9.

29. (a) 3.9×10^4 J; (b) 3.0 min.

31. 76 L.

33. 0.15 J/K.

35. $+11$ kcal/K.

37. $+0.0104$ cal/K·s.

39. 1.7×10^2 J/K.

43. (a) 0.312 kcal/K;
(b) > -0.312 kcal/K.

45. (a) Adiabatic process;
(b) $\Delta S_i = -nR \ln 2, \Delta S_a = 0$;
(c) $\Delta S_{surr,i} = nR \ln 2, \Delta S_{surr,a} = 0$.

47. (a) Entropy is a state function.

51. (a) 5/16; (b) 1/64.

53. (b), (a), (c), (d).

55. 69%.

57. 2.6×10^3 J/K.

59. (a) 17; (b) 5.9×10^7 J/h.

61. (a) 5.3 C°; (b) $+77$ J/kg·K.

63. $\Delta S = K/T$.

65. (a)

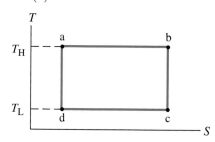

(b) area $= Q_{net} = W_{net}$.

67. $e_{Stirling} = (T_H - T_L) \ln(V_b/V_a)/$
$\left[T_H \ln(V_b/V_a) \right.$
$+ \frac{3}{2}(T_H - T_L) \right]$,
$e_{Stirling} < e_{Carnot}$.

69. 0.091 hp.

71. (a) 1/379; (b) $1/1.59 \times 10^{11}$.

CHAPTER 21

1. 6.3×10^9 N.

3. 2.7×10^{-3} N.

5. 5.5×10^3 N.

7. 8.66 cm.

9. -5.4×10^7 C.

11. 83.8 N away from the center of the triangle.

13. 2.96×10^7 N toward the center of the square.

15. $\mathbf{F}_1 = (kQ^2/\ell^2)[(-2 + 3\sqrt{2}/4)\mathbf{i}$
$+ (4 - 3\sqrt{2}/4)\mathbf{j}], \mathbf{F}_2 = (kQ^2/\ell^2)$
$[(2 + 2\sqrt{2})\mathbf{i} + (-6 + 2\sqrt{2})\mathbf{j}]$,
$\mathbf{F}_3 = (kQ^2/\ell^2)[(-12 - 3\sqrt{2}/4)\mathbf{i}$
$+ (6 + 3\sqrt{2}/4)\mathbf{j}], \mathbf{F}_4 = (kQ^2/\ell^2)$
$[(12 - 2\sqrt{2})\mathbf{i} + (-4 - 2\sqrt{2})\mathbf{j}]$.

17. (a) $Q_1 = Q_2 = \frac{1}{2}Q_T$;
(b) Q_1 (or Q_2) $= 0$.

19. $0.402Q_0, 0.366\ell$ from Q_0.

21. 60.2×10^{-6} C, 29.8×10^{-6} C;
-16.8×10^{-6} C, 106.8×10^{-6} C.

23. $\mathbf{F} = -(1.90kQ^2/\ell^2)(\mathbf{i} + \mathbf{j} + \mathbf{k})$.

25. 2.18×10^{-16} N (west).

27. 7.43×10^6 N/C (up).

29. $(1.39 \times 10^2$ N/C)$\mathbf{j}$.

33.

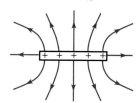

35. 8.26×10^{-10} N/C (south).

37. 4.5×10^6 N/C up, 1.2×10^7 N/C,
56° above the horizontal.

39. 5.61×10^4 N/C away from the opposite corner.

41. $Q_1/Q_2 = \frac{1}{4}$.

43. (a) $2Qy/4\pi\epsilon_0(y^2 + \ell^2)^{3/2}$ $\mathbf{j}$.

45. $\dfrac{Q}{4\pi\epsilon_0} \left[\dfrac{x\mathbf{i} - (2a/\pi)\mathbf{j}}{(x^2 + a^2)^{3/2}} \right]$.

49. $\dfrac{-2\lambda \sin\theta_0}{4\pi\epsilon_0 R}$ $\mathbf{i}$.

51. (a) $\dfrac{\lambda}{4\pi\epsilon_0 x(x^2 + L^2)^{1/2}}$
$\{L\mathbf{i} + [x - (x^2 + L^2)^{1/2}]\mathbf{j}\}$.

53. $(\sigma/2\epsilon_0)\mathbf{k}$.

55. (a) $\mathbf{a} = -(3.5 \times 10^{15}$ m/s$^2)$ $\mathbf{i}$
$-(1.41 \times 10^{16}$ m/s$^2)$ $\mathbf{j}$; (b) $\theta = -104$°.

57. $\theta = -28$°.

59. (b) $2\pi(4\pi\epsilon_0 mR^3/qQ)^{1/2}$.

61. (a) 3.4×10^{-20} C; (b) No;
(c) 8.5×10^{-26} m·N;
(d) 2.5×10^{-26} J.

63. (a) $\theta \ll 1$;
(b) $(pE/I)^{1/2}/2\pi$.

65. (b) Direction of the dipole.

67. 6.8×10^3 C.

69. 5.7×10^{13} C.

71. $\mathbf{F}_1 = 0.30$ N, 265° from x-axis,
$\mathbf{F}_2 = 0.26$ N, 139° from x-axis,
$\mathbf{F}_3 = 0.26$ N, 30° from x-axis.

73. 4.2×10^5 N/C up.

75. $0.444Q_0, 0.333\ell$ from Q_0.

77. 5.60 m from the positive charge, and 3.60 m from the negative charge.

79. (a) In the direction of the velocity, to the right; (b) 2.1×10^2 N/C.

81. $\theta_0 = 18$°.

83. $(1.08 \times 10^7$ N/C)$/$
$\{3.00 - \cos[(12.5$ s$^{-1})t]\}^2$, up.

85. $E_A = 4.2 \times 10^4$ N/C (right),
$E_B = -1.4 \times 10^4$ N/C (left),
$E_C = -2.8 \times 10^3$ N/C (left),
$E_D = -4.2 \times 10^4$ N/C (left).

87. $d(1 + \sqrt{2})$ from the negative charge, and $d(2 + \sqrt{2})$ from the positive charge.

CHAPTER 22

1. (a) 41 N·m^2/C; (b) 29 N·m^2/C;
(c) 0.

3. $\Phi_{net} = 0$,
$\Phi_{x=0} = -(6.50 \times 10^3$ N/C$)\ell^2$,
$\Phi_{x=\ell} = +(6.50 \times 10^3$ N/C$)\ell^2$,
$\Phi_{all\ others} = 0$.

5. 12.8 nC.

7. -1.2 μC.

9. -3.75×10^{-11} C.

11. (a) -1.0×10^4 N/C (toward wire);
(b) -2.5×10^4 N/C (toward wire).

13.

15. (a) 5.5×10^7 N/C (away from center);
(b) 0.

17. (a) -8.00 μC;
(b) $+1.00$ μC

19. (a) 0; (b) σ/ϵ_0;
(c) unaffected.

21. (a) 0; (b) $\sigma_1 r_1^2/\epsilon_0 r^2$;
  (c) $(\sigma_1 r_1^2 + \sigma_2 r_2^2)/\epsilon_0 r^2$;
  (d) $\sigma_2/\sigma_1 = -(r_1/r_2)^2$; (e) $\sigma_1 = 0$.

23. (a) $q/4\pi\epsilon_0 r^2$;
  (b) $(1/4\pi\epsilon_0)[Q(r^3 - r_1^3)$
    $+ q(r_0^3 - r_1^3)]/(r_0^3 - r_1^3)r^2$;
  (c) $(q + Q)/4\pi\epsilon_0 r^2$.

25. (a) $q/4\pi\epsilon_0 r^2$; (b) $(q + Q)/4\pi\epsilon_0 r^2$;
  (c) $E(r < r_0) = Q/4\pi\epsilon_0 r^2$,
    $E(r > r_0) = 2Q/4\pi\epsilon_0 r^2$;
  (d) $E(r < r_0) = -Q/4\pi\epsilon_0 r^2$,
    $E(r > r_0) = 0$.

27. (a) $\sigma R_0/\epsilon_0 r$; (b) 0; (c) same.

29. (a) 0; (b) $Q/2\pi\epsilon_0 Lr$;
  (c) 0; (d) $eQ/4\pi\epsilon_0 L$.

31. (a) 0;
  (b) -2.3×10^5 N/C (toward the axis);
  (c) -1.8×10^4 N/C (toward the axis).

33. (a) $\rho_E r/2\epsilon_0$; (b) $\rho_E R_1^2/2\epsilon_0 r$;
  (c) $\rho_E(r^2 + R_1^2 - R_2^2)/2\epsilon_0 r$;
  (d) $\rho_E(R_3^2 + R_1^2 - R_2^2)/2\epsilon_0 r$;
  (e)

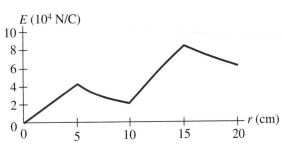

E (10^4 N/C)

35. $Q/\epsilon_0\sqrt{2}$.

37. $\oint \mathbf{g}\cdot d\mathbf{A} = -4\pi GM$.

39. $Q_{enclosed} = \epsilon_0 b\ell^3$.

41. 3.95×10^2 N·m²/C,
  -1.69×10^2 N·m²/C.

43. (a) 0; (b) $Q/25\pi\epsilon_0 r_0^2 \leq E \leq Q/\pi\epsilon_0 r_0^2$;
  (c) not perpendicular;
  (d) not useful.

45. (a) $0.677e = +1.08 \times 10^{-19}$ C;
  (b) 3.5×10^{11} N/C.

47. (a) $\rho_E r_0/6\epsilon_0$ (right);
  (b) $-17\rho_E r_0/54\epsilon_0$ (left).

49. (a) 0; (b) 5.65×10^5 N/C (right);
  (c) 5.65×10^5 N/C (right);
  (d) -5.00×10^{-6} C/m²;
  (e) $+5.00 \times 10^{-6}$ C/m².

CHAPTER 23

1. -4.2×10^{-5} J (done by the field).

3. 3.4×10^{-15} J.

5. $V_a - V_b = +72.8$ V.

7. 7.04 V.

9. 0.8 μC.

11. (a) $V_{BA} = 0$; (b) $V_{CB} = -2100$ V;
  (c) $V_{CA} = -2100$ V.

13. (a) -9.6×10^8 V;
  (b) $V(\infty) = +9.6 \times 10^8$ V.

15. (a) The same;
  (b) $Q_2 = r_2 Q/(r_1 + r_2)$.

17. (a) $Q/4\pi\epsilon_0 r$;
  (b) $(Q/8\pi\epsilon_0 r_0)[3 - (r^2/r_0^2)]$;
  (c)

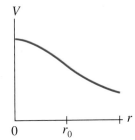

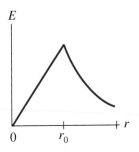

19. (a) $V_0 + (\sigma R_0/\epsilon_0) \ln(R_0/r)$;
  (b) $V = V_0$; (c) $V \neq 0$.

21. (a) 29 V;
  (b) -29 eV (-4.6×10^{-18} J).

23. $+0.19$ J.

25. 4.2 MV.

27. 2.33×10^7 m/s.

29. $V_{BA} = (1/2\pi\epsilon_0)q(2b - d)/b(d - b)$.

31. $\dfrac{\sigma}{2\epsilon_0}[(x^2 + R_2^2)^{1/2} - (x^2 + R_1^2)^{1/2}]$.

33. $\dfrac{Q}{8\pi\epsilon_0 L} \ln\left(\dfrac{x + L}{x - L}\right)$, $x > L$.

35. $\dfrac{a}{6\epsilon_0}[(x^2 + R^2)^{1/2}(R^2 - 2x^2) + 2x^3]$.

37. 3.2 mm.

39. (a) 8.5×10^{-30} C·m; (b) zero.

41. (a) -0.088 V; (b) 1%.

43. (a) 5.2×10^{-20} C.

47. $\mathbf{E} = 2y(2z - 1)\mathbf{i} - 2(y + x - 2xz)\mathbf{j}$
    $+ (4xy)\mathbf{k}$.

49. (a) 9.6×10^4 eV; (b) 1.9×10^5 eV.

51. -2.4×10^4 V.

53. (a) $U = (1/4\pi\epsilon_0)(Q_1 Q_2/r_{12}$
    $+ Q_1 Q_3/r_{13} + Q_1 Q_4/r_{14} + Q_2 Q_3/r_{23}$
    $+ Q_2 Q_4/r_{24} + Q_3 Q_4/r_{34})$.
  (b) $U = (1/4\pi\epsilon_0)(Q_1 Q_2/r_{12}$
    $+ Q_1 Q_3/r_{13} + Q_1 Q_4/r_{14} + Q_1 Q_5/r_{15}$
    $+ Q_2 Q_3/r_{23} + Q_2 Q_4/r_{24} + Q_2 Q_5/r_{25}$
    $+ Q_3 Q_4/r_{34} + Q_3 Q_5/r_{35} + Q_4 Q_5/r_{45})$.

55. (a) 2.0 keV; (b) 42.8.

57. (a) $(-4 + \sqrt{2})Q^2/4\pi\epsilon_0 b$; (b) 0.

59. $3Q^2/20\pi\epsilon_0 r$.

61. 5.4×10^5 V/m.

63. 9×10^2 V.

65. (a) 1.1 MV; (b) 13 kg.

67. 7.2 MV.

69. 1.58×10^{12} electrons.

71. 1.7×10^{-12} V.

73. 1.03×10^6 m/s.

75. $V_a = -3.5\, Q/4\pi\epsilon_0 L$,
    $V_b = -5.2\, Q/4\pi\epsilon_0 L$,
    $V_c = -6.8\, Q/4\pi\epsilon_0 L$.

77. (a) 5.8×10^5 V; (b) 9.2×10^{-14} J.

79. $V_a - V_b = (\lambda/2\pi\epsilon_0) \ln(R_b/R_a)$.

83. (a) $\rho_E(r_2^3 - r_1^3)/3\epsilon_0 r$;
  (b) $(\rho_E/6\epsilon_0)[3r_2^2 - r^2 - (2r_1^3/r)]$;
  (c) $(\rho_E/2\epsilon_0)(r_2^2 - r_1^2)$, potential is
    continuous at r_1 and r_2.

CHAPTER 24

1. 2.6 μF.

3. 6.3 pF.

5. 0.80 μF.

7. 2.0 C.

9. 1.8×10^2 m².

11. 7.1×10^{-4} F.

13. 23 nC.

17. 4.5×10^4 V/m.

19. (a) $\epsilon_0 A/(d - \ell)$; (b) 3.

21. 2880 pF, yes.

23. (a) $(C_1 C_2 + C_1 C_3 + C_2 C_3)/$
      $(C_2 + C_3)$;
  (b) $Q_1 = 350$ μC, $Q_2 = 117$ μC.

25. (a) 3.71 μF; (b) $V_{ab} = V_1 = 26.0$ V,
    $V_2 = 14.9$ V, $V_3 = 11.1$ V.

27. 18 nF (parallel), 1.6 nF (series).

29. (a) $3C/5$; (b) $Q_1 = Q_2 = CV/5$,
    $Q_3 = 2CV/5$, $Q_4 = 3CV/5$,
    $V_1 = V_2 = V/5$, $V_3 = 2V/5$,
    $V_4 = 3V/5$.

31. $Q_1' = C_1 C_2 V_0/(C_1 + C_2)$,
    $Q_2' = C_2^2 V_0/(C_1 + C_2)$.

33. (a) $Q_1 = Q_3 = 30\ \mu C$;
$\quad\ Q_2 = Q_4 = 60\ \mu C$;
(b) $V_1 = V_2 = V_3 = V_4 = 3.75\ V$;
(c) 7.5 V.

35. 3.0 μF.

37. $C \approx (\epsilon_0 A/d)[1 - \frac{1}{2}(\theta\sqrt{A}/d)]$.

39. 2.0×10^{-3} J.

41. 2.3×10^3 J.

43. 1.65×10^{-7} J.

45. (a) 2.5×10^{-5} J; (b) 6.2×10^{-6} J;
(c) $Q_{par} = 4.2\ \mu C$, $Q_{ser} = 1.0\ \mu C$.

47. (a) 2.2×10^{-4} J; (b) 8.1×10^{-5} J;
(c) -1.4×10^{-4} J; (d) stored
potential energy is not conserved.

51. 1.5×10^{-10} F.

53. 0.46 μC.

55. 3.3×10^2 J.

57. $C = 2\epsilon_0 A K_1 K_2/d(K_1 + K_2)$.

59. (a) $0.40Q_0, 1.60Q_0$; (b) $0.40V_0$.

61. (a) 111 pF; (b) 1.66×10^{-8} C;
(c) 1.84×10^{-8} C;
(d) 1.17×10^5 V/m; (e) 3.34×10^4 V/m;
(f) 150 V;
(g) 172 pF; (h) 2.58×10^{-8} C.

63. 22%.

65. $Q \approx 4.41 \times 10^{-7}$ C,
$Q_{ind} = 3.65 \times 10^{-7}$ C,
$E_{air} = 2.69 \times 10^4$ V/m,
$E_{glass} = 4.64 \times 10^3$ V/m;

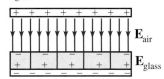

67. 11 μF.

69. (a) $4\times$; (b) $4\times$; (c) $\frac{1}{2}\times$.

71. 10.9 V.

73. (b) 1.5×10^{-10} F/m.

75. $U_2/U_1 = 1/2K$, $E_2/E_1 = 1/K$.

77. (a) 19 J; (b) 0.19 MW.

79. (a) 0.10 MV; (b) voltage will
decrease exponentially.

81. 660 pF in parallel.

83. $Q_1 = 11\ \mu C$, $Q_2 = Q_3 = 13\ \mu C$,
$V_1 = 11\ V$, $V_2 = 6.5\ V$, $V_3 = 4.4\ V$.

85. $Q^2 x/2A\epsilon_0$.

87. (a) 7.4 nF, 0.33 μC, 1.5×10^4 V/m,
7.5×10^{-6} J; (b) 27 nF, 1.2 μC,
1.5×10^4 V/m, 2.7×10^{-5} J.

89. (a) 66 pF; (b) 30 μC; (c) 7.0 mm; (d)
450 V.

CHAPTER 25

1. 9.38×10^{18} electron/s.

3. 2.1×10^{-11} A.

5. 7.5×10^2 V.

7. 2.1×10^{21} electron/min.

9. (a) 16 Ω; (b) 6.8×10^3 C.

11. 0.57 mm.

13. $R_{Al} = 1.2R_{Cu}$.

15. 1/6 the length, 8.3 Ω, 1.7 Ω.

17. 58.3°C.

19. 1.8×10^3 °C.

21. $R_2 = \frac{1}{4}R_1$.

25. $R = (r_2 - r_1)/4\pi\sigma r_1 r_2$.

27. 3.2 W.

29. 37 V.

31. (a) 240 Ω, 0.50 A; (b) 96 Ω, 1.25 A.

33. 0.092 kWh, 22¢/month.

35. 1.1 kWh.

37. 3.

39. 5.3 kW.

41. 0.128 kg/s.

43. 0.094 A.

45. (a) Infinite; (b) 1.9×10^2 Ω.

47. 636 V, 5.66 A.

49. 1.5 kW, 3.0 kW, 0.

51. (a) 7.8×10^{-10} m/s; (b) 10.5 A/m²
along the wire; (c) 1.8×10^{-7} V/m.

53. 2.7 A/m² north.

55. 12 h.

57. 6.67×10^{-2} S.

59. (a) 8.6 Ω, 1.1 W; (b) $4\times$.

61. (a) $44; (b) 1.8×10^3 kg/yr.

63. (a) -19.5%; (b) percentage decrease
in the power output would be less.

65. (a) 1.44×10^3 W; (b) 17 W; (c) 11 W;
(d) 0.8¢/day.

67. (a) 1.5 kW; (b) 12.5 A.

69. 2.

71. 0.303 mm, 28.0 m.

73. (a) 1.2 kW; (b) 100 W.

75. 1.4×10^{12} protons.

77. $j_a = 2.8 \times 10^5$ A/m²,
$j_b = 1.6 \times 10^5$ A/m².

CHAPTER 26

1. (a) 8.39 V; (b) 8.49 V.

3. 0.060 Ω.

5. 360 Ω, 23 Ω.

7. 25 Ω, 70 Ω, 95 Ω, 18 Ω.

9. Series connection.

11. 4.6 kΩ.

13. 310 Ω, 3.7%.

15. 960 Ω in parallel.

17. 105 Ω.

19. (a) V_1 and V_2 increase; V_3 and V_4
decrease; (b) I_1 ($= I$) and I_2
increase; I_3 and I_4 decrease;
(c) increases; (d) $I = I_1 = 0.300$ A,
$I_2 = 0, I_3 = I_4 = 0.150$ A; $I = 0.338$ A,
$I_2 = I_3 = I_4 = 0.113$ A.

21. 0.4 Ω.

23. 0.41 A.

25. $I_1 = 0.68$ A, $I_2 = -0.40$ A.

27. $I_1 = 0.18$ A right, $I_2 = 0.32$ A left,
$I_3 = 0.14$ A up.

29. $I_1 = 0.274$ A, $I_2 = 0.222$ A,
$I_3 = 0.266$ A, $I_4 = 0.229$ A,
$I_5 = 0.007$ A, $I = 0.496$ A.

31. 52 V, −28 V.
The negative value means the
battery is facing the other direction.

33. 70 V.

35. $I_1 = 0.783$ A.

39. (a) $R(3R + 5R')/8(R + R')$; (b) $R/2$.

41. (a) 3.7 nF; (b) 22 μs.

43. $t = 1.23\tau$.

45. (a) $\tau = R_1 R_2 C/(R_1 + R_2)$;
(b) $Q_{max} = \mathscr{E} R_2 C/(R_1 + R_2)$.

47. 2.1 μs.

49. 50 μA.

51. (a) 2.9×10^{-4} Ω in parallel;
(b) 35 kΩ in series.

53. 22 V, 17 V, 14% low.

55. 0.85 mA, 4.3 V.

57. 9.6 V.

61. 3.6×10^{-2} C°.

63. Two resistors in series.

65. 2.2 V, 116 V.

67. 0.19 MΩ.

69. (a) 0.10 A; (b) 0.10 A; (c) 53 mA.

71. 2.5 V.

73. 46.1 V, 0.71 Ω.

75. (a) 72.0 W; (b) 14.2 W; (c) 3.76 W.

77. (a) 40 kΩ; (b) between b and c.

79. 375 cells, 3.8 m × 0.090 m.

81. (a) 0.50 A; (b) 0.17 A; (c) 3.3 μC;
(d) 32 μs.

83. (a) + 6.8 V, 10.2 μC; (b) 28 μs.

CHAPTER 27

1. (a) 6.7 N/m; (b) 4.7 N/m.

3. 2.68 A.

5. 0.243 T.

7. (a) South pole; (b) 3.5×10^2 A;
(c) 5.22 N.

9. 5.5×10^3 A.

13. 1.05×10^{-13} N north.

15. (a) Down; (b) in; (c) right.

21. 1.6 m.

23. (a) 2.7 cm; (b) 3.8×10^{-7} s.

25. 1.034×10^8 m/s (west), gravity can be ignored.

27. $(6.4\mathbf{i} - 10.3\mathbf{j} - 0.24\mathbf{k})] \times 10^{-16}$ N.

29. 5.3×10^{-5} m, 3.3×10^{-4} m.

31. (a) $45°$; (b) 3.5×10^{-3} m.

33. (a) $2\mu B$; (b) 0.

35. (a) 4.33×10^{-5} m $\cdot$ N; (b) north edge.

39. 29 μA.

41. 1.2×10^5 C/kg.

43. (a) 2.2×10^{-4} V/m; (b) 2.7×10^{-4} m/s; (c) 6.4×10^{28} electrons/m^3.

45. (a) Determine polarity of the emf; (b) 0.43 m/s.

47. 1.53 mm, 0.76 mm.

51. 3.0 T up.

53. 1.1×10^{-6} m/s west.

55. 0.17 N, $68°$ above the horizontal toward the north.

57. 0.20 T, $26.6°$ from the vertical.

59. (c) 48 MeV.

61. Slower protons will deflect more, and faster protons will deflect less, $\theta = 12°$.

63. 2.0 A, down.

65. 7.3×10^{-3} T.

67. -2.1×10^{-20} J.

CHAPTER 28

1. 1.7×10^{-4} T, $3.1\times$.

3. 0.18 N attraction.

5.

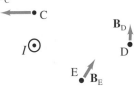

7. 8.9×10^{-5} T, $70°$ above horizontal.

9. 4.0×10^{-5} T, $15°$ below the horizontal.

11. (a) $(2.0 \times 10^{-5}$ T/A$)(15$ A $- I)$ up; (b) $(2.0 \times 10^{-5}$ T/A$)(15$ A $+ I)$ down.

13. 21 A down.

15. $[(\mu_0/4\pi)2I(d - 2x)/x(d - x)]\mathbf{j}$.

17. 4.12×10^{-5} T.

19. (b) $(\mu_0/4\pi)(2I/y)$.

21. 0.123 A.

23. (a) 6.4×10^{-3} T; (b) 3.8×10^{-3} T; (c) 2.1×10^{-3} T.

25. (a) 51 cm; (b) 1.3×10^{-2} T.

27. (a) $(\mu_0 I_0/2\pi R_1^2)r$ circular CCW; (b) $\mu_0 I_0/2\pi r$ circular CCW; (c) $(\mu_0 I_0/2\pi r)(R_3^2 - r^2)/(R_3^2 - R_2^2)$ circular CCW; (d) 0.

29. 3.6×10^{-6} T.

31. $\mu_0 I/8R$ out of the page.

33. (a) $\mu_0 I(R_1 + R_2)/4R_1 R_2$ into the page; (b) $\frac{1}{2}\pi I(R_1^2 + R_2^2)$ into the page.

35. (a) $\dfrac{Q\omega R^2}{4}\mathbf{i}$;

(b)
$$\frac{\mu_0 Q\omega}{2\pi R^2}\left[\frac{R^2 + 2x^2 - 2x\sqrt{R^2 + x^2}}{\sqrt{R^2 + x^2}}\right]\mathbf{i};$$
(c) yes.

37. (b) $B = \mu_0 IL/4\pi y(L^2 + y^2)^{1/2}$ circular.

39. (a) $(\mu_0 I_0/2\pi R)n\tan(\pi/n)$ into the page.

41. $B = \dfrac{\mu_0 I}{4\pi}\Bigg\{\dfrac{(x^2 + y^2)^{1/2}}{xy}$

$+ \dfrac{[(b - x)^2 + y^2]^{1/2}}{y(b - x)}$

$+ \dfrac{[(a - y)^2 + (b - x)^2]^{1/2}}{(a - y)(b - x)}$

$+ \dfrac{[x^2 + (a - y)^2]^{1/2}}{x(a - y)}\Bigg\}$,

out of the page.

43. (a) 26 A $\cdot$ m^2; (b) 31 m $\cdot$ N.

45. 30 T.

47. $F_M/L = 5.84 \times 10^{-5}$ up, $F_N/L = 3.37 \times 10^{-5}$ N/m $60°$ below the line toward P, $F_P/L = 3.37 \times 10^{-5}$ N/m $60°$ below the line toward N.

49. 0.27 mm, 1.4 cm.

51. $B = \mu_0 jt/2$ parallel to the sheet, perpendicular to the current (opposite directions on the two sides).

53. Between long, thin and short, fat.

55. 3×10^9 A.

57. B will decrease.

59. 2.1×10^{-6} g.

61. $2\mu_0 I/L\pi$ (left).

63. 4×10^{-6} T, about 10% of the Earth's field.

CHAPTER 29

1. -3.8×10^2 V.

3. Counterclockwise.

5. 0.026 V.

7. (a) Counterclockwise; (b) clockwise; (c) zero; (d) counterclockwise.

9. Counterclockwise.

11. (a) Clockwise; (b) 43 mV; (c) 17 mA.

13. 1.1×10^{-5} J.

15. 4.21 C.

17. (a) 5.2×10^{-2} A; (b) 0.32 mW.

19. 1.7×10^{-2} V.

21. $(\mu_0 Ia/2\pi)\ln 2$.

23. (a) 0.15 V; (b) 5.4×10^{-3} A; (c) 4.5×10^{-4} N.

25. (a) Will move at constant speed; (b) $v = v_0 e^{-B^2\ell^2 t/mR}$.

27. (a) $\dfrac{\mu_0 Iv}{2\pi}\ln\left(\dfrac{a + b}{b}\right)$ toward long wire;

(b) $\dfrac{\mu_0 Iv}{2\pi}\ln\left(\dfrac{a + b}{b}\right)$ away from long wire.

31. 0.33 kV, 120 rev/s.

33. 100 V.

35. 13 A.

37. 3.54×10^4 turns.

39. 0.18.

41. (a) 5.2 V; (b) step-down transformer.

43. 549 V, 68.6 A.

45. 56.8 kW.

47. 0.188 V/m.

49. (b) Clockwise; (c) $dB/dt > 0$.

51. (a) IR/ℓ (constant); (b) $\dfrac{\mathscr{E}_0}{\ell}e^{-B^2\ell^2 t/mR}$.

53. 31 turns.

55. $v = 0.76$ m/s.

57. 184 kV.

59. 1.5×10^{17}.

61. (a) 23 A; (b) 90 V; (c) 6.9×10^2 W; (d) 75%.

63. (a) 0.85 A; (b) 8.2.

65. $\frac{1}{2}B\omega L^2$ toward the center.

71. $B\omega r$, radially out from the axis.

CHAPTER 30

1. $M = \mu_0 N_1 N_2 A/\ell$.

3. $M/\ell = \mu_0 n_1 n_2 \pi r_2^2$.

5. $M = (\mu_0 w/2\pi)\ln(\ell_2/\ell_1)$.

7. 1.2 H.

9. 2.5×10^{-6} H.

11. $r_1 \geq 2.5$ mm.

13. 3.

15. (a) $L_1 + L_2$; (b) $= L_1L_2/(L_1 + L_2)$.

17. 15.9 J.

19. (a) $u_E = 4.4 \times 10^{-4}$ J/m^3,
$u_B = 1.6 \times 10^6$ J/m^3, $u_B \gg u_E$;
(b) $E = 6.0 \times 10^8$ V/m.

21. 4.4 J/m^3, 1.6×10^{-14} J/m^3.

23. $(\mu_0 I^2/4\pi) \ln(r_2/r_1)$.

25. $t/\tau = 4.6$.

27. $(dI/dt)_0 = V_0/L$.

29. (a) $(LV^2/2R^2)(1 - 2e^{-t/\tau} + e^{-2t/\tau})$;
(b) $t/\tau = 5.3$.

31. (a) 213 pF; (b) 46.5 μH.

35. (a) $Q = Q_0/\sqrt{2}$; (b) $T/8$.

37. $R = 2.30\ \Omega$.

41. Decrease, 1.15 kΩ.

43. 20 mH, 95 turns.

45. (a) 21 mH; (b) 45 mA;
(c) 2.2×10^{-5} J.

47. 3.0×10^3 turns, 95 turns.

51. (b) Positioning one coil
perpendicular to the other;
(c) $L_1L_2/(L_1 + L_2)$;
$(L_1L_2 - M^2)/(L_1 + L_2 \mp 2M)$.

55. (a) $\frac{1}{2}(Q_0^2/C)e^{-Rt/L}$.

57.

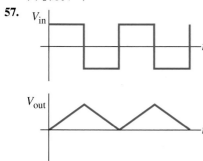

CHAPTER 31

1. (a) $3.7 \times 10^2\ \Omega$; (b) $2.2 \times 10^{-2}\ \Omega$.

3. 9.90 Hz.

5.

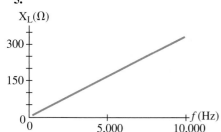

7. 0.13 H.

11. (a) 5.0%; (b) 98%.

13. (a) 9.0 kΩ; (b) 10.2 kΩ.

15. (a) 18 mA; (b) $-29°$; (c) 1.8 W;
(d) $V_R = 105$ V, $V_C = 58$ V.

17. (a) 0.38 A; (b) $-89°$; (c) 0.29 W.

19. 332 Ω.

21. (a) 0; (b) $\dfrac{2}{\pi}V_0$, $\overline{V}_{1/2} = \dfrac{2\sqrt{2}}{\pi}V_{\text{rms}}$.

23. 8.78 kΩ, $-7.66°$, 91.1 mA.

25. 265 Hz, 324 W.

27. 52.5 mA.

29. (b) $\omega'^2 = [(1/LC) - (R^2/2L^2)]$;
(c) $k \leftrightarrow 1/C, m \leftrightarrow L, b \leftrightarrow R$.

31. (a) $V_0^2R/[2R^2 + 2(\omega L - 1/\omega C)^2]$;
(b) $\omega'^2 = 1/LC$; (c) $\Delta\omega = R/L$.

33. 4 Ω.

35. 9.76 nF.

37. 27.9 mH.

39. 1.6 kHz.

41. 14 Ω, 75 mH.

43. 2.2×10^3 Hz, 1.1×10^4 Hz.

45. (a) 23.6 kΩ, 10.8°; (b) 1.88×10^{-5} W;
(c) 2.8×10^{-5} A, 0.66 V, 4.7×10^{-4} V,
0.126 V.

49. $\{(R_1 + R_2)^2$
$+ [\omega(L_1 + L_2) -$
$(C_1 + C_2)/\omega C_1 C_2]^2\}^{1/2}$.

51. 19 Ω, 62 mH.

53. $I = \left(\dfrac{V_0}{Z}\right)\sin(\omega t + \phi)$,

$I_C = \left(\dfrac{V_0}{X_C}\right)\left[-\left(\dfrac{R}{Z}\right)\cos(\omega t + \phi) + \cos(\omega t)\right]$,

$I_L = \left(\dfrac{V_0}{X_L}\right)\left[\left(\dfrac{R}{Z}\right)\cos(\omega t + \phi) - \cos(\omega t)\right]$,

$Z = \sqrt{R^2 + \left(\dfrac{X_C X_L}{X_L - X_C}\right)^2}$,

$\tan\phi = \dfrac{X_C X_L}{(X_L - X_C)R}$.

CHAPTER 32

1. 9.2×10^4 V/m $\cdot$ s.

3. 7.9×10^{14} V/m $\cdot$ s.

7. $\oint \mathbf{B} \cdot d\mathbf{A} = \mu_0 Q_m$,
$\oint \mathbf{E} \cdot d\boldsymbol{\ell} = \mu_0\, dQ_m/dt - d\Phi_B/dt$.

9. 1.4×10^{-13} T.

11. (a) $B_0 = E_0/c$, $-y$-direction;
(b) $-z$-direction.

13. (a) 1.08 cm; (b) 3.0×10^{18} Hz.

15. 314 nm, ultraviolet.

17. (a) 4.3 min; (b) 71 min.

19. 1.77×10^{-6} W/m^2.

21. 7.82×10^{-7} J/h.

23. 4.50×10^{-6} J.

25. 3.8×10^{26} W.

29. $r < 3 \times 10^{-7}$ m.

31. 302 pF.

33. 2.59 nH $\leq L \leq$ 3.89 nH.

35. (a) 441 m; (b) 2.979 m.

37. 5.56 m, 0.372 m.

39. (a) 1.28 s; (b) 4.3 min.

43. 469 V/m.

45. Person at the radio hears the voice
0.14 s sooner.

47. (a) 0.40 W; (b) 12 V/m; (c) 12 V.

49. 1.5×10^{11} W.

51. (a) Parallel;
(b) 8.9 pF $\leq C \leq$ 80 pF;
(c) 1.05 mH $\leq L \leq$ 1.12 mH.

CHAPTER 33

1. (a) 2.21×10^8 m/s;
(b) 1.99×10^8 m/s.

3. 8.33 min.

5. 3 m.

7. 3.4×10^3 rad/s.

9. I_3 is the desired image:

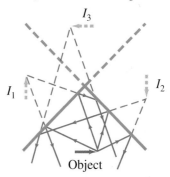

Depending on where you put your eye,
two other images may also be visible.

11. 5°.

15. 36.4 cm.

19. 4.5 m.

21. Convex, -20 m.

23. (a) Center of curvature; (b) real;
(c) inverted; (d) -1.

29. (a) Convex mirror; (b) 22 cm behind
surface; (c) -98 cm; (d) -196 cm.

33. 45.6°.

35. 24.9°.

37. 4.6 m.

43. 3.0%.

45. 0.22°.

47. 61.7°, lucite.

49. 93.5 cm.

51. $n_{\text{liquid}} \geq 1.5$.

55. 17.0 cm below the surface of the glass.

59. (a) 3.0 m, 4.0 m, 7.0 m; (b) toward you, away from you, toward you.

61. −3.80 m.

63. Chose different signs for the magnification; 13.3 cm, 26.7 cm.

65. ≥56.1°.

69. 81 cm (inside the glass).

CHAPTER 34

1. (a)

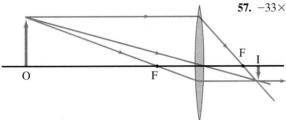

(b) 24.9 cm.

3. (a) 3.64 D, converging; (b) −16.0 cm, diverging.

5. (a) −0.26 mm; (b) −0.47 mm; (c) −1.9 mm.

7. (a) 81 mm; (b) 82 mm; (c) 87 mm; (d) 24 cm.

9. (a) Virtual; (b) converging lens; (c) 7.5 D.

11. (a) −10.5 cm (diverging), virtual; (b) +203 cm (converging).

13. 22.9 cm, 53.1 cm.

15. Real and upright.

17. Real, 21.3 cm beyond second lens, −0.708 (inverted).

19. (a) +7.14 cm; (b) −0.357 (inverted); (c)

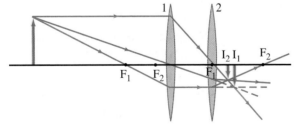

23. 1.54.

25. 8.1 cm.

27. −1.87 m (concave).

29. +1.15 D.

31. $f/2.3$.

35. 41 mm.

37. +2.3 D.

39. Glasses would be better.

41. (a) −1.33 D; (b) 38 cm.

43. −26.8 cm.

45. 17 cm, 100 cm.

47. 8.3 cm.

49. 6.3 cm from the lens, 3.9×.

51. (a) −49.4 cm; (b) 4.7×.

53. (a) 7.2×; (b) 2.2×.

55. 3.2 cm, 83 cm.

57. −33×.

59. 12×.

61. $f_o = 4.0$ m, $r = 8.0$ m.

63. 7.5×.

65. 1.7 cm.

67. (a) 0.85 cm; (b) 230×.

69. (a) 14.4 cm; (b) 137×.

71. (a) 15.9 cm; (b) 14.3 cm; (c) 1.6 cm; (d) $r = 0.46$ cm.

73. $6.87 \text{ m} \leq d_o \leq \infty$.

75. 100 mm, 200 mm.

77. 79.4 cm, 75.5 cm.

79. 0.101 m, −2.7 m.

81. $0 < -d_o < -f$.

83. (c) $\Delta d = \sqrt{d_T^2 - 4d_T f}$,
$$m = \left(\frac{d_T + \sqrt{d_T^2 - 4d_T f}}{d_T - \sqrt{d_T^2 - 4d_T f}} \right)^2.$$

85. $1/f' = [(n/n') - 1][(1/R_1) + (1/R_2)]$;
$(1/d_o) + (1/d_i) = 1/f'$, where
$1/f' = [(n/n') - 1]/f(n - 1)$;
$m = -d_i/d_o$.

87. +3.6 D.

89. 2.9×, 4.1×.

91. (a) −2.5×; (b) 5.0 D.

93. −20×.

CHAPTER 35

3. 5.9 μm.

5. 3.9 cm.

7. 0.21 mm.

9. 613 nm.

11. 533 nm.

15. (a) $I_\theta/I_0 = (1 + 4\cos\delta + 4\cos^2\delta)/9$;
(b) $\sin\theta_{\max} = m\lambda/d$, $m = 0, \pm1, \pm2, \ldots$;
$\sin\theta_{\min} = (m + \frac{1}{3}k)\lambda/d$; $k = 1, 2$;
$m = 0, \pm1, \pm2, \ldots$.

17. Orange-red.

19. 179 nm.

21. 9.1 μm.

23. 120 nm, 240 nm.

25. 1.26.

29. 0.47, 0.23.

31. 0.221 mm.

33. 0.289 mm.

35. (a) 17 lm/W; (b) 156.

37. (a) Constructive; (b) destructive.

39. 464 nm.

41. 646 nm.

43. (a) 81.5 nm; (b) 127 nm.

45. 0.5 cm.

47. $\theta = 63.3°$.

49. $\sin\theta_{\max} = (m + \frac{1}{2})\lambda/2S$, $m = 0, 1, 2, \ldots$,
$\sin\theta_{\min} = m\lambda/2S$, $m = 0, 1, 2, \ldots$.

51. $I/I_0 = \cos^2(2\pi x/\lambda)$.

CHAPTER 36

1. 2.26°.

3. 2.4 m.

5. 5.8 cm.

7. 4.28 cm.

9. (b) Average intensity is 2×.

11. 10.7°.

13. $d = 4a$.

15. (a) 1.8 cm; (b) 11.0 cm.

17.

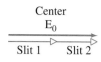

19. 2.4×10^{-7} rad $= (1.4 \times 10^{-5})°$
= $0.050''$.

21. 820 lines/mm, 102 lines/mm.

23. $5.61°$.

25. Two full orders.

27. 497 nm, 612 nm, 637 nm, 754 nm.

29. 600 nm to 750 nm of second order overlaps with 400 nm to 500 nm of third order.

31. 621 nm, 909 nm.

35. (a) Two orders; (b) 6.44×10^{-5} rad
= $13.3''$, 7.36×10^{-5} rad = $15.2''$,
2.52×10^{-4} rad = $52.0''$.

37. (a) 1.60×10^4, 3.20×10^4;
(b) 0.026 nm, 0.013 nm.

39. $\Delta f = f/mN$.

41. (a) $62.0°$; (b) 0.21 nm.

43. 0.033.

45. $57.3°$.

47. (a) $35°$; (b) $63°$.

49. $36.9°$, $53.1°$.

51. $I_0/32$.

55. $31°$ on either side of the normal.

57. 12,500 lines/cm.

59. $\sin \theta = \sin 20° - (m\lambda/a)$,
$m = \pm 1, \pm 2, \ldots$.

61. Two orders.

63. $11.7°$.

65. (a) 16 km; (b) $0.42'$.

67. (a) 0; (b) $0.094I_0$;
(c) no light gets transmitted.

69. (a) $30°$; (b) $18°$; (c) $5.7°$.

73. 0.245 nm.

CHAPTER 37

1. (a) 1.00; (b) 0.99995; (c) 0.995;
(d) 0.436; (e) 0.141; (f) 0.0447.

3. 2.07×10^{-6} s.

5. $0.773c$.

7. $0.141c$.

9. (a) 99.0 yr; (b) 27.7 yr; (c) 26.6 ly;
(d) $0.960c$.

11. $0.89c = 2.7 \times 10^8$ m/s.

13. (a) (470 m, 20 m, 0);
(b) (1820 m, 20 m, 0).

15. (a) $0.80c$; (b) $-0.80c$.

17. 60 m/s, $24°$.

19. 2.7×10^8 m/s, $43°$.

21. (a) $L_0\sqrt{1 - (v/c)^2 \cos^2\theta}$;
(b) $\tan \theta' = \tan \theta/\sqrt{1 - (v/c)^2}$.

23. Not possible in the boy's frame.

25. $0.866c$.

27. (a) 0.5%; (b) 13%.

29. 5.36×10^{-13} kg.

31. 8.20×10^{-14} J, 0.511 MeV.

33. 9×10^2 kg.

37. (a) 11.2 GeV (1.79×10^{-9} J);
(b) 6.45×10^{-18} kg · m/s.

39. 7.49×10^{-19} kg · m/s.

41. $0.941c$.

43. $M = 2m/\sqrt{1 - (v^2/c^2)}$,
$K_{loss} = 2mc^2\{[1/\sqrt{1 - (v^2/c^2)}] - 1\}$.

45. $0.866c$, 4.73×10^{-22} kg · m/s.

47. 39 MeV (6.3×10^{-12} J),
1.5×10^{-19} kg · m/s, -6%, -4%.

49. $0.804c$.

51. 3.0 T.

57. (a) $0.80c$; (b) $2.0c$.

61. 3.8×10^{-5} s.

63. $\rho = \rho_0/[1 - (v^2/c^2)]$.

65. (a) 1.5 m/s less than c; (b) 30 cm.

67. 1.02 MeV (1.64×10^{-13} J).

69. 2.2 mm.

71. 0.78 MeV.

73. Electron.

75. 5.19×10^{-13}%.

81. (a) $\alpha = \tan^{-1}[(c^2/v^2) - 1]^{1/2}$;
(c) $\tan \theta = c/v$, $u = \sqrt{v^2 + c^2}$.

CHAPTER 38

1. 6.59×10^3 K.

3. 5.4×10^{-20} J, 0.34 eV.

5. (a) 114 J; (b) 228 J; (c) 342 J;
(d) $114n$ J; (e) -456 J.

7. (b) $h = 6.63 \times 10^{-34}$ J · s.

9. 3.67×10^{-7} eV.

11. 2.4×10^{13} Hz, 1.2×10^{-5} m.

13. 400 nm.

15. (a) 2.18 eV; (b) 0.92 V.

17. 3.46 eV.

19. 1.88 eV, 43.3 kcal/mol.

21. (a) 2.43×10^{-12} m;
(b) 1.32×10^{-15} m.

23. (a) 5.90×10^{-3}, 1.98×10^{-2},
3.89×10^{-2};
(b) 60.8 eV, 204 eV, 401 eV.

27. 1.82 MeV.

29. 212 MeV, 5.85×10^{-15} m.

31. 3.2×10^{-32} m.

33. 19 V.

35. (a) 0.39 nm; (b) 0.12 nm;
(c) 0.039 nm.

37. 1.84×10^3.

39. 3.3×10^{-38} m/s, no diffraction.

41. 0.026 nm.

43. 3.4 eV.

45. 122 eV.

49. 52.5 nm.

51.

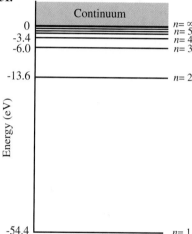

53. $U = -27.2$ eV, $K = +13.6$ eV.

55. Justified.

61. 3.28×10^{15} Hz.

65. 2.78×10^{21} photons/s · m^2.

67. 8.3×10^6 photons/s.

69. $\theta = 89.4°$.

71. 4.7×10^{-14} m.

73. 10.2 eV.

75. 4.4×10^{-40}, yes.

77. 653 nm, 102 nm, 122 nm.

79. 0.64 V.

83. 5×10^{-12} m.

85. 3.

CHAPTER 39

1. 1.8×10^{-7} m.

3. $\pm 1.3 \times 10^{-11}$ m.

5. 7.2×10^3 m/s.

7. 2.4×10^{-3} m, 1.4×10^{-32} m.

9. (a) 6.6×10^{-8} eV; (b) 6.5×10^{-9};
(c) 7.9×10^{-7} nm.

13. (a) $\psi = A \sin (3.5 \times 10^9 \text{ m}^{-1})x + B \cos (3.5 \times 10^9 \text{ m}^{-1})x$;
(b) $\psi = A \sin (6.3 \times 10^{12} \text{ m}^{-1})x + B \cos (6.3 \times 10^{12} \text{ m}^{-1})x$.

17. 3.6×10^6 m/s.

19. (a) 52 nm; (b) 0.22 nm.

23. $E_1 = 0.094$ eV,
$\psi_1 = (1.00 \text{ nm}^{-1/2}) \sin (1.57 \text{ nm}^{-1} \, x)$;
$E_2 = 0.38$ eV,
$\psi_2 = (1.00 \text{ nm}^{-1/2}) \sin (3.14 \text{ nm}^{-1} \, x)$;
$E_3 = 0.85$ eV,
$\psi_3 = (1.00 \text{ nm}^{-1/2}) \sin (4.71 \text{ nm}^{-1} \, x)$;
$E_4 = 1.51$ eV,
$\psi_4 = (1.00 \text{ nm}^{-1/2}) \sin (6.28 \text{ nm}^{-1} \, x)$.

25. (a) 4 GeV; (b) 2 MeV; (c) 2 MeV.

27. (a) 0.18; (b) 0.50; (c) 0.50.

29.

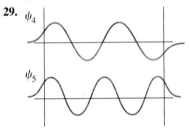

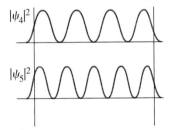

31. 0.03 nm.

33. 9.2 eV.

35. (a) Decreases by 8%;
(b) decreases by 5%.

37. (a) 32 MeV; (b) 56 fm;
(c) 8.8×10^{20} s^{-1}, 10^{10} yr.

39. 21 MeV.

41. 3.00×10^{-10} eV/c^2.

43. r_1, the Bohr radius.

45. 0.5 MeV, 5×10^6 m/s.

47. (b) 4 s.

49. 14% decrease.

CHAPTER 40

1. $\ell = 0, 1, 2, 3, 4, 5$.

3. 32 states, $(4, 0, 0, -\frac{1}{2})$, $(4, 0, 0, +\frac{1}{2})$,
$(4, 1, -1, -\frac{1}{2})$, $(4, 1, -1, +\frac{1}{2})$, $(4, 1, 0, -\frac{1}{2})$,
$(4, 1, 0, +\frac{1}{2})$, $(4, 1, 1, -\frac{1}{2})$, $(4, 1, 1, +\frac{1}{2})$,
$(4, 2, -2, -\frac{1}{2})$, $(4, 2, -2, +\frac{1}{2})$,
$(4, 2, -1, -\frac{1}{2})$, $(4, 2, -1, +\frac{1}{2})$,
$(4, 2, 0, -\frac{1}{2})$, $(4, 2, 0, +\frac{1}{2})$, $(4, 2, 1, -\frac{1}{2})$,
$(4, 2, 1, +\frac{1}{2})$, $(4, 2, 2, -\frac{1}{2})$, $(4, 2, 2, +\frac{1}{2})$,
$(4, 3, -3, -\frac{1}{2})$, $(4, 3, -3, +\frac{1}{2})$,
$(4, 3, -2, -\frac{1}{2})$, $(4, 3, -2, +\frac{1}{2})$,
$(4, 3, -1, -\frac{1}{2})$, $(4, 3, -1, +\frac{1}{2})$, $(4, 3, 0, -\frac{1}{2})$,
$(4, 3, 0, +\frac{1}{2})$, $(4, 3, 1, -\frac{1}{2})$, $(4, 3, 1, +\frac{1}{2})$,
$(4, 3, 2, -\frac{1}{2})$, $(4, 3, 2, +\frac{1}{2})$, $(4, 3, 3, -\frac{1}{2})$,
$(4, 3, 3, +\frac{1}{2})$.

5. $n \geq 5, m_\ell = -4, -3, -2, -1, 0, 1, 2, 3, 4$,
$m_s = -\frac{1}{2}, +\frac{1}{2}$.

7. (a) 6; (b) -0.378 eV; (c) $\ell = 4$,
$\sqrt{20}\hbar = 4,72 \times 10^{-34}$ kg · m^2/s;
(d) $m_\ell = -4, -3, -2, -1, 0, 1, 2, 3, 4$.

13. (a) $-[3/(32\pi r_0^3)^{1/2}] \, e^{-5/2}$;
(b) $(9/32\pi r_0^3) \, e^{-5}$; (c) $(225/8r_0) \, e^{-5}$.

15. 1.85.

17. 17.3%.

19. (a) $1.34 r_0$; (b) $2.7 r_0$; (c) $4.2 r_0$.

21. $\dfrac{r^4}{24 r_0^5} e^{-r/r_0}$.

27. (a) $\dfrac{4r^2}{27 r_0^3} \left(1 - \dfrac{2r}{3r_0} + \dfrac{2r^2}{27 r_0^2} \right)^2 e^{-2r/3r_0}$;

(b)

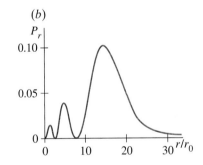

(c) $r = 13 r_0$.

29. (a) $(1, 0, 0, -\frac{1}{2})$, $(1, 0, 0, +\frac{1}{2})$,
$(2, 0, 0, -\frac{1}{2})$, $(2, 0, 0, +\frac{1}{2})$, $(2, 1, -1, -\frac{1}{2})$,
$(2, 1, -1, +\frac{1}{2})$; (b) $(1, 0, 0, -\frac{1}{2})$,
$(1, 0, 0, +\frac{1}{2})$, $(2, 0, 0, -\frac{1}{2})$, $(2, 0, 0, +\frac{1}{2})$,
$(2, 1, -1, -\frac{1}{2})$, $(2, 1, -1, +\frac{1}{2})$, $(2, 1, 0, -\frac{1}{2})$,
$(2, 1, 0, +\frac{1}{2})$, $(2, 1, -1, -\frac{1}{2})$, $(2, 1, -1, +\frac{1}{2})$,
$(3, 0, 0, -\frac{1}{2})$, $(3, 0, 0, +\frac{1}{2})$.

31. (a) $1s^2 2s^2 2p^6 3s^2 3p^6 3d^{10} 4s^2 4p^4$; (b) $1s^2$
$2s^2 2p^6 3s^2 3p^6 3d^{10} 4s^2 4p^6 4d^{10} 4f^{14} 5s^2 5p^6$
$5d^{10} 6s^1$; (c) $1s^2 2s^2 2p^6 3s^2 3p^6 3d^{10} 4s^2 4p^6$
$4d^{10} 4f^{14} 5s^2 5p^6 5d^{10} 6s^2 6p^6 5f^3 6d^1 7s^2$.

33. 5.8×10^{-13} m, 0.115 MeV.

37. 0.041 nm, 1 nm.

41. 0.19 nm.

43. Chromium.

47. (a) 0.25 mm; (b) 0.13 mm.

49. (a) $\frac{1}{2}, \frac{3}{2}, \frac{1}{2}\sqrt{3}\,\hbar, \frac{1}{2}\sqrt{15}\hbar$;
(b) $\frac{5}{2}, \frac{7}{2}, \frac{1}{2}\sqrt{35}\hbar, \frac{1}{2}\sqrt{63}\hbar$;
(c) $\frac{3}{2}, \frac{5}{2}, \frac{1}{2}\sqrt{15}\hbar, \frac{1}{2}\sqrt{35}\hbar$.

51. (a) 0.4 T; (b) 0.4 T.

53. 5.6×10^{-4} rad, 1.7×10^2 m.

55. 3.7×10^4 K.

57. (a) 1.56; (b) 1.4×10^{-10} m.

59. (a) $1s^2 2s^2 2p^6 3s^2 3p^6 3d^7 4s^2$;
(b) $1s^2 2s^2 2p^6 3s^2 3p^6 3d^{10} 4s^2 4p^6$;
(c) $1s^2 2s^2 2p^6 3s^2 3p^6 3d^{10} 4s^2 4p^6 5s^2$.

61. (a) 2.5×10^{74}; (b) 5.0×10^{74}.

63. $r = 5.24 r_0$.

65. (a) ϕ is unknown;
(b) L_x and L_y are unknown.

67. (a) 1.2×10^{-4} eV;
(b) 1.1 cm; (c) no difference.

69. (a) $3 \times 10^{-171}, 7 \times 10^{-203}$;
(b) $1.1 \times 10^{-8}, 6.3 \times 10^{-10}$;
(c) $6.6 \times 10^{15}, 3.8 \times 10^{14}$;
(d) 7×10^{23} photons/s,
4×10^{22} photons/s.

71. 182.

73. 2.25.

CHAPTER 41

1. 5.1 eV.

3. 0.7 eV.

7. (a) 13.941 u; (b) 7.0034 u;
(c) 0.9801 u.

9. (a) 6.86 u; (b) 1.85×10^3 N/m.

11. (a) 1.79×10^{-4} eV;
(b) 7.16×10^{-4} eV, 1.73 mm.

13. 2.36×10^{-10} m.

15. -7.9 eV.

17. 0.283 nm.

19. (b) -6.9 eV; (c) -10.8 eV; (d) 3%.

21. 1.8×10^{21}.

23. (a) 6.9 eV; (b) 6.8 eV.

25. 6.3%.

27. 3.2 eV, 1.05×10^6 m/s.

29. (a) 1.79×10^{29} m^{-3}; (b) 3.

33. (a) 0.021, reasonable; (b) 0.979;
(c) 0.021.

37. A large energy is required to create
a conduction electron by raising an
electron from the valence band to
the conduction band.

39. 1.1 μm.

41. 5×10^6.

43. 1.91 eV.

45. 13 mA.

47. (*a*) 2.1 mA; (*b*) 4.3 mA.

49. (*a*) 9.4 mA (smooth);
(*b*) 6.7 mA (rippled).

51. 4.0 kΩ.

53. 0.43 mA.

55. (*a*) 3.5×10^4 K; (*b*) 1.2×10^3 K.

57. (*a*) 0.9801 u; (*b*) 482 N/m, 88% of the constant for H_2.

59. States with higher values of L are less likely to be occupied, so less likely to absorb a photon; I will depend on L.

61. 2.8×10^4 J/kg.

63. 1.24 eV.

65. 1.09 μm, could be used.

67. (*a*) 0.094 eV; (*b*) 0.63 nm.

69. (*a*) 145 V ≤ V ≤ 343 V;
(*b*) 3.34 kΩ ≤ R_{load} < ∞.

CHAPTER 42

1. 3729 MeV/c^2.

3. 1.9×10^{-15} m.

5. (*a*) 2.3×10^{17} kg/m^3; (*b*) 184 m;
(*c*) 2.6×10^{-10} m.

7. 28 MeV.

9. $^{31}_{15}$P.

11. (*a*) 1.8×10^3 MeV;
(*b*) 7.3×10^2 MeV.

13. 7.48 MeV.

15. (*a*) 32.0 MeV, 5.33 MeV;
(*b*) 1636 MeV, 7.87 MeV.

17. 12.4 MeV, 7.0 MeV, neutron is more closely bound in ^{23}Na.

19. (*b*) Stable.

21. 0.782 MeV.

23. β^+ emitter, 1.82 MeV.

25. $^{228}_{90}$Th, 228.02883 u.

27. (*a*) $^{32}_{16}$S; (*b*) 31.97207 u.

29. 0.862 MeV.

31. 5.31 MeV.

33. (*b*) 0.961 MeV, 0.961 MeV to 0.

35. 3.0 h.

37. 1.2×10^9 decays/s.

39. 1.78×10^{20} nuclei.

41. 7 α particles, 4 β^- particles.

43. (*a*) 4.8×10^{16} nuclei;
(*b*) 3.2×10^{15} nuclei;
(*c*) 7.2×10^{13} decays/s; (*d*) 26 min.

45. 1.68×10^{-10} g.

47. 2.6 min.

49. 3.4 mg.

51. 8.6×10^{-7}.

53. $^{211}_{82}$Pb.

55. $N_D = N_0(1 - e^{-\lambda t})$.

57. (*a*) 0.99946; (*b*) 1.2×10^{-14};
(*c*) 2.3×10^{17} kg/m^3, $10^{14}\times$.

59. 28.6 eV.

61. (*a*) 7.2×10^{55}; (*b*) 1.2×10^{29} kg;
(*c*) 3.2×10^{11} m/s^2.

63. 6×10^3 yr.

65. 6.64 $T_{1/2}$.

69. Calcium, stored by body in bones, 193 yr, $^{90}_{38}$SR → $^{90}_{39}$Y + $^{0}_{-1}$e + $\bar{\nu}$,
$^{90}_{39}$Y is radioactive,
$^{90}_{39}$Y → $^{90}_{40}$Zr + $^{0}_{-1}$e + $\bar{\nu}$,
$^{90}_{40}$Zr is stable.

73. (*a*) 0.002603 u, 2.425 MeV/c^2; (*b*) 0;
(*c*) −0.094909 u, −88.41 MeV/c^2;
(*d*) 0.043924 u, 40.92 MeV/c^2;
(*e*) Δ ≥ 0 for 0 ≤ Z ≤ 8
and Z ≥ 85, Δ < 0 for 9 ≤ Z ≤ 84.

75. 0.083%.

77. $^{228}_{88}$RA, $^{228}_{89}$Ac, $^{228}_{90}$Th, $^{224}_{88}$Ra, $^{220}_{87}$Rn, $^{231}_{90}$Th, $^{231}_{91}$Pa, $^{227}_{89}$Ac, $^{227}_{90}$Th, $^{223}_{88}$Ra.

79. (*b*) ≈ 10^{17} yr; (*c*) ≈ 60 yr; (*d*) 0.4.

CHAPTER 43

1. $^{28}_{13}$Al, β^- emitter, $^{28}_{14}$Si.

3. Possible.

5. 5.701 MeV is released.

7. (*a*) Can occur; (*b*) 19.85 MeV.

9. +4.730 MeV.

11. (*a*) $^{7}_{3}$Li; (*b*) neutron is stripped from the deuteron; (*c*) +5.025 MeV, exothermic.

13. (*a*) $^{31}_{15}$P(p, γ)$^{32}_{16}$S; (*b*) +8.864 MeV.

15. $\sigma = \pi(R_1 + R_2)^2$.

17. Rate at which incident particles pass through target without scattering.

19. (*a*) 0.7 μm; (*b*) 1 mm.

21. 173.2 MeV.

23. 0.116 g.

25. 25.

27. 0.11.

29. 1.3 keV.

33. 6.1×10^{23} MeV/g, 4.9×10^{23} MeV/g,
2.1×10^{24} MeV/g,
5.1×10^{23} MeV/g.

35. Not independent.

37. 3.23×10^9 J, 65$\times$.

39. (*a*) $4.9\times$; (*b*) 1.5×10^9 K.

41. 400 rad.

43. 167 rad.

45. 1.7×10^2 counts/s.

49. 0.225 μg.

51. (*a*) 0.03 mrem ≈ 0.006% of allowed dose; (*b*) 0.3 mrem ≈ 0.06% of allowed dose.

53. (*a*) 1; (*b*) 1 ≤ m ≤ 2.7.

55. (*a*) $^{12}_{6}$C; (*b*) +5.70 MeV.

57. 1.004.

59. 51 mrem/yr.

61. 4.6 m.

63. 18.000953 u.

65. 6.31×10^{14} J/kg, ≈ $10^7\times$ the heat of combustion of coal.

67. 1×10^{24} neutrinos/yr.

69. (*a*) 6.8 bn; (*b*) 3×10^{-14} m.

71. (*a*) 3.7×10^3 decays/s;
(*b*) 5.2×10^{-4} Sv/yr ≈ 0.15 background.

CHAPTER 44

1. 7.29 GeV.

3. 1.8 T.

5. 13 MHz.

7. λ_α = 2.6×10^{-15} m ≈ size of nucleon, λ_p = 5.2×10^{-15} m ≈ 2(size of nucleon), α particle is better.

9. 1.4×10^{-18} m.

11. 2.2×10^6 km, 7.5 s.

15. 33.9 MeV.

17. 1.879 GeV.

19. 67.5 MeV.

21. 2.3×10^{-18} m.

23. (*b*) Uncertainty principle allows energy to not be conserved.

25. 69.3 MeV.

27. 8.6 MeV, 57.4 MeV.

29. 52.3 MeV.

31. 7.5×10^{-21} s.

33. (*a*) 1.3 keV; (*b*) 8.9 keV.

35. (*a*) n = d d u; (*b*) $\bar{n}$ = $\bar{d}$ $\bar{d}$ $\bar{u}$;
(*c*) Λ^0 = u d s; (*d*) Σ^0 = $\bar{u}$ $\bar{d}$ $\bar{s}$.

37. D^0 = c$\bar{u}$.

39. (*a*)

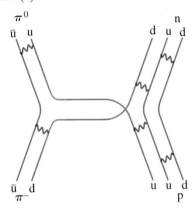

(*b*)

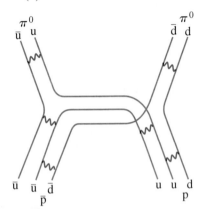

41. 26 GeV, 4.8×10^{-17} m.

43. 5.5 T.

45. (*a*) Possible, through the strong interaction; (*b*) possible, through the strong interaction; (*c*) possible, through the strong interaction; (*d*) forbidden, charge is not conserved; (*e*) possible, through the weak interaction; (*f*) forbidden, charge is not conserved; (*g*) possible, through the strong interaction; (*h*) forbidden, strangeness is not conserved; (*i*) possible, through the weak interaction; (*j*) possible, through the weak interaction.

49. -135.0 MeV, -140.9 MeV.

51. 64.

53. (*b*) 10^{27} K.

55. 6.58×10^{-5} m.

57. $\bar{u}\,\bar{u}\,\bar{d} + u\,d\,d \rightarrow \bar{u}\,d + d\,\bar{d}$.

CHAPTER 45

3. 4.8 ly.

5. $0.059''$, 17 pc.

7. Less, $\phi_1/\phi_2 = \frac{1}{2}$.

9. 48 W/m².

11. 1.4×10^{-4} kg/m³.

13. 1.8×10^9 kg/m³, 3.3×10^5.

15. -92.2 keV, 7.366 MeV.

17. (*a*) 9.594 MeV is released; (*b*) 7.6 MeV; (*c*) 6×10^{10} K.

19. $d_2/d_1 = 6.5$.

23. 540°.

25. 1.4×10^8 ly.

27. 0.86*c*.

29. 0.328*c*, 4.6×10^9 ly.

31. 1.1 mm.

33. (*a*) 10^{-3}; (*b*) 10^{-10}; (*c*) 10^{-13}; (*d*) 10^{-27}.

35. (*a*) Temperature increases, luminosity is constant, size decreases; (*b*) temperature is constant, luminosity decreases, size decreases; (*c*) temperature decreases, luminosity increases, size increases.

37. 8×10^3 rev/s.

39. 7×10^{24} W.

41. (*a*) 46 eV; (*b*) 18 eV.

43. $\approx (2 \times 10^{-5})°$.

45. 1.4×10^{16} K, hadron era.

47. Venus is brighter, $\ell_V/\ell_S = 16$.

49. $R \geq GM/c^2$.

51. $R = \dfrac{h^2}{16 m_n^{8/3} G M^{1/3}} \left(\dfrac{18}{\pi^2} \right)^{2/3}$.

Index

Note: The abbreviation *defn* means the page cited gives the definition of the term; *fn* means the reference is in a footnote; *pr* means it is found in a problem or question; *ff* means also the following pages. Page numbers such as A-3 refer to the appendices, after the main text.

A

Aberration:
  chromatic, 856 *fn*, 859
  of lenses, 858–59
  spherical, 817, 829, 858
Absolute luminosity (stars & galaxies), 1145–46
Absolute space, 918, 923
Absolute temperature scale, 447 *ff*, 460, 538
Absolute zero, 455, 538
Absorbed dose, 1102
Absorption lines, 902, 964, 1039, 1042–43
Absorption spectra, 902, 964, 968

Abundancies, natural, 1062
Acceleration, 23–37, 54–55, 64–65, 80–82
  angular, 241–42
  average, 23–24, 54
  centripetal, 64–65, 144 *ff*, 242
  Coriolis, 292–94
  in *g*'s, 36
  of gravity, 31 *ff*, 85, 137–39
  instantaneous, 25, 54
  of the moon, 134
  motion at constant, 26 *ff*, 54–55
  radial, 64–65, 69, 114 *ff*, 242
  related to force, 80–82
  tangential, 121–22, 242
  uniform, 26 *ff*, 55, 243–44
  variable, 36–37
Accelerators, 1115–20
Accelerometer, 94
Acceptor level, 1052
AC circuits, 658 *fn*, 772–83
Accommodation of eye, 851
AC generator, 740–41
Achromatic doublet, 859
Achromatic lens, 859
AC motor, 698
Action at a distance, 147, 554

Action-reaction (Newton's third law), 82–85
Activation energy, 471, 1034
Activity, 1075
Addition of velocities, 66–68
  relativistic, 935–36
Adhesion, 352
Adiabatic process, 495–96, 502–3, 531, 537
Air bags, 30
Air columns, vibrations of, 424–29
Air conditioner, 525–28
Airfoil, 348
Airplane wing, 348
Air resistance, 31–33, 122–23
Airy disc, 896
Alkali metals, 1013
Allowed transition, 1007, 1039, 1042–43
Alpha decay, 1067–70, 1073
  and tunneling, 998, 1069
Alpha particles (or rays), 1068–70
Alternating current (ac), 645–46, 772–83
Alternators, 742
AM, 804
Ammeter, 674–76, 697

Amorphous solids, 1044
Ampère, Andrè, 637, 712, 722, 787
Ampère (unit), 637, 712
  operational definition of, 712
Ampère's law, 712–15, 788–90
Amplifier, 1054–55
Amplitude:
  pressure, 420
  of vibration, 363
  of wave, 390, 396, 419, 422
Amplitude modulation (AM), 804
Analog information, 749
Analyzer (of polarized light), 908
Andromeda, 1143
Aneroid gauge, 338–39
Angle:
  Brewster's, 910
  of dip, 688
  of incidence, 403, 408, 813, 822
  phase, 366, 398, 778
  polarizing, 910
  radian measure of, 240
  of reflection, 403, 813
  of refraction, 408–9, 822
Angstrom (unit), 824 fn
Angular acceleration, 241–42
Angular displacement, 246–47
Angular frequency, 366
Angular magnification, 853
Angular momentum, 256–60, 281–89
  in atoms, 965, 1005–7, 1015–18
  conservation, law of, 257–59,
    288–89
  of nuclei, 1064
  total, 1017–18
  vector, nature of, 259–60, 281–90
Angular position, 240
Angular quantities, 240–43, 246–47
Angular velocity, 241–44
Annihilation, 957, 1124, 1163
Anode, 605
Antenna, 787, 805, 876–77
Anthropic principle, 1169
Antilock brakes, 119
Antimatter, 1137 pr, 1163 fn
Antineutrino, 1071
Antineutron, 1124
Antinodes, 405, 427
Antiparticles, 1072, 1124–28, 1163 fn
Antiproton, 1124
Antiquark, 1130–31
Apparent brightness (stars and
  galaxies), 1145
Apparent magnitude, 1172 pr

Apparent weightlessness, 142
Approximate calculations, 9–12
Arago, F., 887
Arch, 319–21
Archimedes, 340–41
Archimedes' principle, 340–43
Area:
  formulas for, A-1
  under a curve or graph, 161–64
Aristotle, 2, 78–79
Armature, 698, 744
Asymptotic freedom, 1133
Astigmatism, 851
Astronomical telescope, 854–56
Astrophysics, 1140 ff
Atmosphere (unit), 337–40
Atmosphere, scattering of light by,
  911
Atmospheric pressure, 336–40
Atomic bomb, 1095
Atomic force microscope, 999
Atomic mass, 446
Atomic mass number, 1062
Atomic mass unit, 7, 446, 1063
Atomic number, 1010, 1012–15, 1062
Atomic spectra, 863–65
Atomic structure:
  Bohr model of, 965–71, 977, 1003–5
  early models of, 862–63
  quantum mechanics of, 1003–24
Atomic weight, 446 fn
Atoms, 446–47, 965–73 (see also
  Atomic structure, Kinetic theory)
  angular momentum in, 967
Atoms (cont.)
  binding energy in, 968
  complex, 1010–15
  electric charge in, 547
  energy levels in, 967
  hydrogen, 964–71, 1004–10
  ionization energy in, 968, 970
  probability distribution in, 1004,
    1008–10
  shells and subshells in, 1012–13
  vector model of, 1028 pr
ATP, 1036
Atwood machine, 92, 277 pr, 286
Audible range, 418
Aurora borealis, 694
Autoclave, 482 pr
Autofocusing camera, 419
Average binding energy per nucleon,
  1065
Average lifetime, 1084 pr

Average speed, 18
Avogadro, Amedeo, 459
Avogadro's hypothesis, 459
Avogadro's number, 459
Axis, instantaneous, 246
Axis of lens, 837
Axis of rotation (defn), 240

B

Back, forces in, 331 pr
Back emf, 742–43
Background radiation, cosmic
  microwave, 1159–61, 1164
Balance, 308
Ballistic galvanometer, 755 pr
Ballistic pendulum, 218
Balmer, J. J., 964
Balmer formula, 964, 969
Balmer series, 964, 968–69
Band gap, 1049–50
Band spectra, 1038
Band theory of solids, 1048–50
Banking of curves, 118–20
Bar (unit), 337
Bar-code reader, 1022
Barn (unit), 1089
Barometer, 339
Barrier penetration, 996–99,
  1069–70, 1155 fn, 1168
Baryon (defn), 1127
Baryon number, 1124–26, 1132, 1162
Base (of transistor), 1054
Base and derived quantities and
  units, 8
Battery, 635–36, 659, 668
Beam splitter, 881
Beat frequency, 431
Beats, 429–32
Becquerel, H., 1067
Becquerel (unit), 1101
Bel, 421
Bell, Alexander Graham, 421
Bernoulli, Daniel, 345–46
Bernoulli's equation, 345–47
Bernoulli's principle, 345–49
Beta decay, 1067, 1070–73, 1125
Beta particle, 1067, 1070
Betatron, 754 pr
Biasing and bias voltage, 1052,
  1054–55
Big Bang, 1136, 1159–64
Big crunch, 1165

Bimetallic-strip thermometer, 448
Binary system, 1146, 1155
Binding energy:
    in atoms, 968
    in molecules, 1031, 1033, 1036
    in nuclei, 1064–66
    in solids, 1044
    total, 1064
Binding energy per nucleon, 1065
Binoculars, 827, 856
Binomial expansion, A-1
Biological damage by radiation,
    1101–2
Biot, Jean Baptist, 719
Biot-Savart law, 719–21
Blackbody, 950
Blackbody radiation, 950–51, 1160
Black dwarf, 1149
Black hole, 149, 1144, 1151, 1155,
    1158, 1166, 1168
Blood flow, 344–45
Blueshift, 1156
Blue sky, 911
Bohr, Niels, 958, 965, 969, 985
Bohr magneton, 1016, 1064
Bohr model of the atom, 965–71,
    977, 1003–5
Bohr radius, 590 pr, 966, 1004
Boiling, 475–76, 490–93 (see also
    Phase, changes of)
Boiling point, 475–76
Boltzmann, Ludwig, 535–36
Boltzmann's constant, 459
Boltzmann distribution, 1020
Boltzmann factor, 1020, 1046
Bomb calorimeter, 490
Bond:
    covalent, 1030–33, 1044
    dipole, 1036–37
    hydrogen, 1036
    ionic, 1031–33, 1044
    metallic, 1044
    molecular, 1030–33
    partially covalent/ionic, 1032–33,
        1044
    strong (defn), 1036
    van der Waals, 1036–37
    weak, 1036–37, 1045
Bond energy, 1031, 1036
Bonding:
    in molecules, 1030–33
    in solids, 1044–45
Bone density, 957
Bose, S. N., 1011 fn

Bose-Einstein statistics, 1046 fn
Bosons, 1011 fn, 1046 fn, 1126–27,
    1132–34
Bottomness and bottom quark,
    1130–31
Boundary conditions, 990, 995
Bound charge, 625
Bound state, 994
Bow wave, 436
Boyle, Robert, 455
Boyle's law, 455–56, 467
Bragg, W. H., 906
Bragg, W. L., 906
Bragg equation, 906
Brahe, Tycho, 143
Brake, hydraulic, 338
Braking distance of car, 29, 167
Brayton cycle, 544 pr
Breakdown voltage, 596
Break-even (fusion), 1099
Breaking point, 309
Breeder reactor, 1094
Bremsstrahlung, 1014–15
Brewster, D., 910
Brewster's angle and law, 910
Bridges, 315–18
Brightness, apparent, 1145
British system of units, 8
Broglie, Louis de (see de Broglie)
Bronchoscope, 827
Brown, Robert, 446
Brown dwarfs, 1166
Brownian movement, 446
Brushes, 698, 740
BSCCO, 650
Btu, 486
Bubble chamber, 1120, 1124
Bulk modulus, 310, 312
Buoyancy, 340–43
Buoyant force, 340–43
Burglar alarm, 954

C

Calculator errors, 5
Caloric theory, 486
Calorie (unit), 486–87
    related to the joule, 486
Calorimeter, 489–90
Calorimetry, 489–90
Camera, 848–50
    autofocusing, 419
    gamma, 1104

Camera flash unit, 621
Cancer, 1101, 1104
Candela (unit), 882
Cantilever, 304–5
Capacitance, 614–21
Capacitance bridge, 630 pr
Capacitive reactance, 775
Capacitor discharge, 672
Capacitor microphone, 677
Capacitors, 613–21
    charging of, 789–90
    in circuits, 669–73, 764–65, 774 ff
    as filter, 776
    reactance of, 775
    in series and parallel, 617–20
    uses of, 776
Capacity (see Capacitance)
Capillaries, 352
Capillarity, 351–52
Car, stopping distance, 29, 167
Carbon cycle, 1097, 1112 pr, 1148
Carbon dating, 1061, 1078–79
Carburetor of car, 348
Carnot, S., 520
Carnot cycle, 520–25
Carnot efficiency, 522–23
Carnot engine, 520–25
Carnot's theorem, 522–23
Carrier frequency, 804
Carrier of forces, 1121 ff
Cassegrainian focus, 856
Cathedrals, 319
Cathode, 605
Cathode rays, 605, 699 (see also
    Electron)
Cathode ray tube (CRT), 605–6, 700,
    805
CAT scan, 1105–6
Causality, 146
Cavendish, Henry, 135, 137, 148
CD player, 1022, 1054
Cell, electric, 635
Cellular telephone, 806
Celsius temperature scale, 448–49
Center of buoyancy, 357 pr
Center of gravity (CG), 223
Center of mass (CM), 221–24
    for human body, 221–22
    translational motion and, 225–27,
        262–65
Centigrade (see Celsius temperature
    scale)
Centipoise (unit), 350
Centrifugal (pseudo) force, 115, 291

Centrifugal pump, 353
Centripetal acceleration, 64–65, 114
    ff, 242
Centripetal force, 114
Cepheid variables, 1170 *pr*
Čerenkov radiation, 886 *pr*
CERN, 1117–18, 1123
CGS system of units, 8
Chadwick, J., 1062
Chain reaction, 1091–95
Chandrasekhar limit, 1149
Change of phase (or state), 473–76,
    490–93
Characteristic expansion time, 1158
Characteristic X-rays, 1013
Charge (*see* Electric charge)
Charge density, 582
Charging a battery, 659 *fn*, 668
Charles, Jacques, 455
Charles' law, 455
Charm, 1130–31
Chemical bonds, 1030–37
Chemical reaction, rate of, 471
Chernobyl, 1094
Chip, 1030, 1052–55
Christmas tree lights, 658
Chromatic aberration, 856 *fn*, 859
Circle of confusion, 849
Circle of least confusion, 858
Circuit (*see* Electric circuits)
Circuit breaker, 644
Circular motion, 63–65, 114–22
    nonuniform, 121–22
    uniform, 63–65, 114 *ff*, 371
Circulating pump, 353
Classical physics (*defn*), 1, 916, 978
Clausius, R. J. E., 517, 533
Clausius equation of state, 477
Clausius statement of second law of
    thermodynamics, 517, 526, 533
Clock paradox (*see* Twin paradox)
Closed system, 493
Cloud chamber, 1120
Coating of lenses, optical, 880–81
Coaxial cable, 632 *pr*, 715, 760–61,
    800
Coefficient (*see* name of)
Coefficient of performance, 526
Coherent light, 873–74, 877, 1019
Coherent sources of light, 873–74,
    877
Cohesion, 352
Coil (*see* Inductor)
Collector, 1054

Colliding beams, 1118–19
Collimated beam, 1105
Collision, 211–20
    elastic, 214–17
    inelastic, 214, 217–18
Collision frequency, 483 *pr*
Colonoscope, 827
Color:
    of light related to frequency and
        wavelength, 872, 878
    of quarks, 1132–33
    of sky, 911
    of stars, 950, 1146
Color charge, 1133–34
Color force, 1133–34
Coma, 859
Commutative property, 160
Commutator, 698
Compass, magnetic, 688, 710
Complementarity, principle of,
    958–59
Completely inelastic collision,
    217–18
Complex atoms, 1010–15
Complex quantities, 980 *fn*, 986 *fn*, 989
Components of vector, 48–52
Composite wave, 401
Compound lenses, 859, 899
Compound microscope, 856–58
Compound nucleus, 1090
Compression (longitudinal wave),
    391
Compressive strength, 313
Compressive stress, 311
Compton, A. H., 955–57, 1092
Compton effect, 955–57
Compton shift, 956
Compton wavelength, 956, 975 *pr*
Computer:
    chip, 1030
    digital information and, 749
    disks, 749
    hard drive, 243–44
    keyboards, 616
    monitor, 606
Computerized axial tomography,
    1105–6
Computerized tomography, 1105–6
Concave mirror, 816, 820
Concentration gradient, 480
Concrete, prestressed and reinforced,
    313
Condensation, 475
Condensed-matter, physics, 1044–50

Conductance, 656 *pr*
Conduction, of heat, 503–5
Conduction band, 1049–50
Conduction current (*defn*), 791
Conductivity:
    electrical, 640, 649
    thermal, 503–5
Conductors:
    electric, 547–48, 562–63, 640
    heat, 504
    quantum theory of, 1048–50
Confinement (fusion), 1098–99
Conical pendulum, 116
Conservation laws:
    of angular momentum, 257–59,
        288–89
    of baryon number, 1124–26, 1135,
        1162
    of electric charge, 546–47
    of energy, 182–92, 493–95
    of lepton number, 1124–26, 1162
    of linear momentum, 208–11
    of nucleon number, 1073, 1124–26
    in nuclear and particle physics,
        1073, 1124–26
    of strangeness, 1129
Conservative force, 177–78, 749
Conserved quantity, 183
Constants, values of, inside front
    cover
Constant-volume gas thermometer,
    449, 460
Constructive interference, 404, 430,
    871, 1031
Contact force, 86, 148
Contact lenses, 852
Contact potential, 677 *fn*
Continental drift, 342
Continuity, equation of, 344
Continuous laser, 1022
Continuous spectrum, 902
Control rods, 1093
Convection, 505
Conventional current (*defn*), 637
Conventions, sign (geometric optics),
    819, 822
Converging lens, 836 *ff*
Conversion factors, 8, inside front
    cover
Convex mirror, 816
Conveyor belt, 228
Coordinate systems, 17
Copenhagen interpretation of
    quantum mechanics, 985

Copier, electrostatic, 555
Core, of reactor, 1093
Coriolis force, 292–94
Cornea, 850
Corrective lenses, 850–52
Correspondence principle, 943, 971, 976 pr, 978
Cosmic microwave background radiation, 1159–61, 1164
Cosmic rays, 1102, 1114
Cosmological principle, 1157 ff
"perfect," 1159 fn
Cosmology, 1136, 1140–69
Coulomb, Charles, 549–50
Coulomb (unit), 550, 712
operational definition of, 712
Coulomb barrier, 998, 1098, 1148
Coulomb's law, 549–53, 575–86, 966
vector form of, 553
Coulomb potential, 1034
Counter emf, 742–43
Counter torque, 743
Covalent bond, 1030–33, 1044
Crane, 304
Creativity in science, 3
Crick, F.H.C., 906
Critical damping, 376
Critical density, of universe, 1165–68
Critical mass, 1092–95
Critical point, 473
Critical temperature, 473
Crossed Polaroids, 908
Cross product, 279–80
Cross section, 1088–90
CRT, 605–6, 700, 805
Crystal lattice, 446–47, 1044
Crystallography, 906
CT scan, 1105–6
Curie, M. and P., 726, 1067
Curie (unit), 1101
Curie temperature, 722
Curie's law, 726
Current (see Electric current)
Current density, 647–49
Current gain, 1055
Current sensitivity, 674
Curvature of field, 859
Curvature of universe (space-time), 149, 1151–55, 1165
Cycle (defn), 363
Cyclic universe, 1168
Cyclotron, 707 pr, 1116–17
Cyclotron frequency, 694, 1116

D

Damped harmonic motion, 374–77
Dark matter, 1166
Dating, radioactive, 1078–79
Daughter nucleus (defn), 1068
Davisson, C. J., 960
dB (unit), 421
DC circuits, 658–77
DC generator, 741
DC motor, 698
de Broglie, Louis, 959–61, 971, 977–78
de Broglie wavelength, 959, 971–72
Debye's law, 514 pr
Decay:
of elementary particles, 1124–34
radioactive (see Radioactivity)
Decay constant, 1073–76
Decay series, 1077–78
Deceleration (defn), 24
Deceleration parameter, 1168
Decibel (db), 421
Declination, magnetic, 688
Decommissioning nuclear plant, 1094
Dee, 1116
Defects of the eye, 851
Defibrillator, 632 pr, 651 fn
Degrees of freedom, 500–501
Del, 603 fn
Delayed neutron, 1093
Delta particle, 1128
Demagnetization, 725
Density, 332–33
and floating, 342
Density of occupied states, 1047
Density of states, 1045–48
Depth of field, 849
Derivatives, 21–22, A-4–A-5
partial, 399
Derived quantities, 8
Descartes, R., 146
Destructive interference, 404, 430, 871, 1031
Detection of radiation, 1080
Determinism, 147, 984–85
Deuterium, 1086, 1092, 1095–1100
Deuteron, 1086
Dew point, 476
Diamagnetism, 725–26
Diamond, 826–27
Dielectric constant, 621–22
Dielectrics, 621–26
molecular description of, 624–26

Dielectric strength, 622
Diesel engine, 540 pr
Differential cross section, 1089
Diffraction, 867, 887–911, 1020
by circular opening, 896–97
as distinguished from interference, 896
in double slit experiment, 896
of electrons, 960
Fraunhofer, 888 fn
Fresnel, 888 fn
of light, 867, 887–911
as limit to resolution, 896–97
by single slit, 888–93
of water waves, 410
X-ray, 905–6
Diffraction grating, 900–905
resolving power of, 903–5
Diffraction limit of lens resolution, 896–97
Diffraction patterns, 888 ff
of circular opening, 896–97
of single slit, 888–93
X-ray, 905–6
Diffraction spot or disc, 896
Diffuse reflection, 813
Diffusion, 479–80
Fick's law of, 480
Diffusion constant, 480
Diffusion equation, 480
Digital information, 749
Digital voltmeter, 675
Dimensional analysis, 12–13
Dimensions, 12–13
Diodes, 1052–54
Diopter, 838
Dip, angle of, 688
Dipole antenna, 793
Dipole bonds, 1036–37
Dipoles and dipole moments:
of atoms, 1015–18
electric, 565–67, 578, 601–3
magnetic, 695–97, 721
Dirac, P. A. M., 1005, 1046 fn
Direct current (dc) (see Electric current)
Discharge tube, 963
"Discovery" in science, 700
Disintegration energy (defn), 1068–69
Disorder and order, 533–34
Dispersion, 402, 824–25
Displacement, 17–18, 46–47
angular, 246–47

Displacement (*continued*)
  vector, 46–47, 53
  in vibrational motion, 363
  of wave, 396
Displacement current, 791
Dissipative force, 189–92
Dissociation energy, 1031
Distance:
  image, 814, 819
  object, 814, 819
Distortion, by lenses, 859
Distributive property, 160
Diverging lens, 836 *ff*
DNA, 573 *pr*, 906, 1036–37
Domains, magnetic, 722
Dome, 319–21
Donor level, 1052
Doorbell, 724
Door opener, automatic, 954
Doping of semiconductors, 1051 *ff*
Doppler, J.C., 432 *fn*
Doppler effect:
  for light, 435, 942–43, 1156–57
  for sound, 432–35
Dose, 1101–3
Dosimetry, 1101–3
Dot product, 159–61
Double-slit experiment (light),
    870–73
  for electrons, 978–79
  intensity in pattern, 874–77, 895
Drag force, 122–23
Drift velocity, 647–49, 701
Dry cell, 636
Duality, wave-particle, 958–59, 972
Dulong and Petit value, 501
Dynamic lift, 348
Dynamics, 16, 77 *ff*
  of rotational motion, 247 *ff*
Dynamo (*see* Electric generator)
Dyne (unit), 81
Dynode, 1080

E

$E = mc^2$, 938–42
Ear, response of, 422
Earth, rocks and earliest life, 1079
Earthquake waves, 394, 396, 409
ECG, 606
Eddy currents, 744
Edison, Thomas, 605
Effective cross section, 1089

Effective dose, 1102
Efficiency of heat engine, 519,
    522–23
Einstein, A., 148, 169, 446, 916,
    922–23, 940–41, 965, 977, 985,
    1095, 1134, 1151–55
Einstein cross, 1153
EKG, 606
Elastic collision, 214–17
Elasticity, 309–12
Elastic limit, 309
Elastic moduli, 309–12
  and speed of sound waves, 393
Elastic potential energy, 181, 369
Elastic region, 309
Elastic scattering, 1089
Elastic spring, 181
Electric battery, 635–36, 659, 668
Electric car, 592, 657 *pr*
Electric cell, 635
Electric charge, 545 *ff*
  in atom, 547
  bound, 625
  charge density, 582
  conservation of, 546–47
  of electron, 550, 722–23
  elementary, 550
  free, 625
  induced, 548–49, 625
  motion of, in electric field, 564–65
  motion of, in magnetic field, 692–95
  point (*defn*), 551
  quantization of, 550
  "test," 554
Electric circuits, 636, 658–77, 772–83
  ac, 658 *fn*, 772–83
  containing capacitors, 669–73,
    764–67, 774 *ff*
  dc, 658–77
  grounding, 651–52
  household, 644–45
  impedance matching of, 781–82
  integrated, 1054–55
  and Kirchhoff's rules, 664–69
  LC, 764–65
  LR, 762–63
  LRC, 766–67, 776–79
  RC, 669–73
  rectifier, 1053–54
  resonant, 780
  time constants of, 670, 762
Electric current, 634–52 (*see also*
    Electric circuits)
  ac, 645–46, 772–83

conduction (*defn*), 791
conventional, 637
dc (*defn*), 645
density, 647–49
displacement, 791
hazards of, 651–52
induced, 735
leakage, 652
magnetic force on, 689–92
microscopic view, 647–49
and Ohm's law, 638–39
peak, 645
produced by magnetic field, 747–49
produces magnetic field, 689,
    719–21
rms, 646
Electric dipole, 565–67, 578, 601–3
Electric energy, 603–5, 620–21
  stored in capacitor, 620–21
  stored in electric field, 321
Electric field, 545 *ff*, 595–97, 602–3
  calculation of, 558–61, 595, 602–3
  and conductors, 562–63
  in dielectric, 622–25
  of dipole, 565–67, 603
  in EM wave, 792–803
  energy stored in, 621
  motion of charged particle in,
    564–65, 694
  produced by changing magnetic
    field, 747–49
  produced by dipole, 566
  produces magnetic field, 788–91
  relation to electric potential,
    595–97, 602–3
Electric field lines, 561–62, 576
Electric flux, 575–78, 789
Electric force, 545 *ff*, 694
  Coulomb's law for, 549–53
Electric generator, 740–42
Electric hazards, 651–52
Electric heater, 643
Electricity, static, 545
Electric motor, 698, 742–43
  counter emf in, 742–43
Electric potential, 591–604 (*see also*
    Potential difference)
  of dipole, 601–2
  due to any charge distribution, 599
  relation to electric field, 595–97,
    602–3
  of single point charge, 597–98
Electric potential energy, 591–94,
    603–5

Electric power, 642–44
  in ac circuits, 778
  generation, 741
  in household circuits, 644–45
  and impedance matching, 781–82
  transmission of, 746–47
Electric shocks, 651–52
Electrocardiogram (ECG, EKG), 606
Electrode, 635
Electrolyte, 635
Electromagnet, 723–24
Electromagnetic force, 1121–22,
    1126–27, 1134–36
Electromagnetic induction, 734 ff
Electromagnetic oscillations, 764–65
Electromagnetic pumping, 703 pr
Electromagnetic spectrum, 798–800,
    874–75
Electromagnetic (EM) waves,
    792–806, 866 (see also Light)
Electrometer, 548
Electromotive force (emf), 659–60,
    734–35, 739–40, 742–43
  back, 742–43
  counter, 742–43
  of generator, 740–42
  Hall, 701
Electromotive force (emf) (cont.)
  induced, 734–35, 739–40
  motional, 739
  series and parallel, 668
  sources of, 659–60, 734–35
Electron:
  as beta particle, 1067, 1070
  charge on, 547, 550, 722–23
  discovery of, 699–700
  in double-slit experiment, 978–79
  as elementary particle, 1126
  free, 548
  mass of, 1063
  in pair production, 957–58
  wavelength of, 959–60
Electron band theory, 1048–50
Electron capture, 1072
Electron cloud, 1004, 1030–31
Electron configuration, 925–26
Electron gun, 606
Electronically steered phase array, 877
Electron microscope, 899, 961, 999
Electron sharing, 1030
Electron spin, 723, 1005–6, 1017–18
Electron spin resonance, 1029 pr
Electron volt (unit), 603–5, 1063
Electroscope, 548–49, 635 fn

Electrostatic copier, 555
Electrostatic potential energy, 603–5
Electrostatics, 551 ff
Electrostatic unit, 550 fn
Electroweak force, 148, 545 fn
Electroweak theory, 1132–34
Elementary charge, 550
Elementary particle physics, 1114–36
Elements:
  origin of in universe, 1149–50
  periodic table of, 1011–13, 1062 fn,
    inside rear cover
  transmutation of, 1068, 1070,
    1085–88
Elevator and counterweight, 92
Ellipse, 143
EMF (see Electromotive force)
Emission spectra, 902, 949–51,
    963–65, 968–69
Emission tomography, 1006–7
Emissivity, 506
Emitter, 1054
EM waves (see Electromagnetic
    waves)
Endoergic reaction (defn), 1087
Endoscope, 827
Endothermic reaction, 1086–87
Energy, 155, 165–69, 176–97, 260–65,
    487–88, 493–94
  activation, 471, 1034
  binding, 968, 1031, 1033, 1036,
    1064–66
  chemical, 190
  conservation of, 182–92, 493–95,
    1086
  degradation of, 535
  disintegration, 1068
  distinguished from heat and
    temperature, 487
  electric (see Electric energy)
  in EM waves, 800–802
  equipartition of, 500–501
  equivalence to mass, 938–42
  and first law of thermodynamics,
    493–94
  internal, 189–90, 487–88
  kinetic, 164–69, 182 ff, 260–65
  mechanical, 182–89
  molecular rotational and
    vibrational, 500–501
  nuclear, 990–1016
  potential (see Potential energy)
  quantization of, 951, 965–71,
    991–92

reaction, 1086
  relation to work, 164–69, 178–82,
    190–91, 195, 260–61
  rest, 939, 1063
  rotational, 260–65
  in simple harmonic motion, 369–70
  thermal, 189–90, 487–88
  threshold, 1088, 1113 pr
  total mechanical energy (defn), 183
  transformation of, 189–90
  unavailability of, 534–35
  and uncertainty principle, 982–83,
    996
  units of, 167
  vibrational, 369–70
  of waves, 395–96
  zero point, 992, 1041
Energy bands, 1048–50
Energy density:
  in electric field, 621
  in magnetic field, 761
Energy gap, 1049–50
Energy levels:
  in atoms, 967
  for fluorescence, 1019
  for lasers, 1019–22
Energy states, in atoms, 967
Energy transfer, heat as, 189–90,
    485 ff
Engine, heat (see Heat engine)
Enriched uranium, 1092
Entropy, 528–37
  in life processes, 534
  statistics and, 535–37
Environmental pollution, 525
Enzyme, 1037
Equally-tempered chromatic scale,
    424
Equation of continuity, 344
Equation of motion, oscillations, 364,
    375, 377
Equation of state, 454, 457, 477–78
  Clausius, 477
  ideal gas, 457
  van der Waals, 477–78
Equilibrium, 197–98, 300–303, 308
  conditions for, 301–3
  stable, unstable, neutral, 197–98,
    308
  thermal, 449–50
Equilibrium position (vibrational
    motion), 363
Equilibrium state (defn), 454
Equipartition of energy, 500–501

Equipotential lines, 600–601
Equipotential surface, 600–601
Equivalence, principle of, 148, 1151–52
Erg (unit), 156
Escape velocity, 192–94
Estimated uncertainty, 4
Estimating, 9–12
Ether, 919–22
Euclidean space, 1153
Evaporation, 474–75, 493
Event horizon, 1155
Everest, Mt., 138
Evolution and entropy, 534
Exact differential, 494 *fn*
Exchange coupling, 723
Excited state, of atom, 957 *ff*
Exclusion principle, 1010–12, 1046, 1132–33, 1150
Exoergic reaction (*defn*), 1086
Exothermic reaction, 1086
Expansion, thermal (*see* Thermal expansion)
Expansion of universe, 1156–59
Expansions, in waves, 391
Expansions, mathematical, A-1
Exponential curves, 1074–75
Exponential decay, 1074–75
Extragalactic (*defn*), 1143
Eye:
   accommodation, 851
   defects of, 851
   far and near points of, 851
   normal (*defn*), 851
   resolution of, 899
   structure and function of, 850–52
Eyeglass lenses, 851–52
Eyepiece, 854

F

Fahrenheit temperature scale, 448–49
Falling bodies, 31–36
Fallout, radioactive, 1095
Farad (unit of capacitance), 614
Faraday, Michael, 147, 554, 622, 734, 736, 787
Faraday's law, 736–38, 747–48, 756, 792
Far point of eye, 851
Farsighted eye, 851–52
Fermat's principle, 835 *pr*

Fermi, E., 11, 958, 978, 1011 *fn*, 1046 *fn*, 1071, 1088, 1090–95, 1128
Fermi-Dirac statistics, 1046–48
Fermi energy, 1046–48
Fermi factor, 1046–47
Fermi gas, 1046
Fermilab, 1114, 1117
Fermi level, 1046–48
Fermions, 1011 *fn*, 1046, 1132–33
Fermi speed, 1047
Fermi temperature, 1060 *pr*
Ferris wheel, 124 *pr*
Ferromagnetism and ferromagnetic materials, 687, 722–23
Feynman, R., 1121
Feynman diagram, 1121–22
Fiber optics, 826–27
Fick's law of diffusion, 480
Field, 147 (*see also* Electric field, Gravitational field, Magnetic field)
   conservative and nonconservative, 748–49
   in elementary particles, 1121
Film badge, 1103
Filter circuit, 776, 785 *pr*
Fine structure, 977, 1003, 1006, 1018
Fine structure constant, 1018
First law of motion, 78 *ff*
First law of thermodynamics, 493–98
Fission, nuclear, 1090–95
Fission bomb, 1095
Fission fragments, 1090–95
Fitzgerald, G. F., 922
Flashlight battery, 636
Flashlight bulb, 639
Flavor (of elementary particles), 1132–33
Floating, 342
Flow of fluids, 343–45
   laminar, 343–45
   streamline, 343–45
   in tubes, 351
   turbulent, 343
Flow rate, 343–45
Fluids, 332–61 (*see also* Gases, Liquids)
Fluorescence, 1019
Fluorescent lightbulbs, 1019
Flux:
   electric, 575–78, 789
   magnetic, 736, 791
Flying buttress, 319
FM radio, 804

Focal length:
   of lens, 838, 840–41
   of spherical mirror, 817, 821
Focal plane, 838
Focal point, 817, 821, 837–38
Focus, 817
Focusing, of camera, 849
Foot-candle (*defn*), 882 *fn*
Foot-pound (unit), 156, 167
Forbidden transition, 1007, 1020 *fn*, 1041 *fn*
Force, in general, 77–85, 147–48, 177–78 (*see also* Electric force, Gravitational force, Magnetic force)
   buoyant, 340–43
   centrifugal, 115
   centripetal, 114
   color, 1133–34
   conservative and nonconservative, 177–78
   contact, 86, 148
   Coriolis, 292–94
   damping, 374–77 (*see also* Drag force)
   dissipative, 189–92
   drag, 122–23
   elastic, 162
   electromagnetic, 1121–22, 1126–27, 1134–36
   electroweak, 148
   in equilibrium, 300–303
   exerted by inanimate object, 83–84
   fictitious, 291
   of friction, 78, 106–14, 118–20, 178, 190–92
   of gravity, 85–86, 133 *ff*, 1123, 1150–55
   impulsive, 212
   inertial, 291–92
   long and short range, 1066
   measurement of, 78
   in muscles and joints, 306
   net, 80, 88
   in Newton's laws, 78–85, 207–9
   nonconservative, 178
   normal, 85–87, 107
   nuclear, 1064–66, 1072, 1121–36
   pseudoforce, 291
   relation of momentum to, 206–8
   resistive, 122–23
   restoring, 162, 363
   strong, 1064–66, 1121–36

types of, in nature, 147–48, 545 *fn*, 1123, 1136
units of, 81
van der Waals, 1036–37
velocity-dependent, 122–23
viscous, 344, 350–51
weak, 1066, 1072, 1121–36
work done by, 156 *ff*
Forced convection, 505
Force diagram (*see* Free-body diagram)
Forced vibrations, 378–79
Force pump, 353
Fossil-fuel power plants, 741
Foucault, J., 868
Four-dimensional space-time, 932
Fourier integral, 401
Fourier's theorem, 401
Fovea, 850, 899
Fractal behavior, 446
Fracture, 312–15
Frames of reference, 17, 79, 291
Franklin, Benjamin, 546, 586
Fraunhofer diffraction, 888 *fn*
Free-body diagram, 88–96, 303
Free charge, 625
Free electrons, 548
Free-electron theory of metals, 1045–48
Free expansion, 498, 531, 537
Free fall, 31–33
Free particle, and Schrödinger equation, 989–94
Freezing (*see* Phase, changes of)
Frequency, 65, 242
  angular, 366
  of audible sound, 422
  beat, 429–32
  carrier, 804
  of circular motion, 65
  cyclotron, 694
  fundamental, 406, 424–26
  infrasonic, 419
  of light, 799, 869
  natural, 366, 378, 405
  resonant, 378–79, 405, 780
  of rotation, 242–43
  ultrasonic, 418, 437
  of vibration, 363, 366
  of wave, 390
Frequency modulation (FM), 804
Fresnel, A., 887
Fresnel diffraction, 888 *fn*
Friction, 78, 106–14, 118–20, 178, 184

coefficients of, 107, 112
  kinetic, 106 *ff*
  in rolling, 106, 263–64, 267
  static, 107 *ff*
Fringes, interference, 871
Frisch, O., 1090
f-stop, 848
Full-wave rectifier, 1053
Fundamental frequency, 406, 424–26
Fuse, 644–45
Fusion, heat of, 490–93
Fusion, nuclear, 1095–1100
  in stars, 1148–49, 1163
Fusion reactor, 1097

G

g-factor, 1017
Galaxies, 1141–45, 1155 *ff*
Galaxy clusters, 1143
Galilean-Newton relativity, 917–19, 932–36
Galilean telescope, 856
Galilean transformation, 932–36
Galileo, 2–3, 16, 31–32, 56–57, 78–79, 146, 339, 448, 811, 854–56, 854 *fn*, 917–18, 932, 1141, 1150
Galvani, Luigi, 635
Galvanometer, 674, 697, 755 *pr*
Gamma camera, 1104
Gamma decay, 1067, 1072–73
Gamma rays, 957, 1067, 1072–73
Gamow, G., 1160
Gas constant, 457
Gases, 332, 445–46, 454–81
  adiabatic expansion of, 502–3
  change of phase, 473–76, 490–93
  heat capacities of, 498–99
  ideal, 456–60, 466–72
  real, 473–74
  work done by, 496–98
Gas laws, 454–59
Gauge bosons, 1026–27, 1132, 1134
Gauge pressure, 337
Gauges, pressure, 338–40
Gauge theory, 1134
Gauss, C. F., 1154
Gauss (unit), 691
Gauss's law, 575–86, 713
  for magnetism, 791
Gay-Lussac, Joseph, 456
Gay-Lussac's law, 456

Geiger counter, 612 *pr*, 1080
Gell-Mann, M., 1130
General motion, 262–68, 283–84
General theory of relativity, 148, 149, 169, 922, 1140–41, 1151–55
Generator, electric, 740–42
Genetic damage, 1101
Geodesic, 1153–54
Geological dating, 1079
Geometric optics, 811 *ff*
Geophone, 759–60
Geosynchronous satellite, 140, 145
Germer, L. H., 960
Glaser, D., 1120
Glashow, S., 1134
Glasses, eye, 851–52
Glueballs, 1133 *fn*
Gluino, 1136
Gluons, 1123, 1133
Gradient:
  concentration, 480
  electric potential, 602
  pressure, 351
  velocity, 350
Gradient operator, 603 *fn*
Gram (unit), 7
Grand unified era, 1162
Grand unified theories, 148, 1134–36
Grating, diffraction (*see* Diffraction grating)
Gravitation, universal law of, 133 *ff*
Gravitational collapse, 1155
Gravitational constant (*G*), 135
Gravitational field, 146–47, 562, 1151–55
Gravitational force, 85–86, 133 *ff*
  due to spherical mass distribution, A-6–A-8
Gravitational mass, 148, 1152
Gravitational potential energy, 178–80, 192–94
Gravitino, 1164 *fn*
Graviton, 1122–23
Gravity, 32–33, 85, 133 *ff*, 1123, 1150–55
  acceleration of, 32–33, 85, 137–39
  center of, 223
  free fall under, 31–36
  specific, 333
Gravity anomalies, 138
Gray (unit), 1102
Greek alphabet, inside front cover
Grimaldi, F., 867, 872
Grounding, electrical, 651–52

Ground state, of atom, 967 *ff*
GUT, 148, 1134–36
Gyromagnetic ratio, 1017
Gyroscope, 289

H

h-bar (*h*), 1013
Hadron era, 1162
Hadrons, 1126–36, 1162
Hahn, O., 1090
Hair dryer, 646
Halflife, 1073–76
Half-wave rectification, 1053
Hall, E. A., 701
Hall effect, Hall emf, Hall field, Hall probe, 701
Halogen, 1013
Hard drive, 243–44
Harmonic motion:
    damped, 374–77
    forced, 378–79
    simple, 364–71, 390
Harmonics, 406, 424, 426–27
Harmonic wave, 398
Hazards of electricity, 651–52
Headlights, 643
Hearing, threshold of, 422
Heart, 353
    defibrillator, 632 *pr*, 651 *fn*
    pacemaker, 673, 758
Heat, 485–544
    compared to work, 493
    conduction, convection, radiation, 503–8
    distinguished from internal energy and temperature, 487
    as energy transfer, 189–90, 485 *ff*
    in first law of thermodynamics, 493–98
    as flow of energy, 485–86
    of fusion, 490–93
    latent, 490–93
    mechanical equivalent of, 486
    specific, 488 *ff*, 498–99
    of vaporization, 491
Heat capacity, 515 *pr* (*see also* Specific heat)
Heat conduction, 503–5
Heat death, 534–35
Heat engine, 517–25, 1093
    Carnot, 520–25
    efficiency of, 519, 522–23

internal combustion, 518–19
    steam, 518–19
Heating element, 642
Heat of fusion, 490–93
Heat pump, 527–28
Heat reservoir, 495
Heat transfer, 503–8
Heat of vaporization, 491
Heavy water, 1092
Heisenberg, W., 949, 978, 981–84
Heisenberg uncertainty principle, 981–84, 996, 1031
Helium:
    liquid, 474
    primordial production, 1164
Helium-neon laser, 1021
Helmholtz coils, 733 *pr*
Henry, Joseph, 734
Henry (unit), 757
Hertz, Heinrich, 798
Hertz (unit of frequency), 243, 363, 418
Hertzsprung-Russell diagram, 1146–49
Higgs boson, 1134
Higgs field, 1134
High energy accelerators, 1115–20
High energy physics, 1114–36
Highway curves, banked and unbanked, 118–20
Holes (in semiconductors), 1050–52
Holograms, 1023–24
Hooke, Robert, 309, 878 *fn*
Hooke's law, 162, 181, 309
Horsepower (unit), 195
H-R diagram, 1146–49
Hubble, E., 943, 1143, 1156–57
Hubble age, 1158
Hubble's law, 1156–58
Hubble parameter, 1156–58
Hubble Space Telescope, 897, 1140
Humidity, 476
Huygens, C., 867
Huygens' principle, 867–69
Hydraulic brakes, 338
Hydraulic lift, 338
Hydrodynamics, 343 *ff*
Hydroelectric power, 741
Hydrogen atoms, 604–5
    Bohr theory of, 965–71
    magnetic moment of, 697
    quantum mechanics of, 1004–10
    spectrum of, 968–70
Hydrogen bomb, 1098

Hydrogen bond, 1036
Hydrogen-like atoms, 966 *fn*
Hydrogen molecule, 1030–31, 1033, 1040–42
Hydrometer, 342
Hyperopia, 851
Hysteresis, 724–25
Hysteresis loop, 725

I

Ideal gas, 456–60, 466 *ff*
    internal energy of, 487–88
Ideal gas law, 456–60, 466 *ff*
Ideal gas temperature scale, 460
Ignition (fusion), 1099–1100
Illuminance, 882
Image:
    as tiny diffraction pattern, 896–97
    CAT scan, 1105–6
    fiber optic, 826–27
    formed by lens, 836 *ff*
    formed by plane mirror, 812–15
    formed by spherical mirror, 816–22
    NMR, 1064, 1107–8
    PET and SPET, 1106–7
    real, 814, 818, 839
    ultrasound, 437
    virtual, 814, 821, 840
Image distance, 814, 819
Imaging, medical, 437, 1105–9
Imbalance, rotational, 287–88
Impedance, 759, 774–75, 781–82
Impedance matching, 781–82
Impulse, 211–13
Impulsive forces, 212
Incidence, angle of, 403, 408, 813, 822
Inclines, motion on, 94–95, 262–64
Incoherent source of light, 873–74, 877
Indeterminacy principle (*see* Heisenberg uncertainty principle)
Index of refraction, 811–12
    in Snell's law, 822–24
Induced electric charge, 548–49, 625
Induced emf, 734–35, 739–40
    counter, 742–43
    in electric generator, 740–42
    in transformer, 744–47
Inductance, 756–60
    in ac circuits, 758–63, 772–83
    mutual, 756–58
    self-, 758–60

Induction, electromagnetic, 734 *ff*
Induction, Faraday's law, 736–38, 792
Inductive reactance, 774
Inductor, 758–59
  in circuits, 758–63, 772–83
  energy stored in, 760–61
  reactance of, 774
Inelastic collision, 214, 217–18
Inelastic scattering, 1089
Inertia, 79
  law of, 78–79
  moment of, 249–56, 374
  rotational (*defn*), 249–50
Inertial confinement, 1098–99
Inertial forces, 291–92
Inertial mass, 148, 1152
Inertial reference frame, 79, 291,
  917 *ff*
  equivalence of all, 918–19, 922
  transformations, 932–36
Inflationary universe, 1164
Infrared radiation (IR), 506, 799,
  825
Infrasonic waves, 419
Initial conditions, 365
Instantaneous axis, 246
Instruments, wind and stringed,
  424–29
Insulators:
  electrical, 547–48, 640, 1049–50
  thermal, 504, 1049–50
Integrals and integration, 36–37, 161
  ff, A-4–A-5, B-5
Integrated circuits, 1054–55
Intensity:
  of coherent and incoherent sources,
    877
  in interference and diffraction
    patterns, 874–77, 890–93
  of light, 877, 882
  of sound, 420–23
  of waves, 395–96, 420–23
Intensity level (*see* Sound level)
Interference, 404–5
  constructive, 404, 430, 871, 1031
  destructive, 404, 430, 871, 1031
  as distinguished from diffraction,
    895
  of light waves, 870–81, 894–96, 900
  of sound waves, 429–32
  by thin films, 877–81
  of water waves, 404–5
  of waves on a string, 404
Interference fringes, 871

Interference pattern:
  double slit, 870–73, 978–79
  double slit, including diffraction,
    893–95
  multiple slit, 900–905
Interferometers, 881
Internal combustion engine, 518–19
Internal conversion, 1073
Internal energy, 189–90, 487–88
Internal reflection, total, 415 *pr*,
  826–27
Internal resistance, 659
Intrinsic semiconductor (*defn*), 1049
Inverted population, 1020
Ion, 547
Ionic bonds, 1031–33, 1044
Ionic cohesive energy, 1044
Ionization energy, 968, 970
Ionizing radiation (*defn*), 1100
IR radiation, 506, 799, 825
Irreversible process, 520–25
Isobaric process (*defn*), 496–97
Isochoric process (*defn*), 496–97
Isolated system, 209, 493
Isomer, 1073
Isotherm, 495
Isothermal process (*defn*), 495
Isotopes, 702, 1062
  table of, A-9–A-13
Isotropic material, 867

J

J (total angular momentum),
  1017–18
J/ψ particle, 1131
Jeweler's loupe, 854
Joint, 315
Joule, James Prescott, 486
Joule (unit), 156, 167, 248 *fn*
  relation to calorie, 486
Jump start, 668–69
Junction diode, 1054–55
Junction transistor, 1054–55
Jupiter, moons of, 152, 854

K

Kant, Immanuel, 1143
Kaon, 1127
K-capture, 1072
K lines, 1013

Kelvin (unit), 455
Kelvin-Planck statement of second
  law of thermodynamics, 520, 523,
  533
Kelvin temperature scale, 455, 460,
  538
Kepler, Johannes, 143, 854 *fn*, 1150
Keplerian telescope, 854
Kepler's laws, 143–46, 288–89
Keyboard, computer, 616
Kilocalorie (unit), 486–87
Kilogram (unit), 7, 79
Kilowatt-hour (unit), 643
Kinematics, 16–67, 240–44 (*see also*
  Motion)
  for rotational motion, 240–44
  for translational motion, 16–67
  for uniform circular motion, 63–65
  vector kinematics, 53–55
Kinetic energy, 164–69, 182 *ff*,
  260–65
  in collisions, 214–17, 219
  molecular, relation to temperature,
    469
  relativistic, 938–42
  rotational, 260–65
  total, 262–65
  translational, 165–69
Kinetic friction, 106 *ff*
Kinetic theory, 445, 466 *ff*
Kirchhoff, G. R., 665
Kirchhoff's rules, 664–69

L

Ladder, 306–7
Laminar flow, 343–45
Land, Edwin, 907
Large Magellanic Cloud, 1150
Laser light, 873
Lasers, 873, 887, 1019–22, 1100
Latent heats, 490–93
Lateral magnification, 819, 841
Lattice structure, 446–47, 1044,
  1051–52
Laue, Max von, 905
Lawrence, E. O., 1116
Laws, 3–4 (*see also* specific name of
  law)
Lawson, J. D., 1099
Lawson criterion, 1099–1100
LC circuit, 764–65
LCD, 909

LC oscillation, 766–67
Leakage current, 652
LED, 1054
Length, standard of, 6, 881
Length contraction, 930–31
Lens, 836–60
  achromatic, 859
  camera, 848–50
  coating of, 80–81
  color corrected, 859
  compound, 859, 899
  contact, 852
  converging, 836 *ff*
  corrective, 851–52
  cylindrical, 851
  diverging, 836 *ff*
  of eye, 850–51
  eyeglass, 851–52
  eyepiece, 854
  focal length of, 838, 840–41
  magnetic, 961
  magnification of, 841–43
  normal, 850
  objective, 854–55, 898–99
  ocular, 856
  positive and negative, 841
  power of (diopters), 838
  resolution of, 896–99
  telephoto, 850
  thin (*defn*), 837
  wide angle, 850
  zoom, 850
Lens aberrations, 858–59
Lens elements, 859
Lens equation, 840–43, 862 *pr*
Lensmaker's equation, 845–47
Lenz's law, 736–38, 758
LEP collider, 1118–19
Lepton era, 1162–63
Leptons and Lepton number, 1121,
  1124–27, 1130–32, 1134, 1162
Level range formula, 60
Lever, 303
Lever arm, 247–48
LHC collider, 1118
Lifetime, 1075, 1127 (*see also*
  Halflife)
Lift, dynamic, 348
Light, 810–912 (*see also* Diffraction,
  Intensity, Interference,
  Interference pattern, Reflection,
  Refraction)
  coherent sources of, 873–74, 877
  color of, and wavelength, 872

dispersion of, 824–25
  as electromagnetic wave, 798–800,
    810
  frequencies of, 798, 869
  gravitational deflection of, 1152
  incoherent sources of, 873–74, 877
  infrared (IR), 506, 799, 825
  intensity of, 877, 882
  monochromatic (*defn*), 870
  photon (particle) theory of, 952–55
  polarized, 907–10
  ray model of, 811 *ff*
  reflection of, 813–22
  refraction of, 822–24, 867–69
  scattering, 911
  from sky, 911
  speed of, 6, 797, 811–12, 867–68,
    919–23, 938
  ultraviolet (UV), 799, 825
  unpolarized (*defn*), 907
  velocity of, 6, 797, 811–12, 867–68,
    919–23, 938
  visible, 811, 867 *ff*
  wavelengths of, 799, 868–69,
    871–73
  wave-particle duality of, 958–59
  wave theory of, 866–911
  white, 825
Light bulb, 634, 642, 664, 882, 954,
  1019
Light-emitting diode, 1054
Light-gathering power, 855
Light meter, 954
Lightning, 591, 643
Light pipe, 827
Light-year (unit), 14 *pr*, 1141
Linac, 1118
Linear accelerator, 1118
Linear expansion, coefficient of, 450
Linear momentum, 206 *ff*
Lines of force, 562
Line spectrum, 902, 963 *ff*, 977
Line voltage, 646, 646 *fn*
Liquefaction, 473
Liquid crystal display (LCD), 909
Liquid drop model, 1090
Liquids, 332 *ff*, 446–47 (*see also*
  Phase, changes of)
Liquid scintillation, 1080
Lloyd's mirror, 886 *pr*
Logarithms, A-3
Longitudinal magnification, 833 *pr*
Longitudinal wave, 391 *ff*, 419–20
  (*see also* Sound waves)

Long-range force, 1066
Lorentz, H. A., 922
Lorentz equation, 694
Lorentz transformation, 932–36
Los Alamos laboratory, 1095
Loudness, 418, 420–23 (*see also*
  Intensity)
Loudness level, 422
Loudspeakers, 367, 421, 698, 776,
  954
LRC circuit, 766–67, 776–79
LR circuit, 762–63
Lumen (unit), 882
Luminosity (stars and galaxies),
  1145–46
Luminous efficiency, 885 *pr*
Luminous flux, 882
Luminous intensity, 882
Lyman series, 964, 968–69

M

Mach, E., 435 *fn*
Mach number, 435
MACHOS, 1166
Macroscopic description of system,
  445
Macrostate, 535–36
Madelung constant, 1044
Magnet, 686–88
  domains of, 722
  electro-, 723–24
  permanent, 722
  superconducting, 650, 723
Magnetic confinement, 1098–99
Magnetic damping, 751 *pr*
Magnetic declination, 688
Magnetic dipoles and dipole
  moments, 695–97, 721
Magnetic domains, 722
Magnetic field, 686–702, 709–26
  of circular loop, 720–21
  definition of, 687–88
  determination of, 691
  direction of, 689–92
  of Earth, 688
  energy stored in, 760–61
  induces emf, 736–38
  motion charged particle in, 692–95
  produced by changing electric field,
    788–91
  produced by electric current, 689,
    719–21 (*see also* Ampère's law)

produces electric field and current, 747–49
of solenoid, 716–18
sources of, 709–26
of straight wire, 710, 720
of toroid, 708
Magnetic field lines, 687
Magnetic flux, 736, 791
Magnetic flux density (*see* Magnetic field)
Magnetic force, 686, 689–95
on electric current, 689–92
on moving electric charges, 692–95
Magnetic induction (*see* Magnetic field)
Magnetic lens, 961
Magnetic moment, 695–97, 1015–18, 1064
Magnetic permeability, 710, 724–25
Magnetic poles, 686–87
of Earth, 688
Magnetic quantum number, 1005–6, 1016
Magnetic resonance imaging, 1064, 1108–9
Magnetic tape and discs, 749
Magnetism, 686–726
Magnetization (vector), 726
Magnification:
angular, 853
of electron microscope, 961
lateral, 819, 841
of lens, 841–43
of lens combination, 843–45
longitudinal, 833 *pr*
of magnifying glass, 853–54
of microscope, 856–58, 898–99
of mirror, 819
sign conventions for, 819, 841
of telescope, 855, 898–99
useful, 899, 961
Magnifier, simple, 853–54
Magnifying glass, 836, 853–54
Magnifying power, 853, 855 (*see also* Magnification)
Magnitude (stars and galaxies), 1172 *pr*
Main-sequence, stars, 1146–49
Manhattan Project, 1095
Marconi, Guglielmo, 803
Manometer, 338
Mass, 7, 79, 82
atomic, 446
center of, 221–24

critical, 1092–95
gravitational vs. inertial, 148, 1152
inertial, 148, 1152
molecular, 446
in relativity theory, 936–38
rest, 938–40, 1063
standard of, 7
units of, 7, 79
variable, 227–29
Mass-energy transformation, 938–42
Mass excess, 1084 *pr*
Mass increase in relativity, 936–38
Mass spectrometer (spectrograph), 702
Matter, states of, 332, 446–47
Matter-dominated universe, 1161, 1164
Maxwell, James Clerk, 470, 787, 792, 797–98, 918
Maxwell distribution of molecular speeds, 470–72
Maxwell's equations, 787, 792–806, 918–19, 934
Mean free path, 478–79
Mean life, 1075, 1084 *pr*
Mean speed, 469
Measurement, 4–5, 981
Mechanical advantage, 93, 303
Mechanical energy, 182–89
Mechanical equivalent of heat, 486
Mechanical waves, 388–411
Mechanics (*defn*), 16
Medical imaging, 437, 1105–9
Meitner, L., 978, 1090
Melting point, 491 (*see also* Phase, changes of)
Mendeleev, D., 1012
Meson exchange, 1122
Mesons, 1122–36
Metallic bond, 1044
Metals, free electron theory of, 1045–48
Metastable state, 1019–20, 1073
Meter (unit), 6, 812 *fn*, 881
Meters, 674–76, 697
correction for resistance of, 676
Metric system, 6–8
MeV (million electron volts) (*see* Electron volt)
Mho (unit), 656 *pr*
Michelson, A., 811–12, 881, 919–22
Michelson interferometer, 881, 919–22

Michelson-Morley experiment, 919–23, 923 *fn*
Micrometer, 10
Microphones:
capacitor, 677
magnetic, 749
ribbon, 749
Microscope:
compound, 856–58
electron, 899, 961, 999
magnification of, 856–58, 898–99
resolving power of, 898–99
useful magnification, 899
Microscopic description of a system, 445
Microstate, 535–36
Microwaves, 798
Milky Way, 854, 1141
Millikan, R. A., 700, 953
Millikan oil-drop experiment, 700
Mirage, 869
Mirror equation, 819–22
Mirrors:
concave and convex, 819 *ff*
focal length of, 817, 821
plane, 812–15
spherical, 816–22
used in telescope, 856
Missing mass, 1166–68
MKS units (*see* SI units)
mm-Hg (unit), 339
Models, 3
Moderator, 1092–94
Modern physics (*defn*), 1, 916
Modulation, 804
Moduli of elasticity, 309–12
Molar specific heat, 498–99
Mole (*defn*), 456
volume of, for ideal gas, 457
Molecular mass, 446
Molecular rotation, 1038–40
Molecular spectra, 1037–43
Molecular speeds, 466–72
Molecular vibrations, 501, 1040–42
Molecules, 1030–43
bonding in, 1030–33
polar, 1032–33
P E diagrams for, 1033–36
spectra of, 1037–43
Moment arm, 247
Moment of a force, 247
Moment of inertia, 249–56, 374
Momentum, 206 *ff*, 936–38, 941
angular, 256–60, 281–89, 965

Momentum (*continued*)
  in collisions, 208–20
  conservation of angular, 257–59,
    288–89
  conservation of linear, 208–11
  linear, 206 *ff*
  relation of force to, 206–8
  relativistic, 936–38, 941
  total, of systems of particles, 222–23
Monochromatic aberration, 859
Monochromatic light (*defn*), 870
Moon, 134, 1141
Morley, E. W., 919–22
Moseley, H. G. J., 1014
Moseley plot, 1014
Motion, 16–299, 916–44
  of charged particle in electric field,
    564–65
  at constant acceleration, 26 *ff*
  damped harmonic, 374–77
  description of (kinematics), 16–67
  dynamics of, 77 *ff*
  general, 262–68, 283–84
  under gravity (free fall), 31–36
  harmonic, 364 *ff*
  Kepler's laws of planetary, 143–46,
    288–89
  linear, 16–37
  Newton's laws of, 78–85 and *ff*, 207,
    209, 225, 226, 250, 256, 282,
    283–85, 916, 938, 978, 985–86
  nonuniform circular, 121–22
  oscillatory, 362–80
  periodic (*defn*), 362
  projectile, 55–63
  rectilinear, 16–37
Motion (*cont.*)
  relative, 66–68, 916–44
  rolling, 244–45, 262–68
  rotational, 239–94
  simple harmonic, 364 *ff*
  translational, 16–238, 262–65
  uniform circular, 63–65, 114 *ff*, 371
  uniformly accelerated, 26 *ff*, 54–55
  uniform rotational, 243–44
  vibrational, 362–80
  of waves, 388 *ff*
Motional emf, 739
Motor, electric, 698, 742–43
  counter emf in, 742–43
Movie projector, 954
MRI, 1064, 1108–9
Multimeter, 675
Multiplication factor, 1093

Muon, 928, 1114, 1125–27
Muscles and joints, forces in, 306
Musical instruments, 424–29
Mutation, 1101
Mutual inductance, 756–58
Myopic eye, 851

N

n-type semiconductor, 1051 *ff*
Natural abundance, 1062
Natural convection, 505
Natural frequency, 366, 378, 405 (*see also* Resonant frequency)
Natural radioactive background, 1102
Near field, 793
Near point, of eye, 851
Nearsighted eye, 851–52
Nebulae, 1143
Negative, 848 *fn*
Negative electric charge, 546, 593
Negative lens, 841
Neon tubes, 1003
Neptune, 146
Neptunium, 1088
Net force, 80, 88
Neutral equilibrium, 198, 308
Neutralino, 1166
Neutrino, 1071, 1126–28, 1121, 1163,
  1166–68
  mass of, 1166–67
Neutron, 547, 1062–63
  delayed, 1093
  in fission, 1090–95
  thermal, 1089
Neutron cross section, 1089
Neutron number, 1062
Neutron star, 1144, 1150
Newton, Isaac, 16, 79, 133–34,
  146–48, 207, 856 *fn*, 917, 1151,
  1154 *fn*
Newton (unit), 81
Newtonian focus, 862
Newtonian mechanics, 79–146
Newton's law of cooling, 515 *pr*
Newton's law of universal
  gravitation, 133–37, 1151–52
Newton's laws of motion, 78–85 and
  ff, 207, 209, 225, 226, 916, 938, 978,
  985–86
  for rotational motion, 250, 256, 282,
    283–85
Newton's rings, 878

Newton's synthesis, 143–46
NMR, 1064, 1107–8
Noble gases, 1012–13, 1045
Nodes, 405, 427
Noise, 429
Nonconductors, 547–48, 640
Nonconservative field, 748–49
Nonconservative force, 178
Non-Euclidean space, 1153
Noninductive winding, 758
Noninertial reference frame, 79, 291
Nonlinear device, 1054
Nonohmic device, 638
Nonreflecting glass, 880
Nonuniform circular motion, 121–22
Normal eye, 851
Normal force, 85–87, 107
Normalization, 987, 990 *fn*, 993
Normal lens, 850
North pole, 687
Novae, 1144
NOVA laser, 1100
Nuclear angular momentum, 1064
Nuclear binding energy, 1064–66
Nuclear collision, 217, 220
Nuclear fission, 1090–95
Nuclear forces, 1064–66, 1072, 1121–36
Nuclear fusion, 1095–1100, 1148–51,
  1163
Nuclear magnetic resonance, 1064,
  1107–8
Nuclear magneton, 1064
Nuclear medicine, 1104
Nuclear physics, 1061–1113
Nuclear power, 741, 1093–94
Nuclear reactions, 1085–88
Nuclear reactors, 1090–95
Nuclear spin, 1064
Nuclear weapons testing, 1095
Nucleon (*defn*), 1062
Nucleon number, 1073
Nucleosynthesis, 1140, 1149–51, 1163
Nucleus, 1061 *ff*
  compound, 1090
  daughter and parent (*defn*), 1068
Nuclide (*defn*), 1062
Null result, 922
Numerical integration, 36–37

O

Object distance, 814, 819, 828–29
Objective lens, 854

Occhialini, G., 1122
Ocular lens (see Eyepiece)
Oersted, H. C., 688–89, 709, 787–88
Off-axis astigmatism, 859
Ohm, G. S., 638
Ohm (unit), 638
Ohmmeter, 675, 697
Ohm's law, 638–39
Oil-drop experiment, 700
Onnes, H. K., 650
Open system, 493
Operating temperatures, 517
Operational definitions, 8, 712
Oppenheimer, J. R., 1095
Optical coating, 880–81
Optical illusion, 869
Optical instruments, 827, 836–60
Optical pumping, 1021
Optical tweezers, 803
Optics, geometric, 811 ff
Orbital angular momentum, in
    atoms, 1005–7, 1015–18
Orbital quantum number, 1005–6
Order and disorder, 533–34
Order of interference or diffraction
    pattern, 871
Order of magnitude and rapid
    estimating, 9, 96
Organ pipe, 427–28
Orion, 1143
Oscillations (see Vibrations)
Oscillator, sawtooth, 685 pr
Oscilloscope, 605–6, 782
Osteoporosis, 957
Otto cycle, 524
Overdamping, 376
Overexposure, 848
Overtone, 406, 426–27

P

P waves, 394
p-type semiconductor, 1051 ff
p-n junction, 1052–54, 1080
Pacemaker, 673, 758
Pair production, 957–58
Parabola (projectile), 63
Parabolic reflector, 817
Parallax, 1144–45
Parallel axis theorem, 255–56
Parallelogram method of adding
    vectors, 47
Paramagnetism, 725–26

Paraxial rays (defn), 817
Parent nucleus, 1068
Parsec (unit), 1144
Partial ionic character, 1032–33
Partial pressure, 476
Particle (defn), 16
Particle acclerators, 1115–20
Particle exchange, 1121–23
Particle interactions, 1124–26
Particle physics, 1114–36
Particle resonance, 1127–28
Pascal, Blaise, 333, 337–38
Pascal (unit of pressure), 333
Pascal's principle, 337–38
Paschen series, 964, 968–69
Pauli, W., 978, 1011, 1071
Pauli exclusion principle, 1010–12,
    1046, 1132–33, 1150
Peak current, 645
Peak voltage, 645
Pendulum:
    ballistic, 218
    conical, 116
    physical, 373–74
    simple, 13, 188, 371–73
    torsion, 374
Pentium chip, 1030
Penzias, A., 1160
Percent uncertainty, 4
Perfect cosmological principle,
    1159 fn
Performance, coefficient of, 526
Perfume atomizer, 348
Period, 65, 243
    of circular motion, 65
    of pendulums, 13, 372, 374
    of planets, 143
    of rotation, 243
    of vibration, 363, 366
Periodic motion, 362 ff
Periodic table, 1011–13, 1062 fn,
    inside rear cover
Periodic wave, 389–90
Permeability, magnetic, 710, 724–25
Permittivity, 551, 622
Perpendicular-axis theorem, 255–56
Perturbation, 145
PET, 1106–7
Phase:
    in ac circuit, 772–80
    changes of, 473–76, 490–93
    of matter, 332, 446–47
    of waves, 397
Phase angle, 366, 398, 778

Phase diagram, 474
Phase-echo technique, 437
Phase transition, 473–76, 490–93
Phase velocity, 397
Phasor diagram:
    ac circuits, 777–78
    interference and diffraction of
        light, 874–75, 903–4
Phons (unit), 422
Phosphor, 1080
Phosphorescence, 1019
Photino, 1136, 1164 fn
Photocathode, 1080
Photocell, 952
Photodiode, 954, 1054
Photoelectric effect, 952–55, 957
Photographic emulsion, 1120
Photographic film, 848
Photomultiplier tube, 1080
Photon, 952–57, 978–79, 1121,
    1126–28, 1162–64 (see also
    Gamma rays, X-rays)
Photon exchange, 1121–23
Photon interactions, 957–58
Photon theory of light, 952–55
Photosynthesis, 955
Physical pendulum, 373–74
Piano tuner example, 12
"Pick-up" nuclear reaction, 1111 pr
Piezoelectric effect, 677
Pin, structural, 305, 315
Pion, 940, 1122–27
Pipe, vibrating air columns in, 427–28
Pitch of a sound, 418
Pixel, 1106
Planck, Max, 949, 951, 965
Planck's constant, 951
Planck's quantum hypothesis, 949–51
Plane geometry, A-2
Plane-polarized light, 907–10
Planetary motion, 143–46
Plane waves, 403, 793–94, 989–90
Plasma, 1085, 1098–99
Plastic region, 309
Plate tectonics, 342
Plum-pudding model of atom, 962
Pluto, 146, 1141
Plutonium, 1088, 1094
Point charge (defn), 551
    potential, 597–98
Poise (unit), 350–51
Poiseuille, J. L., 351
Poiseuille's equation, 351
Poisson, S., 887

Polarization of light, 907–10
  plane, 907–10
  by reflection, 910, 911
  of skylight, 911
Polarized light (*see* Polarization of light)
Polarizer, 908
Polarizing angle, 910
Polar molecules, 547, 565, 624–25, 1032–33
Polaroid, 907–9
Poles, magnetic, 686–87
  of Earth, 688
Pole vault, 176, 185–86
Pollution, 525
  thermal, 525
Position vector, 53, 55, 240, 243–44
Positive electric charge, 546
Positive lens, 841
Positron, 1072, 1106, 1121
Positron emission tomography, 1106–7
Post-and-beam construction, 319
Potential difference, electric, 591–94
  (*see also* Electric potential)
Potential energy, 178–82, 192–93, 197–98
  diagrams, 197–98
  elastic, 181, 369
  electric, 591–94, 603–5
  gravitational, 178–80, 192–94
  for molecular bonds, 1033–36, 1041
  for square well and barriers, 994–99
Potential well, 994–96, 1009 *fn*
Potentiometer, 685 *pr*
Pound (unit), 81
Powell, C.F., 1122
Power, 195–97, 642–44, 778 (*see also* Electric power)
Power factor (ac circuit), 778
Power generation, 741
Power of a lens, 838
Power plants, 741, 1094
Power reactor, 1093
Powers of ten, 5
Power transmission, 746–47, 782
Poynting, J. H., 801 *fn*
Poynting vector, 800–802
Precession, 290
Presbyopia, 851
Pressure, 333–40
  absolute, 337
  atmospheric, 336–40
  in fluids, 333–36
  in a gas, 337, 454–56, 468

gauge, 337
hydraulic, 338
measurement of, 338–40
partial, 476
radiation, 802–3
units for and conversions, 333, 337
vapor, 475–76
Pressure amplitude, 420
Pressure cooker, 476
Pressure gradient, 351
Pressure head, 335
Pressure transducers, 677
Pressure waves, 417 *ff*
Prestressed concrete, 313
Principal axis, 817
Principle of complementarity, 958–59
Principle of correspondence, 943, 971, 976 *pr*
Principle of equivalence, 148, 1151–52
Principle of superposition, 400–402
Principle quantum number, 1005–6
Principles (*see* names of)
Prism, 825
Prism binoculars, 856
Probability, in quantum mechanics, 980, 984–85
Probability and entropy, 536–37
Probability density, 980, 989, 992, 995, 1004, 1008–10, 1031–32
Problem solving techniques, 28, 89, 96, 159
Process, reversible and irreversible (*defn*), 520–25
Projectile, horizontal range of, 60–61
Projectile motion, 55–63
Proper length, 930
Proper time, 928, 1139 *pr*
Proportional limit, 309
Proton, 1062–63, 1127
  decay (?) of, 1135
  electric charge on, 547
Proton-proton collision, 220
Proton-proton cycle, 999–1100, 1148
Protostar, 1148
Proxima Centauri, 1141
Pseudoforce, 291
Pseudovector, 246 *fn*
psi (unit), 333
PT diagram, 474
Pulley, 93
Pulsar, 1150
Pulsating universe, 1168
Pulsed laser, 1021–22

Pulse-echo technique, 437
Pumps, 353
  heart as, 353
  heat, 527–28
Pupil, 850, 899
PV diagram, 473
P waves, 394
Pythagorean theorem, A-2

Q

Q-value, 380, 771 *pr*, 785 *pr*, 1068–69, 1086
QCD, 1123, 1133–35
QED, 1121
Quadratic formula, 35, A-1
Quadrupole, 573 *pr*
Quality factor, 1102
Quality of a sound, 429
Quality value (*Q*-value) of a resonant system, 380, 771 *pr*, 785 *pr*, 1068–69, 1086
Quantities, base and derived, 8
Quantization:
  of angular momentum, 965
  of electric charge, 550
  of energy, 951, 965–71, 991–92
Quantum chromodynamics, 1123, 1133–35
Quantum condition, Bohr's, 965, 972
Quantum electrodynamics, 1121
Quantum hypothesis, Planck's, 949–51
Quantum mechanics, 972, 977–1060
Quantum numbers, 965 *ff*, 992, 1004–7
  in H atom, 1004–7
  for complex atoms, 1010–11
  for molecules, 1038–43
Quantum theory, 916, 949–72
  of atoms, 965–71
  of blackbody radiation, 951
  of light, 949–58
Quarks, 550 *fn*, 1121, 1123, 1130–32, 1134, 1162
Quasars, 1144, 1157–58
Quasistatic process (*defn*), 495

R

Rad (unit), 1102
Radar, 435, 876–77

Radial acceleration, 64–65, 69, 114 *ff*, 240
Radial probability density, 1008–10
Radian measure for angles, 240
Radiant flux, 882
Radiation:
  blackbody, 950–51, 1160
  Čerenkov, 886 *pr*
  cosmic, 1102, 1114
  detection of, 1080, 1102
  from hot objects, 506–8, 950–51
  infrared (IR), 506, 799, 825
  ionizing (*defn*), 1100
  measurement of, 1101–3
  microwave, 798
  nuclear, 1067–80, 1100–1103
  synchrotron, 1117
  thermal, 506–8
  ultraviolet (UV), 799, 825
  X (*see* X-rays)
Radiation damage, 1100–1101
Radiation-dominated universe, 1161, 1163
Radiation dose, 1101–3
Radiation era, 1162–63
Radiation field, 793
Radiation film badge, 1103
Radiation pressure, 802–3
Radiation sickness, 1103–4
Radiation therapy, 1104
Radio, 632, 745, 780, 803–6
Radioactive dating, 1078–79
Radioactive decay (*see* Radioactivity)
Radioactive decay constant, 1073–76
Radioactive decay law, 1074
Radioactive decay series, 1077–78
Radioactive fallout, 1095
Radioactive tracers, 1104
Radioactive waste, 1093–94
Radioactivity, 998, 1067–80
  alpha, 1067–70, 1073
  artificial, 1067
  beta, 1067, 1070–73
  dosage of, 1101–3
  gamma, 1037, 1072–73
  natural, 1067, 1102
  probabilistic nature of, 1073
  rate of decay, 1073–78
Radionuclide (*defn*), 1101
Radio waves, 787, 798
Radius of gyration, 273 *pr*
Radius of nuclei, 1062
Radon, 1102

Rainbow, 825
Random access memory (RAM), 613
Range of projectile, 60–61
Rapid estimating, 9–12
Rarefaction, in wave, 391
Ray, 403, 811
  paraxial (*defn*), 817
Ray diagramming, 837 *ff*
Rayleigh, Lord, 897, 951
Rayleigh criterion, 897
Rayleigh-Jeans theory, 951
Ray model of light, 811 *ff*
RBE, 1102
RC circuit, 669–73
Reactance, 759, 774–77 (*see also* Impedance)
Reaction energy, 1086
Reactions, nuclear, 1085–88
Reactors, nuclear, 1090–95
Real image, 814, 818, 839
Rearview mirror, 822
Receivers, radio and television, 805
Recombination, 1170 *pr*
Rectifiers, 1053–54
Red giant star, 1144, 1146, 1148
Redshift, 435, 943, 1145, 1156 *ff*
Reduced mass, 1039
Reference frames, 17, 917 *ff*
  accelerating, 79, 148, 291
  inertial, 79, 291, 917 *ff*
  noninertial, 79, 291
  rotating, 291
  transformations between, 932–36
Reflecting telescope, 856
Reflection:
  angle of, 403, 813
  diffuse, 813
  law of, 403, 813
  of light, 813–22
  phase changes during, 877–81
  polarization by, 910
  specular, 813
  from thin films, 877–81
  total internal, 415 *pr*, 826–27
  of water waves, 402–3
  of waves on a string, 402
Reflection coefficient, 997
Reflection grating, 900
Refracting telescope, 854
Refraction, 408–9, 822–24
  angle of, 408–9, 822
  index of, 811–12
  law of, 409, 823–24, 867–69
  of light, 822–24, 867–69

and Snell's law, 822–24, 868
  at spherical surface, 828–30
  by thin lenses, 837–40
  of water waves, 408–9
Refrigerator, 525–26
Reinforced concrete, 313
Relative biological effectiveness (RBE), 1102
Relative humidity, 476
Relative motion, 66–68, 916–41
Relative permeability, 725
Relative velocity, 66–68
Relativistic mass, 936–38
Relativistic momentum, 936–38, 941
Relativity, Galilean-Newtonian, 917–19, 932–36
Relativity, general theory of, 148, 149, 169, 922, 1140–41, 1151–55
Relativity, special theory of, 916–44
  and appearance of object, 931
  constancy of speed of light, 923
  four-dimensional space-time, 932
  and length, 930–31
  and Lorentz transformation, 932–36
  and mass, 936–38
  mass-energy relation in, 938–42
  postulates of, 922–23
  simultaneity in, 924–26
  and time, 926–29, 932
Relativity principle, 917–19, 923 *ff*
Relay, 727
Rem (unit), 1102–3
Research reactor, 1093
Resistance and resistors, 638–39
  in ac circuit, 773, 776–79
  with capacitor, 669–73, 776–79
  color code, 639
  electric currents and, 634–52
  with inductor, 762–63, 766–67, 776–79
  internal, 659
  in series and parallel, 660–64
  shunt, 674
Resistance thermometer, 641, 676
Resistive force, 122–23
Resistivity, 640–42
  temperature coefficient of, 641–42
Resolution:
  of diffraction grating, 903–5
  of electron microscope, 961
  of eye, 899
  of lens, 896–99
  of light microscope, 898–99

Resolution (*continued*)
of telescope, 898–99
of vectors, 48–52
Resolving power, 898–99, 903–5 (*see also* Resolution)
Resonance, 378–79, 405–8, 780
in ac circuit, 780
particle, 1127–28
Resonant frequency, 378–79, 405, 780
Resources, energy, 741
Rest energy, 939, 1063
Restitution, coefficient of, 234 *pr*
Rest mass, 938–40
Restoring force, 162, 363
Resultant vector, 47, 50
Retentivity (magnetic), 725
Retina, 850
Reversible process, 520–25
Reynolds' number, 359 *pr*
Right-hand rule, 246, 280, 689, 690, 692, 711
Rigid body (*defn*), 239
rotational motion of, 239–68, 285–90
Rigid box, particle in, 990–94
Ripple voltage, 1060 *pr*
rms (root-mean-square):
current, 646
velocity, 469
voltage, 646
Rockets, 83, 210, 227–29
Roemer, Olaf, 811
Roentgen, W. C., 905
Roentgen (unit), 1102
Rolling friction, 106, 263–64, 267
Rolling motion, 244–45, 262–68
Root-mean-square (rms) current, 646
Root-mean-square (rms) speed, 469
Root-mean-square (rms) voltage, 646
Rotation, frequency of, 242–43
Rotational angular momentum quantum number, 1038–39, 1042–43
Rotational imbalance, 287–88
Rotational inertia (*defn*), 249–50
Rotational kinetic energy, 260–65
molecular, 500–501
Rotational transitions, 1038–40
Rotation and rotational motion, 239–94
Rotor, 698, 742
Ruby laser, 1021
Russell, Bertrand, 960

Rutherford, Ernest, 962, 1061, 1067, 1085
Rutherford's model of the atom, 962–63
R-value, 505
Rydberg constant, 964, 969
Rydberg states, 1029 *pr*

S

S waves, 394
Safety factor, 312
Sailboat and Bernoulli's principle, 348
Salam, A., 1134
Satellites, 139–43, 145
Saturated vapor pressure, 475
Saturation (magnetic), 724
Savart, Felix, 719
Sawtooth oscillator, 685 *pr*
Sawtooth voltage, 673
Scalar (*defn*), 45
Scalar product, 159–61
Scales, musical, 424
Scanning electron microscope, 961
Scanning tunneling electron microscope, 999
Scattering of light, 911
Schrödinger, Erwin, 869, 978
Schrödinger equation, 978, 985–95, 1004–7
Schwarzschild radius, 1155
Scientific notation, 5
Scintillation counter, 1080
Scintillator, 1080, 1120
Search coil, 754 *pr*
Second (unit), 7
Second law of motion, 80 *ff*, 250 *ff*
Second law of thermodynamics, 516–37
Clausius statement of, 517, 526, 533
entropy and, 529–33
general statements of, 532–33
Kelvin-Planck statement of, 520, 523, 533
statistical interpretation of, 535–37
Seismograph, 749–50
Selection rules, 1007, 1042
Self-inductance, 758–60
Self-sustaining chain reaction, 1091–95
Semiconductor detector, 1080, 1120

Semiconductor doping, 1051 *ff*
Semiconductors, 548, 640, 954, 1049–55
n and *p* types, 1051 *ff*
resistivity of, 640
Sensitivity of meters, 675–76
Separation of variables, 988
Shear modulus, 312
Shear strength, 313
Shear stress, 311–12
Shells and subshells, 1012–13
SHM (*see* Simple harmonic motion)
Shock waves, 435–36
Short circuit, 644
Short-range forces, 1066
Shunt resistor, 674
Shutter speed, 848, 850
Sievert (unit), 1102
Sign conventions (geometric optics), 819, 822
Significant figures, 4–5
Silicon, 1051 *ff*
Simple harmonic motion (SHM), 364–71, 390
applied to pendulum, 371–73
Simple harmonic oscillator (SHO), 364–71
molecular vibration as, 1040–42
Simple pendulum, 13, 188, 371–73
Simultaneity, 924–26
Single photon emission tomography, 1106–7
Single slit diffraction, 888–93
Singularity, 1155
Sinusoidal curve, 364 *ff*
Siphon, 354 *pr*
SI units (Système Internationale), 6–8, 81
Sky, color of, 911
SLAC, 1118
Slepton, 1136
Slope, of a curve, 21
Slow neutron reaction, 1087
Slug (unit), 81
Smoke detector, 954, 1070
SN 1987a, 1150, 1167
Snell, W., 823
Snell's law, 822–24, 826
Soap bubble, 877, 880
Soaps, 352
Sodium chloride bond, 628 *pr*, 1031–32, 1034–35, 1044
Solar cell, 1054
Solar constant, 507

Solenoid, 709, 716–18, 723–24, 757, 759
Solids, 309 *ff*, 332, 446–47, 474 (*see also* Phase, changes of)
Solid state physics, 1044–50
Somatic damage, 1101
Sonar, 437
Sonic boom, 435–36
Sound "barrier," 435
Sounding board, 425
Sounding box, 425
Sound level, 421–22
Sound spectrum, 429
Sound track, 954
Sound waves, 417–37
  Doppler shift of, 432–35
  infrasonic, 419
  intensity of (and dB), 420–23
  interference of, 429–32
  quality of, 429
  sources of, 424–29
  speed of, 418
  supersonic, 418
  ultrasonic, 418
Source activity, 1101
Source of emf, 659–60, 734–35
South pole, 687
Space:
  absolute, 918, 923
  relativity of, 930–36
Space quantization, 1005, 1016–17
Space-time, 932
  curvature of, 149, 1151–55, 1165
Special theory of relativity (*see* Relativity, special theory of)
Specific gravity, 333
Specific heat, 488 *ff*, 498–99
Spectrograph, mass, 702
Spectrometer:
  light, 901–2
  mass, 702
Spectrophotometer, 901 *fn*
Spectroscope and spectroscopy, 901–2
Spectroscopic notation, 1018
Spectrum, 901
  absorption, 902, 964, 968
  atomic emission, 902, 963–65
  band, 1038
  continuous, 902
  electromagnetic, 798–800, 874–75
  emitted by hot object, 902, 949–51
  line, 902, 963 *ff*
  molecular, 1037–43
  visible light, 799, 824–25
  X-ray, 1013–15

Specular reflection, 813
Speed, 18–22 (*see also* Velocity)
  average, 18
  of EM waves, 794–97
  mean (molecular), 469
  molecular, 466–72
  most probable, 471–72
  rms, 469
Speed of light, 6, 797, 811–12, 867–68, 919–23, 938
  constancy of, 923
  measurement of, 811–12
  as ultimate speed, 938
Speed of sound, 418
  supersonic, 418, 435–36
  ultrasonic, 418
SPET, 1106–7
Spherical aberration, 817, 829, 858
Spherical mirrors, image formed by, 816–22
Spherical waves, 395
Spin, electron, 723, 1005–6, 1017–18
Spin-echo technique, 1109
Spinning top, 290
Spin-orbit interaction, 1018
Spin quantum number, 1005
Spiral galaxy, 1143
Spring:
  potential energy of, 181, 187, 369
  vibration of, 363 *ff*
Spring constant, 162, 363
Spring equation, 162
Spyglass, 856
Squark, 1136
Stable and unstable equilibrium, 197–98, 308
Standard conditions (STP), 457
Standard length, 6, 881
Standard mass, 7, 79
Standard model:
  cosmology, 1161–64
  elementary particles, 1132–34
Standards and units, 6–8
Standard temperature and pressure (STP), 457
Standing waves, 405–8, 424–28, 971
Star clusters, 1143
Stars, 1141–51
Starter, automobile, 724
State:
  changes of, 473–76, 490–93
  energy (*see* Energy states)
  equation of, 454, 457, 459, 477–78
  of matter, 332, 445–47

  as physical condition of system, 445
State variable, 445, 494, 529
Static electricity, 545
Static equilibrium, 300–331
Static friction, 107 *ff*
Statics, 300–321
Stationary states in atom, 965 (*see also* Energy states)
Statistics and entropy, 535–37
Stator, 742
Steady state model of universe, 1159
Steam engine, 518–19
Steam power plant, 1094
Stefan-Boltzmann constant, 506
Stefan-Boltzmann law (or equation), 506
Stellar evolution, 1145–51, 1159
Step-down transformer, 745
Step-up transformer, 745
Stern-Gerlach experiment, 1016–18
Stereo broadcasting, FM, 805 *fn*
Stimulated emission, 1019–22
Stirling cycle, 543 *pr*
Stopping distance for car, 29, 167
Stopping potential, 952
Storage rings, 1118–19
STP, 457
Strain, 309–12
Strain factor, 683 *pr*
Strain gauge, 677
Strangeness and strange particles, 1129
Strassmann, F., 1090
Streamline (*defn*), 343
Streamline flow, 343–45
Strength of materials, 309, 312–15
Stress, 309–12
  compressive, 311
  shear, 311–12
  tensile, 311
  thermal, 454
Stringed instruments, 424–25
Strings, vibrating, 405–8, 424–25
String theory, 1136
"Stripping" nuclear reaction, 1111 *pr*
Strong bonds, 1036
Strongly-interacting particles (*defn*), 1126
Strong nuclear force, 545 *fn*, 1066, 1121–36
Sublimation, 474
Subshell, atomic, 1012–13
Suction, 340

Sun, energy source of, 1096–98
Sunglasses, polarized, 908–10
Sunsets, 911
Supercluster, 1143
Superconducting magnets, 650, 723
Superconducting supercollider, 1118
Superconductivity, 650, 723
Superfluidity, 474
Supernovae, 1144, 1150, 1167
Superposition, principle of, 400, 401–2, 552, 555
Supersaturated air, 476
Supersonic speed, 418, 435–36
Superstring theory, 1136
Supersymmetry, 1136
Surface tension, 351–52
Surface waves, 394
Surfactants, 352
Surgery, laser, 1022
Survival equation, 483 *pr*
Suspension bridge, 318
S waves, 394
Symmetry, 10, 35, 93, 557–58, 714, 715, 716
Symmetry breaking, 1134, 1162, 1164
Synchrocyclotron, 1117
Synchrotron, 1117
Synchrotron radiation, 1117
Système Internationale (SI), 6–8, 81
Systems, 445, 493
    closed, 493
    isolated, 209, 493
    open, 493
    as set of objects, 445
    of units, 6–8

T

Tangential acceleration, 121–22, 242
Tape-recorder head, 749
Tau lepton, 1126–27, 1131
Telephone, cellular, 806
Telephoto lens, 850
Telescopes, 854–56
    astronomical, 854–56
    field-lens, 856
    Galilean, 856
    Keplerian, 854
    magnification of, 855, 899
    reflecting, 856
    refracting, 854
    resolution of, 898–99
    terrestrial, 856

Television, 595, 605–6, 803–6
Temperature, 447 *ff*, 460, 537–38
    absolute, 455, 460, 538
    Celsius (or centigrade), 448–49
    critical, 473
    Curie, 722
    distinguished from heat and
        internal energy, 487
    Fahrenheit, 448–49
    Fermi, 1060 *pr*
    human body, 449
    Kelvin, 455, 460, 538
    molecular interpretation of, 466–70
    operating (of heat engine), 517
    relation to molecular velocities,
        470–72
    scales of, 448, 455, 460, 538
    transition of, 650
Temperature coefficient of resistivity,
    641–42
Tensile strength, 313
Tensile stress, 311
Tension, 90, 311
Terminal, 635
Terminal speed or velocity, 32 *fn*,
    122–23
Terminal voltage, 659–60
Terrestrial telescope, 856
Tesla (unit), 691
Test charge, 554
Tevatron, 1117
Theories (general), 3
Theories of everything, 1136
Theory of relativity (*see* Relativity)
Thermal conductivity, 503–5
Thermal contact, 449
Thermal energy, 189–90, 487–88 (*see
    also* Internal energy)
    distinguished from heat and
        temperature, 487
Thermal equilibrium, 449–50
Thermal expansion, 450–53
    coefficients of, 450–51
    of water, 453
Thermal neutron, 1089
Thermal pollution, 525
Thermal radiation, 506–8
Thermal stress, 454
Thermionic emission, 605–6
Thermistor, 641, 676
Thermochemical calorie, 486–87
Thermocouple, 677
Thermodynamic probability, 536–37
Thermodynamics, 445, 485–544

first law of, 493–98
second law of, 516–37
third law of, 529 *fn*, 537–38
zeroth law of, 449–50
Thermodynamic temperature scale,
    537–38
Thermoelectric effect, 677
Thermography, 507–8
Thermometers, 447–49
    constant-volume gas, 449
    resistance, 641, 676
Thermonuclear device, 1098
Thermostat, 461 *pr*
Thin-film interference, 877–81
Thin lenses, 837–40
Third law of motion, 82 *ff*
Third law of thermodynamics, 529 *fn*,
    537–38
Thomson, G. P., 960
Thomson, J. J., 699–700, 960, 962
Three Mile Island, 1094
Threshold energy, 1088, 1113 *pr*
Thrust, 228
TIA, 349
Timbre, 429
Time:
    absolute, 918
    proper, 928
    relativity of, 924–29, 931–36
    standard of, 7
Time constant, 670, 762
Time dilation, 926–29
Time's arrow, 533
Tire pressure, 458
Tire pressure gauge, 338
Tokamak, 1085, 1099
Tokamak Fusion Test Reactor, 1099
Tomography, 1105–6
Tone color, 429
Top quark and topness, 1130–31
Toroid, 718
Torque, 247–50 and *ff*, 266–67,
    280–81 and *ff*, 302 *ff*
    counter, 743
    on current loop, 695–97
Torr (unit), 339
Torricelli, Evangelista, 339, 347
Torricelli's theorem, 347
Torsion pendulum, 374
Total angular momentum, 1017–18
Total binding energy, 1064
Total cross section, 1089
Total internal reflection, 415 *pr*,
    826–27

Total mechanical energy, 183
Total reaction cross section, 1089
Townsend, J. S., 700
Tracers, 1104
Transducers, 676–77
Transformations:
  Galilean, 932–36
  Lorentz, 932–36
Transformer, 744–47, 758
Transformer equation, 745
Transient ischemic attack (TIA), 349
Transistors, 1054–55
Transition elements, 1013
Transition temperature, 650
Translational motion, 16–238, 262–65
Transmission coefficient, 997
Transmission electron microscope, 961
Transmission grating, 900 ff
Transmission lines, 746–47, 782
Transmission of electricity, 746–47, 782
Transmutation of elements, 1068, 1070, 1085–88
Transverse waves, 391 ff, 794, 907
Trap, sink, 349
Triangulation, 11, 1144 fn, 1153
Trigonometric functions and identities, 49, A-2–A-3, inside back cover
Triple point, 460, 474
Tritium, 1084 pr, 1097–99
Trusses, 315–18
Tube, vibrating column of air in, 425–28
Tubes, flow in, 351
Tunnel diode, 998–99
Tunneling through a barrier, 996–99, 1069–70, 1155 fn, 1168
Turbine, 741
Turbulent flow, 343
Turning points, 197
Turn signal, automobile, 673
Tweeter, 776
Twin paradox, 926–29
Tycho Brahe, 143

U

UHF, 804–5
Ultimate speed, 938
Ultimate strength, 309, 312–13
Ultrasonic waves, 418, 437

Ultrasound, 437
Ultrasound imaging, 437
Ultraviolet (UV) light, 799, 825
Unavailability of energy, 534-35
Uncertainty (in measurements), 4, 981
Uncertainty principle, 981–84, 996, 1031
Underdamping, 376
Underexposure, 848
Unification scale, 1134
Unified atomic mass unit, 7, 446, 1063
Uniform circular motion, 63–65, 114 ff, 371
  dynamics of, 114–17
  kinematics of, 63–65
Uniformly accelerated motion, 26 ff, 54–55
Uniformly accelerated rotational motion, 243–44
Unit conversion, 8–9, inside front cover
Units of measurement, 6–8
Unit vectors, 52–53
Universal gas constant, 457
Universal law of gravitation, 133–37 and ff
Universe (see also Cosmology):
  age of, 1158–59
  Big Bang theory of, 1136, 1159–64
  curvature of, 149, 1151–55, 1165
  expanding, 1156–59
  future of, 1165–69
  inflation scenario of, 1164
  matter-dominater, 1161, 1164
  open or closed, 1151, 1165–69
  pulsating, 1168
  radiation-dominated, 1161, 1163
  standard model of, 1161–64
  steady state model of, 1159
Unpolarized light (defn), 907
Unstable equilibrium, 198, 308
Uranium, 1077, 1079, 1090–95
Uranus, 146
Useful magnification, 899, 961
UV light, 799, 825

V

Vacuum pump, 353
Vacuum tube, 803
Valence, 1013
Valence band, 1049–50

Van de Graaff generator and accelerator, 612 pr, 1115
van der Waals, J. D., 477
van der Waals bonds and forces, 1036–37
van der Waals equation of state, 477–78
Vapor, 473 (see also Gases)
Vaporization, latent heat of, 491
Vapor pressure, 474–76
Variable-mass systems, 227–29
Vector displacement, 46–47, 53
Vector model (atoms) 1028 pr
Vector product, 279–80
Vectors, 17, 46–53, 159–61, 279–80, A-3
  addition of, 46–52
  components of, 48–52
  cross product, 279–80
  multiplication of, 48, 159–61, 279–80
  multiplication by a scalar, 48
  position, 53, 55, 240, 243–44
  resolution of, 48–52
  resultant, 47, 50
  scalar (dot) product, 159–61
  subtraction of, 48
  sum, 46–52
  unit, 52–53
  vector (cross) product, 279–80
Velocity, 18–22 and ff, 36–37, 45, 54–55
  addition of, 66–68
  angular, 241–44
  average, 18–19, 21, 54
  drift, 647–49, 701
  of EM waves, 794–97
  escape, 192–94
  gradient, 350
  instantaneous, 20–22, 26–27, 54–55
  of light, 6, 797, 811–12, 867–68, 919–23, 938
  molecular, and relation to temperature, 466–72
  phase, 397
  relative, 66–68
  rms, 469
  of sound, 418
  supersonic, 418, 435–36
  terminal, 32 fn, 122–23
  of waves, 390–94, 397, 418
Velocity selector, 695
Venturi meter, 348
Venturi tube, 348

Venus, phases of, 854
Vibrational energy, 369–70
    molecular, 500–501, 1040–42
Vibrational motion, 362–80
Vibrational quantum number, 1041
Vibrational transition, 1040–42
Vibrations, 362–80, 405
    of air columns, 425–28
    of atoms and molecules, 500–501
    forced, 378–79
    as source of wave, 388–90, 417, 424–28
    of springs, 363 ff
    of strings, 405–8, 424–25
Virtual image, 814, 821, 840
Virtual particles, 1121
Virtual photon, 1121
Viscosity, 344, 350–51
    coefficient of, 350
Viscous force, 344, 350–51
Visible light, wavelengths of, 799, 824–25, 867 ff
Visible spectrum, 824–25
Volt (unit), 592
Volta, Alessandro, 592, 634–35
Voltage, 592–93 (see also Electric potential)
    bias, 1054–55
    breakdown, 596
    line, 646, 646 fn
    peak, 645
    ripple, 1060 pr
    rms, 646
    sawtooth, 673
    terminal, 659–60
Voltage drop, 665
Voltage gain, 1055
Voltmeter, 674–76, 697
    digital, 675
Volume, formulas for, A-1
Volume expansion, coefficient of, 452

W

W particle, 1123, 1126–27, 1133
Walking, 83
Water:
    anomalous behavior below 4°, 453
    cohesion of, 352
    dipole moment of, 566
    dissolving power of, 628 pr
    expansion of, 453
    heavy, 1092
    latent heats of, 490–93
    molecule, 1032–34
    polar nature of, 547, 565
    saturated vapor pressure, 475
    thermal expansion of, 453
    triple point of, 460, 474
Water equivalent, 490
Watson, J.D., 906
Watt, James, 195 fn
Watt (unit), 195, 642
Wave(s), 388–411 (see also Light, Sound waves)
    amplitude of, 390, 396
    composite, 401
    continuous (defn), 389–90
    diffraction of, 410
    displacement of, 396 ff
    earthquake, 394, 396, 409
    electromagnetic, 792–806, 866
    energy in, 395–96
    frequency, 390
    harmonic (defn), 398
    infrasonic, 419
    intensity, 395–96, 420–23
    interference of, 404–5
    light, 866–911
    longitudinal, 391 ff, 419–20
    mathematical representation of, 396–401, 419–20
    mechanical, 388–411
    P-, 394
    periodic (defn), 389–90
    phase, 397
    plane (defn), 403, 793–94, 989–90
    pressure, 419 ff
    pulse, 389
    radio, 787, 798
    reflection of, 402–3
    refraction of, 408–9
    S-, 394
    shock, 435–36
    sound, 417–37
    source is vibration, 388–90, 417, 424–28
    spherical, 395
    standing, 405–8, 424–28, 971
    surface, 394
    transmission, 402–3
    transverse, 391 ff, 794, 907
    ultrasonic, 418, 437
    velocity of, 390–94, 397, 418
    water, 388 ff
Wave equation, 399–401, 797
    Schrödinger, 978, 985–95, 1004–7
Waveform, 429
Wave front (defn), 403
Wave function, 978–79, 985–95, 1003–4, 1007–10
    for H atom, 1003–4, 1007–10
    for square well, 994–96
    measurement of, 901–15
    of visible light, 798
Wave intensity, 395–96, 420–23 (see also Light)
Wavelength (defn), 390
    Compton, 956, 975 pr
    de Broglie, 959, 971–72
    depending on index of refraction, 869
    as limit to resolution, 898–99
    of material particles, 959
Wave motion, 388 ff
Wave nature of matter, 959 ff
Wave number (defn), 397
Wave packet, 989–90
Wave-particle duality:
    of light, 958–59
    of matter, 959, 972, 978, 981–82
Wave pulse, 389
Wave theory of light, 866–911
Wave velocity, 390–94, 397, 418
Weak bonds, 1036–37, 1045
Weak charge, 1134
Weakly interacting massive particles (WIMPS), 1166
Weak nuclear force, 1066, 1072, 1123–36
Weber (unit), 736
Weight, 79, 85–87, 137–39
Weightlessness, 139–42
Weinberg, S., 1134
Wheatstone bridge, 677
Whirlpool galaxy, 1143
White dwarf, 1144, 1146, 1148–49
White-light holograms, 1024
Whole-body dose, 1103
Wide-angle lens, 850
Wien's displacement law, 950, 1146
Wien's radiation theory, 950
Wilson, R., 1160
WIMPS, 1166
Wind instruments, 425–27
Windshield wipers, 673
Wing of an airplane, lift on, 348
Wire drift chamber, 1120
Woofer, 776
Work, 155–69, 260–61
    compared to heat, 493
    defined, 156

done by a gas, 496–98
in first law of thermodynamics, 494 *ff*
from heat engines, 517 *ff*
relation to energy, 164–69, 177–83, 190–91, 193, 195, 260–61
units of, 156
Work-energy principle, 165–68, 182, 190–91, 261
Work function, 953, 1048
Working substance (*defn*), 519
Wright, Thomas, 1141

X

X-ray crystallography, 906
X-ray diffraction, 905–6
X-rays, 799, 905–6, 1013–15, 1073, 1105–9

and atomic number, 1013–15
in electromagnetic spectrum, 799
spectra, 1013–15

Y

Young, Thomas, 870
Young's double-slit experiment, 870–73
Young's modulus, 309–10
Yukawa, H., 1121–22

Z

Z^0 particle, 1123, 1126–28, 1133
Zeeman effect, 708 *pr*, 1005, 1016, 1018, 1107
Zener diode, 1053

Zero, absolute, 455, 538
Zero-point energy, 992, 1041
Zeroth law of thermodynamics, 449–50
Zoom lens, 850
Zweig, G., 1130

Photo Credits

29-15 Tomas D.W. Friedmann/Photo Researchers, Inc. **29-20** Jon Feingersh/Comstock **29-28b** National Earthquake Information Center, U.S.G.S. **CO-30** Adam Hart-Davis/Science Photo Library/Photo Researchers, Inc. **CO-31v1** Albert J. Copley/Visuals Unlimited **CO-31v2** Mary Teresa Giancoli **CO-32** Richard Megna/Fundamental Photographs **32-1** AIP Emilio Segre' Visual Archives **32-14** The Image Works **CO-33** Douglas C. Giancoli **33-6** Douglas C. Giancoli **33-11a** Mary Teresa Giancoli and Suzanne Saylor **33-11b** Mary Teresa Giancoli **33-21** Mary Teresa Giancoli **33-25** David Parker/Science Photo Library/Photo Researchers, Inc. **33-28b** Michael Giannechini/Photo Researchers, Inc. **33-34b** S. Elleringmann/Bilderberg/Aurora & Quanta Productions **33-40** Douglas C. Giancoli **33-42** Mary Teresa Giancoli **CO-34** Mary Teresa Giancoli **34-1c** Douglas C. Giancoli **34-1d** Douglas C. Giancoli **34-2** Douglas C. Giancoli and Howard Shugat **34-4** Douglas C. Giancoli **34-7a** Douglas C. Giancoli **34-7b** Douglas C. Giancoli **34-18** Mary Teresa Giancoli **34-19a** Mary Teresa Giancoli **34-19b** Mary Teresa Giancoli **34-26** Mary Teresa Giancoli **34-30a** Franca Principe/Istituto e Museo di Storia della Scienza, Florence, Italy **34-32** Yerkes Observatory, University of Chicago **34-33c** Palomar Observatory/California Institute of Technology **34-33d** Joe McNally/Joe McNally Photography **34-35b** Olympus America Inc. **CO-35** Larry Mulvehill/Photo Researchers, Inc. **35-4a** John M. Dunay IV/Fundamental Photographs **35-9a** Bausch & Lomb Incorporated **35-16a** Paul Silverman/Fundamental Photographs **35-16b** Richard Megna/Fundamental Photographs **35-16c** Yoav Levy/Phototake NYC **35-18b** Ken Kay/Fundamental Photographs **35-20b** Bausch & Lomb Incorporated **35-20c** Bausch & Lomb Incorporated **35-22** Kristen Brochmann/Fundamental Photographs **CO-36** Richard Megna/Fundamental Photographs **36-2a** Reprinted with permission from P.M. Rinard, American Journal of Physics, Vol. 44, #1, 1976, p. 70. Copyright 1976 American Association of Physics Teachers. **36-2b** Ken Kay/Fundamental Photographs **36-2c** Ken Kay/Fundamental Photographs **36-11a** Richard Megna/Fundamental Photographs **36-11b** Richard Megna/Fundamental Photographs **36-12a and b** Reproduced by permission from M. Cagnet, M. Francon, and J. Thrier, The Atlas of Optical Phenomena. Berlin: Springer-Verlag, 1962. **36-15** Space Telescope Science Institute **36-16** The Arecibo Observatory is part of the National Astronomy and Ionosphere Center which is operated by Cornell University under a cooperative agreement with the National Science Foundation. **36-21** Wabash Instrument Corp./Fundamental Photographs **36-26** Photo by W. Friedrich/Max von Laue. Burndy Library, Dibner Institute for the History of Science and Technology, Cambridge, Massachusetts. **36-29b** Bausch & Lomb Incorporated **36-36** Diane Schiumo/Fundamental Photographs **36-39a** Douglas C. Giancoli **36-39b** Douglas C. Giancoli **CO-37** Image of Albert Einstein licensed by Einstein Archives, Hebrew University, Jerusalem, represented by Roger Richman Agency, Beverly Hills, California **37-1** AIP Emilio Segre Visual Archives **CO-38** Wabash Instrument Corp./Fundamental Photographs **38-10** Photo by S.A. Goudsmit, AIP Emilio Segre' Visual Archives **38-11** Education Development Center, Inc. **38-20b** Richard Megna/Fundamental Photographs **38-21abc** Wabash Instrument Corp./Fundamental Photographs **CO-39** Institut International de Physique/American Institute of Physics/Emilio Segre Visual Archives **39-01** American Institute of Physics/Emilio Segre Visual Archives **39-02** F.D. Rosetti/American Institute of Physics/Emilio Segre Visual Archives **39-04** Advanced Research Laboratory/Hitachi Metals America, Ltd. **39-18** Driscoll, Youngquist, and Baldeschwieler, Caltech/Science Photo Library/Photo Researchers, Inc. **CO-40** Patricia Peticolas/Fundamental Photographs **40-16** Paul Silverman/Fundamental Photographs **40-22** Yoav Levy/Phototake NYC **40-24b** Paul Silverman/Fundamental Photographs **CO-41** Charles O'Rear/Corbis **CO-42** Reuters Newmedia Inc/Corbis **42-03** Chemical Heritage Foundation **42-07** University of Chicago, Courtesy of AIP Emilio Segre Visual Archives **CO-43** AP/Wide World Photos **43-07** Gary Sheahan "Birth of the Atomic Age" Chicago (Illinois); 1957. Chicago Historical Society. **43-10** Liaison Agency, Inc. **43-11** LeRoy N. Sanchez/Los Alamos National Laboratory **43-12** Corbis **43-16a** Lawrence Livermore National Laboratory/Science Source/Photo Researchers, Inc. **43-16b** Gary Stone/Lawrence Livermore National Laboratory **43-22a** Martin M. Rotker/Martin M. Rotker **43-22b** Simon Fraser/Science Photo Library/Photo Researchers, Inc. **43-26b** Southern Illinois University/Peter Arnold, Inc. **43-28** Mehau Kulyk/Science Photo Library/Photo Researchers, Inc. **CO-44** Fermilab Visual Media Services **44-06** CERN/Science Photo Library/Photo Researchers, Inc. **44-08** Fermilab Visual Media Services **CO-45** Jeff Hester and Paul Scowen, Arizona State University, and NASA **45-01** NASA Headquarters **45-02c** NASA/Johnson Space Center **45-03** U.S. Naval Observatory Photo/NASA Headquarters **45-04** National Optical Astronomy Observatories **45-05a** R.J. Dufour, Rice University **45-05b** U.S. Naval Observatory **45-05c** National Optical Astronomy Observatories **45-10** Space Telescope Science Institute/NASA Headquarters **45-11** National Optical Astronomy Observatories **45-15** European Space Agency/NASA Headquarters **45-22** Courtesy of Lucent Technologies/Bell Laboratories

Table of Contents Photos p. v (left) NOAA/Phil Degginger/Color-Pic, Inc. **p. v** (right) Mark Wagner/Tony Stone Images **p. vi** (left) Jess Stock/Tony Stone Images **p. vi** (right) Photograph by Dr. Harold E. Edgerton, © The Harold E. Edgerton 1992 Trust. Courtesy Palm Press **p. vii** (left) Ch. Russeil/Kipa/Sygma Photo News **p.vii** (right) Tibor Bognar/The Stock Market **p. viii** (left) Richard Megna/Fundamental Photographs **p. viii** (right) Douglas C. Giancoli **p. ix** (top) S. Feval/Le Matin de Lausanne/Sygma Photo News **p. ix** (bottom) Richard A. Cooke III/Tony Stone Images (right) David Woodfall/Tony Stone Images and AP/ Wide World Photos **p. x** (left) Fundamental Photographs **p. x** (right) Richard Kaylin/Tony Stone Images **p. xi** (left) Mahaux Photography/The Image Bank **p. xi** (right) Manfred Cage/Peter Arnold, Inc. **p. xii** Werner H. Muller/Peter Arnold, Inc. **p. xiii** (left) Mary Teresa Giancoli. **p. xiii** (right) Larry Mulvehill/Photo Researchers, Inc. **p. xiv** Image of Albert Einstein licensed by Einstein Archives, Hebrew University, Jerusalem, represented by Roger Richman Agency, Beverly Hills, California **p. xiv** (right) Donna McWilliam/AP/Wide World Photos **p. xv** (left) Charles O'Rear/Corbis (right) Fermilab **p. xvi** (left) Jeff Hester and Paul Scowen, Arizona State University, and NASA (right) R. J. Dufour, Rice University.

Periodic Table of the Elements[§]

Transition Elements

Legend: Symbol — Atomic Number, Atomic Mass[§], Electron Configuration (outer shells only). Example: Cl 17, 35.4527, $3p^5$

Group I	Group II	Transition Elements →										Group III	Group IV	Group V	Group VI	Group VII	Group VIII
H 1 1.00794 $1s^1$																	**He** 2 4.002602 $1s^2$
Li 3 6.941 $2s^1$	**Be** 4 9.012182 $2s^2$											**B** 5 10.811 $2p^1$	**C** 6 12.0107 $2p^2$	**N** 7 14.00674 $2p^3$	**O** 8 15.9994 $2p^4$	**F** 9 18.9984032 $2p^5$	**Ne** 10 20.1797 $2p^6$
Na 11 22.989770 $3s^1$	**Mg** 12 24.3050 $3s^2$											**Al** 13 26.981538 $3p^1$	**Si** 14 28.0855 $3p^2$	**P** 15 30.973761 $3p^3$	**S** 16 32.066 $3p^4$	**Cl** 17 35.4527 $3p^5$	**Ar** 18 39.948 $3p^6$
K 19 39.0983 $4s^1$	**Ca** 20 40.078 $4s^2$	**Sc** 21 44.955910 $3d^14s^2$	**Ti** 22 47.867 $3d^24s^2$	**V** 23 50.9415 $3d^34s^2$	**Cr** 24 51.9961 $3d^54s^1$	**Mn** 25 54.938049 $3d^54s^2$	**Fe** 26 55.845 $3d^64s^2$	**Co** 27 58.933200 $3d^74s^2$	**Ni** 28 58.6934 $3d^84s^2$	**Cu** 29 63.546 $3d^{10}4s^1$	**Zn** 30 65.39 $3d^{10}4s^2$	**Ga** 31 69.723 $4p^1$	**Ge** 32 72.61 $4p^2$	**As** 33 74.92160 $4p^3$	**Se** 34 78.96 $4p^4$	**Br** 35 79.904 $4p^5$	**Kr** 36 83.80 $4p^6$
Rb 37 85.4678 $5s^1$	**Sr** 38 87.62 $5s^2$	**Y** 39 88.90585 $4d^15s^2$	**Zr** 40 91.224 $4d^25s^2$	**Nb** 41 92.90638 $4d^45s^1$	**Mo** 42 95.94 $4d^55s^1$	**Tc** 43 (98) $4d^55s^2$	**Ru** 44 101.07 $4d^75s^1$	**Rh** 45 102.90550 $4d^85s^1$	**Pd** 46 106.42 $4d^{10}5s^0$	**Ag** 47 107.8682 $4d^{10}5s^1$	**Cd** 48 112.411 $4d^{10}5s^2$	**In** 49 114.818 $5p^1$	**Sn** 50 118.710 $5p^2$	**Sb** 51 121.760 $5p^3$	**Te** 52 127.60 $5p^4$	**I** 53 126.90447 $5p^5$	**Xe** 54 131.29 $5p^6$
Cs 55 132.90545 $6s^1$	**Ba** 56 137.327 $6s^2$	57–71[†]	**Hf** 72 178.49 $5d^26s^2$	**Ta** 73 180.9479 $5d^36s^2$	**W** 74 183.84 $5d^46s^2$	**Re** 75 186.207 $5d^56s^2$	**Os** 76 190.23 $5d^66s^2$	**Ir** 77 192.217 $5d^76s^2$	**Pt** 78 195.078 $5d^96s^1$	**Au** 79 196.96655 $5d^{10}6s^1$	**Hg** 80 200.59 $5d^{10}6s^2$	**Tl** 81 204.3833 $6p^1$	**Pb** 82 207.2 $6p^2$	**Bi** 83 208.98038 $6p^3$	**Po** 84 (209) $6p^4$	**At** 85 (210) $6p^5$	**Rn** 86 (222) $6p^6$
Fr 87 (223) $7s^1$	**Ra** 88 (226) $7s^2$	89–103[‡]	**Rf** 104 (261) $6d^27s^2$	**Db** 105 (262) $6d^37s^2$	**Sg** 106 (266) $6d^47s^2$	**Bh** 107 (264) $6d^57s^2$	**Hs** 108 (269)	**Mt** 109 (268)	110 (271)	111 (272)	112 (277)		114 (289)		116 (289)		118 (293)

[†] Lanthanide Series

La 57 138.9055 $5d^16s^2$	**Ce** 58 140.115 $4f^15d^16s^2$	**Pr** 59 140.90765 $4f^35d^06s^2$	**Nd** 60 144.24 $4f^45d^06s^2$	**Pm** 61 (145) $4f^55d^06s^2$	**Sm** 62 150.36 $4f^65d^06s^2$	**Eu** 63 151.964 $4f^75d^06s^2$	**Gd** 64 157.25 $4f^75d^16s^2$	**Tb** 65 158.92534 $4f^95d^06s^2$	**Dy** 66 162.50 $4f^{10}5d^06s^2$	**Ho** 67 164.93032 $4f^{11}5d^06s^2$	**Er** 68 167.26 $4f^{12}5d^06s^2$	**Tm** 69 168.93421 $4f^{13}5d^06s^2$	**Yb** 70 173.04 $4f^{14}5d^06s^2$	**Lu** 71 174.967 $4f^{14}5d^16s^2$

[‡] Actinide Series

Ac 89 (227.02775) $6d^17s^2$	**Th** 90 232.0381 $6d^27s^2$	**Pa** 91 231.03588 $5f^26d^17s^2$	**U** 92 238.0289 $5f^36d^17s^2$	**Np** 93 (237) $5f^46d^17s^2$	**Pu** 94 (244) $5f^66d^07s^2$	**Am** 95 (243) $5f^76d^07s^2$	**Cm** 96 (247) $5f^76d^17s^2$	**Bk** 97 (247) $5f^96d^07s^2$	**Cf** 98 (251) $5f^{10}6d^07s^2$	**Es** 99 (252) $5f^{11}6d^07s^2$	**Fm** 100 (257) $5f^{12}6d^07s^2$	**Md** 101 (258) $5f^{13}6d^07s^2$	**No** 102 (259) $5f^{14}6d^07s^2$	**Lr** 103 (262) $5f^{14}6d^17s^2$

[§] Atomic mass values averaged over isotopes in percentages they occur on Earth's surface. For many unstable elements, mass of the longest-lived known isotope is given in parentheses. 1999 revisions. (See also Appendix D.)

Trigonometric Table

Angle in Degrees	Angle in Radians	Sine	Cosine	Tangent	Angle in Degrees	Angle in Radians	Sine	Cosine	Tangent
0°	0.000	0.000	1.000	0.000					
1°	0.017	0.017	1.000	0.017	46°	0.803	0.719	0.695	1.036
2°	0.035	0.035	0.999	0.035	47°	0.820	0.731	0.682	1.072
3°	0.052	0.052	0.999	0.052	48°	0.838	0.743	0.669	1.111
4°	0.070	0.070	0.998	0.070	49°	0.855	0.755	0.656	1.150
5°	0.087	0.087	0.996	0.087	50°	0.873	0.766	0.643	1.192
6°	0.105	0.105	0.995	0.105	51°	0.890	0.777	0.629	1.235
7°	0.122	0.122	0.993	0.123	52°	0.908	0.788	0.616	1.280
8°	0.140	0.139	0.990	0.141	53°	0.925	0.799	0.602	1.327
9°	0.157	0.156	0.988	0.158	54°	0.942	0.809	0.588	1.376
10°	0.175	0.174	0.985	0.176	55°	0.960	0.819	0.574	1.428
11°	0.192	0.191	0.982	0.194	56°	0.977	0.829	0.559	1.483
12°	0.209	0.208	0.978	0.213	57°	0.995	0.839	0.545	1.540
13°	0.227	0.225	0.974	0.231	58°	1.012	0.848	0.530	1.600
14°	0.244	0.242	0.970	0.249	59°	1.030	0.857	0.515	1.664
15°	0.262	0.259	0.966	0.268	60°	1.047	0.866	0.500	1.732
16°	0.279	0.276	0.961	0.287	61°	1.065	0.875	0.485	1.804
17°	0.297	0.292	0.956	0.306	62°	1.082	0.883	0.469	1.881
18°	0.314	0.309	0.951	0.325	63°	1.100	0.891	0.454	1.963
19°	0.332	0.326	0.946	0.344	64°	1.117	0.899	0.438	2.050
20°	0.349	0.342	0.940	0.364	65°	1.134	0.906	0.423	2.145
21°	0.367	0.358	0.934	0.384	66°	1.152	0.914	0.407	2.246
22°	0.384	0.375	0.927	0.404	67°	1.169	0.921	0.391	2.356
23°	0.401	0.391	0.921	0.424	68°	1.187	0.927	0.375	2.475
24°	0.419	0.407	0.914	0.445	69°	1.204	0.934	0.358	2.605
25°	0.436	0.423	0.906	0.466	70°	1.222	0.940	0.342	2.747
26°	0.454	0.438	0.899	0.488	71°	1.239	0.946	0.326	2.904
27°	0.471	0.454	0.891	0.510	72°	1.257	0.951	0.309	3.078
28°	0.489	0.469	0.883	0.532	73°	1.274	0.956	0.292	3.271
29°	0.506	0.485	0.875	0.554	74°	1.292	0.961	0.276	3.487
30°	0.524	0.500	0.866	0.577	75°	1.309	0.966	0.259	3.732
31°	0.541	0.515	0.857	0.601	76°	1.326	0.970	0.242	4.011
32°	0.559	0.530	0.848	0.625	77°	1.344	0.974	0.225	4.331
33°	0.576	0.545	0.839	0.649	78°	1.361	0.978	0.208	4.705
34°	0.593	0.559	0.829	0.675	79°	1.379	0.982	0.191	5.145
35°	0.611	0.574	0.819	0.700	80°	1.396	0.985	0.174	5.671
36°	0.628	0.588	0.809	0.727	81°	1.414	0.988	0.156	6.314
37°	0.646	0.602	0.799	0.754	82°	1.431	0.990	0.139	7.115
38°	0.663	0.616	0.788	0.781	83°	1.449	0.993	0.122	8.144
39°	0.681	0.629	0.777	0.810	84°	1.466	0.995	0.105	9.514
40°	0.698	0.643	0.766	0.839	85°	1.484	0.996	0.087	11.43
41°	0.716	0.656	0.755	0.869	86°	1.501	0.998	0.070	14.301
42°	0.733	0.669	0.743	0.900	87°	1.518	0.999	0.052	19.081
43°	0.750	0.682	0.731	0.933	88°	1.536	0.999	0.035	28.636
44°	0.768	0.695	0.719	0.966	89°	1.553	1.000	0.017	57.290
45°	0.785	0.707	0.707	1.000	90°	1.571	1.000	0.000	∞